Student Study Art Notebook

to accompany

Biology

Eighth Edition

Sylvia S. Mader

 **Higher Education**

Boston   Burr Ridge, IL   Dubuque, IA   Madison, WI   New York   San Francisco   St. Louis
Bangkok   Bogotá   Caracas   Kuala Lumpur   Lisbon   London   Madrid   Mexico City
Milan   Montreal   New Delhi   Santiago   Seoul   Singapore   Sydney   Taipei   Toronto

The McGraw·Hill Companies

Student Study Art Notebook to accompany
BIOLOGY, EIGHTH EDITION
SYLVIA S. MADER

Published by McGraw-Hill Higher Education, an imprint of The McGraw-Hill Companies, Inc.,
1221 Avenue of the Americas, New York, NY 10020. Copyright © 2004 by The McGraw-Hill
Companies, Inc. All rights reserved.

 This book is printed on recycled, acid-free paper containing
10% postconsumer waste.
RECYCLED

1 2 3 4 5 6 7 8 9 0 QPD/QPD 0 9 8 7 6 5 4 3

ISBN 0-07-297802-3

www.mhhe.com

DIRECTORY OF NOTEBOOK FIGURES
TO ACCOMPANY
MADER
BIOLOGY 8/e

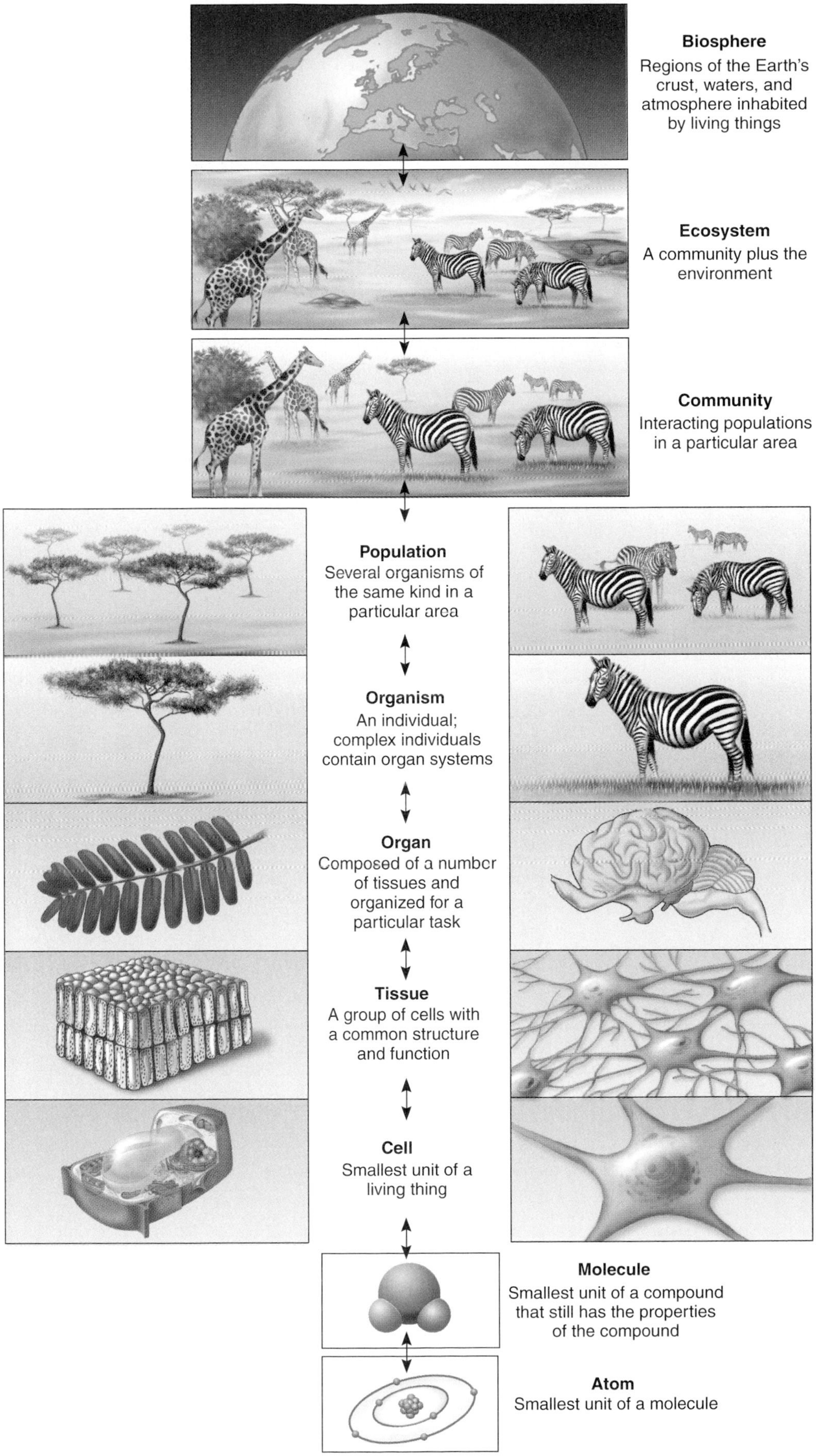

Biosphere
Regions of the Earth's crust, waters, and atmosphere inhabited by living things

Ecosystem
A community plus the environment

Community
Interacting populations in a particular area

Population
Several organisms of the same kind in a particular area

Organism
An individual; complex individuals contain organ systems

Organ
Composed of a number of tissues and organized for a particular task

Tissue
A group of cells with a common structure and function

Cell
Smallest unit of a living thing

Molecule
Smallest unit of a compound that still has the properties of the compound

Atom
Smallest unit of a molecule

Levels of biological organization
Figure 1.2

1

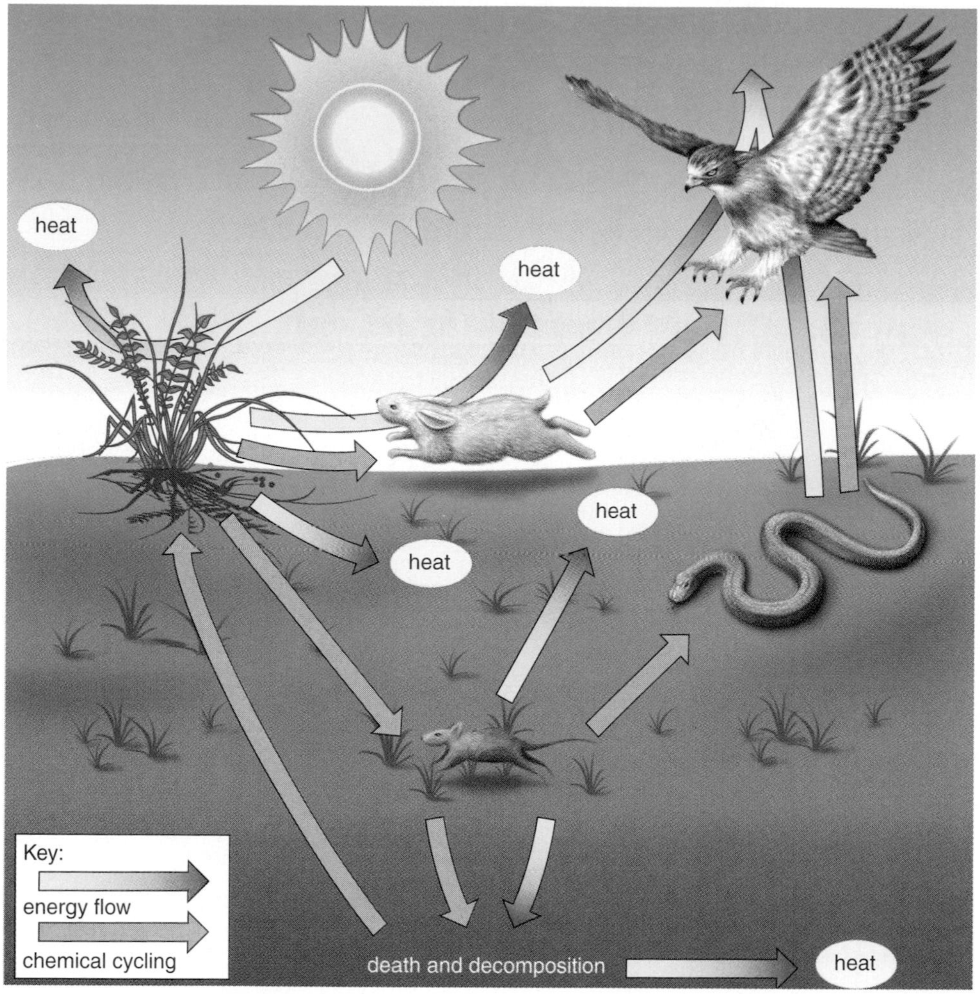
heat
heat
heat
heat
Key:
energy flow
chemical cycling
death and decomposition
heat

A grassland, a terrestrial ecosystem
Figure 1.5

A coral reef, a marine ecosystem
Figure 1.6

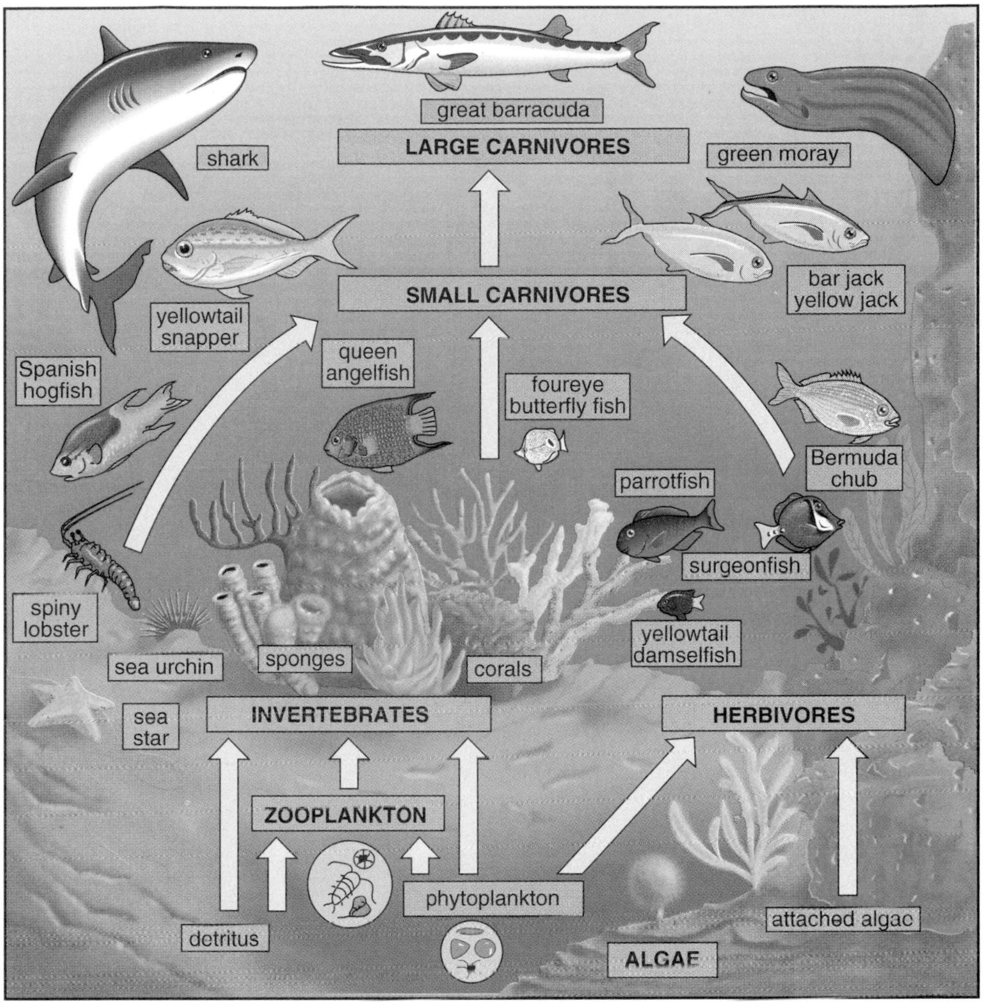

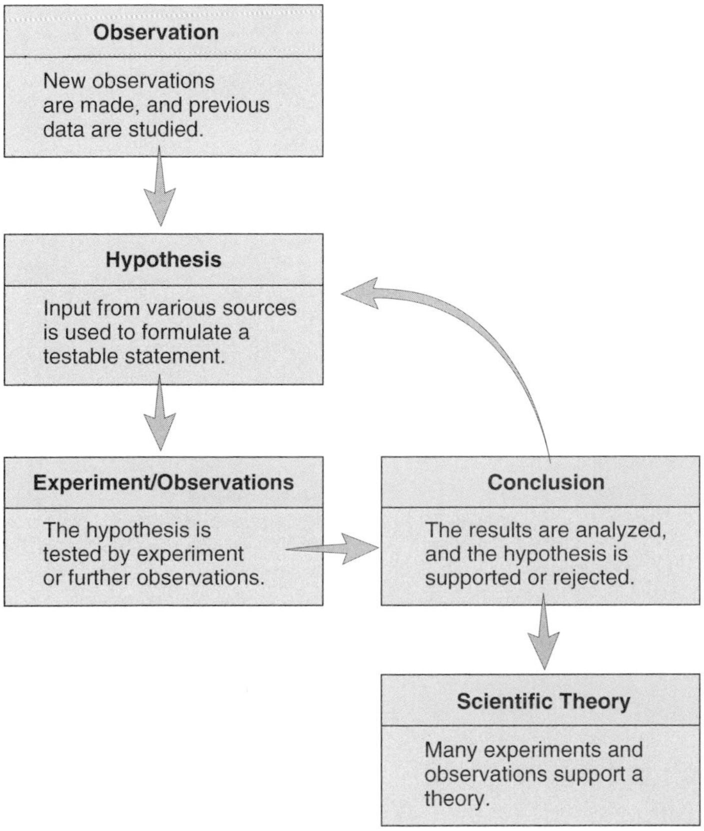

Observation

New observations are made, and previous data are studied.

Hypothesis

Input from various sources is used to formulate a testable statement.

Experiment/Observations

The hypothesis is tested by experiment or further observations.

Conclusion

The results are analyzed, and the hypothesis is supported or rejected.

Scientific Theory

Many experiments and observations support a theory.

Flow diagram for the scientific method
Figure 1.10

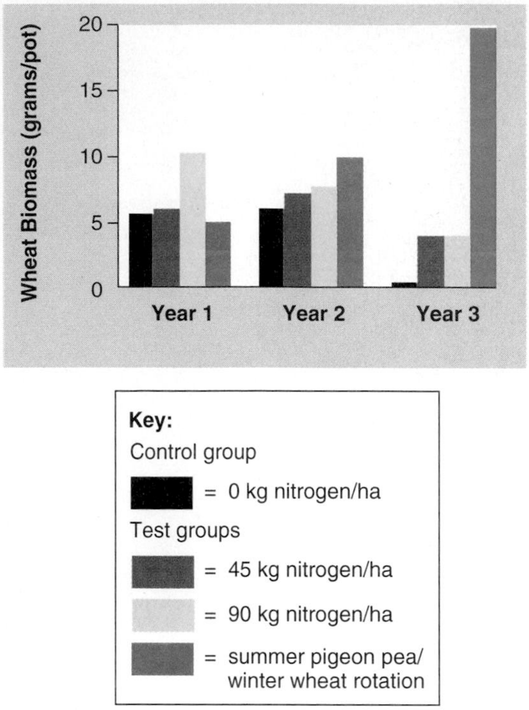

Key:

Control group

■ = 0 kg nitrogen/ha

Test groups

■ = 45 kg nitrogen/ha

□ = 90 kg nitrogen/ha

■ = summer pigeon pea/
winter wheat rotation

Summer pigeon pea/winter wheat rotation study
Figure 1.12

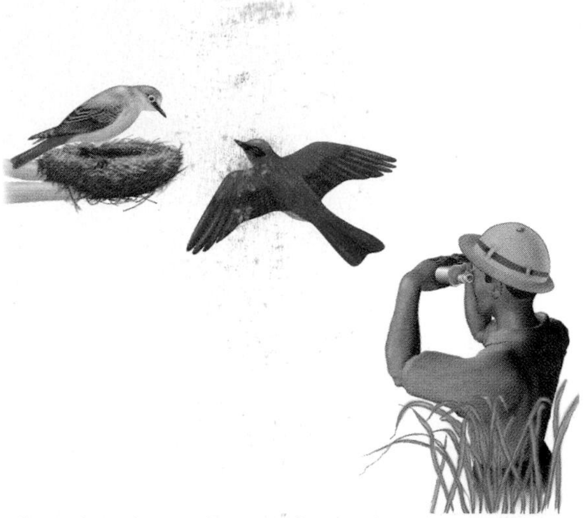

a. Scientist makes observations, studies previous data, and formulates a hypothesis.

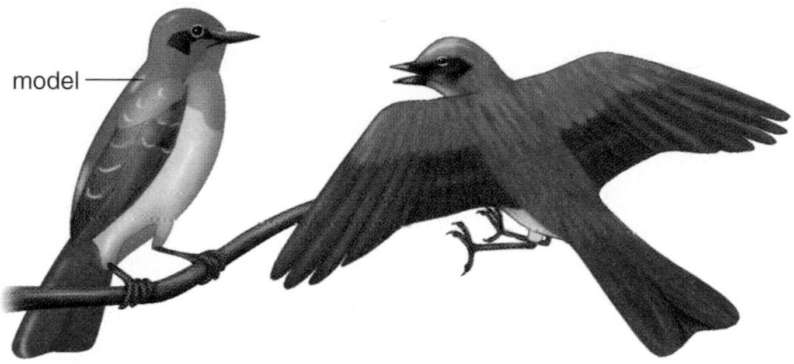

model

b. Scientist performs experiment and collects objective data.

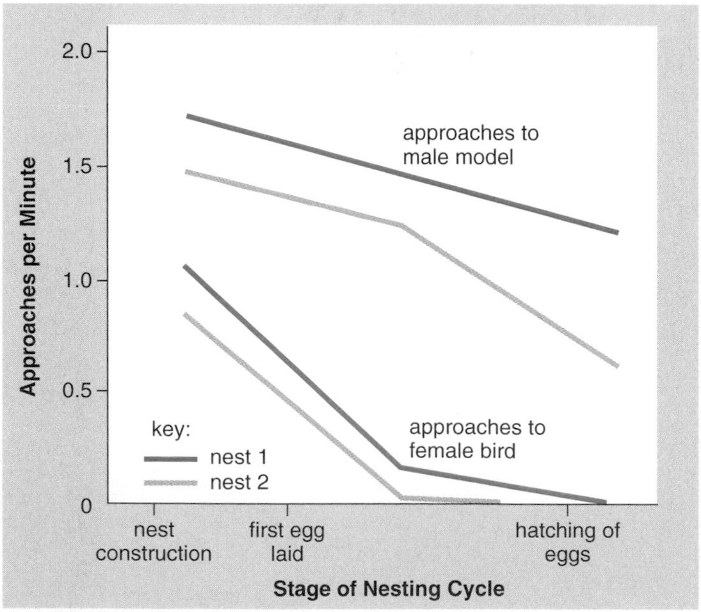

c. Scientist analyzes data and comes to a conclusion.

A field study
Figure 1.13

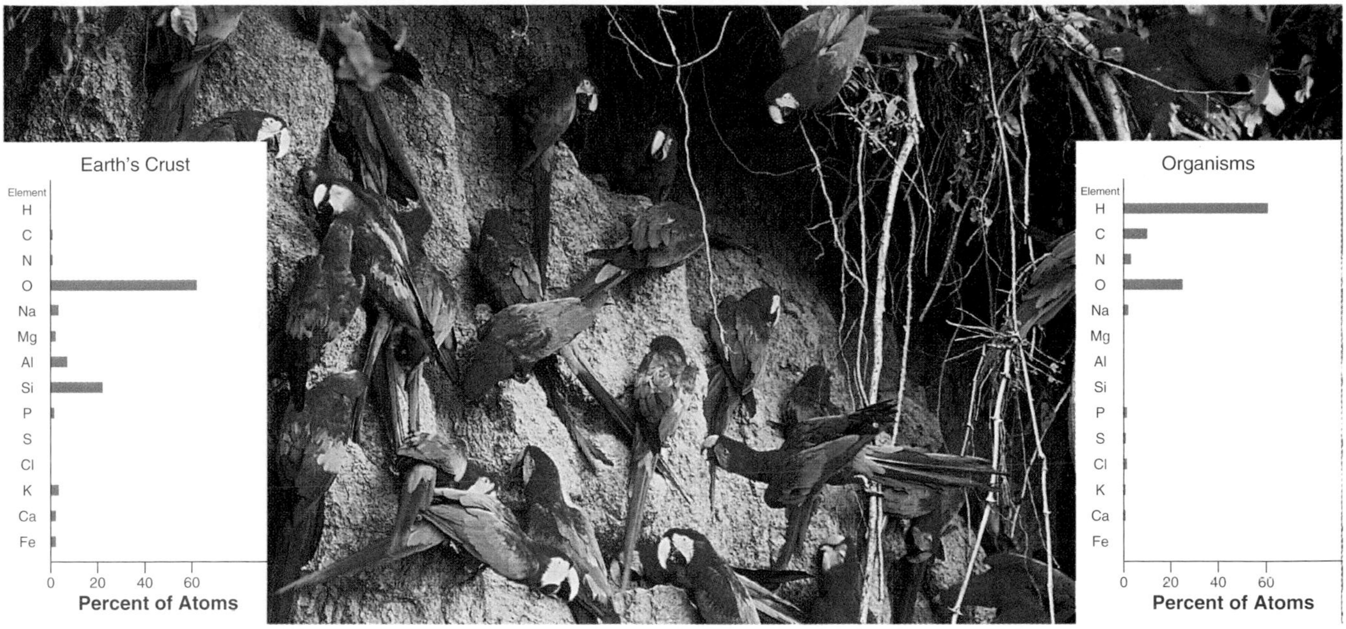

Elements that make up the Earth's crust and its organisms
Figure 2.1

© Gunter Ziesler/Peter Arnold, Inc.

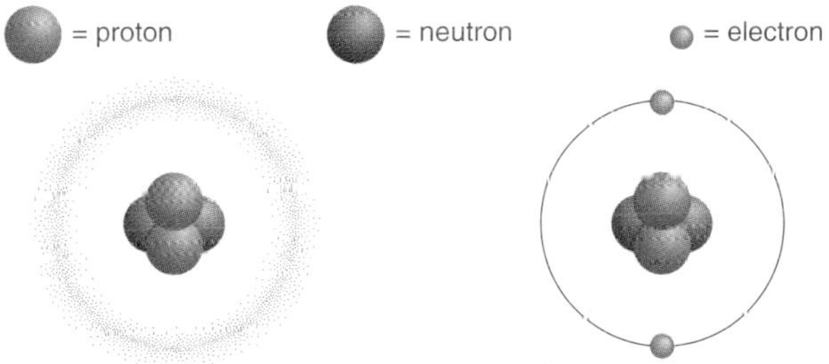

🔴 = proton   🔴 = neutron   🔴 = electron

Subatomic Particles			
Particle	Electric Charge	Atomic Mass	Location
Proton	+1	1	Nucleus
Neutron	0	1	Nucleus
Electron	1	0	Electron shells

a.   b.   c.

Model of helium (He)
Figure 2.2

			Groups				
1							**8**
1							2
H							**He**
1.008	2	3	4	5	6	7	4.003
3	4	5	6	7	8	9	10
Li	**Be**	**B**	**C**	**N**	**O**	**F**	**Ne**
6.941	9.012	10.81	12.01	14.01	16.00	19.00	20.18
11	12	13	14	15	16	17	18
Na	**Mg**	**Al**	**Si**	**P**	**S**	**Cl**	**Ar**
22.99	24.31	26.98	28.09	30.97	32.07	35.45	39.95
19	20	31	32	33	34	35	36
K	**Ca**	**Ga**	**Ge**	**As**	**Se**	**Br**	**Kr**
39.10	40.08	69.72	72.59	74.92	78.96	79.90	83.60

A portion of the periodic table
Figure 2.3

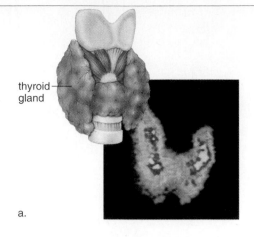

thyroid gland

a.

Low levels of radiation
Figure 2.4

a: © Biomed Comm./Custom Medical Stock Photo

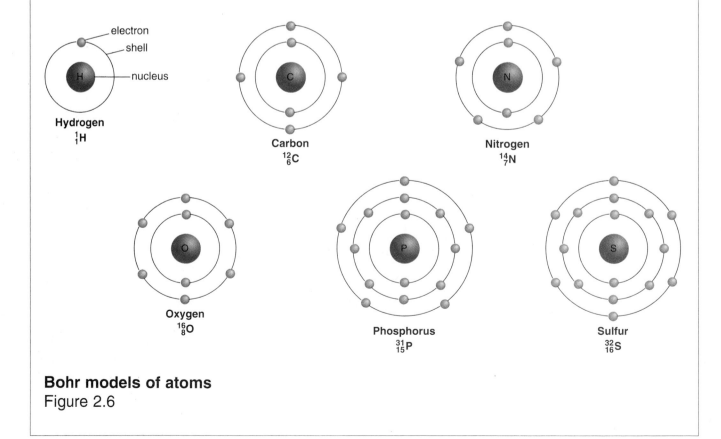

electron
shell
nucleus

Hydrogen
^1_1H

Carbon
$^{12}_6C$

Nitrogen
$^{14}_7N$

Oxygen
$^{16}_8O$

Phosphorus
$^{31}_{15}P$

Sulfur
$^{32}_{16}S$

Bohr models of atoms
Figure 2.6

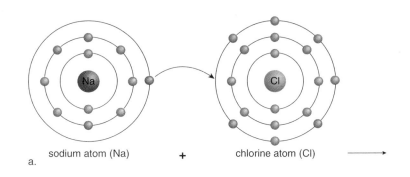

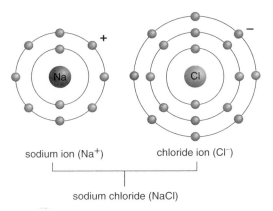

a.

| sodium atom (Na) | + | chlorine atom (Cl) | → | sodium ion (Na⁺) | chloride ion (Cl⁻) |

sodium chloride (NaCl)

Formation of sodium chloride
Figure 2.7

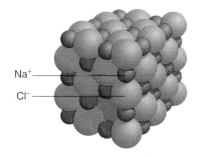

Na^+
Cl^-

Electron Model	Structural Formula	Molecular Formula
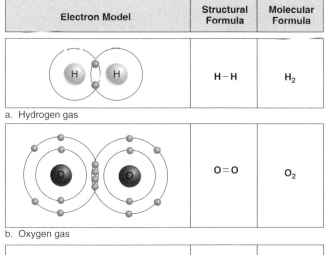 a. Hydrogen gas	H—H	H_2
b. Oxygen gas	O=O	O_2
c. Methane	H—C—H	CH_4

Covalently bonded molecules
Figure 2.8

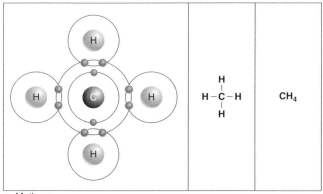

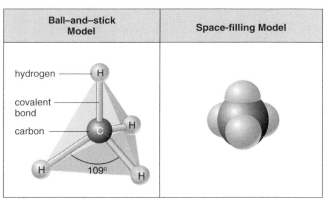

Ball–and–stick Model	Space-filling Model
hydrogen covalent bond carbon 109°	

d. Methane, cont'd.

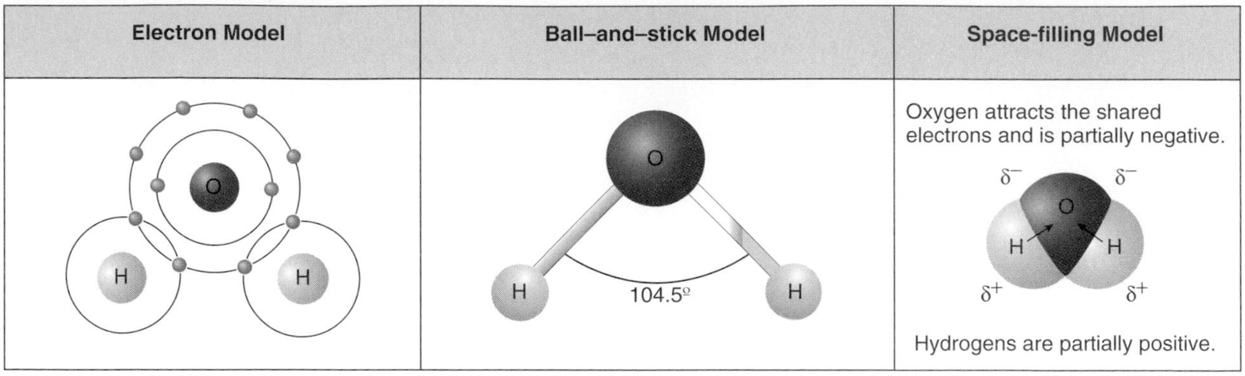

Electron Model	Ball–and–stick Model	Space-filling Model
		Oxygen attracts the shared electrons and is partially negative. Hydrogens are partially positive.

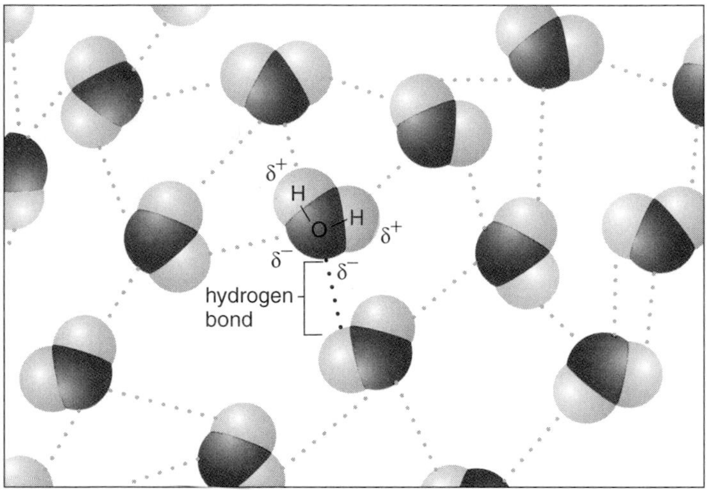

Water molecule
Figure 2.9

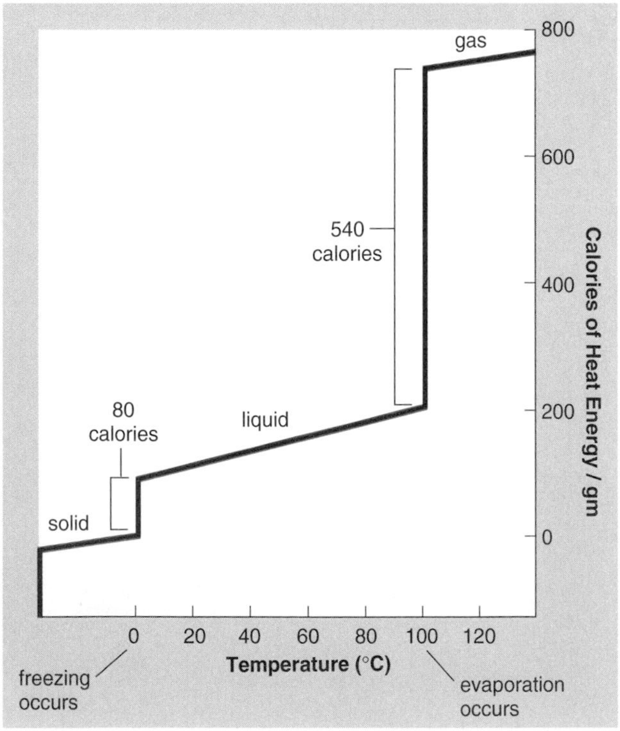

Temperature and water
Figure 2.10

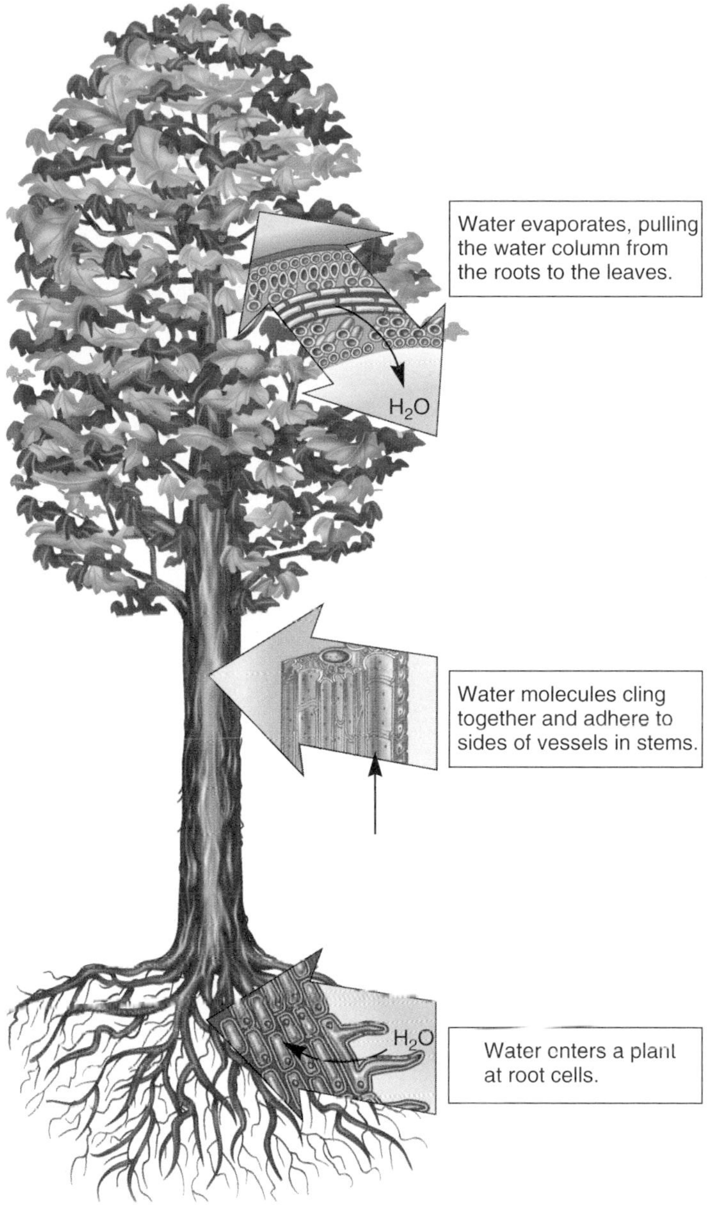

Water evaporates, pulling the water column from the roots to the leaves.

H_2O

Water molecules cling together and adhere to sides of vessels in stems.

Water enters a plant at root cells.

H_2O

Water as a transport medium
Figure 2.11

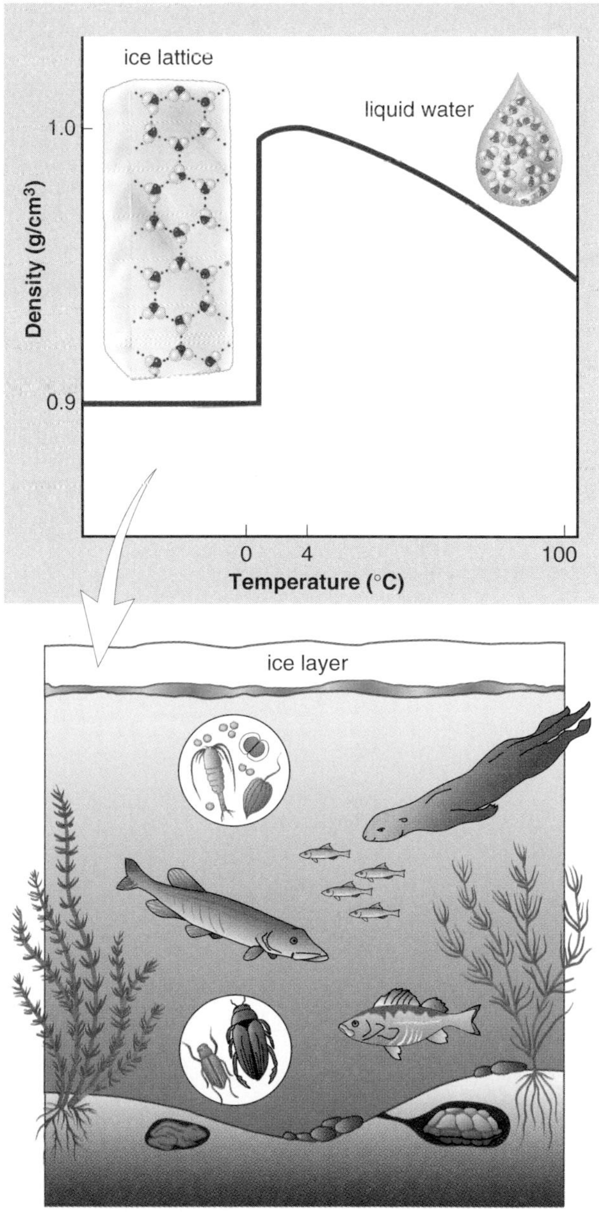

ice lattice

liquid water

1.0

Density (g/cm^3)

0.9

0   4   100
Temperature (°C)

ice layer

Water as ice
Figure 2.12

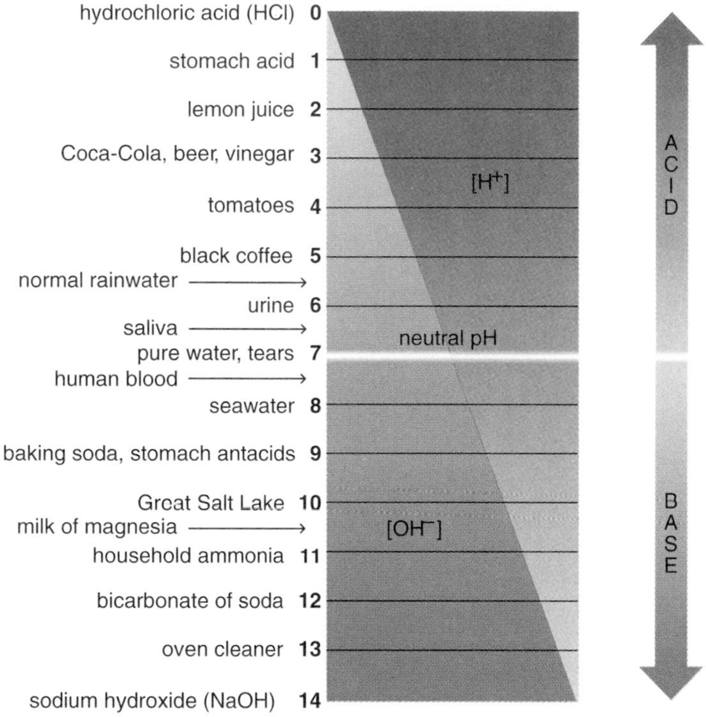
hydrochloric acid (HCl) 0
stomach acid 1
lemon juice 2
Coca-Cola, beer, vinegar 3
tomatoes 4
black coffee 5
normal rainwater
urine 6
saliva
pure water, tears 7
human blood
seawater 8
baking soda, stomach antacids 9
Great Salt Lake 10
milk of magnesia
household ammonia 11
bicarbonate of soda 12
oven cleaner 13
sodium hydroxide (NaOH) 14
[H+]
neutral pH
[OH−]
ACID
BASE

The pH scale
Figure 2.13

Functional Groups			
Group	Structure	Compound	Significance
Hydroxyl	$R-OH$	Alcohol	Polar, forms hydrogen bonds; in sugars and some amino acids
Carbonyl	$R-C{\displaystyle \atop OH}^{O}$	Aldehyde	Polar; in some sugars
	$R-\overset{O}{\overset{\|}{C}}-R$	Ketone	Polar; in some sugars
Carboxyl	$R-C{\displaystyle \atop H}^{O}$	Carboxylic acids	Polar, acidic; in fats and amino acids
Amino	$R-N{\displaystyle \atop H}^{H}$	Amines	Polar, basic; in amino acids
Sulfhydryl	$R-SH$	Thiols	Forms disulfide bonds in some amino acids
Phosphate	$R-O-\overset{O}{\overset{\|}{\underset{\underset{OH}{\|}}{P}}}-OH$	Organic phosphates	Polar, acidic; in some amino acids
R = remainder of molecule			

Functional groups
Figure 3.2

a. Glyceraldehyde   b. Dihydroxyacetone

Isomers
Figure 3.3

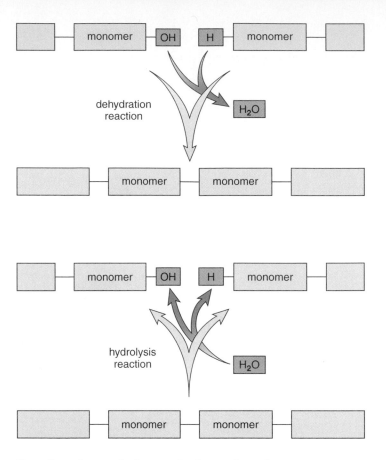

Synthesis and degradation of polymers
Figure 3.5

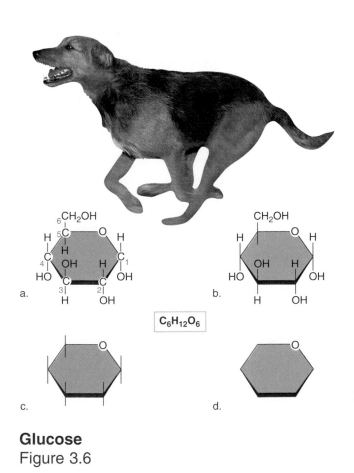

$C_6H_{12}O_6$

a.

b.

c.

d.

Glucose
Figure 3.6

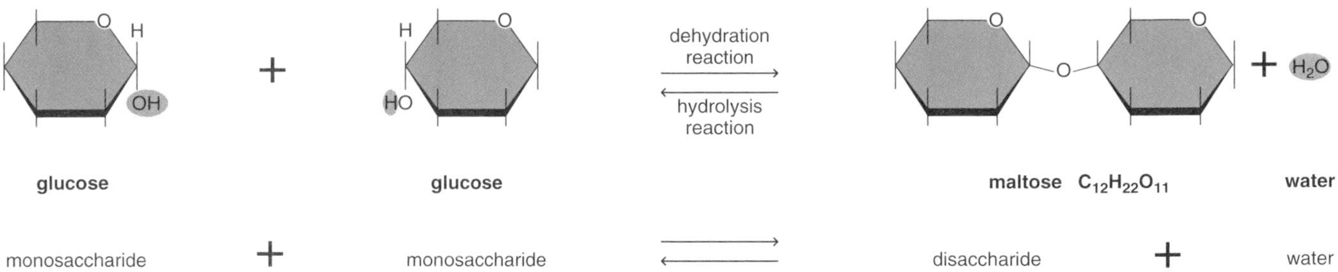

Synthesis and degradation of maltose, a disaccharide
Figure 3.7

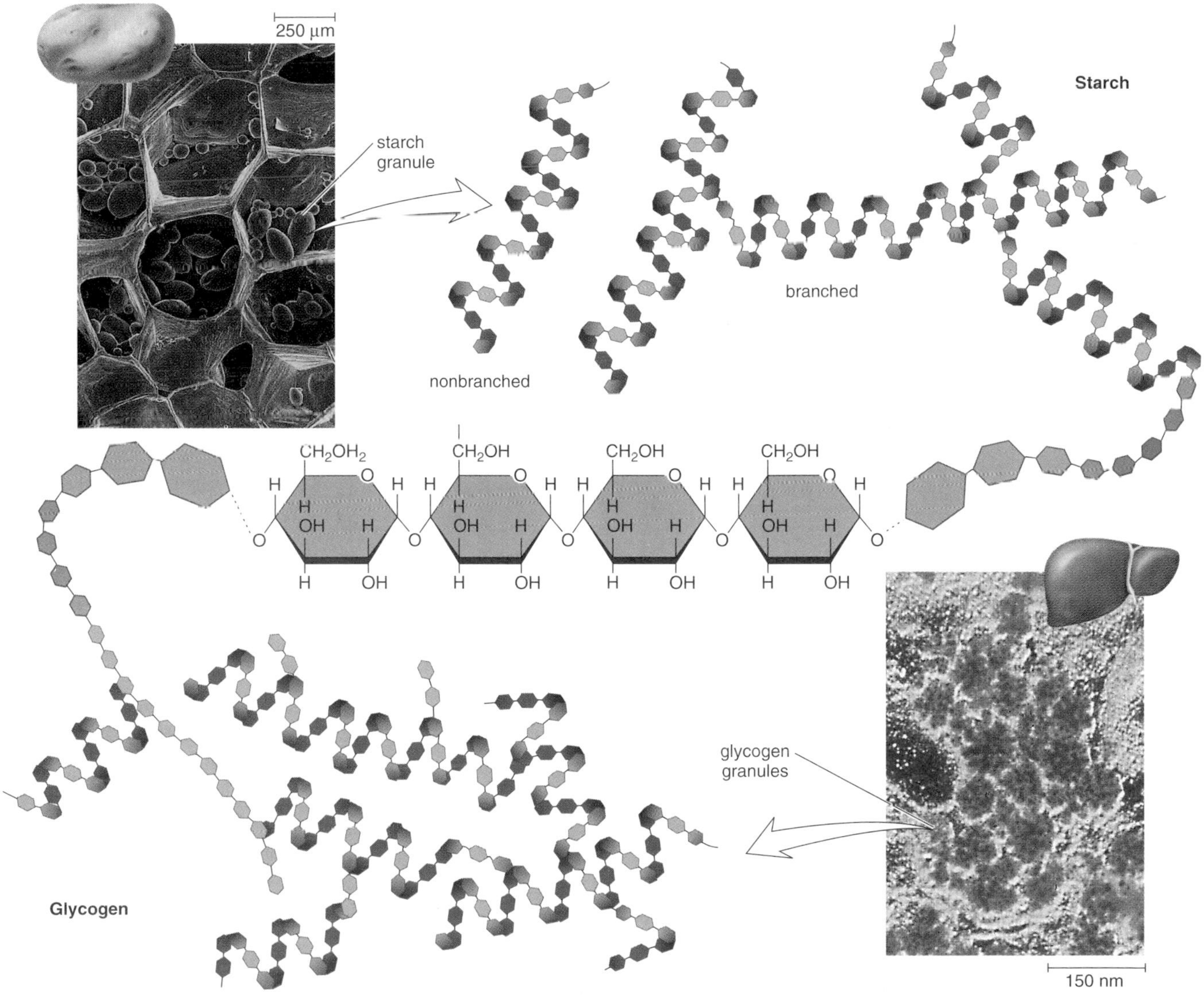

Starch and glycogen structure and function
Figure 3.8

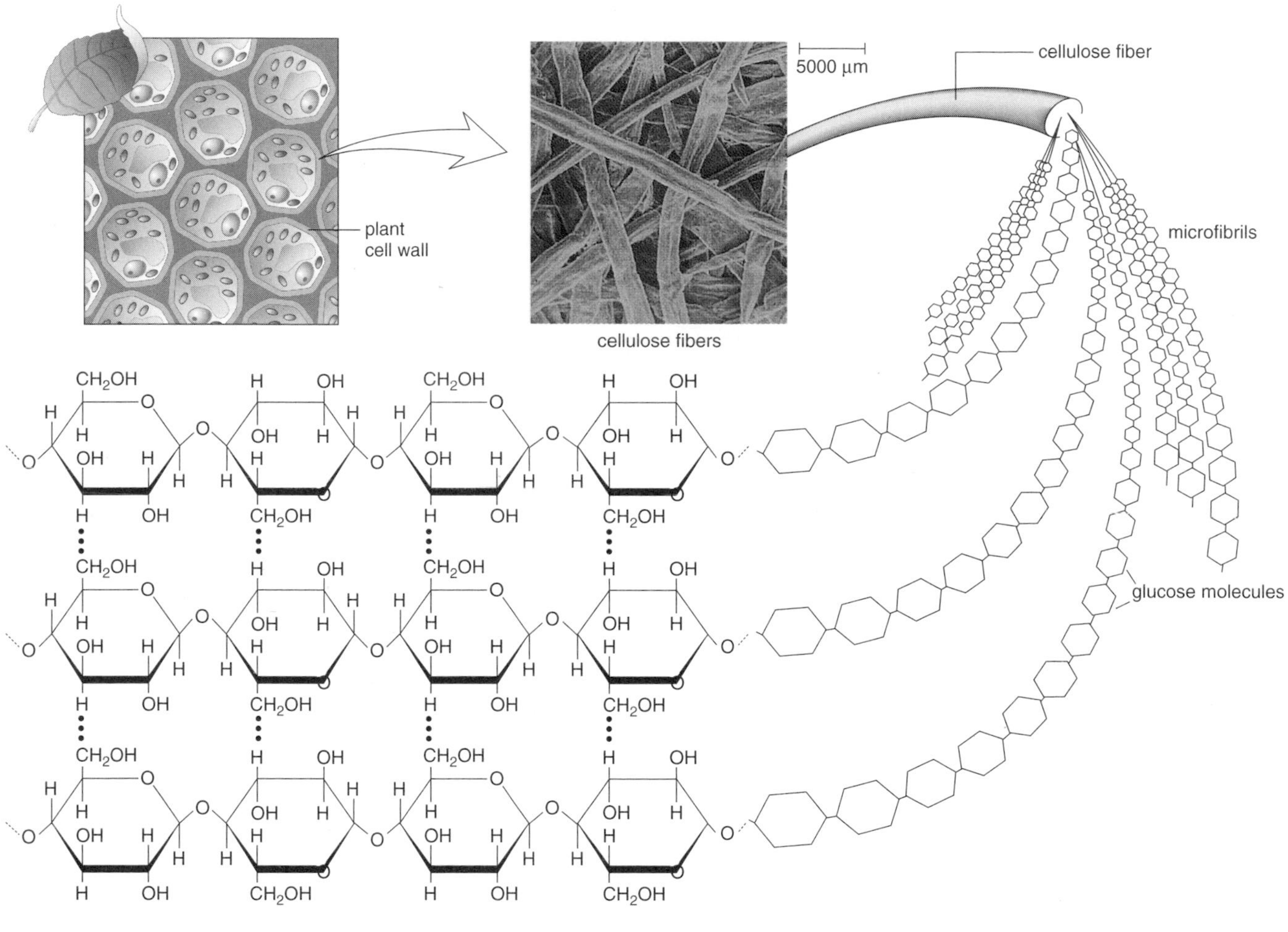

5000 µm

cellulose fiber

microfibrils

glucose molecules

plant
cell wall

cellulose fibers

Cellulose fibrils

Figure 3.9

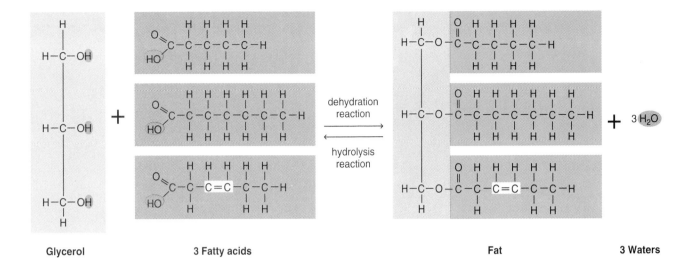

Glycerol **3 Fatty acids** **Fat** **3 Waters**

a. Formation of a fat.

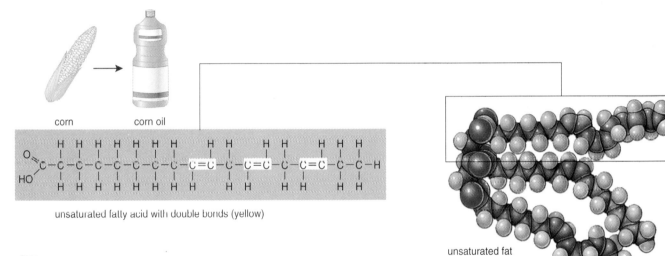

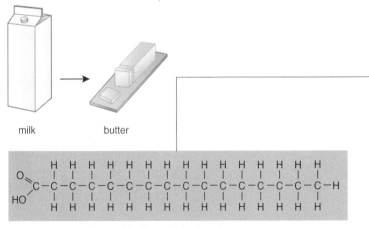

corn corn oil

unsaturated fatty acid with double bonds (yellow)

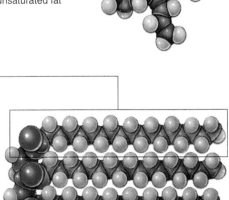

unsaturated fat

milk butter

saturated fatty acid with no double bonds

b. Types of fatty acids

saturated fat

c. Types of fats

Fat and fatty acids
Figure 3.11

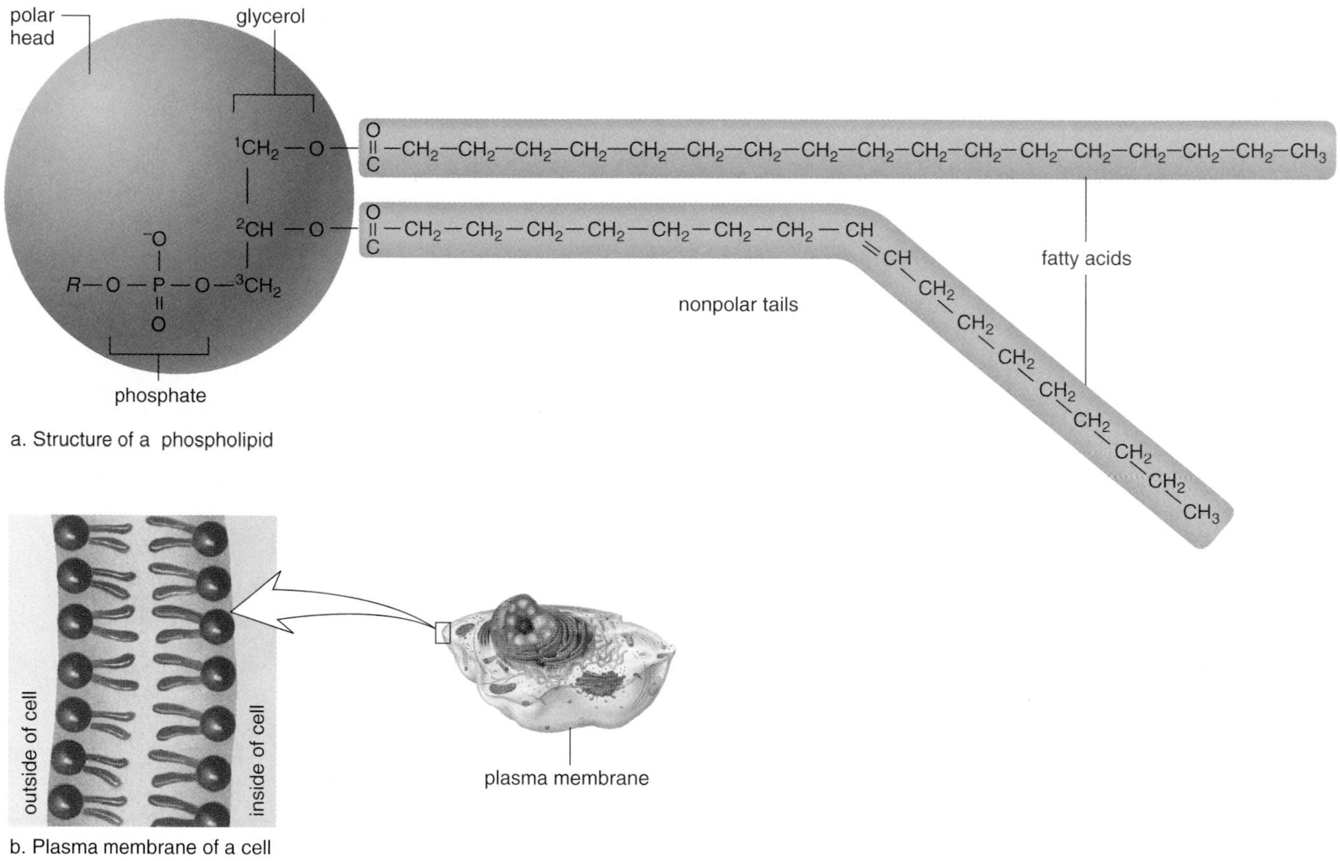

polar head

glycerol

R—O—P—O—^{3}CH$_2$

phosphate

a. Structure of a phospholipid

nonpolar tails

fatty acids

outside of cell

inside of cell

b. Plasma membrane of a cell

plasma membrane

Phospholipids form membranes
Figure 3.12

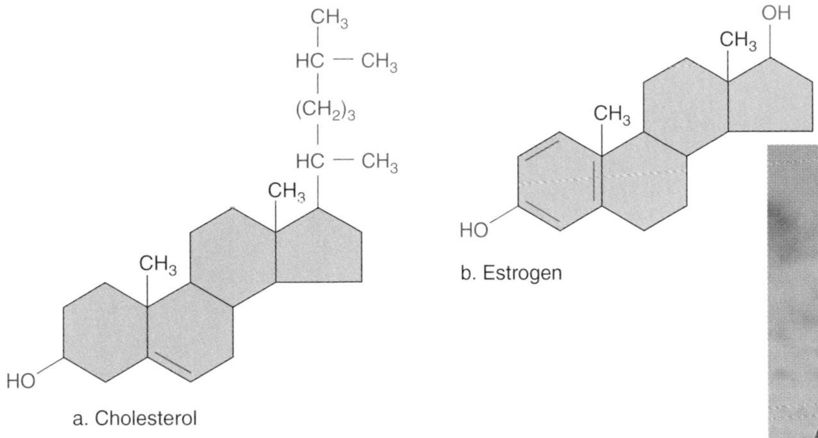

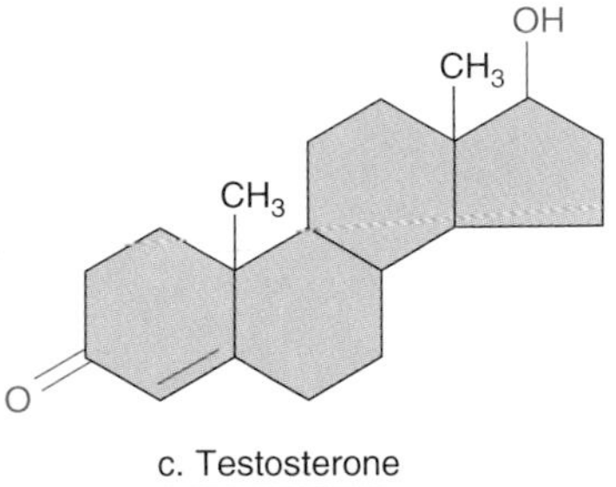

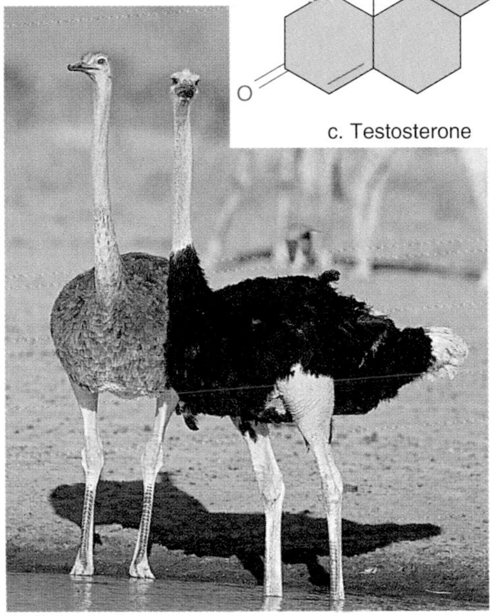

Steroid diversity
Figure 3.13
© Martin Harvey/Peter Arnold, Inc.

a. Cholesterol

b. Estrogen

c. Testosterone

Synthesis and degradation of a peptide
Figure 3.15

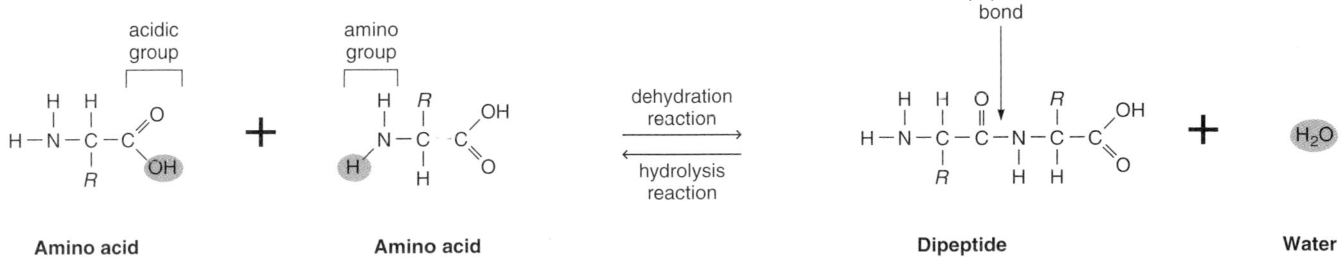

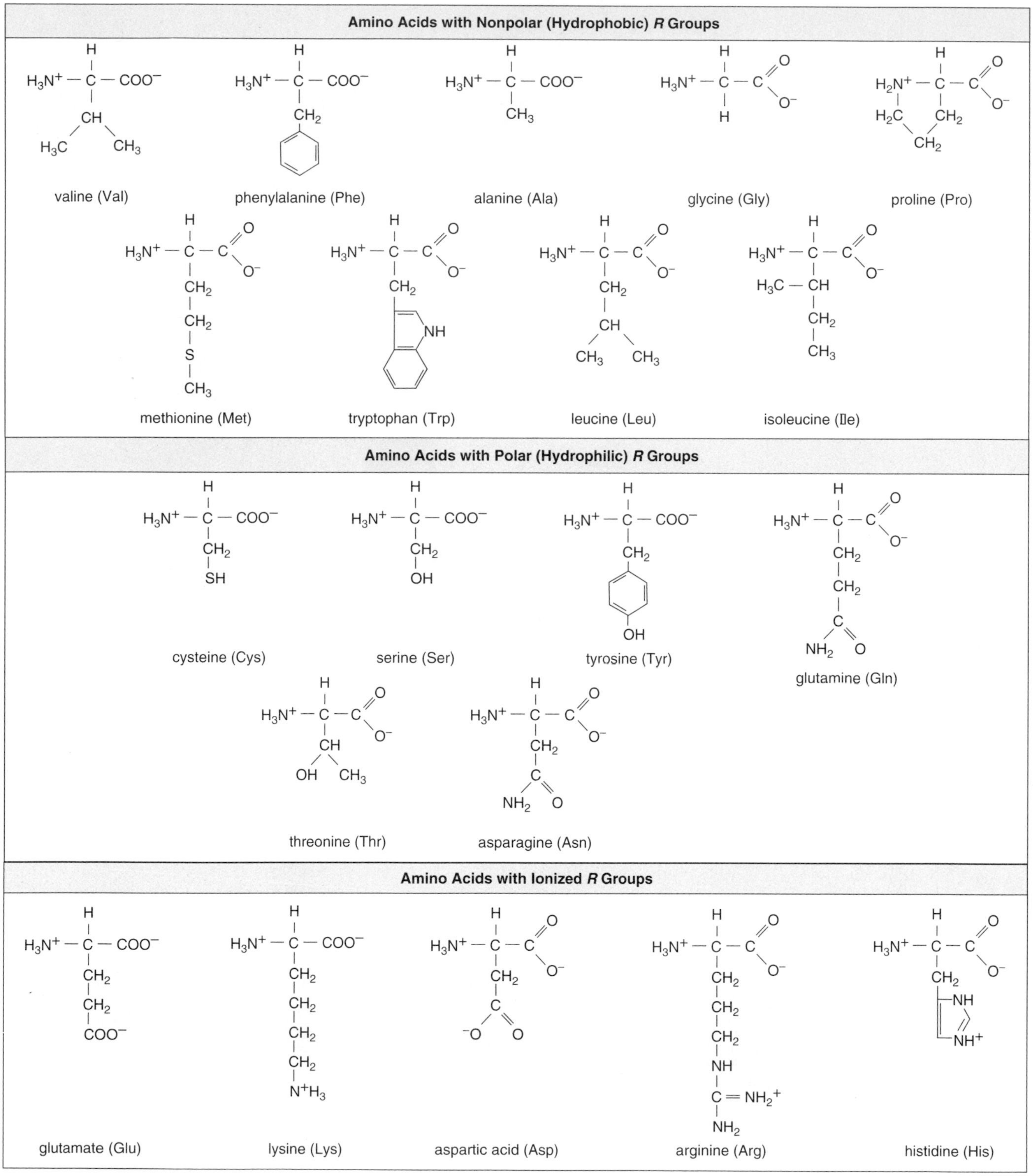
Amino Acids with Nonpolar (Hydrophobic) R Groups
valine (Val)
phenylalanine (Phe)
alanine (Ala)
glycine (Gly)
proline (Pro)
methionine (Met)
tryptophan (Trp)
leucine (Leu)
isoleucine (Ile)
Amino Acids with Polar (Hydrophilic) R Groups
cysteine (Cys)
serine (Ser)
tyrosine (Tyr)
glutamine (Gln)
threonine (Thr)
asparagine (Asn)
Amino Acids with Ionized R Groups
glutamate (Glu)
lysine (Lys)
aspartic acid (Asp)
arginine (Arg)
histidine (His)

Amino acids
Figure 3.16

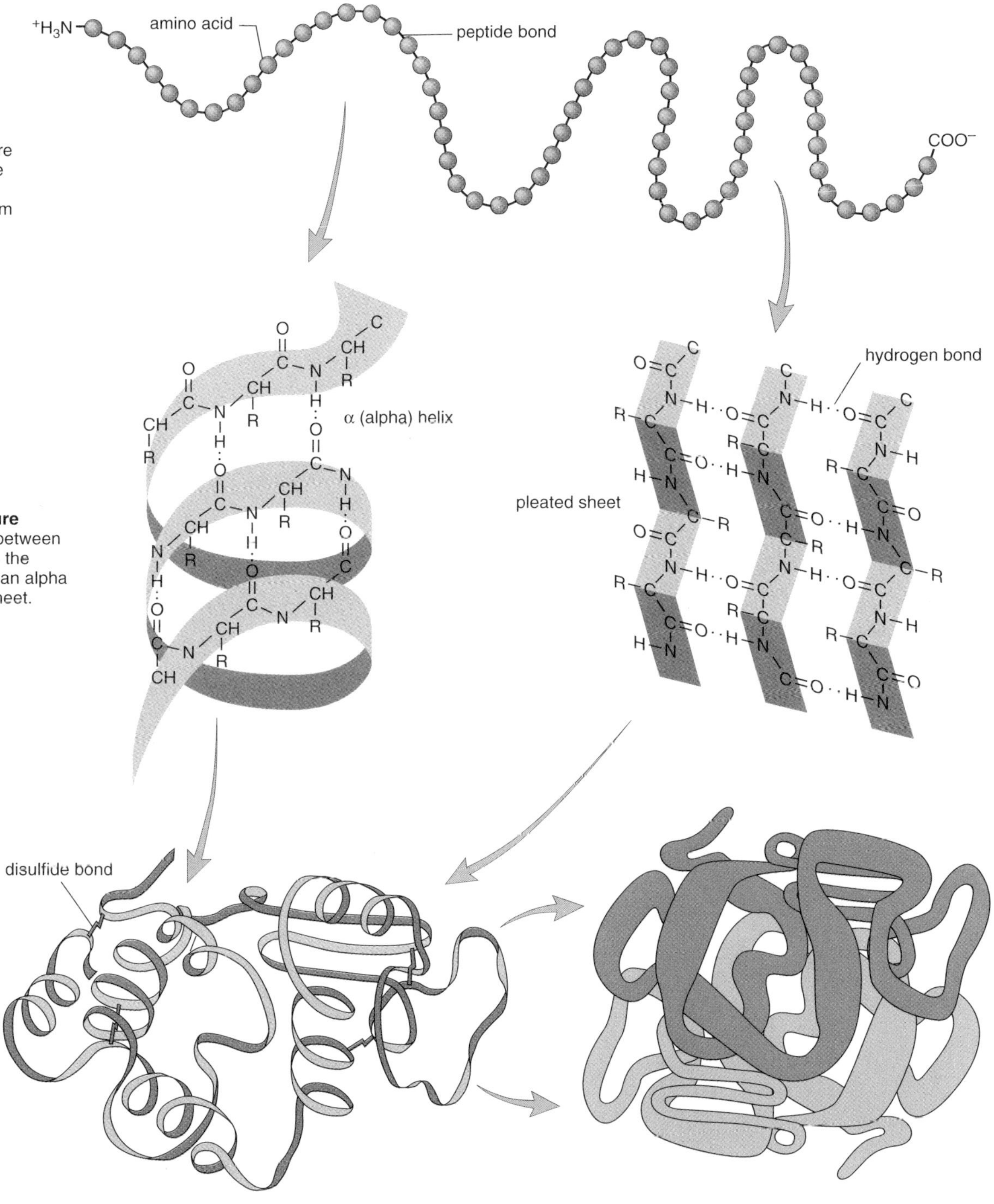

Primary Structure
This level of structure is determined by the sequence of amino acids that join to form a polypeptide.

amino acid

peptide bond

^+H_3N

COO^-

Secondary Structure
Hydrogen bonding between amino acids causes the polypeptide to form an alpha helix or a pleated sheet.

α (alpha) helix

pleated sheet

hydrogen bond

disulfide bond

Tertiary Structure
The helix folds into a characteristic globular shape due in part to covalent bonding between *R* groups.

Quaternary Structure
This level of structure occurs when two or more polypeptides join to form a single protein.

Levels of protein organization
Figure 3.17

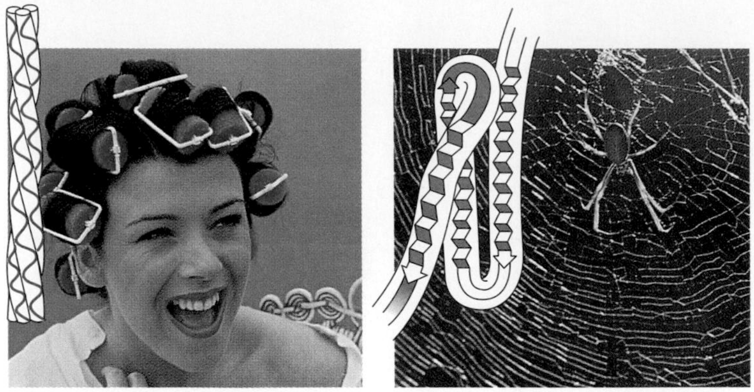

Fibrous proteins
Figure 3.18

a: © Vision MR/Photo Researchers, Inc.;
b: © Terry Whittaker/Photo Researchers, Inc.

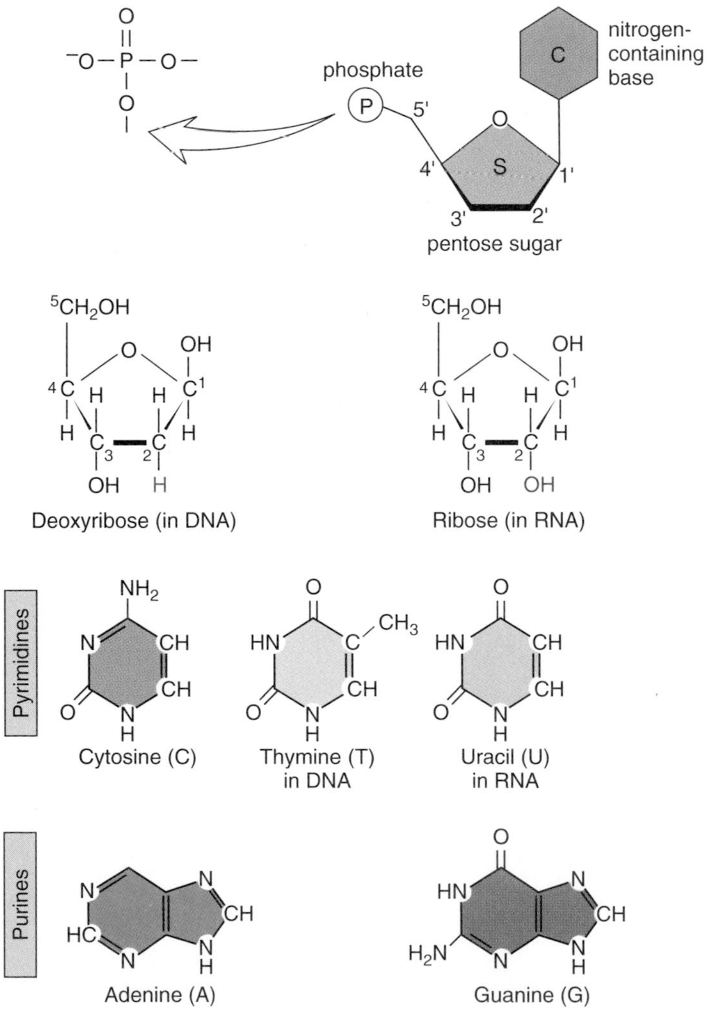

Nucleotides
Figure 3.19

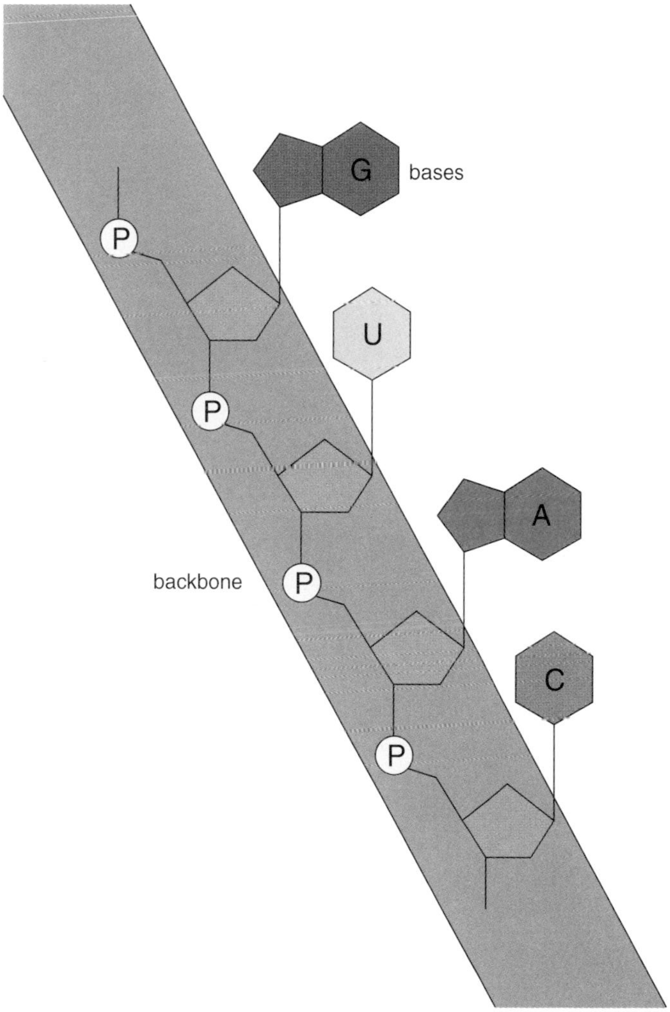

RNA structure
Figure 3.20

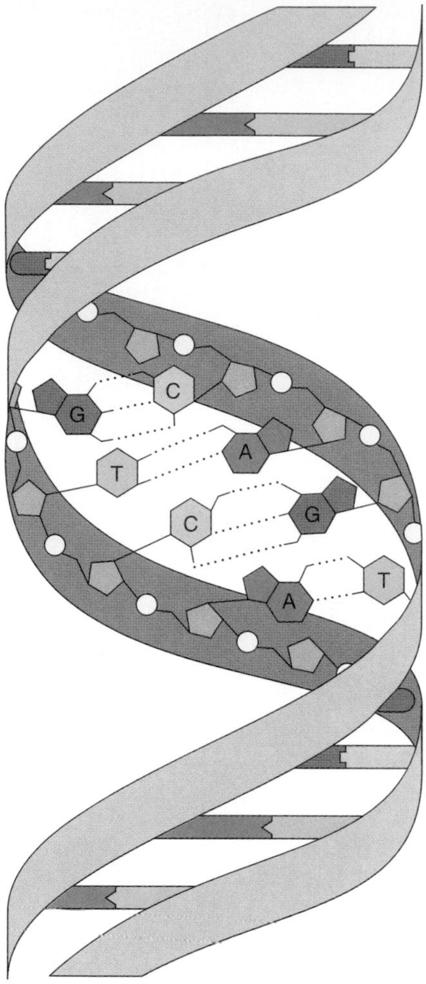

DNA structure
Figure 3.21

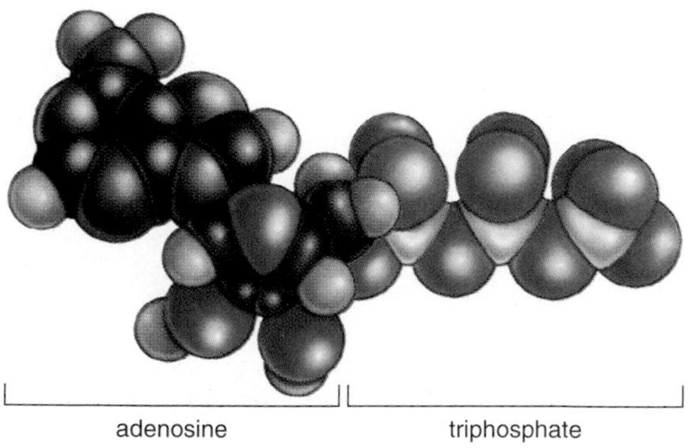

adenosine   triphosphate

ATP
Figure 3.22

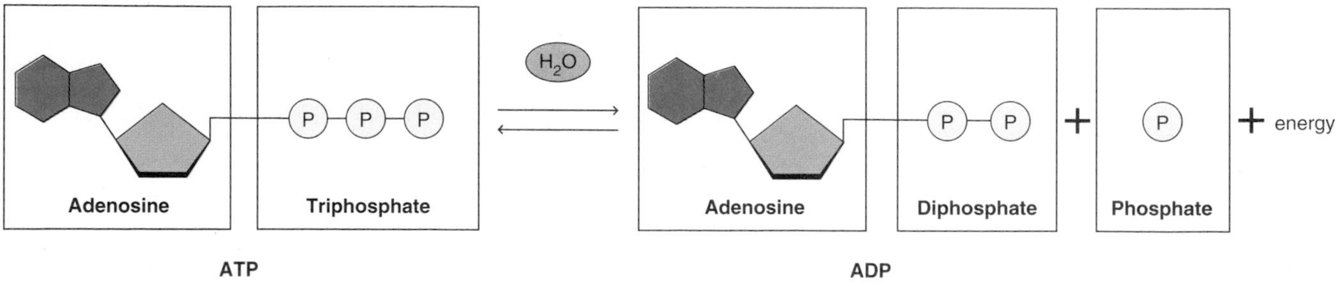

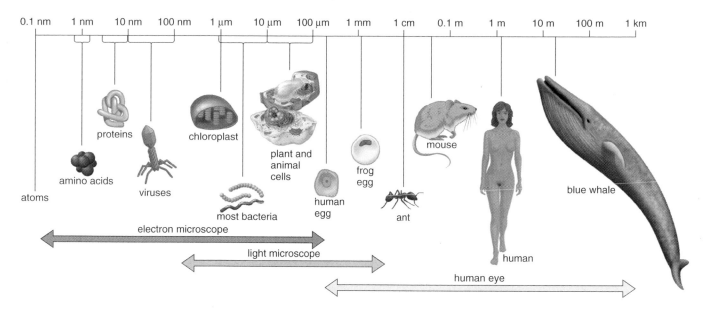

The sizes of living things and their components
Figure 4.2

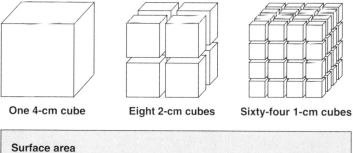

One 4-cm cube   Eight 2-cm cubes   Sixty-four 1-cm cubes

Surface area		
96 cm^2	192 cm^2	384 cm^2
Volume		
64 cm^3	64 cm^3	64 cm^3
Surface area: Volume per cube		
1.5:1	3:1	6:1

Surface-area-to-volume relationships
Figure 4.3

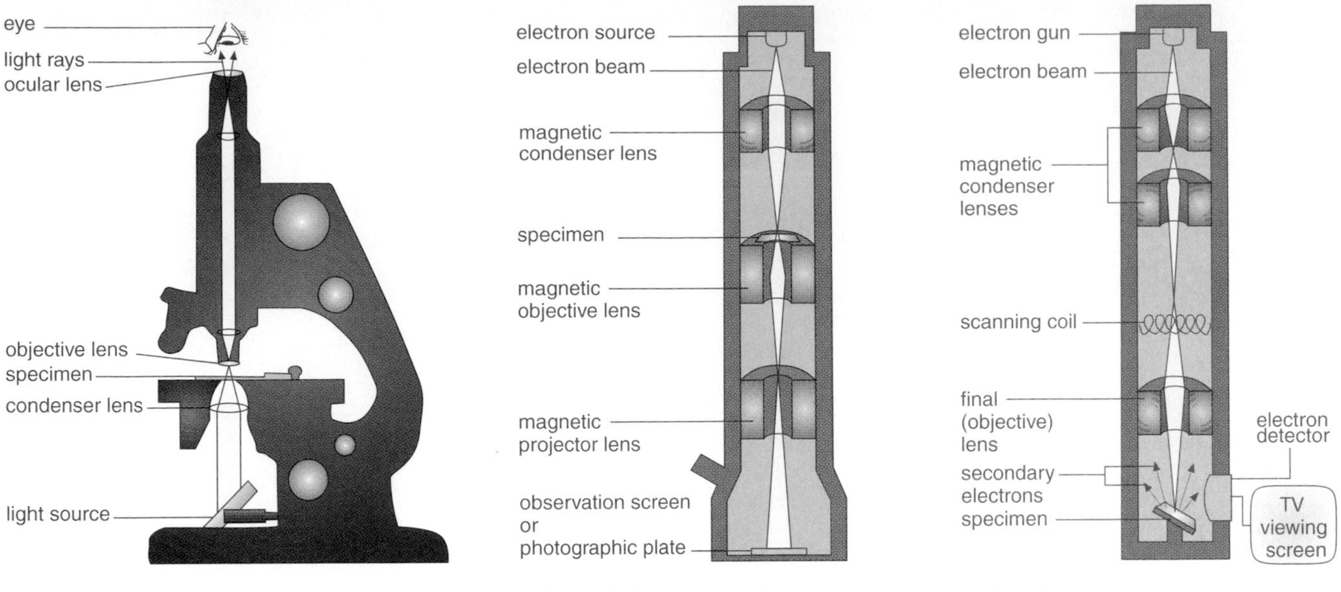

eye
light rays
ocular lens

objective lens
specimen
condenser lens

light source

a. Compound light microscope

electron source
electron beam

magnetic
condenser lens

specimen

magnetic
objective lens

magnetic
projector lens

observation screen
or
photographic plate

b. Transmission electron microscope

electron gun
electron beam

magnetic
condenser
lenses

scanning coil

final
(objective)
lens

secondary
electrons
specimen

electron
detector

TV
viewing
screen

c. Scanning electron microscope

Diagram of microscopes with accompanying micrographs of *Amoeba proteus*
Figure 4A

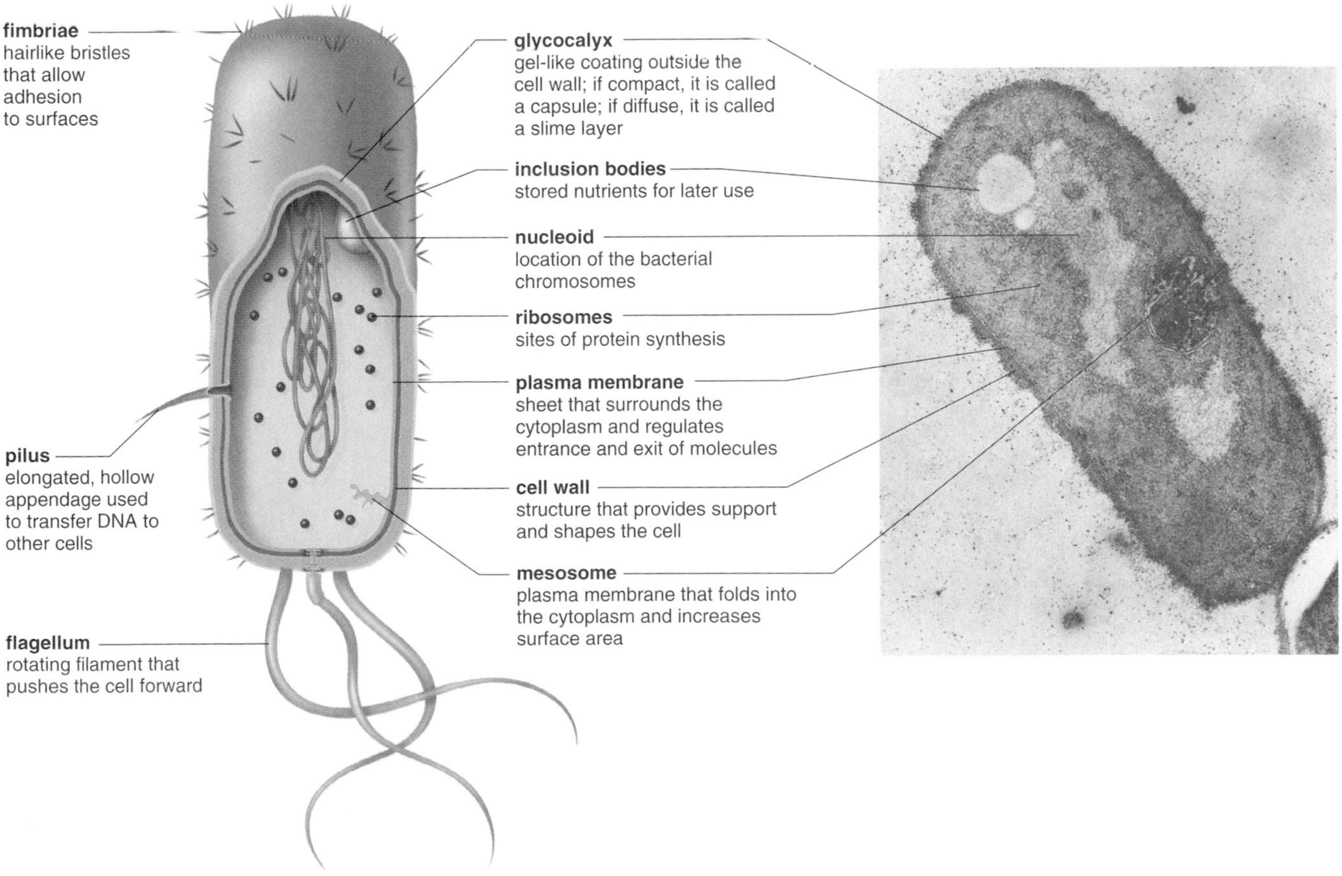

fimbriae
hairlike bristles
that allow
adhesion
to surfaces

glycocalyx
gel-like coating outside the
cell wall; if compact, it is called
a capsule; if diffuse, it is called
a slime layer

inclusion bodies
stored nutrients for later use

nucleoid
location of the bacterial
chromosomes

ribosomes
sites of protein synthesis

plasma membrane
sheet that surrounds the
cytoplasm and regulates
entrance and exit of molecules

cell wall
structure that provides support
and shapes the cell

mesosome
plasma membrane that folds into
the cytoplasm and increases
surface area

pilus
elongated, hollow
appendage used
to transfer DNA to
other cells

flagellum
rotating filament that
pushes the cell forward

Prokaryotic cell
Figure 4.4
© Ralph A. Slepecky/Visuals Unlimited

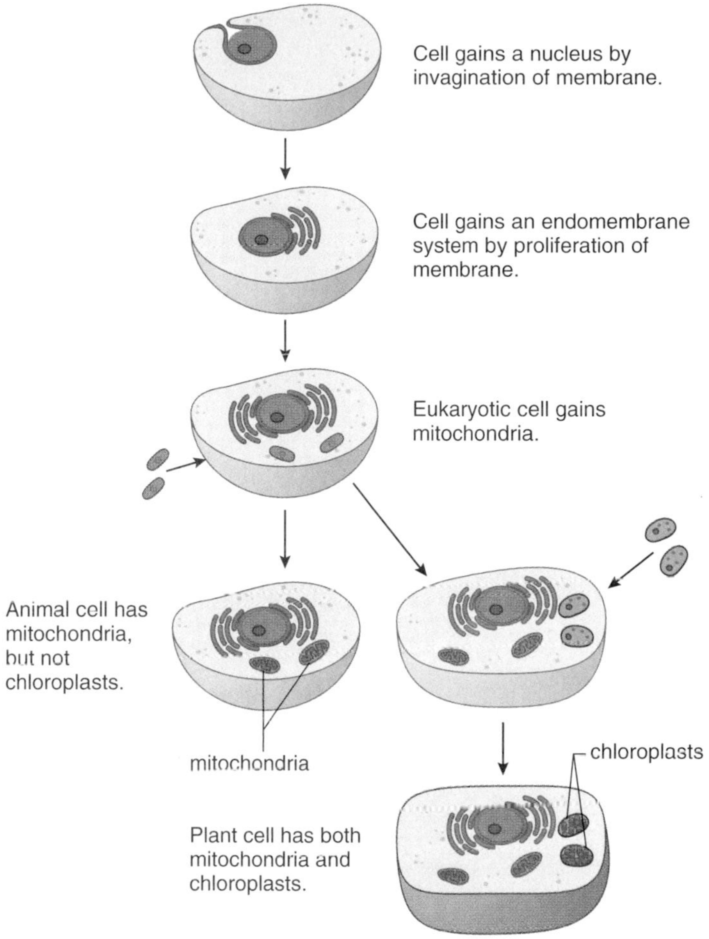

Cell gains a nucleus by invagination of membrane.

Cell gains an endomembrane system by proliferation of membrane.

Eukaryotic cell gains mitochondria.

Animal cell has mitochondria, but not chloroplasts.

mitochondria

chloroplasts

Plant cell has both mitochondria and chloroplasts.

Origin of organelles
Figure 4.5

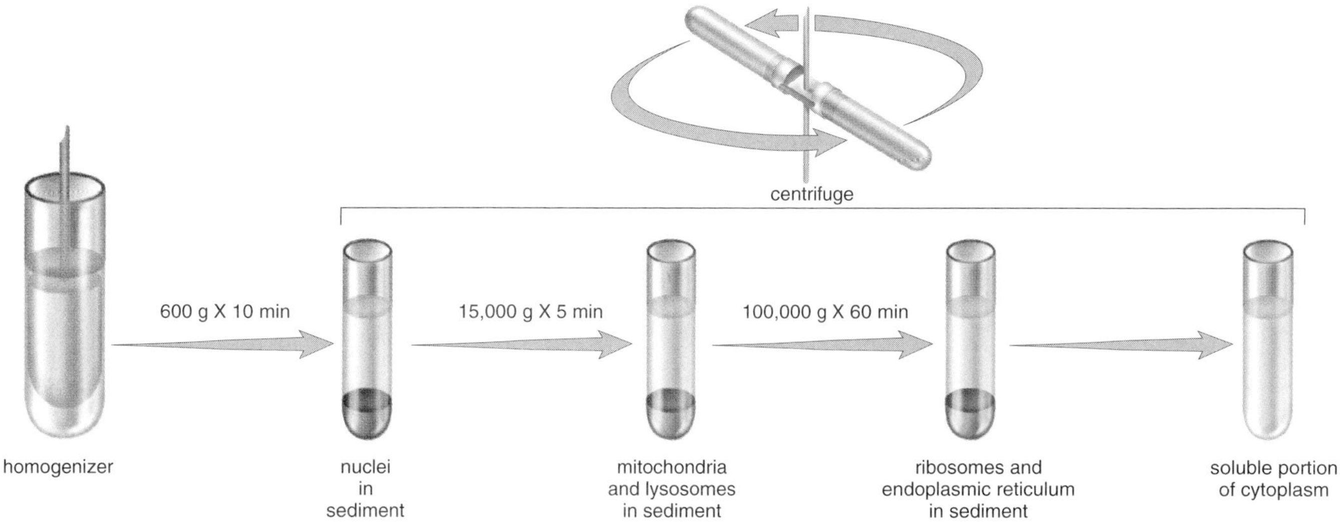

centrifuge

homogenizer

600 g X 10 min

nuclei in sediment

15,000 g X 5 min

mitochondria and lysosomes in sediment

100,000 g X 60 min

ribosomes and endoplasmic reticulum in sediment

soluble portion of cytoplasm

Cell fractionation and differential centrifugation
Figure 4C

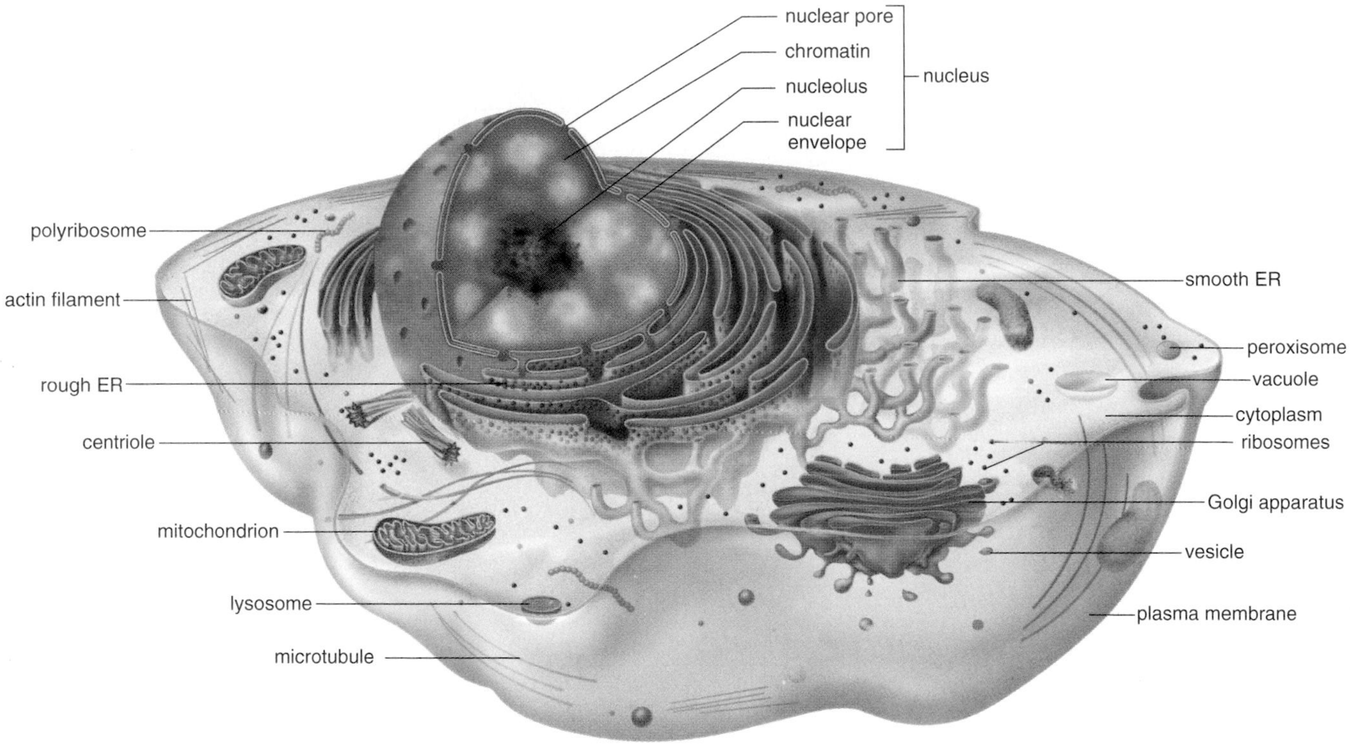

nuclear pore
chromatin
nucleolus
nuclear envelope
nucleus

polyribosome
actin filament
rough ER
centriole
mitochondrion
lysosome
microtubule

smooth ER
peroxisome
vacuole
cytoplasm
ribosomes
Golgi apparatus
vesicle
plasma membrane

Animal cell anatomy
Figure 4.6

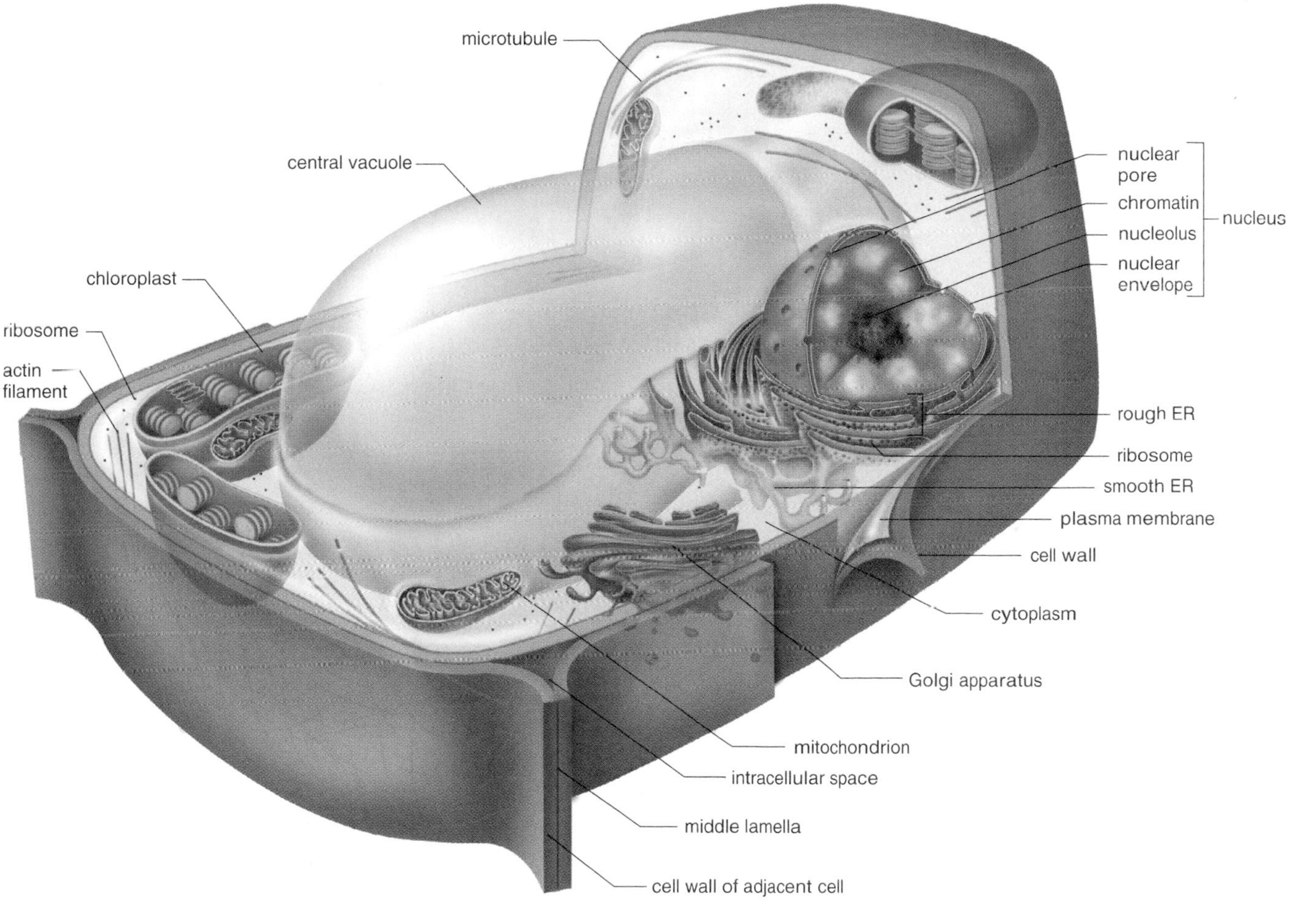

microtubule

central vacuole

chloroplast

ribosome

actin
filament

nuclear
pore

chromatin

nucleolus — nucleus

nuclear
envelope

rough ER

ribosome

smooth ER

plasma membrane

cell wall

cytoplasm

Golgi apparatus

mitochondrion

intracellular space

middle lamella

cell wall of adjacent cell

Plant cell anatomy
Figure 4.7

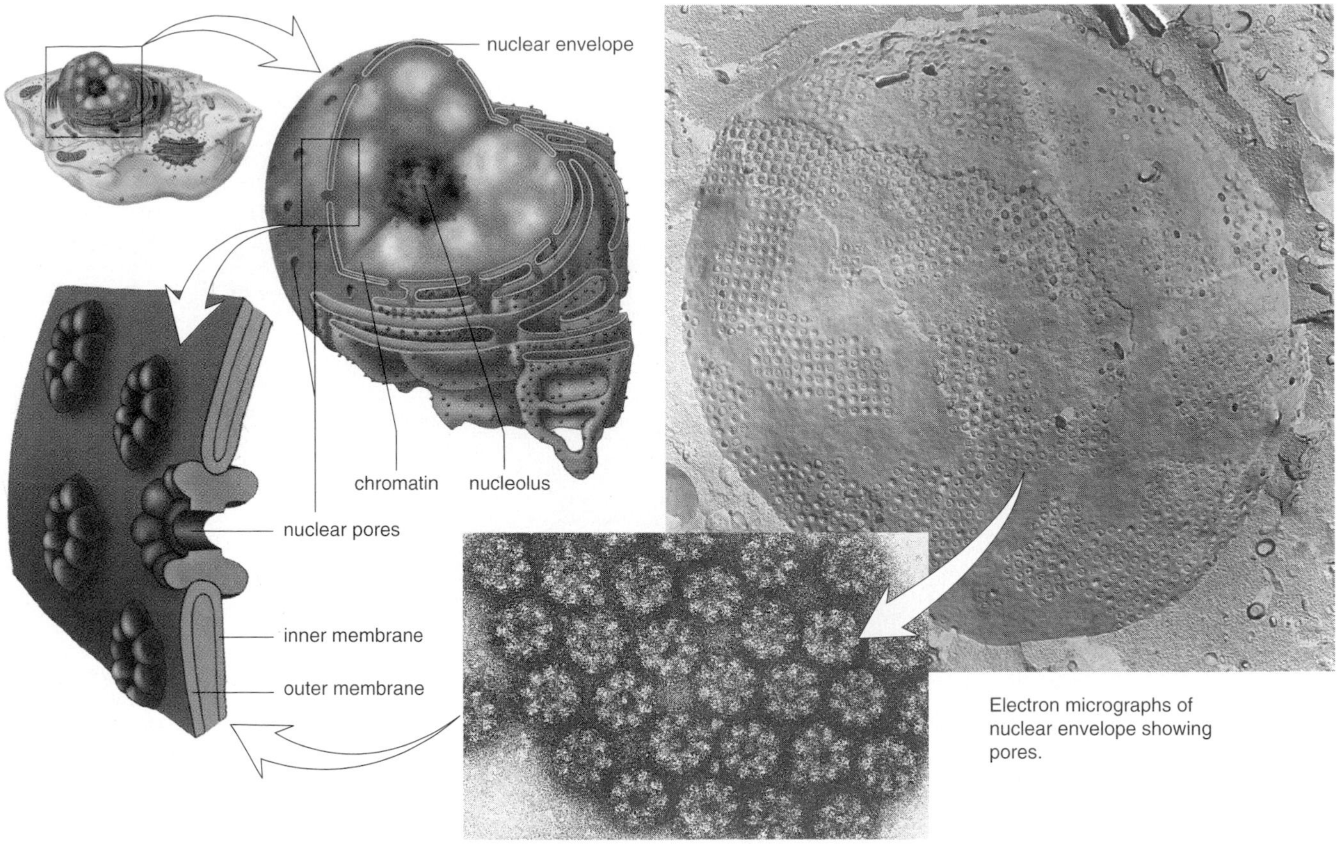

nuclear envelope

chromatin   nucleolus

nuclear pores

inner membrane

outer membrane

Electron micrographs of
nuclear envelope showing
pores.

Anatomy of the nucleus
Figure 4.8

right: Courtesy Ron Milligan/Scripps Research Institute; bottom: Courtesy E.G. Pollock

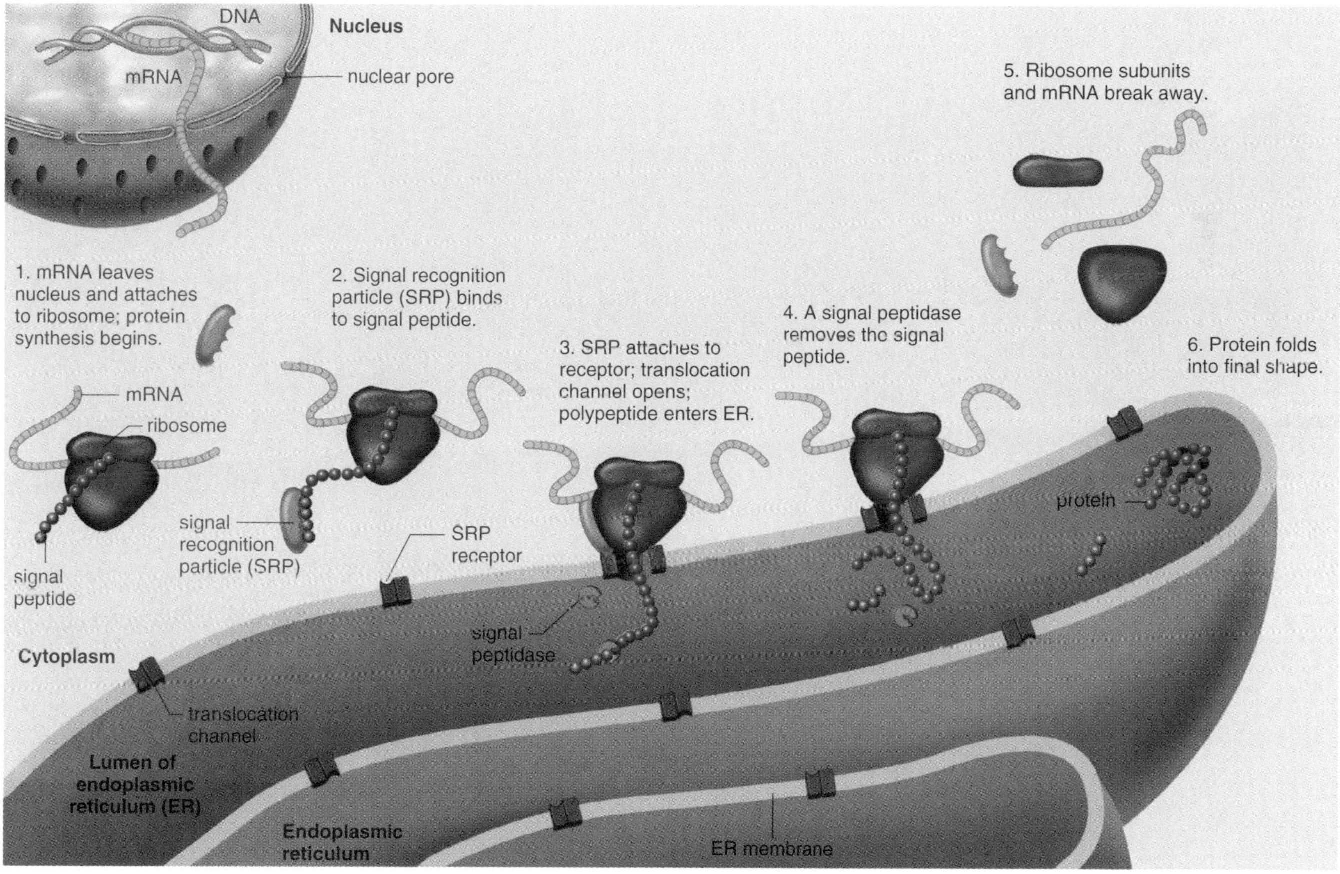

The nucleus, ribosomes, and endoplasmic reticulum (ER)
Figure 4.9

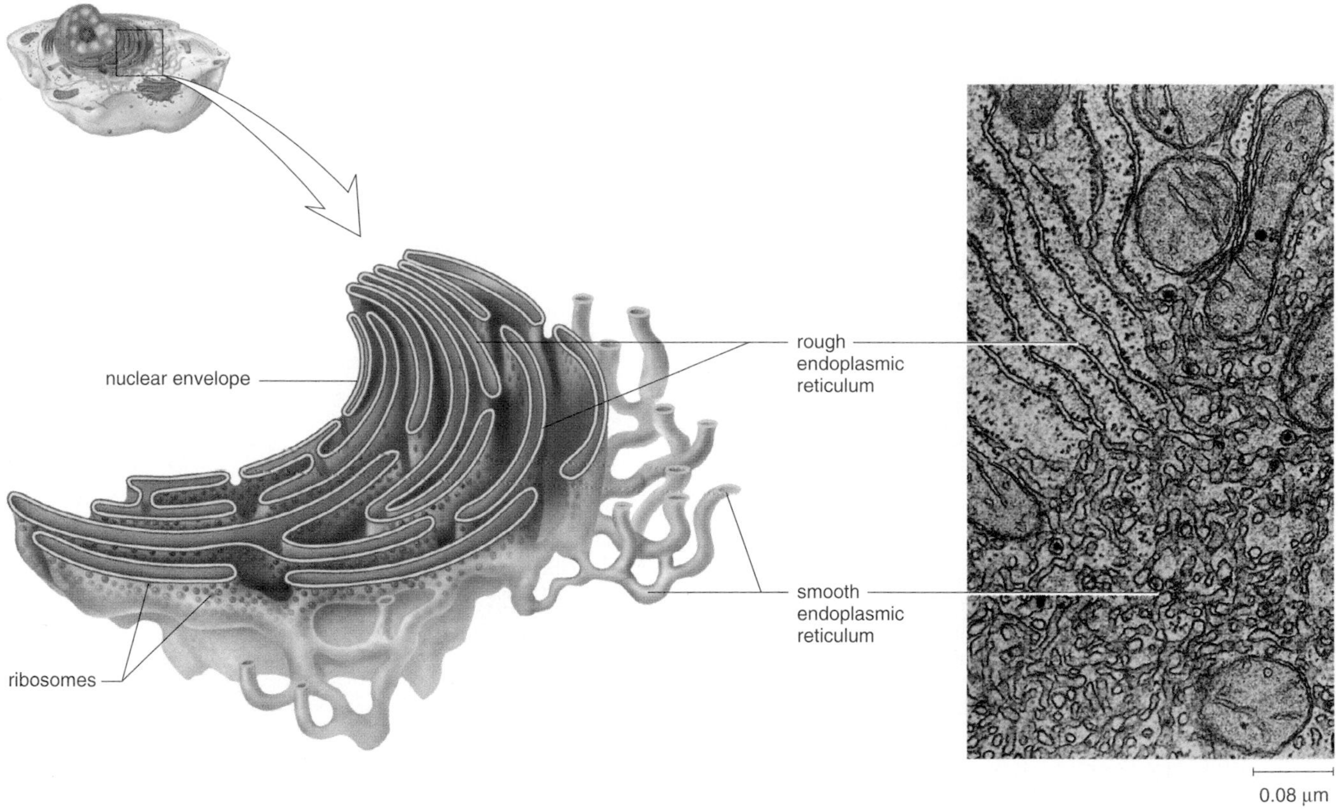

nuclear envelope

ribosomes

rough endoplasmic reticulum

smooth endoplasmic reticulum

0.08 μm

Endoplasmic reticulum (ER)
Figure 4.10
© R. Bolender & D. Fawcett/Visuals Unlimited

lysosome

peroxisome fragment

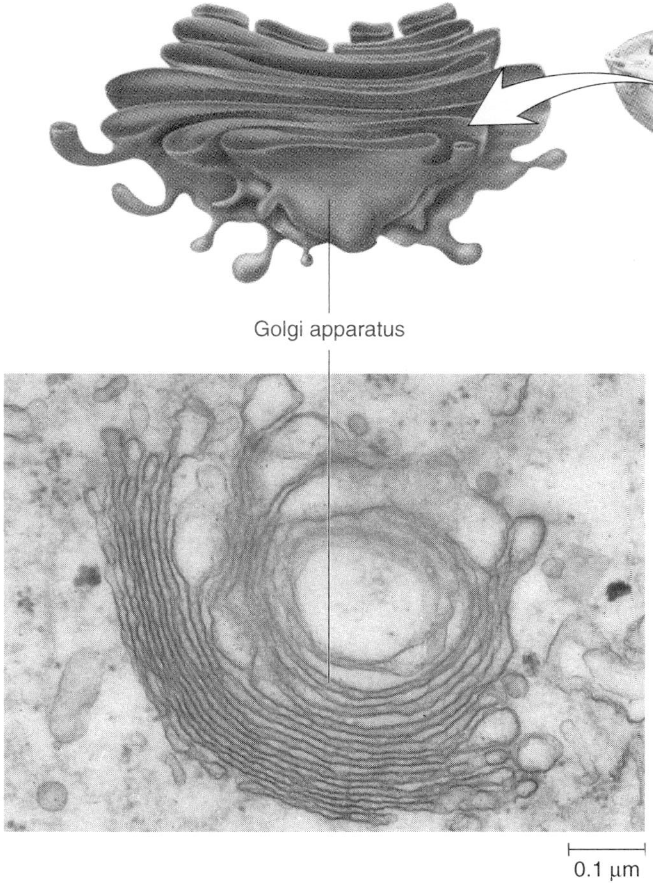

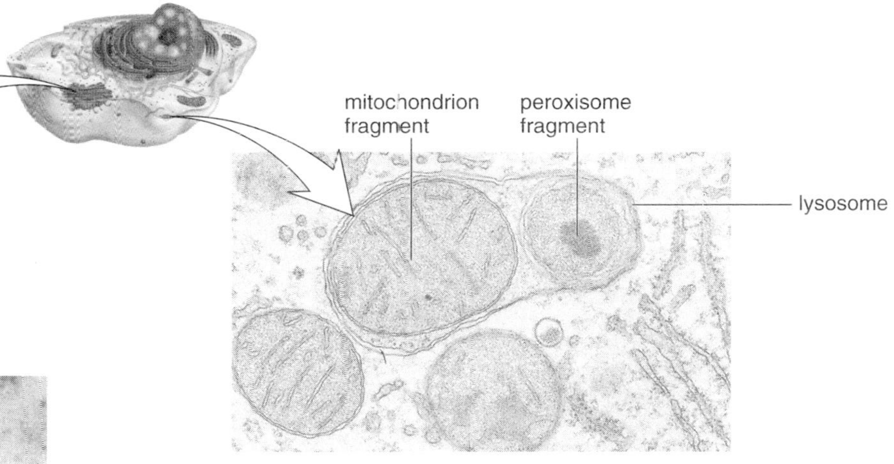

mitochondrion fragment

a. Mitochondrion and a peroxisome in a lysosome.

Golgi apparatus

0.1 μm

Golgi apparatus
Figure 4.11

Charles Courtesy Charles Flickinger, from *Journal of Cell Biology:* 49:221–226, 1971, Fig. 1 page 224

Lysosomes
Figure 4.12

a: Courtesy Daniel S. Friend

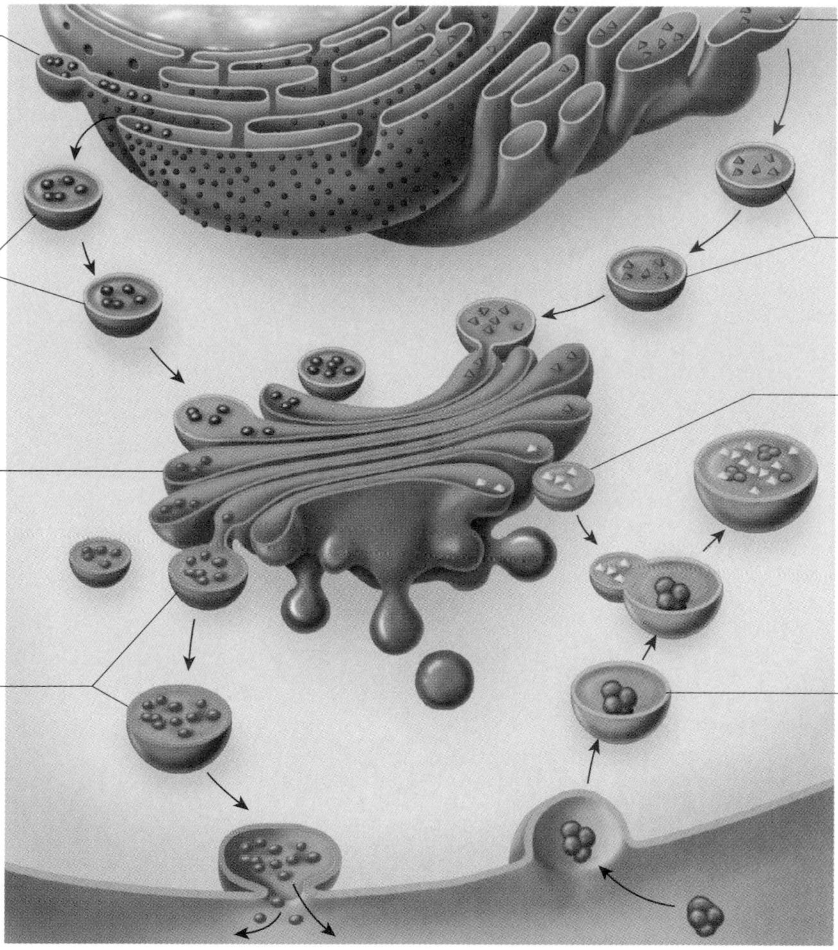

rough endoplasmic reticulum
synthesizes proteins and packages them in vesicles; vesicles commonly go to the Golgi apparatus

transport vesicles
shuttle proteins and lipids to various locations such as the Golgi apparatus

Golgi apparatus
modifies lipids and proteins from the ER; sorts them and packages them in vesicles

secretory vesicles
fuse with the plasma membrane as secretion occurs

smooth endoplasmic reticulum
synthesizes lipids and also performs various other functions

transport vesicles
shuttle proteins and lipids to various locations such as the Golgi apparatus

lysosomes
contain digestive enzymes that break down worn-out cell parts or substances entering the cell at the plasma membrane

incoming vesicles
bring substances into the cell that are digested when vesicle fuses with a lysosome

Endomembrane system
Figure 4.13

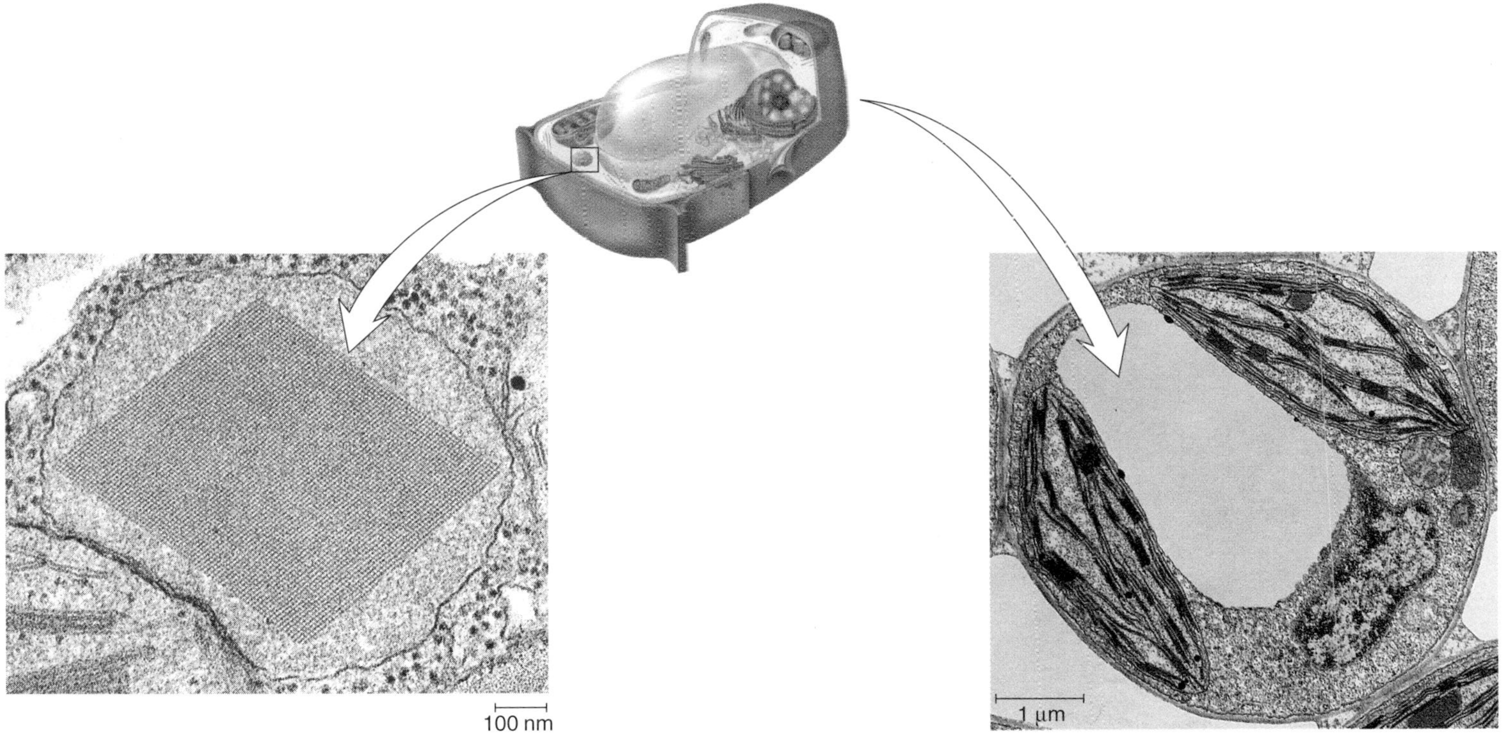

Plant cell vacuole
Figure 4.15
© Newcomb/Wergin/BPS/Tony Stone Images/Getty

Peroxisomes
Figure 4.14
© S.E. Frederick & E.H. Newcomb/Biological Photo Service

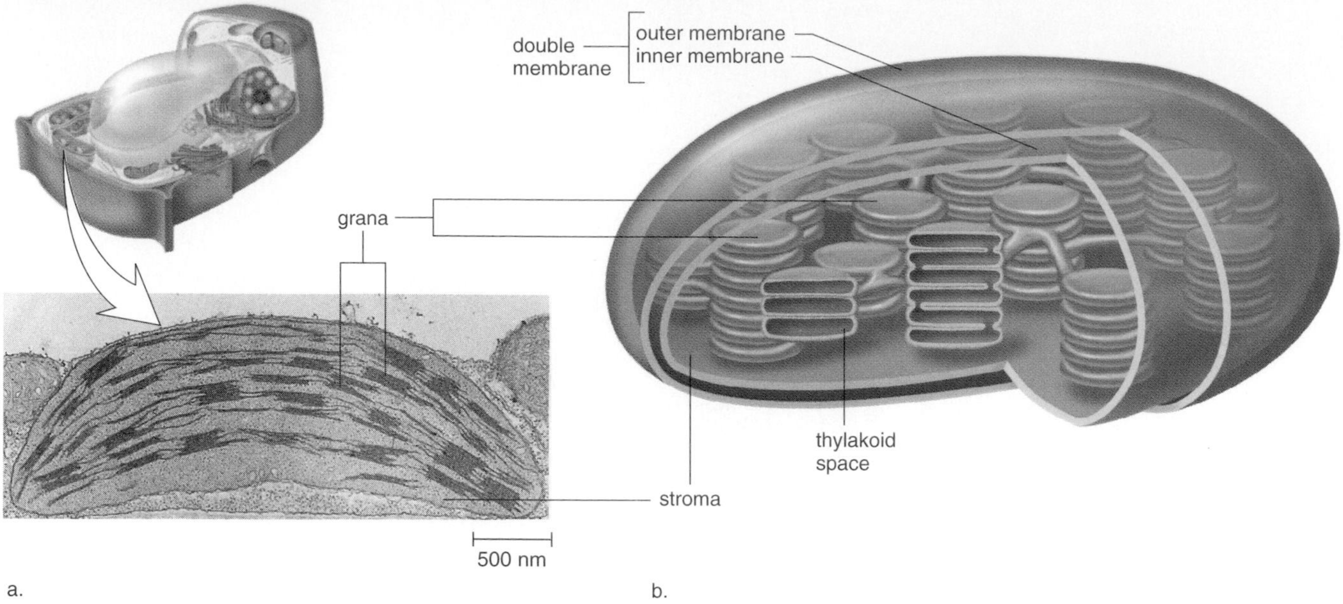

double membrane
- outer membrane
- inner membrane

grana

thylakoid space

stroma

500 nm

a.

b.

Chloroplast structure
Figure 4.16

a: Courtesy Herbert W. Israel, Cornell University

Mitochondrion structure
Figure 4.17

a: Courtesy Dr. Keith Porter

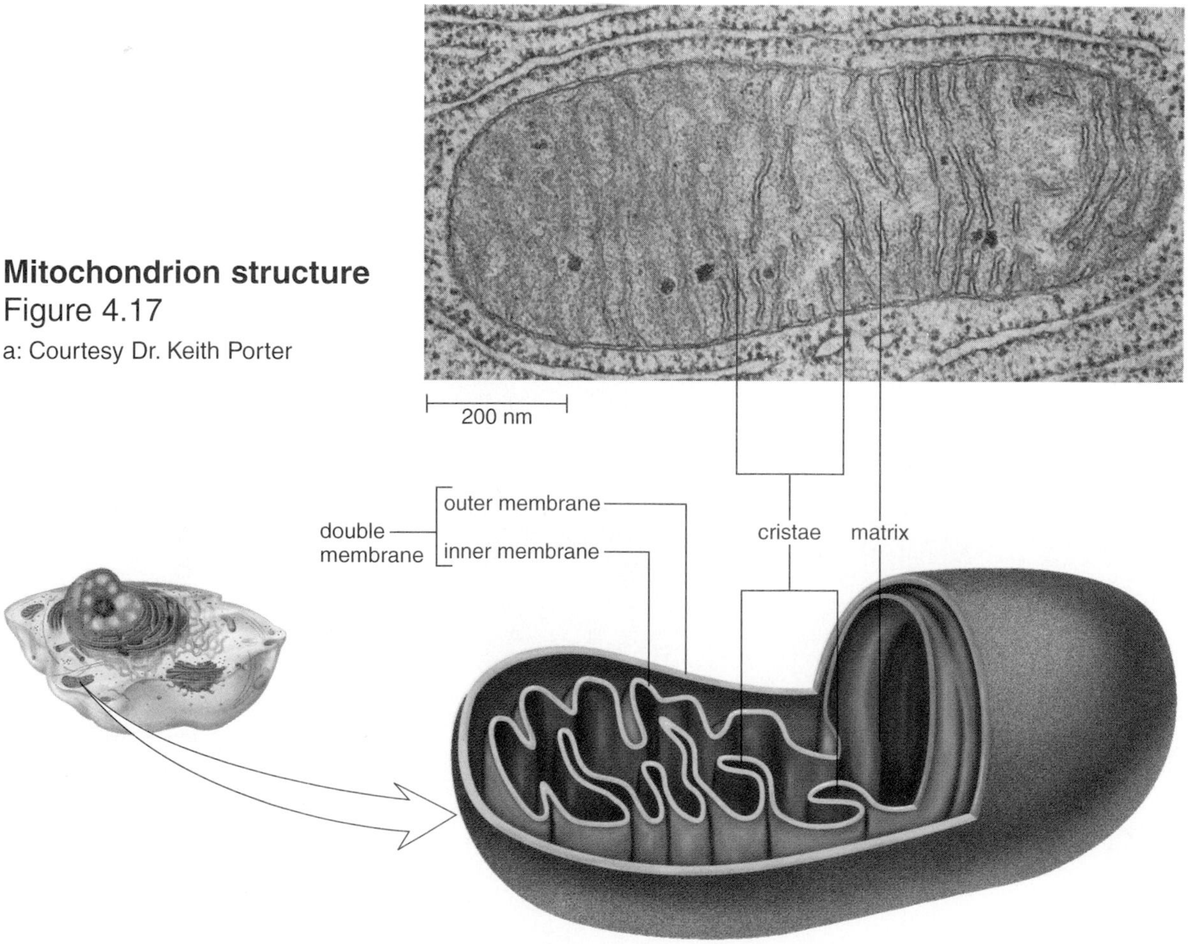

200 nm

double membrane
- outer membrane
- inner membrane

cristae

matrix

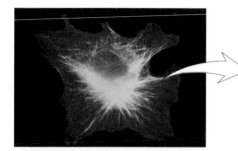

a. Actin filaments

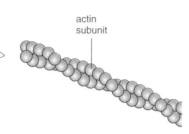

actin
subunit

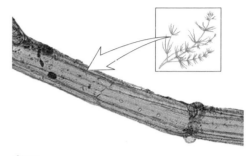

Chara

b. Intermediate filaments

fibrous
subunits

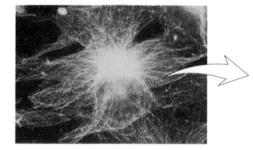

c. Microtubules

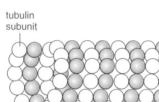

tubulin
subunit

The cytoskeleton
Figure 4.18

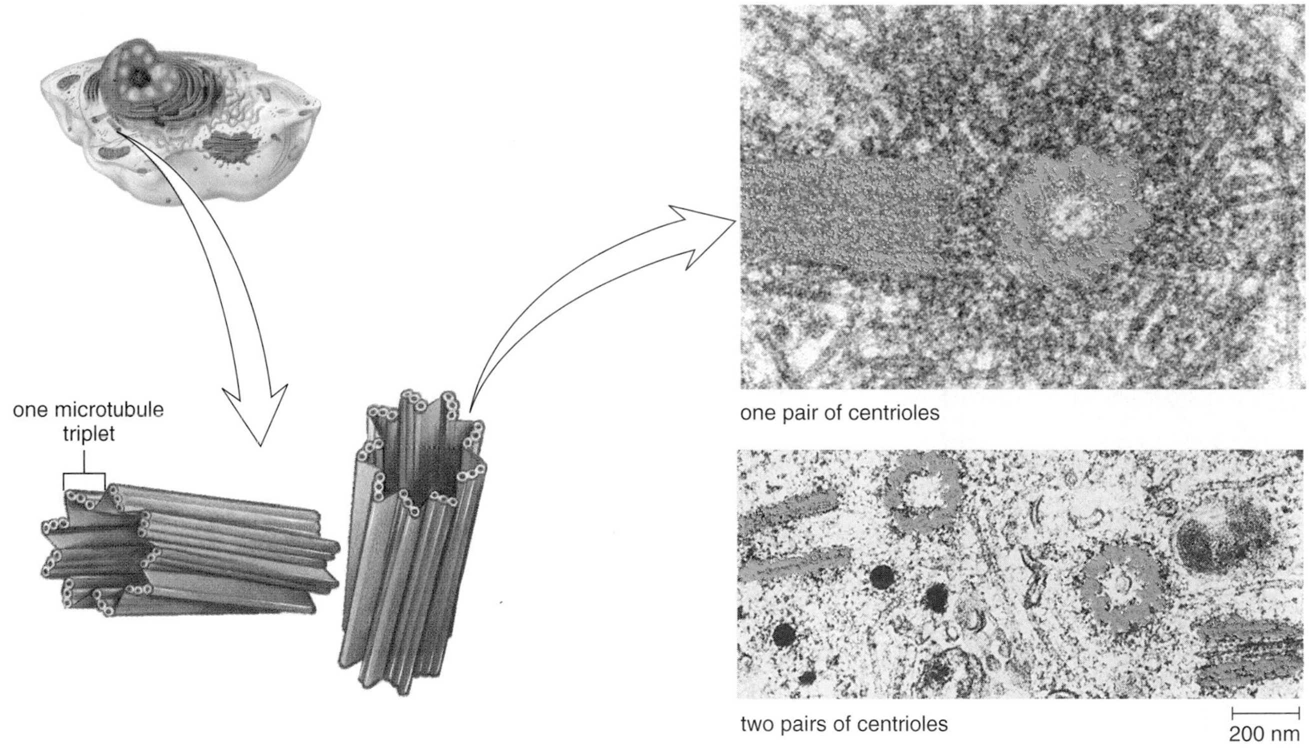

one microtubule triplet

one pair of centrioles

two pairs of centrioles

200 nm

Centrioles

Figure 4.19

top: Courtesy Kent McDonald, University of Colorado Boulder; bottom: From Manley McGill, D.P. Highfield, T.M. Monahan, and B.R. Brinkley, *Journal of Ultrastructure Research 57*, 43-53 pg. 48, fig. 6, (1976) Academic Press

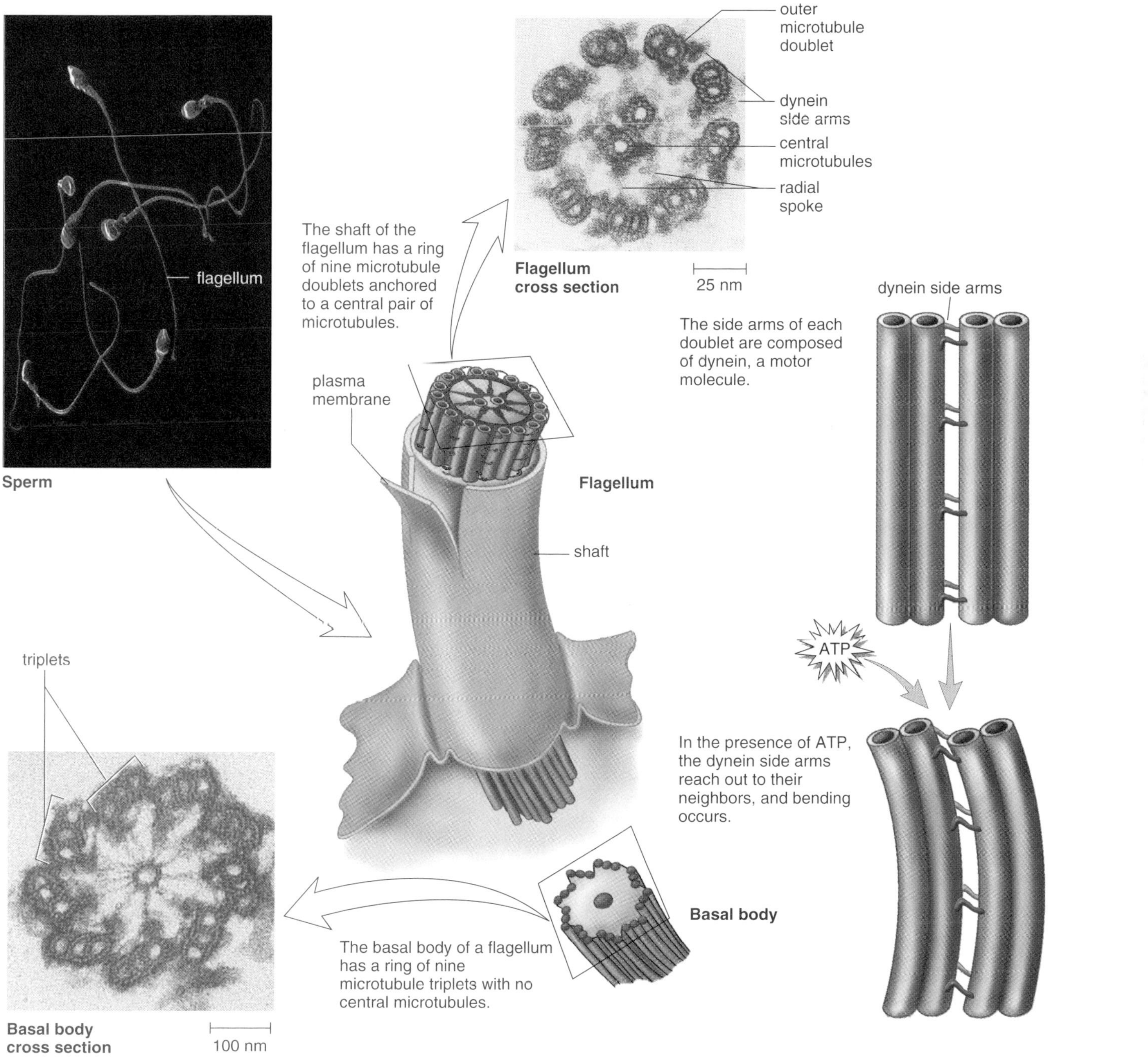

Sperm

flagellum

The shaft of the flagellum has a ring of nine microtubule doublets anchored to a central pair of microtubules.

plasma membrane

outer microtubule doublet

dynein side arms

central microtubules

radial spoke

Flagellum cross section

25 nm

The side arms of each doublet are composed of dynein, a motor molecule.

dynein side arms

Flagellum

shaft

ATP

In the presence of ATP, the dynein side arms reach out to their neighbors, and bending occurs.

triplets

Basal body

The basal body of a flagellum has a ring of nine microtubule triplets with no central microtubules.

Basal body cross section

100 nm

Structure of cilium or flagellum
Figure 4.20

Sperm: © David M. Phillips/Photo Researchers, Inc.; Flagellum, Basal body: © William L. Dentler/Biological Photo Service

b. Two possible models

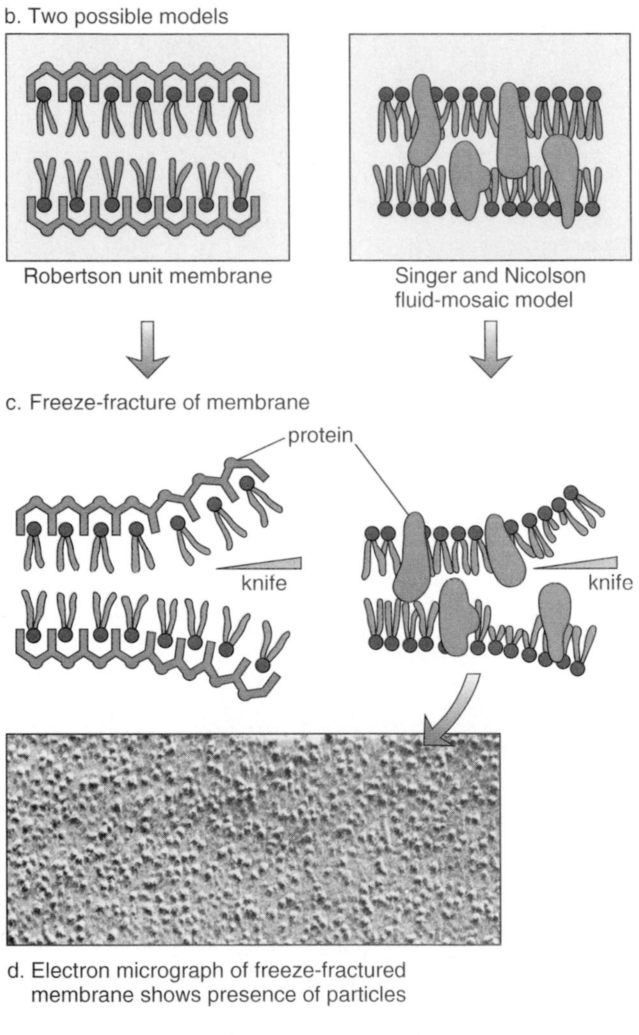

Robertson unit membrane

Singer and Nicolson
fluid-mosaic model

c. Freeze-fracture of membrane

protein

knife

knife

d. Electron micrograph of freeze-fractured
membrane shows presence of particles

Membrane structure
Figure 5.1

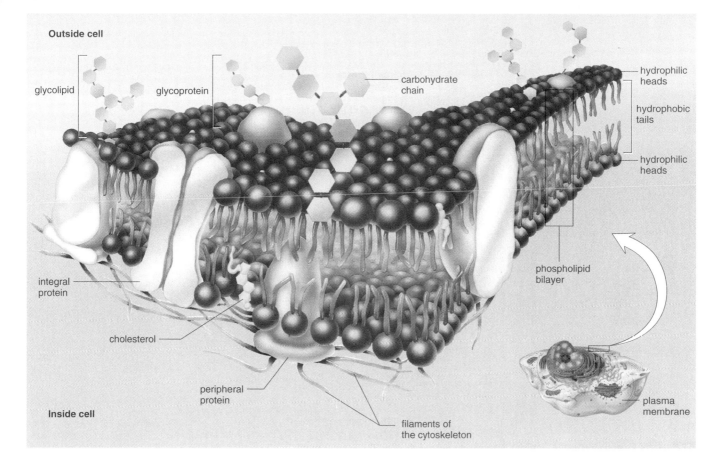

Fluid-mosaic model of plasma membrane structure
Figure 5.2

mouse cell   human cell

cell fusion

immediately after fusion

mixed membrane proteins

Experiment to demonstrate lateral drifting of plasma membrane proteins
Figure 5.3

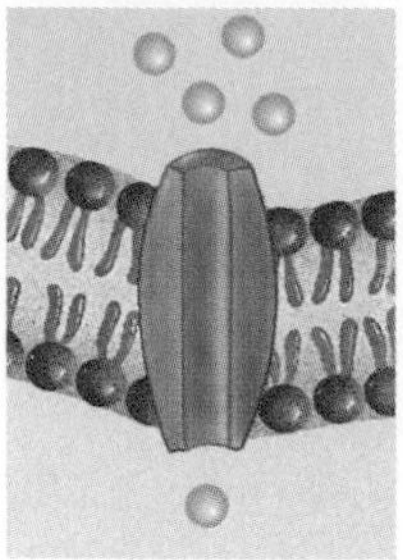

Channel Protein
Allows a particular molecule or ion to cross the plasma membrane freely. Cystic fibrosis, an inherited disorder, is caused by a faulty chloride (Cl⁻) channel; a thick mucus collects in airways and in pancreatic and liver ducts.

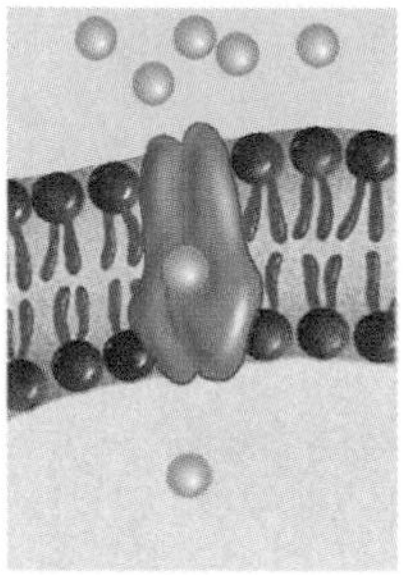

Carrier Protein
Selectively interacts with a specific molecule or ion so that it can cross the plasma membrane. The inability of some persons to use energy for sodium-potassium (Na⁺–K⁺) transport has been suggested as the cause of their obesity.

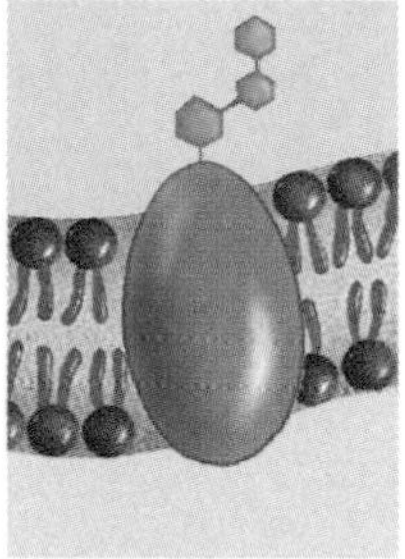

Cell Recognition Protein
The MHC (major histocompatibility complex) glycoproteins are different for each person, so organ transplants are difficult to achieve. Cells with foreign MHC glycoproteins are attacked by blood cells responsible for immunity.

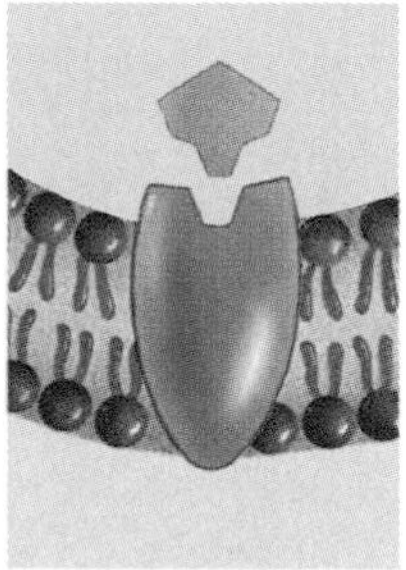

Receptor Protein
Is shaped in such a way that a specific molecule can bind to it. Pygmies are short, not because they do not produce enough growth hormone, but because their plasma membrane growth hormone receptors are faulty and cannot interact with growth hormone.

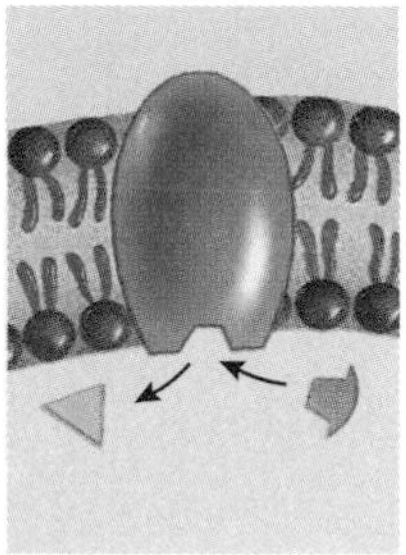

Enzymatic Protein
Catalyzes a specific reaction. The membrane protein, adenylate cyclase, is involved in ATP metabolism. Cholera bacteria release a toxin that interferes with the proper functioning of adenylate cyclase; sodium ions and water leave intestinal cells, and the individual dies from severe diarrhea.

Membrane protein diversity
Figure 5.4

How molecules cross the plasma membrane
Figure 5.5

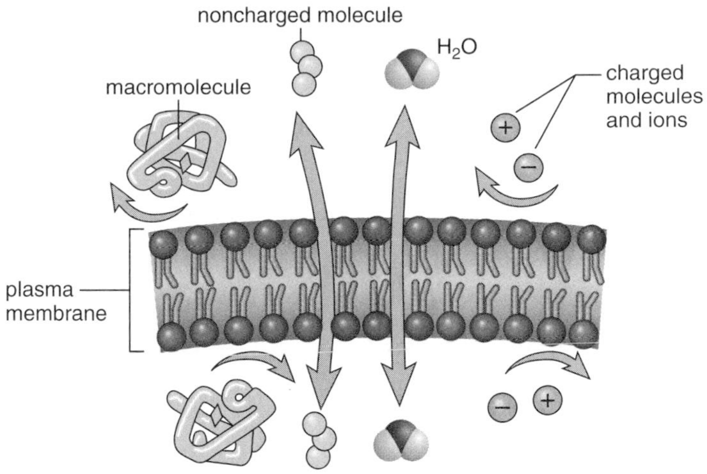

noncharged molecule

macromolecule

H_2O

charged molecules and ions

plasma membrane

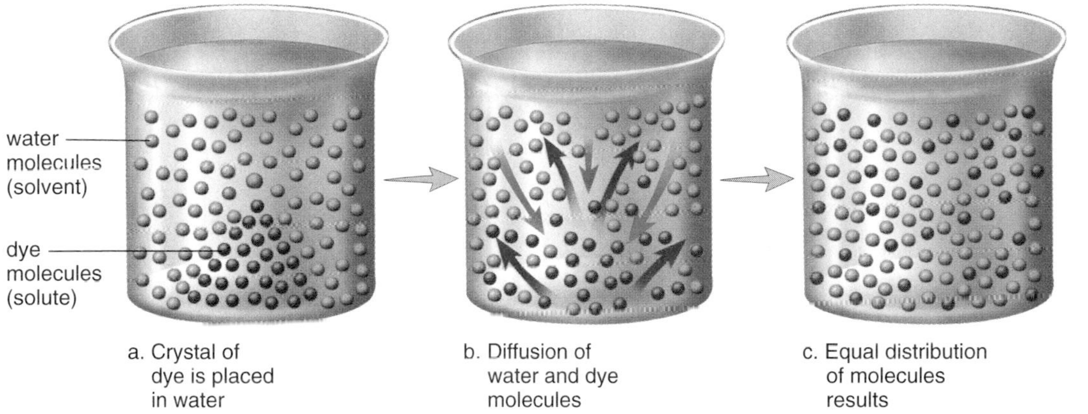

water molecules (solvent)

dye molecules (solute)

a. Crystal of dye is placed in water

b. Diffusion of water and dye molecules

c. Equal distribution of molecules results

Process of diffusion
Figure 5.6

Gas exchange in lungs
Figure 5.7

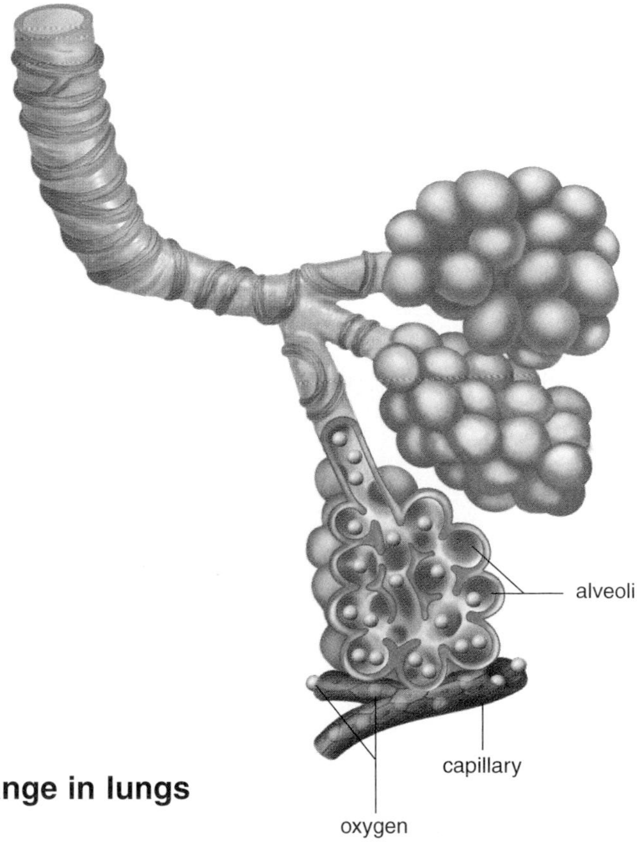

alveoli

capillary

oxygen

41

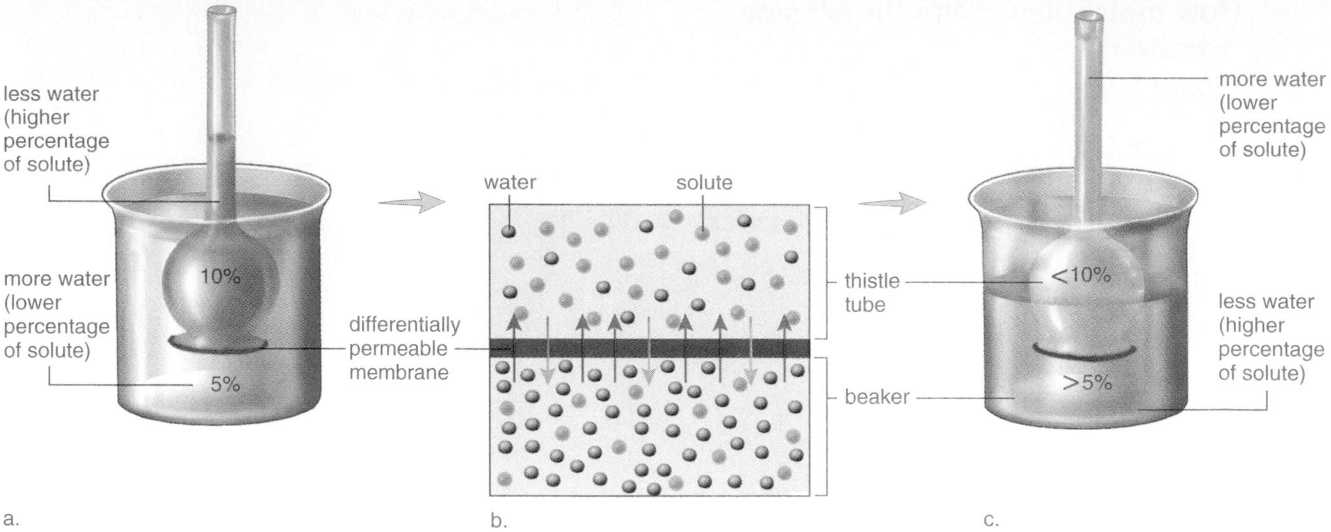

less water
(higher
percentage
of solute)

more water
(lower
percentage
of solute)

10%

5%

a.

differentially
permeable
membrane

water   solute

thistle
tube

beaker

b.

more water
(lower
percentage
of solute)

less water
(higher
percentage
of solute)

<10%

>5%

c.

Osmosis demonstration
Figure 5.8

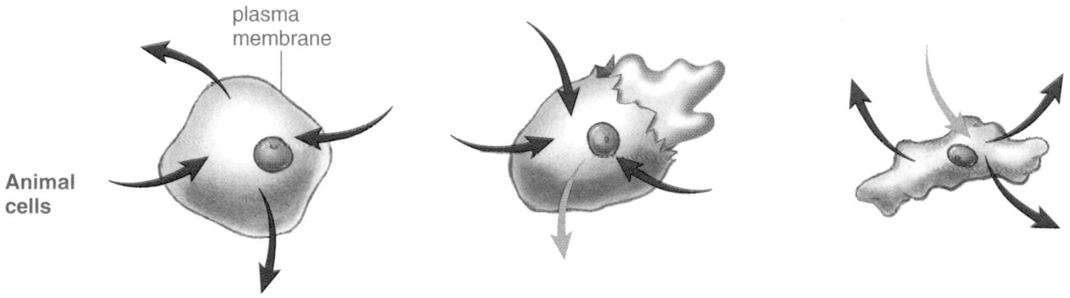

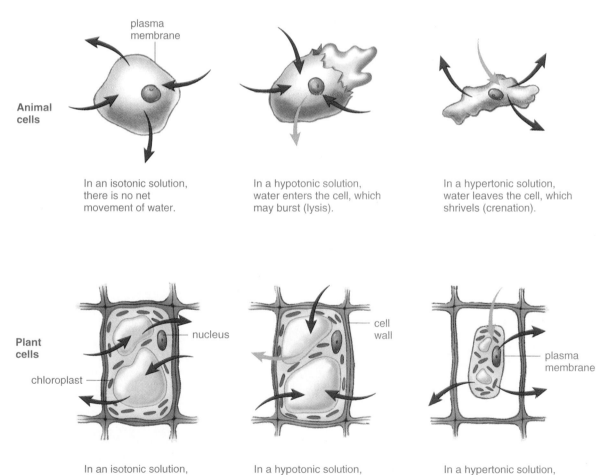

Animal cells

plasma membrane

In an isotonic solution, there is no net movement of water.

In a hypotonic solution, water enters the cell, which may burst (lysis).

In a hypertonic solution, water leaves the cell, which shrivels (crenation).

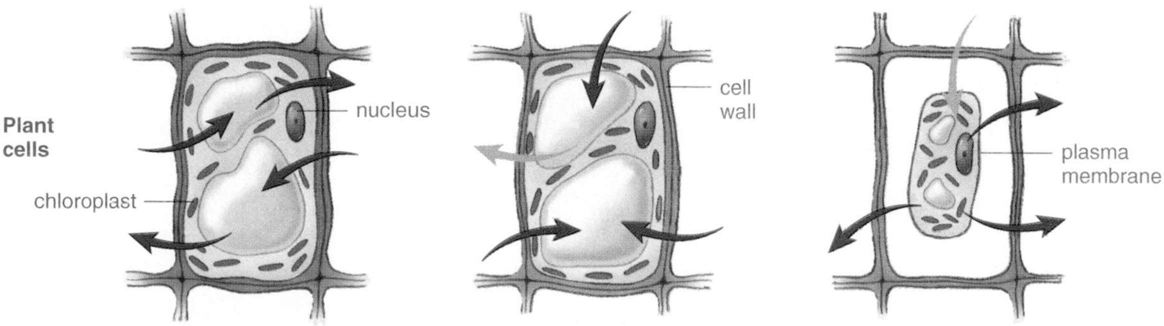

Plant cells

nucleus

chloroplast

cell wall

plasma membrane

In an isotonic solution, there is no net movement of water.

In a hypotonic solution, vacuoles fill with water, turgor pressure develops, and chloroplasts are seen next to the cell wall.

In a hypertonic solution, vacuoles lose water, the cytoplasm shrinks (plasmolysis), and chloroplasts are seen in the center of the cell.

Osmosis in animal and plant cells
Figure 5.9

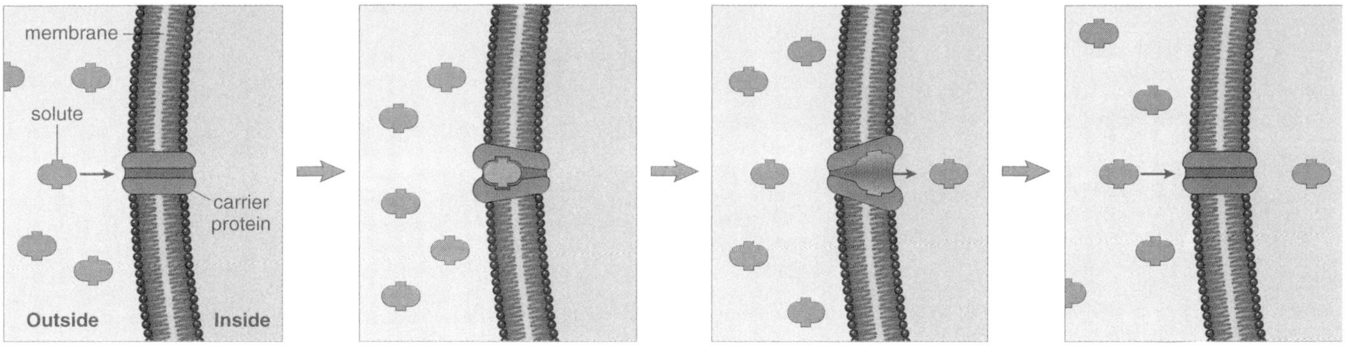

Facilitated transport
Figure 5.10

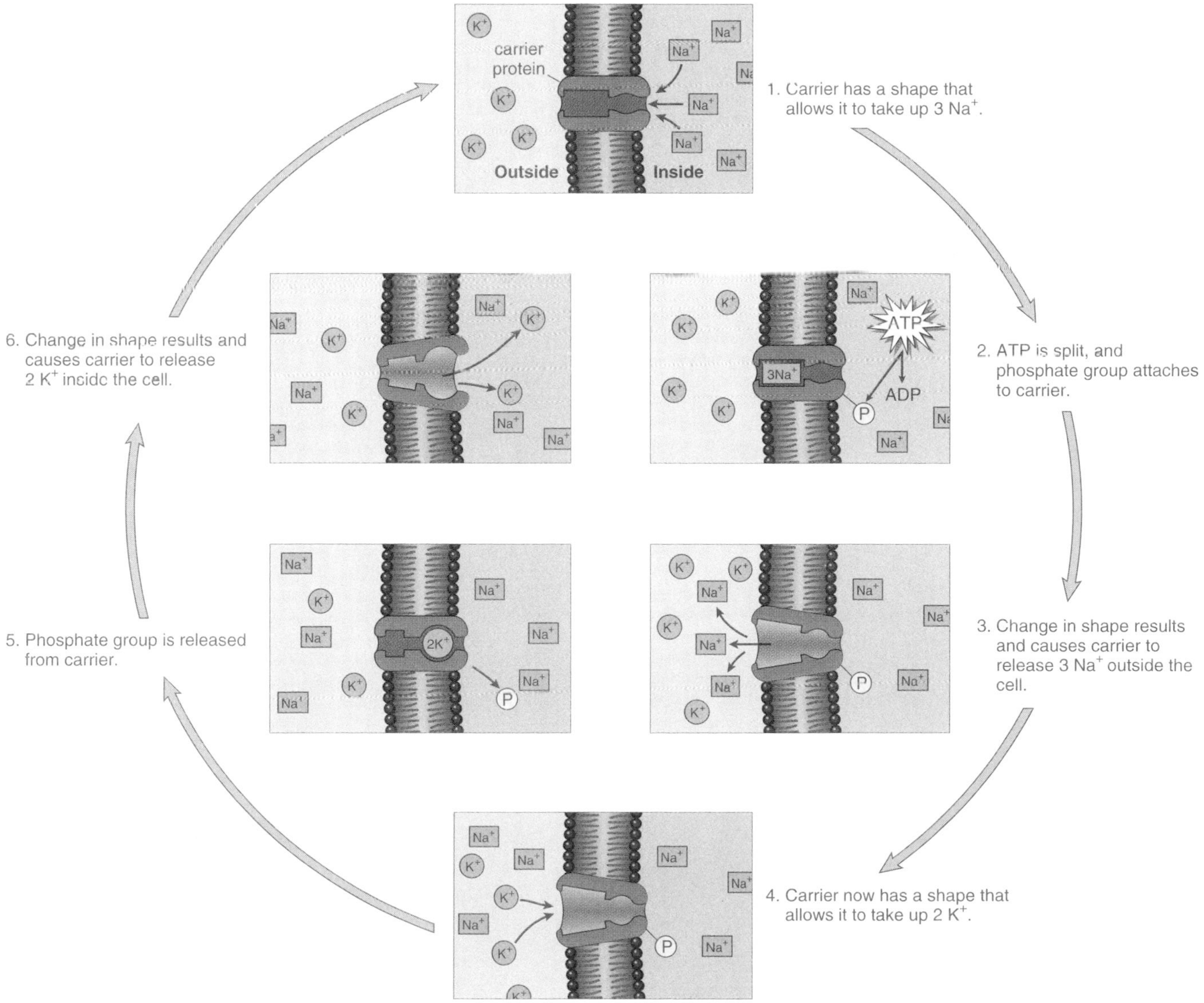

1. Carrier has a shape that allows it to take up 3 Na⁺.

2. ATP is split, and phosphate group attaches to carrier.

3. Change in shape results and causes carrier to release 3 Na⁺ outside the cell.

4. Carrier now has a shape that allows it to take up 2 K⁺.

5. Phosphate group is released from carrier.

6. Change in shape results and causes carrier to release 2 K⁺ inside the cell.

The sodium-potassium pump
Figure 5.11

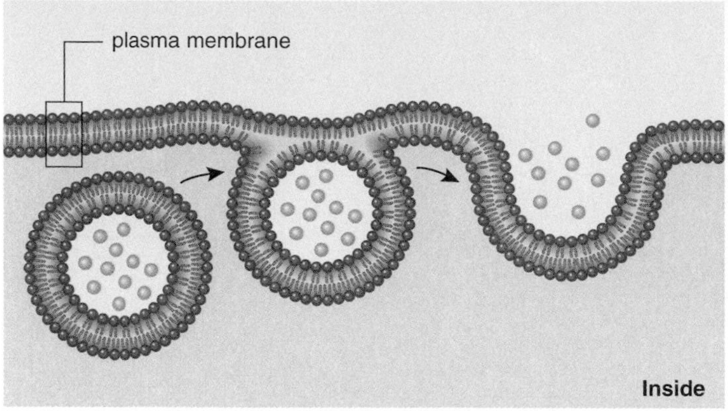

Exocytosis
Figure 5.12

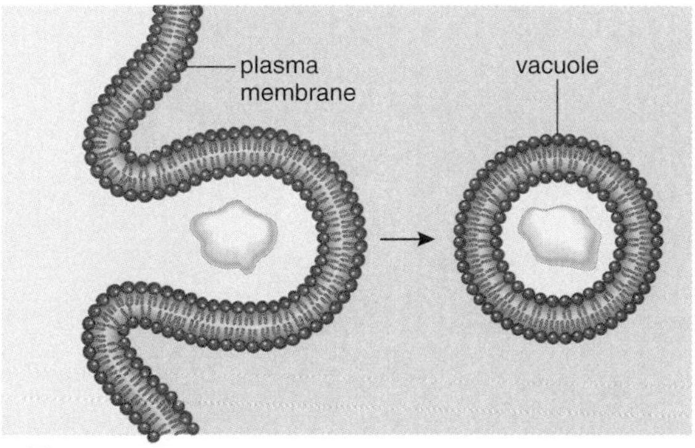

a. Phagocytosis

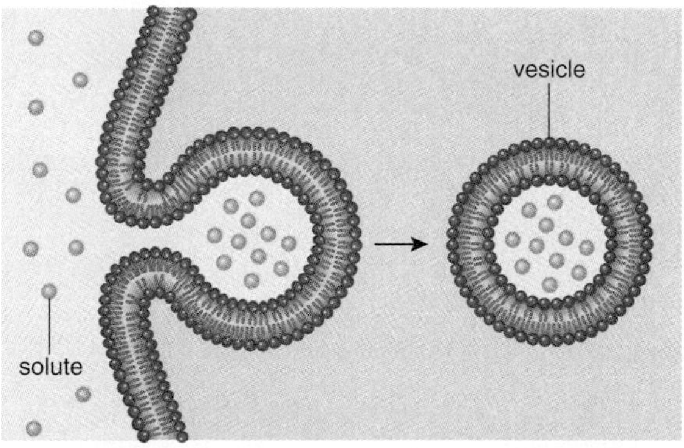

b. Pinocytosis

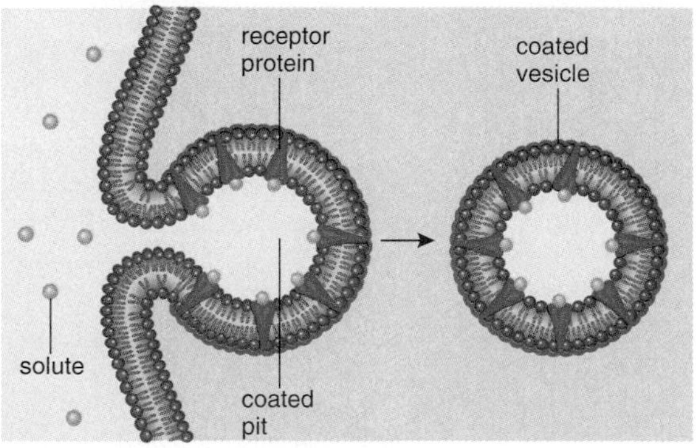

c. Receptor-mediated endocytosis

Three methods of endocytosis
Figure 5.13

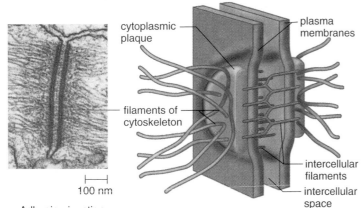

cytoplasmic plaque

plasma membranes

filaments of cytoskeleton

intercellular filaments

intercellular space

100 nm

a. Adhesion junction

Junctions between cells of the intestinal wall

Figure 5.14

a: Courtesy Camillo Peracchia; b: © David M. Phillips/Visuals Unlimited; c: From Douglas E. Kelly, *J. Cell Biol. 28* (1966): 51. Reproduced by copyright permission of The Rockefeller University Press

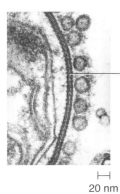

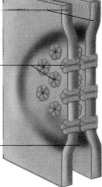

plasma membranes

tight junction proteins

intercellular space

50 nm

b. Tight junction

plasma membranes

membrane channels

intercellular space

20 nm

c. Gap junction

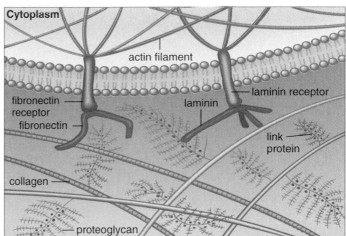

Cytoplasm

actin filament

fibronectin receptor

fibronectin

laminin

laminin receptor

link protein

collagen

proteoglycan

Animal cell extracellular matrix

Figure 5.15

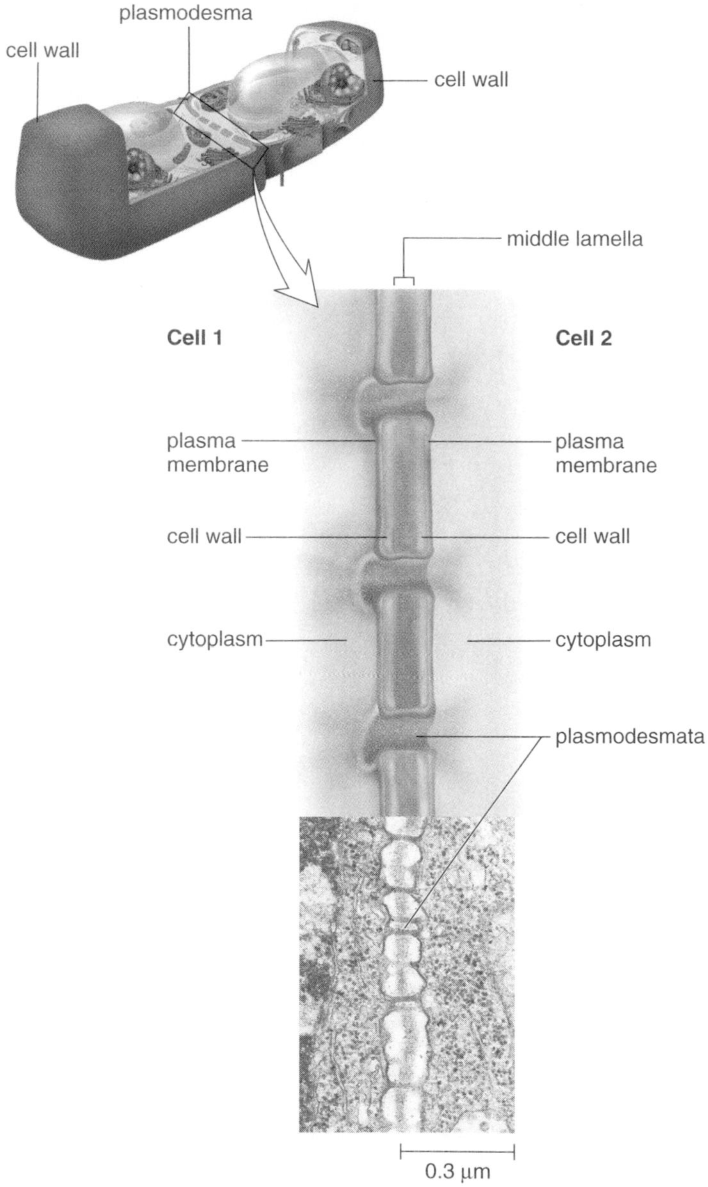

cell wall

plasmodesma

cell wall

middle lamella

Cell 1 **Cell 2**

plasma
membrane

plasma
membrane

cell wall

cell wall

cytoplasm

cytoplasm

plasmodesmata

0.3 μm

Plasmodesmata
Figure 5.16

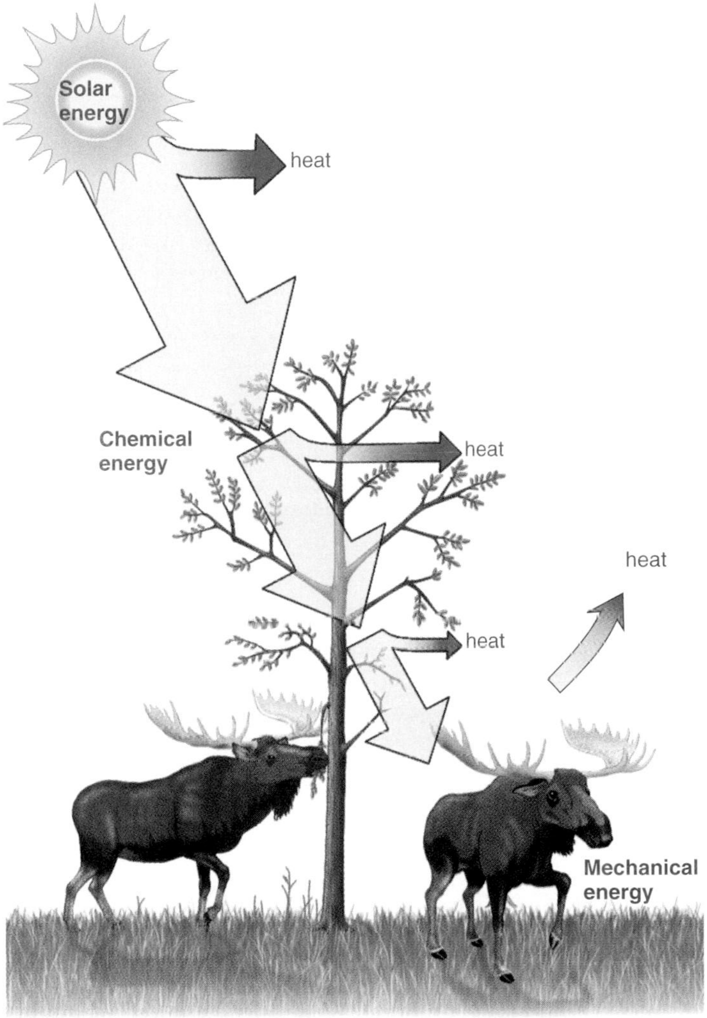

Solar
energy

heat

Chemical
energy

heat

heat

heat

Mechanical
energy

Flow of energy
Figure 6.1

Glucose
- more organized
- more potential energy
- less stable (entropy)

glucose

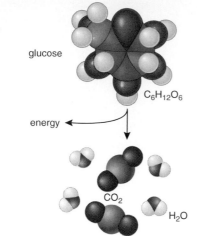

$C_6H_{12}O_6$

energy ←

Carbon dioxide and water
- less organized
- less potential energy
- more stable (entropy)

CO_2

H_2O

a.

Unequal distribution of hydrogen ions
- more organized
- more potential energy
- less stable (entropy)

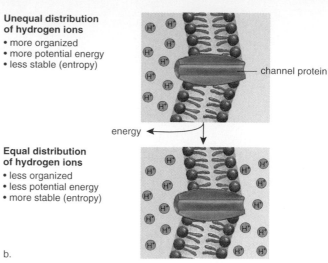

channel protein

energy ←

Equal distribution of hydrogen ions
- less organized
- less potential energy
- more stable (entropy)

b.

Cells and entropy
Figure 6.2

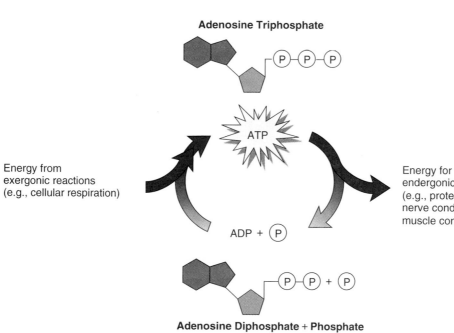

Adenosine Triphosphate

P—P—P

ATP

Energy from exergonic reactions (e.g., cellular respiration)

Energy for endergonic reactions (e.g., protein synthesis, nerve conduction, muscle contraction)

ADP + P

P—P + P

Adenosine Diphosphate + Phosphate

The ATP cycle
Figure 6.3

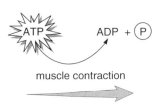

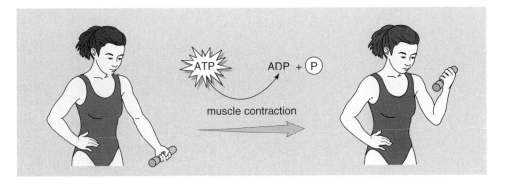

Coupled reactions
Figure 6.4

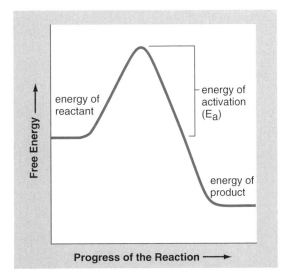

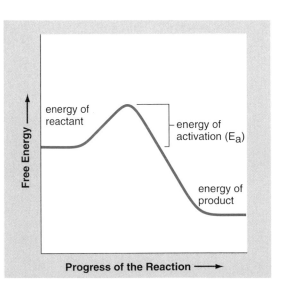

Energy of activation (E$_a$)
Figure 6.5

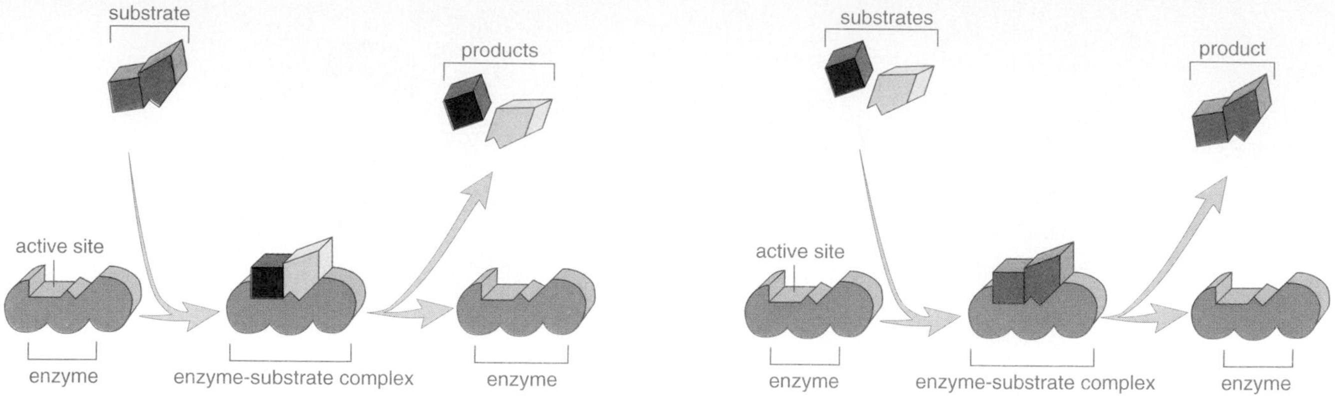

Enzymatic action
Figure 6.6

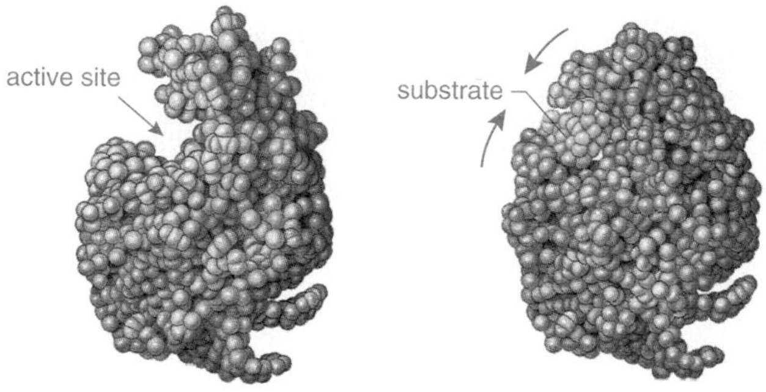

Induced fit model
Figure 6.7

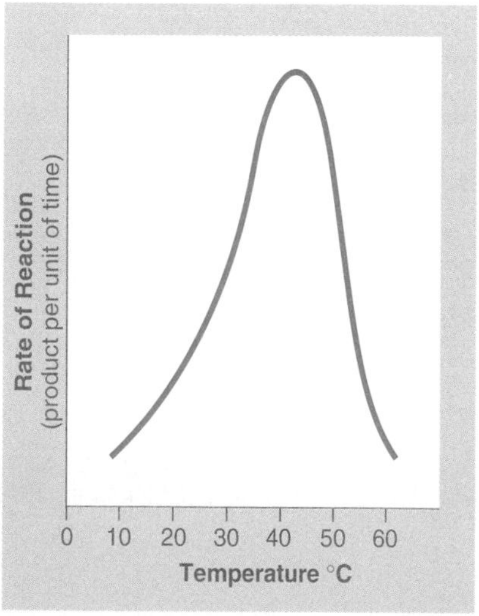

**Effect of temperature on
rate of reaction**
Figure 6.8

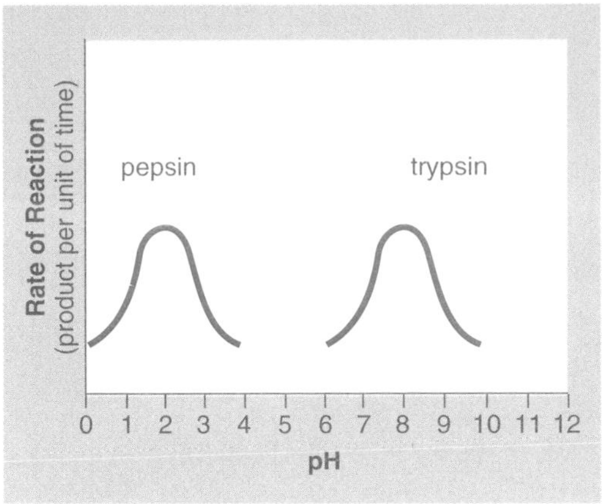

Effect of pH on rate of reaction
Figure 6.9

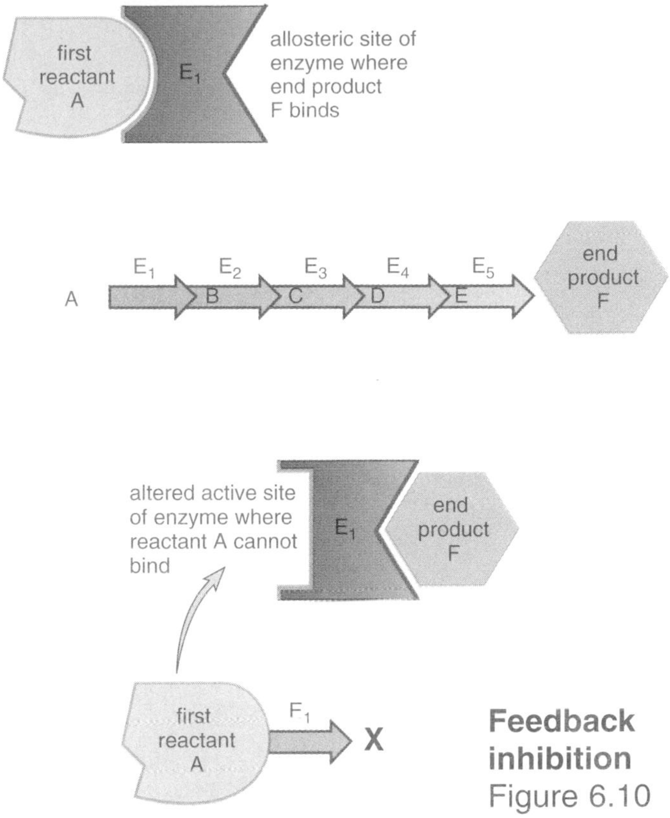

Feedback inhibition
Figure 6.10

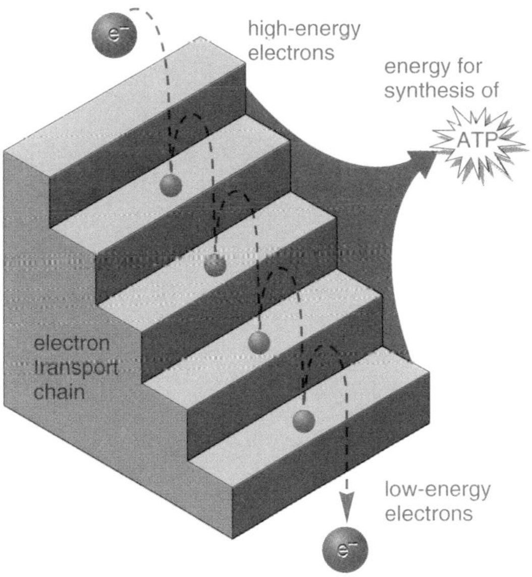

Electron transport system
Figure 6.11

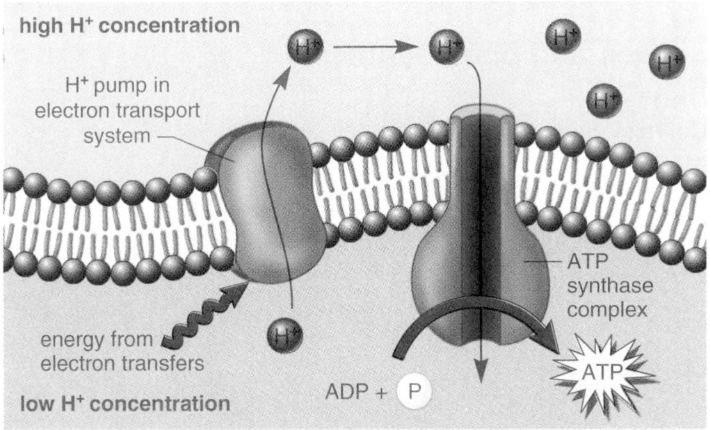

Chemiosmosis
Figure 6.12

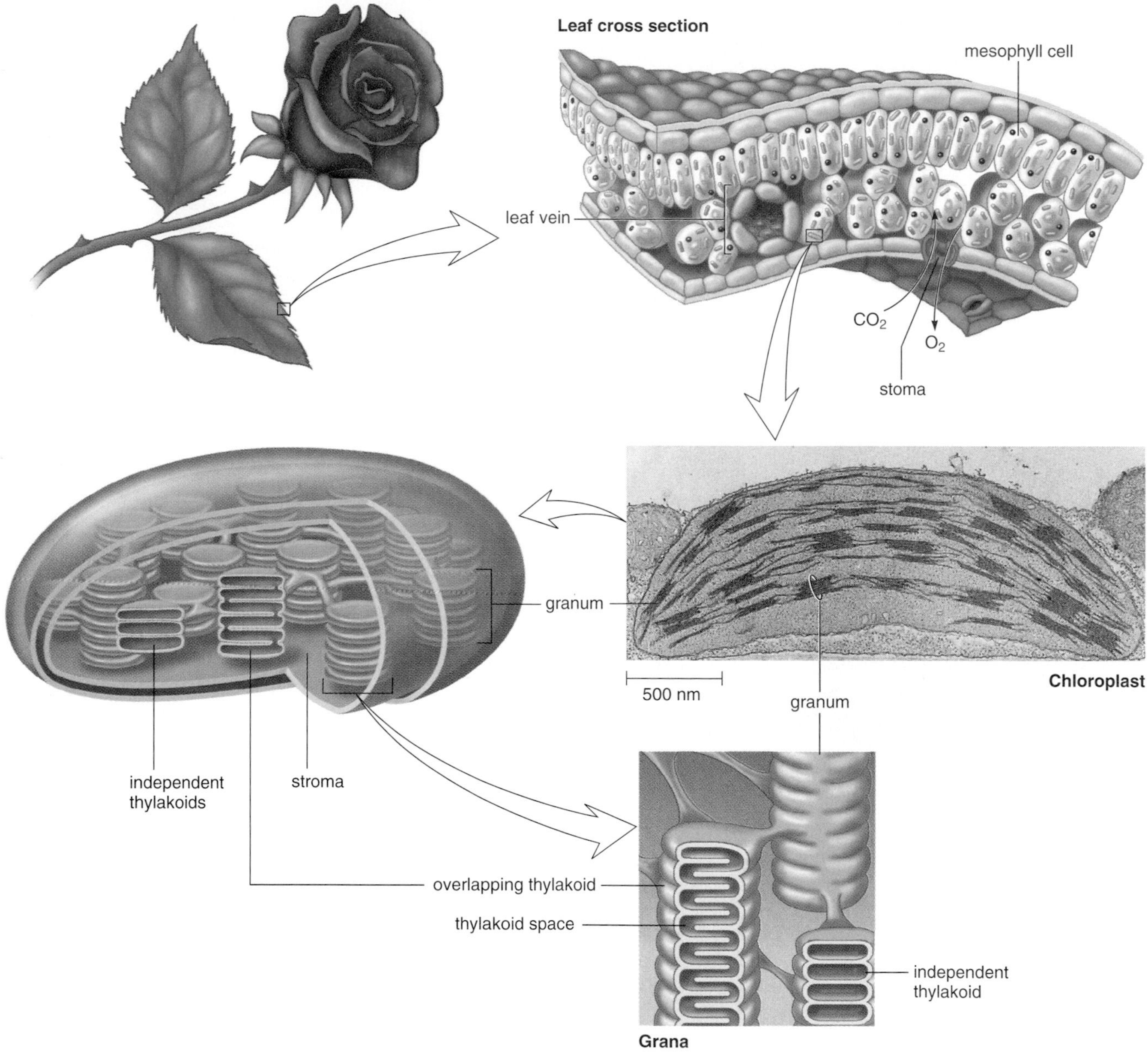

Leaf cross section

mesophyll cell

leaf vein

CO_2

O_2

stoma

granum

500 nm

granum

Chloroplast

independent thylakoids

stroma

overlapping thylakoid

thylakoid space

independent thylakoid

Grana

Leaves and photosynthesis
Figure 7.2
Courtesy Herbert W. Israel, Cornell University

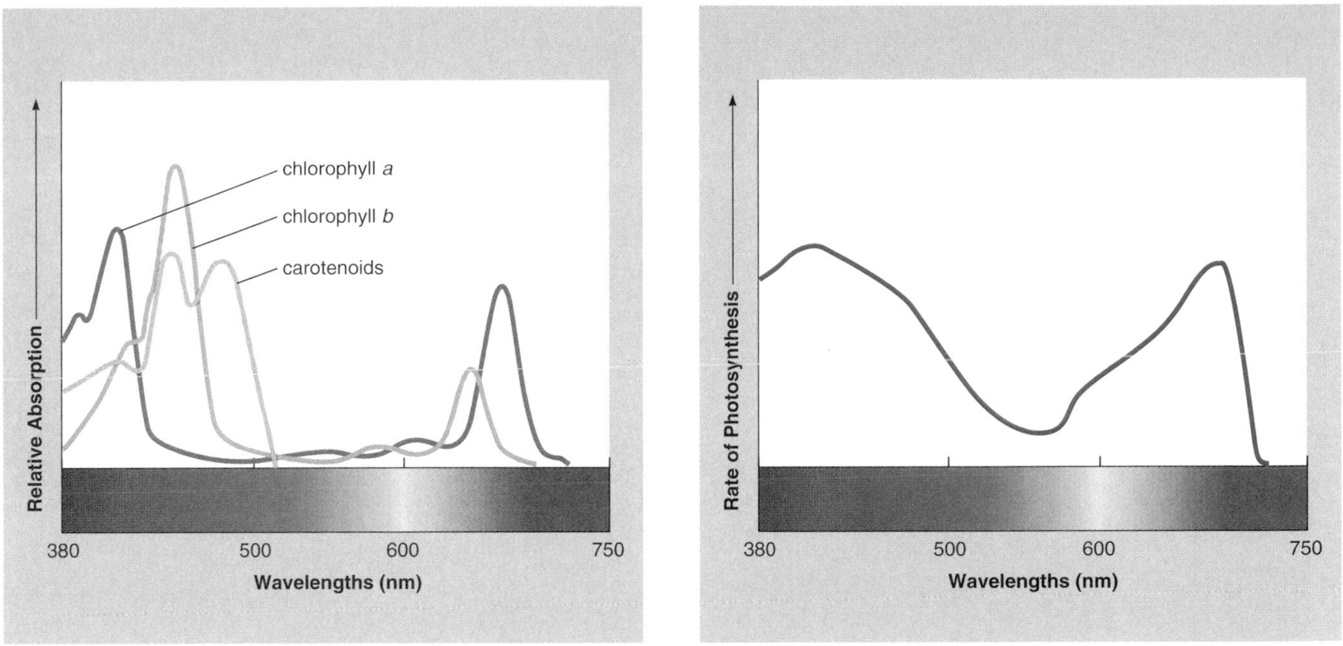

Photosynthetic pigments and photosynthesis
Figure 7.3

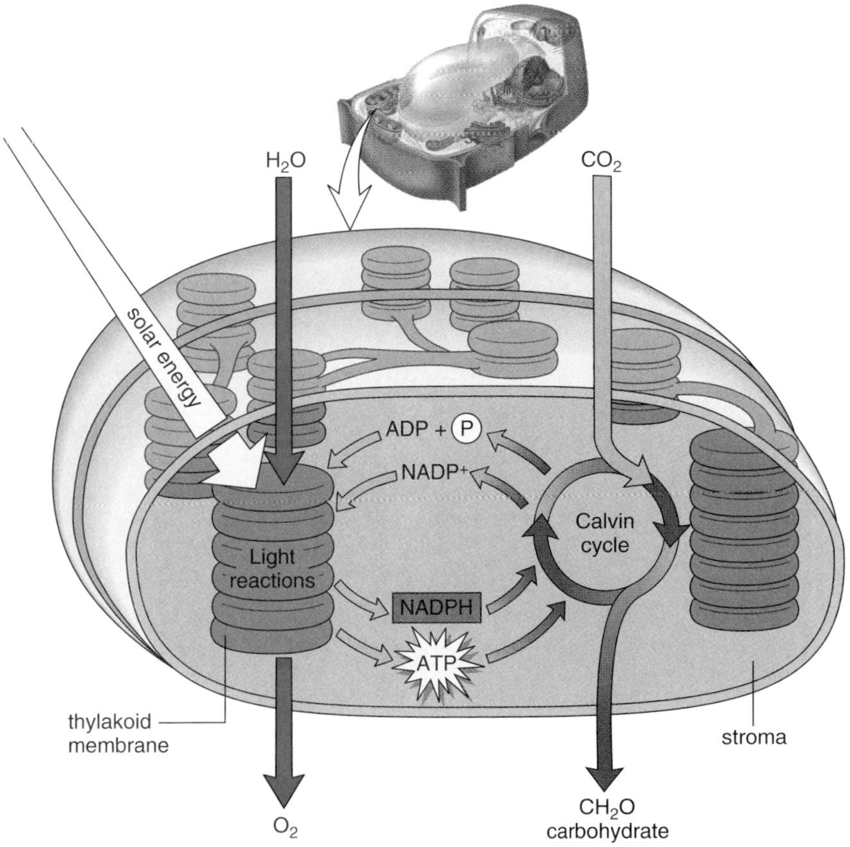

Overview of photosynthesis
Figure 7.4

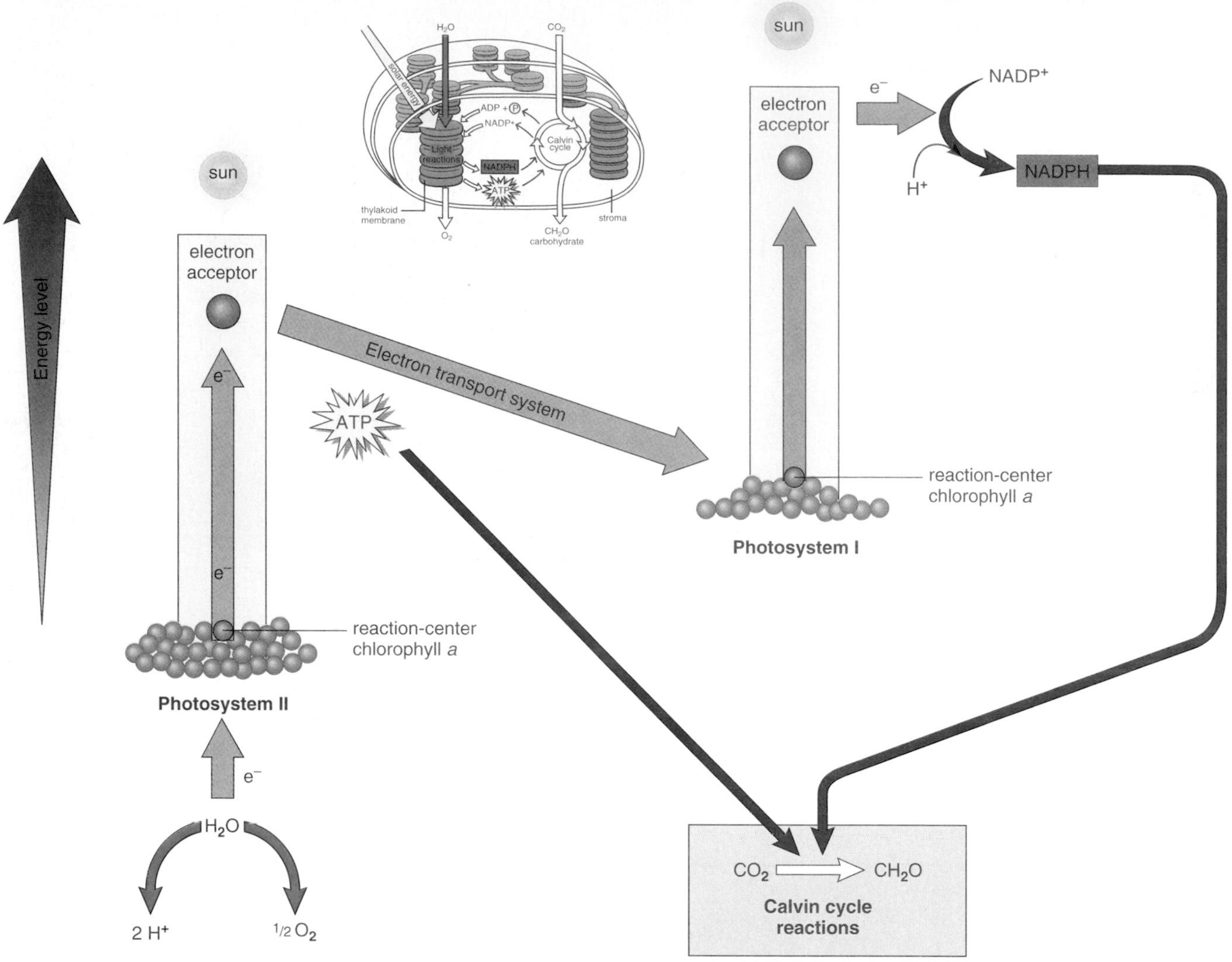

Noncyclic electron pathway:
Electrons move from water to NADP⁺
Figure 7.5

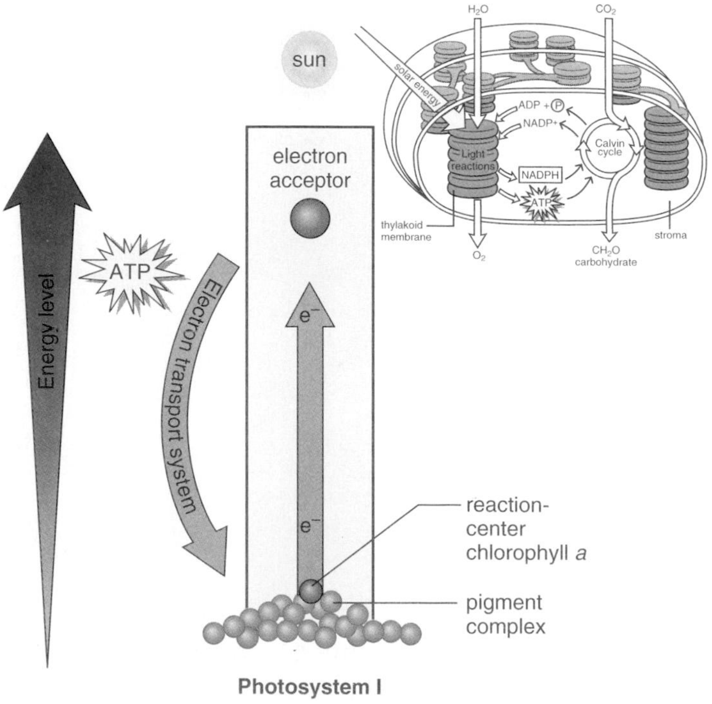

Cyclic electron pathway: Electrons
leave and return to photosystem I.
Figure 7.6

54

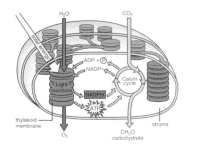

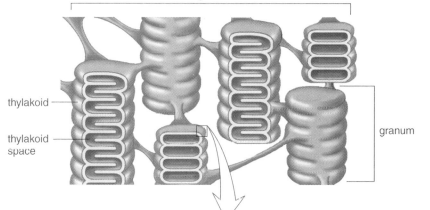

grana

thylakoid

thylakoid space

granum

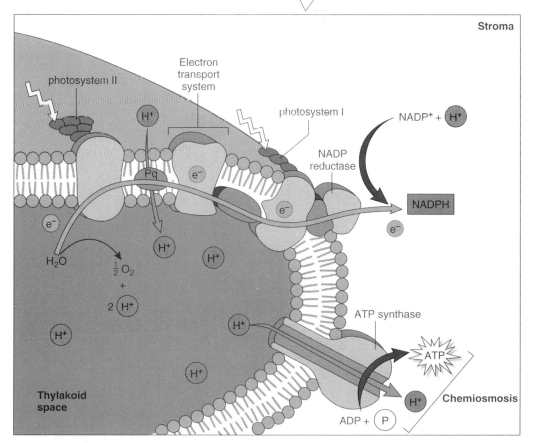

Stroma

photosystem II

Electron transport system

photosystem I

H^+

Pq

e^-

NADP reductase

$NADP^+ + H^+$

e^-

NADPH

e^-

e^-

H^+

H^+

e^-

H_2O

$\frac{1}{2} O_2$

$+$

$2\ H^+$

H^+

H^+

H^+

ATP synthase

ATP

H^+

Chemiosmosis

ADP + P

Thylakoid space

Organization of a thylakoid
Figure 7.7

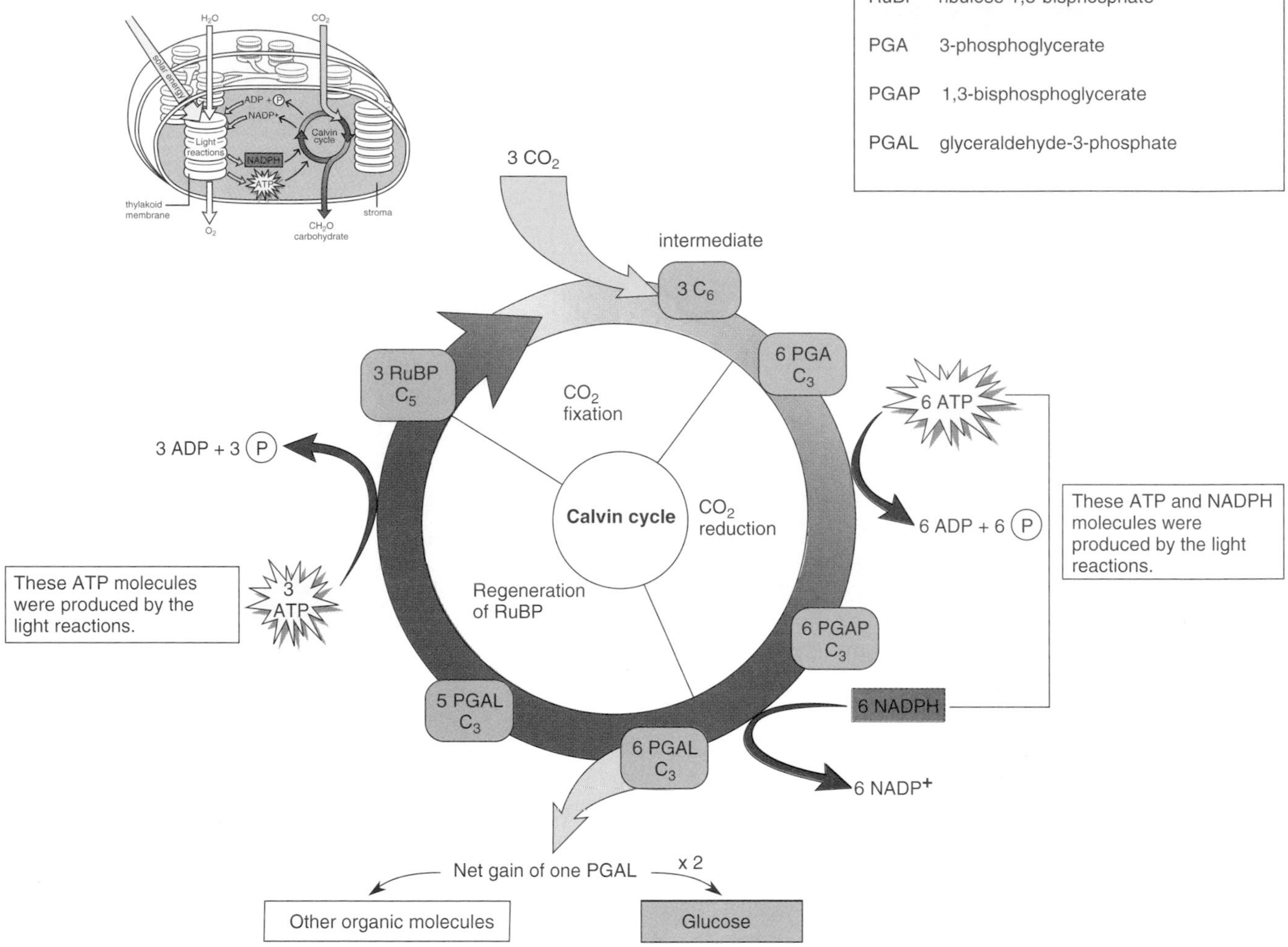

Metabolites of the Calvin Cycle	
RuBP	ribulose 1,5-bisphosphate
PGA	3-phosphoglycerate
PGAP	1,3-bisphosphoglycerate
PGAL	glyceraldehyde-3-phosphate

3 CO_2

intermediate

3 C_6

6 PGA C_3

3 RuBP C_5

CO_2 fixation

6 ATP

6 ADP + 6 (P)

These ATP and NADPH molecules were produced by the light reactions.

3 ADP + 3 (P)

Calvin cycle

CO_2 reduction

These ATP molecules were produced by the light reactions.

3 ATP

Regeneration of RuBP

6 PGAP C_3

5 PGAL C_3

6 PGAL C_3

6 NADPH

6 NADP$^+$

Net gain of one PGAL — x 2

Other organic molecules

Glucose

The Calvin cycle reactions
Figure 7.8

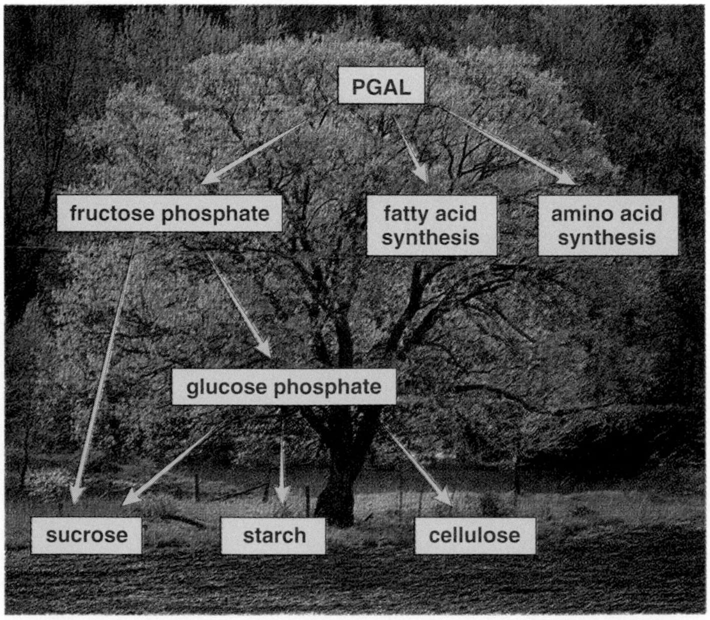

PGAL

fructose phosphate

fatty acid synthesis

amino acid synthesis

glucose phosphate

sucrose   starch   cellulose

Fate of PGAL
Figure 7.9

© The McGraw-Hill Companies, Inc./Bob Coyle, photographer

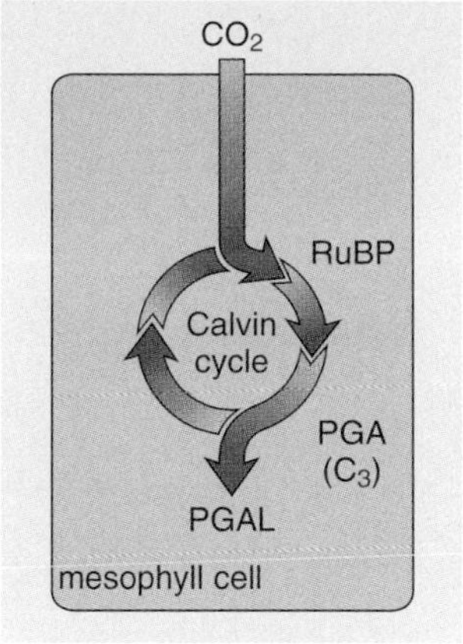

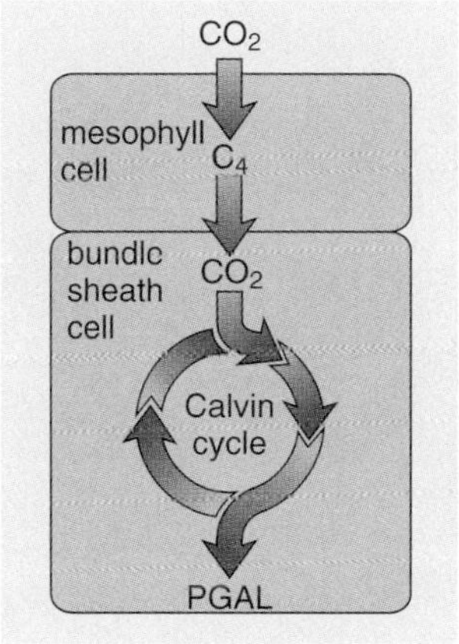

Carbon dioxide fixation in C$_3$ and C$_4$ plants
Figure 7.10

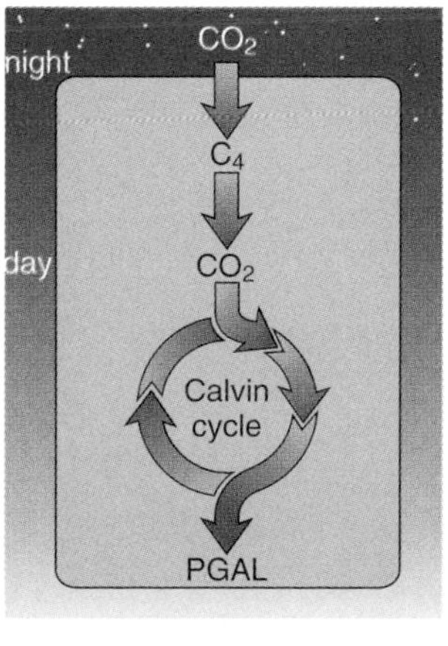

Carbon dioxide fixation in a CAM plant
Figure 7.11

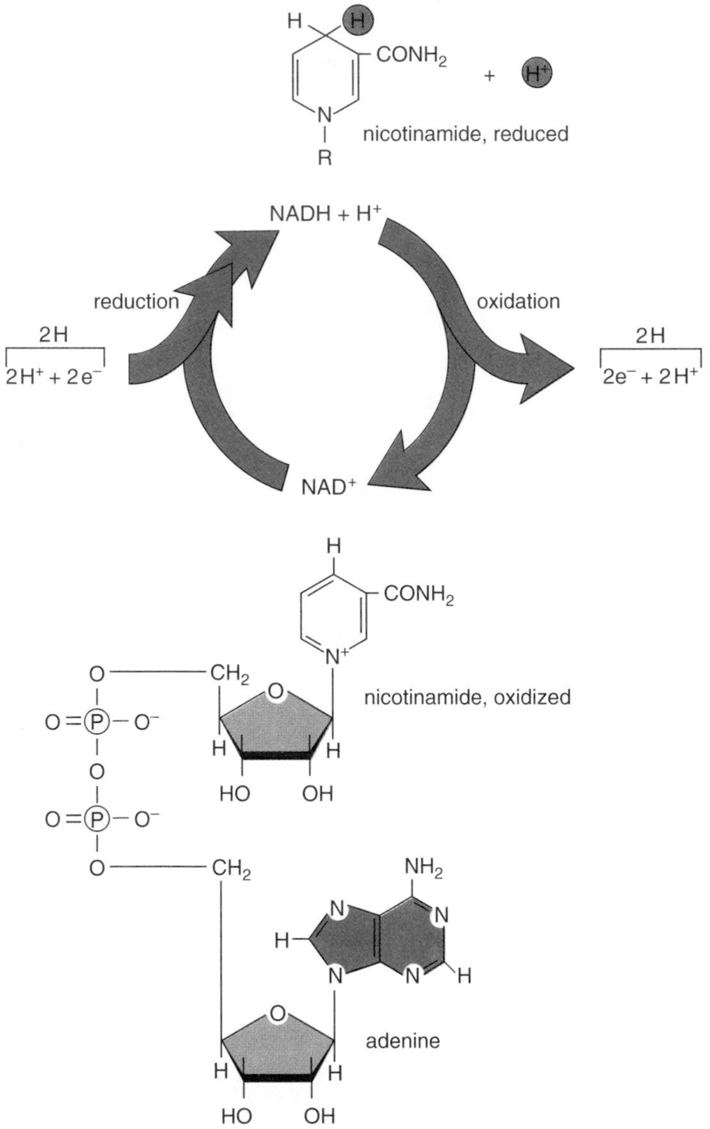
H
H
CONH2
+
H+
N
R
nicotinamide, reduced
NADH + H+
reduction
oxidation
2H
2H+ + 2e−
2H
2e− + 2H+
NAD+
H
CONH2
N+
nicotinamide, oxidized
O
CH2
O
O=P−O−
H
H
O
HO
OH
O=P−O−
O
CH2
NH2
N
N
H
N
N
H
O
adenine
H
H
HO
OH

The NAD+ cycle
Figure 8.1

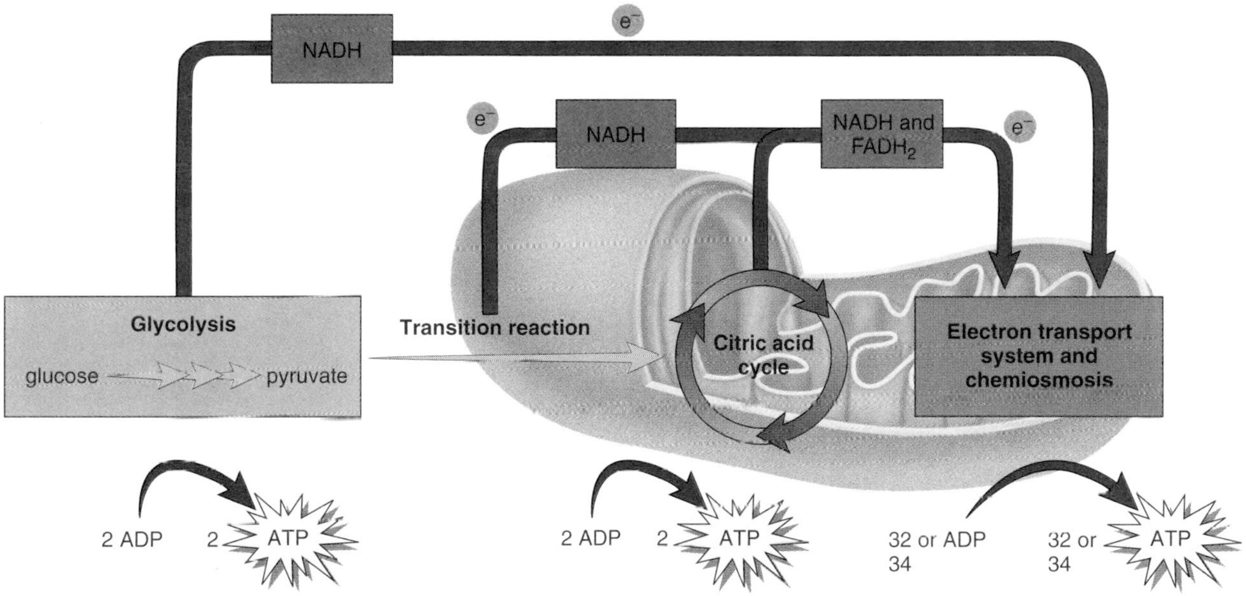

The four phases of complete glucose breakdown
Figure 8.2

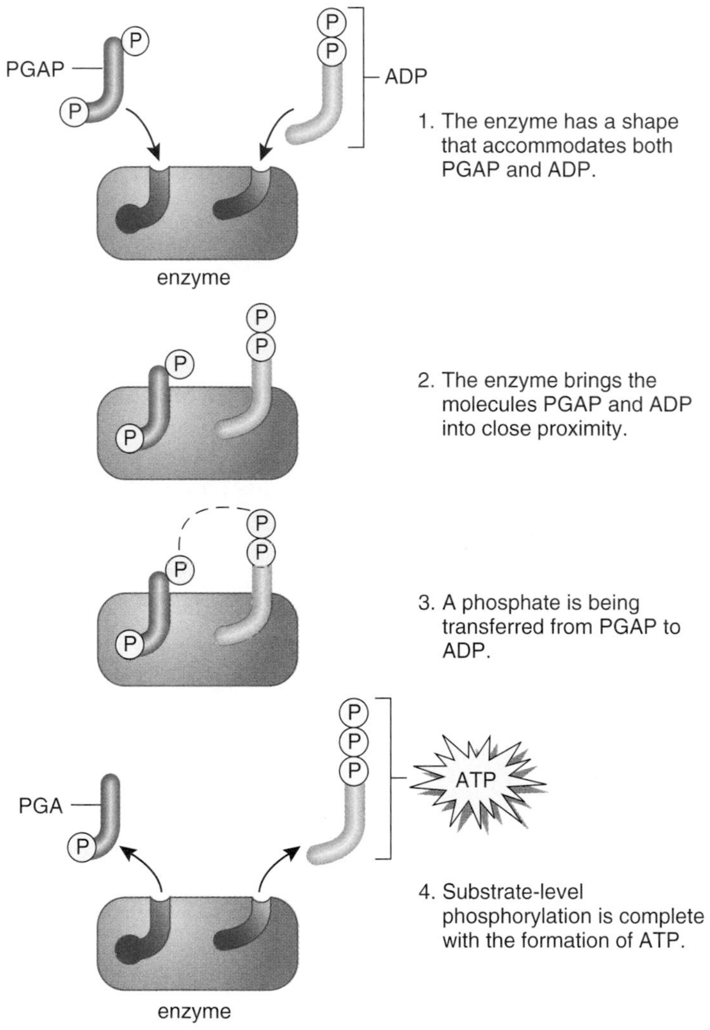

Glycolysis
Figure 8.4

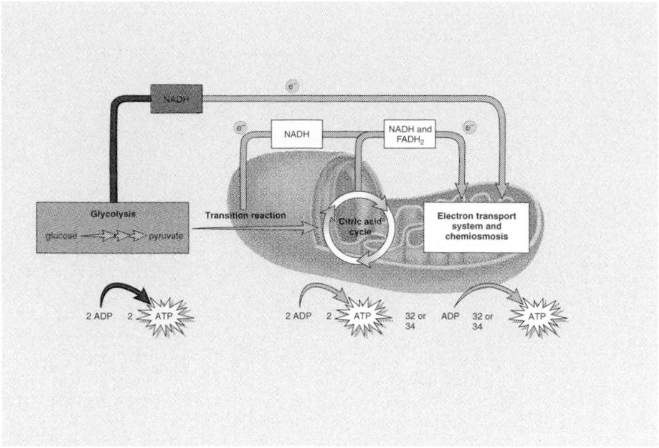

PGAP

ADP

1. The enzyme has a shape that accommodates both PGAP and ADP.

enzyme

2. The enzyme brings the molecules PGAP and ADP into close proximity.

3. A phosphate is being transferred from PGAP to ADP.

PGA

ATP

4. Substrate-level phosphorylation is complete with the formation of ATP.

enzyme

Substrate-level phosphorylation
Figure 8.3

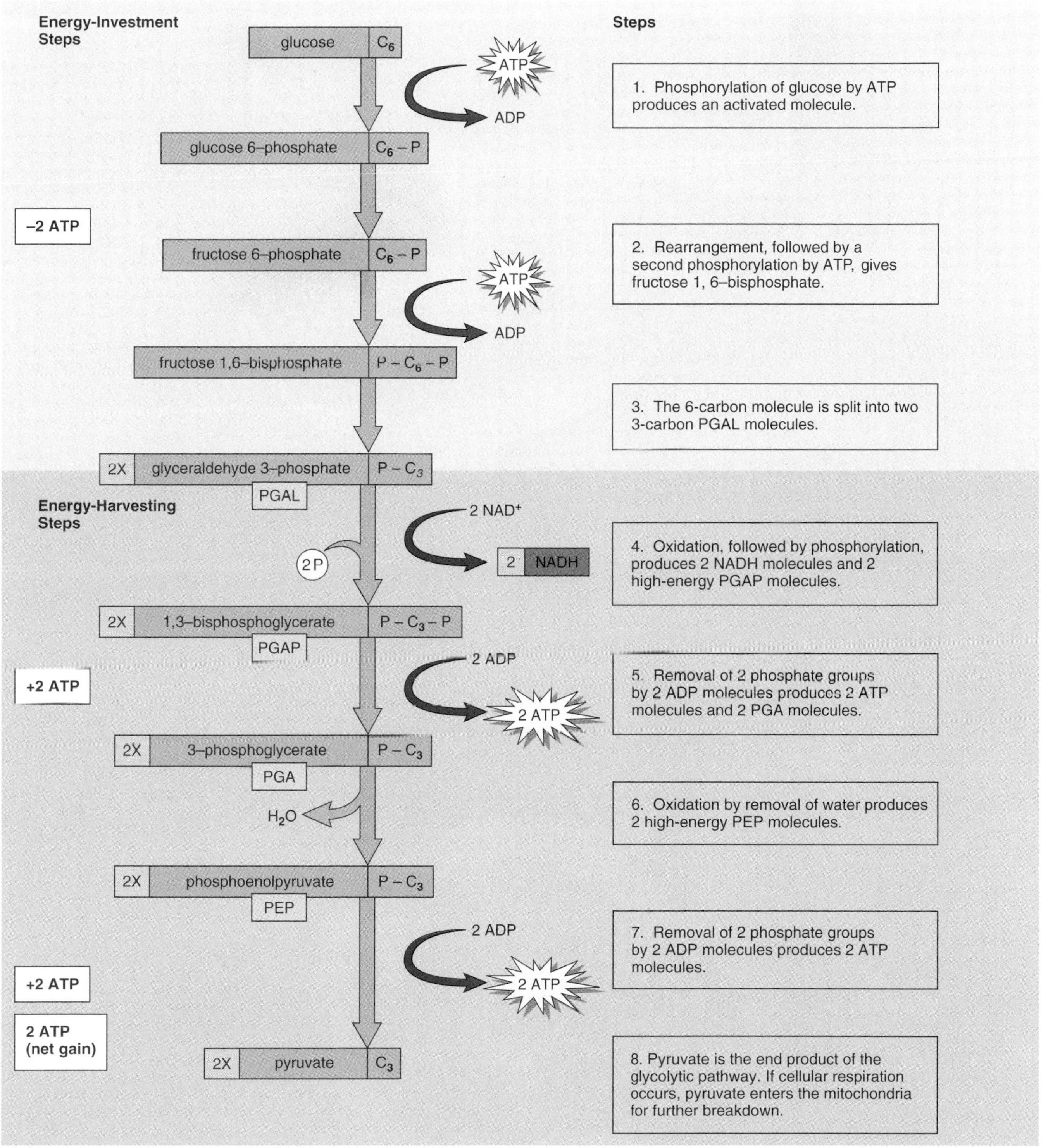

Energy-Investment Steps

glucose | C₆

1. Phosphorylation of glucose by ATP produces an activated molecule.

ATP → ADP

glucose 6–phosphate | C₆ – P

−2 ATP

2. Rearrangement, followed by a second phosphorylation by ATP, gives fructose 1, 6–bisphosphate.

fructose 6–phosphate | C₆ – P

ATP → ADP

fructose 1,6–bisphosphate | P – C₆ – P

3. The 6-carbon molecule is split into two 3-carbon PGAL molecules.

2X glyceraldehyde 3–phosphate | P – C₃ / PGAL

Energy-Harvesting Steps

2 NAD⁺

2P → 2 NADH

4. Oxidation, followed by phosphorylation, produces 2 NADH molecules and 2 high-energy PGAP molecules.

2X 1,3–bisphosphoglycerate | P – C₃ – P / PGAP

2 ADP → 2 ATP

+2 ATP

5. Removal of 2 phosphate groups by 2 ADP molecules produces 2 ATP molecules and 2 PGA molecules.

2X 3–phosphoglycerate | P – C₃ / PGA

H₂O

6. Oxidation by removal of water produces 2 high-energy PEP molecules.

2X phosphoenolpyruvate | P – C₃ / PEP

2 ADP → 2 ATP

+2 ATP

7. Removal of 2 phosphate groups by 2 ADP molecules produces 2 ATP molecules.

2 ATP (net gain)

2X pyruvate | C₃

8. Pyruvate is the end product of the glycolytic pathway. If cellular respiration occurs, pyruvate enters the mitochondria for further breakdown.

Steps

Glycolysis (continued)

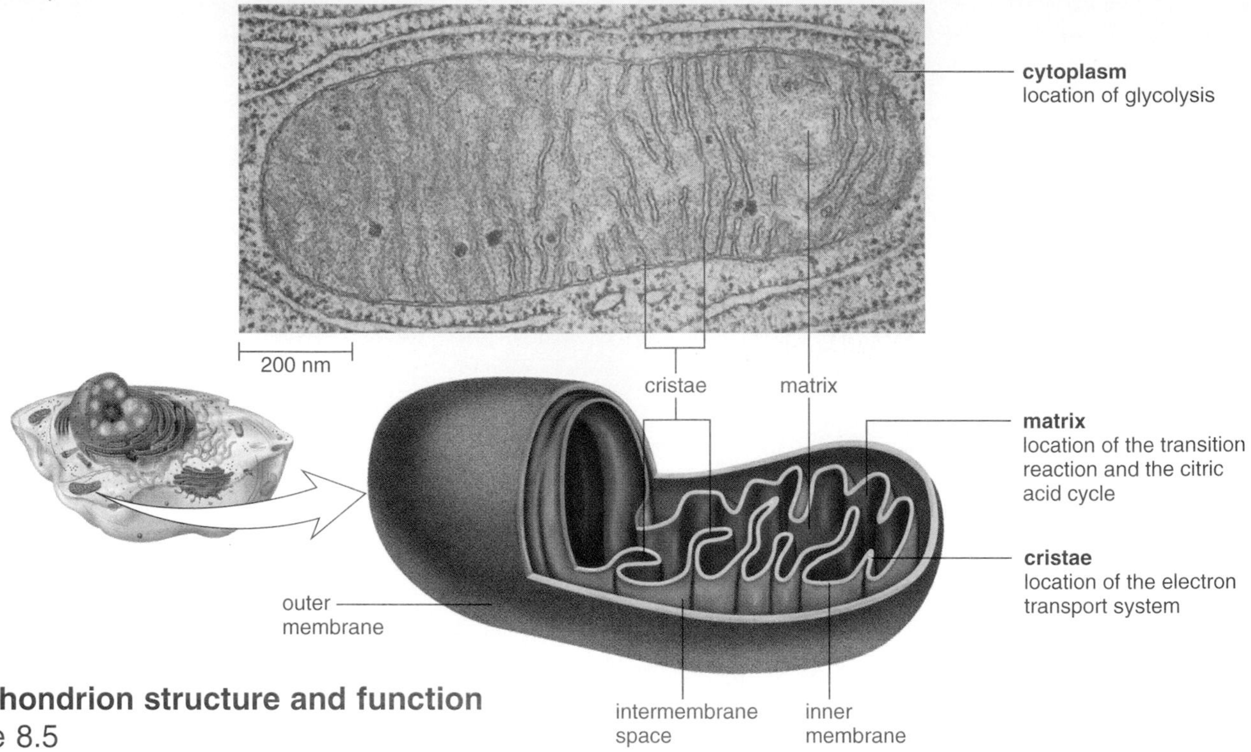

cytoplasm
location of glycolysis

cristae   matrix

matrix
location of the transition
reaction and the citric
acid cycle

cristae
location of the electron
transport system

200 nm

outer
membrane

intermembrane   inner
space   membrane

Mitochondrion structure and function
Figure 8.5

Courtesy Dr. Keith Porter

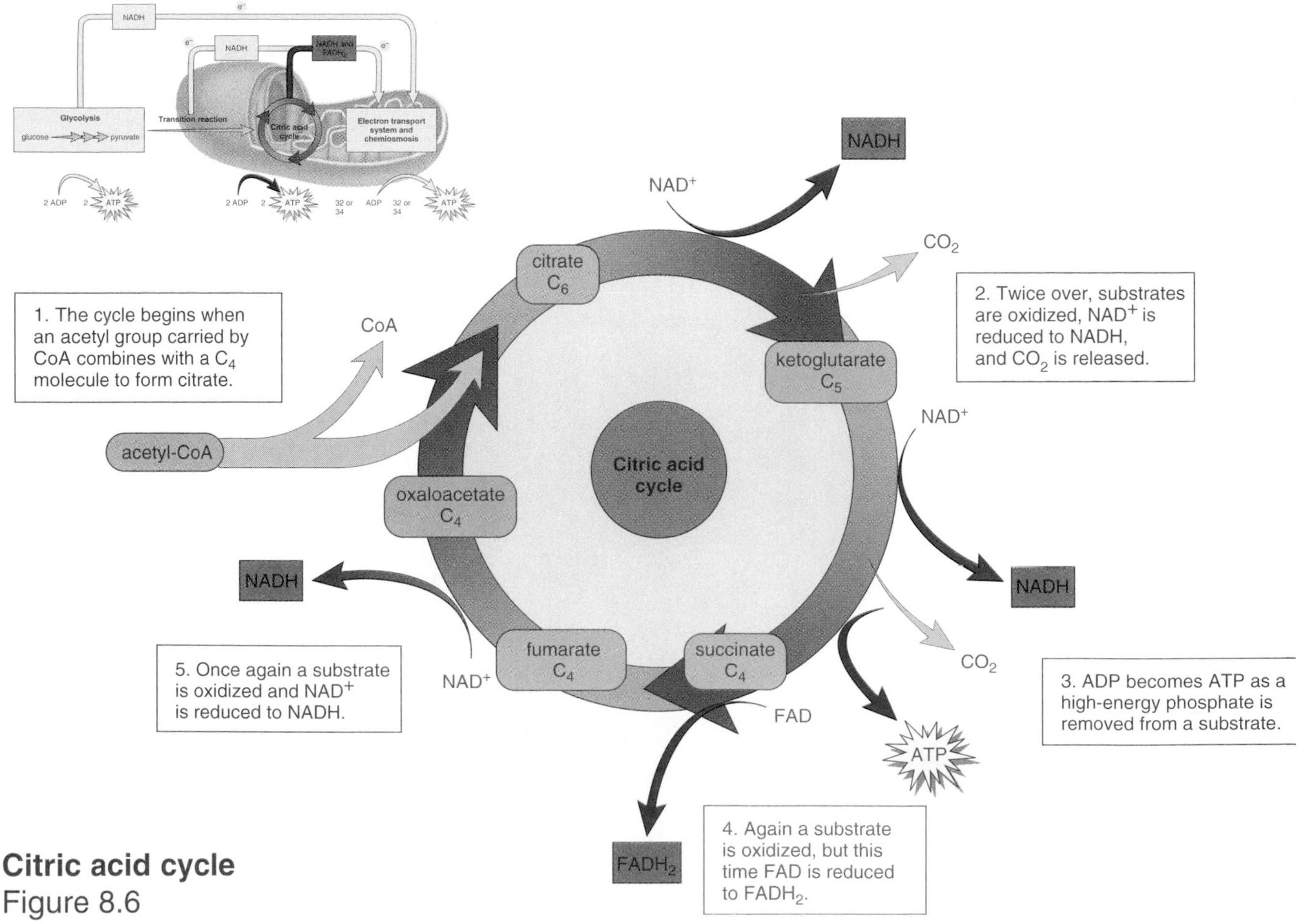

1. The cycle begins when
an acetyl group carried by
CoA combines with a C_4
molecule to form citrate.

CoA

acetyl-CoA

NAD⁺

NADH

CO_2

citrate
C_6

ketoglutarate
C_5

2. Twice over, substrates
are oxidized, NAD^+ is
reduced to NADH,
and CO_2 is released.

NAD⁺

oxaloacetate
C_4

Citric acid
cycle

NADH

NADH

fumarate
C_4

succinate
C_4

CO_2

NAD⁺

FAD

ATP

3. ADP becomes ATP as a
high-energy phosphate is
removed from a substrate.

5. Once again a substrate
is oxidized and NAD^+
is reduced to NADH.

4. Again a substrate
is oxidized, but this
time FAD is reduced
to $FADH_2$.

$FADH_2$

Citric acid cycle
Figure 8.6

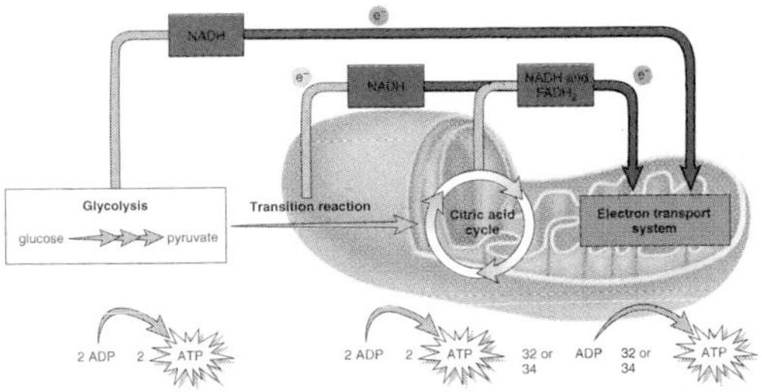
NADH
NADH and FADH2
Glycolysis
glucose
pyruvate
Transition reaction
Citric acid cycle
Electron transport system
2 ADP
2
ATP
32 or 34
ADP

The electron transport system
Figure 8.7

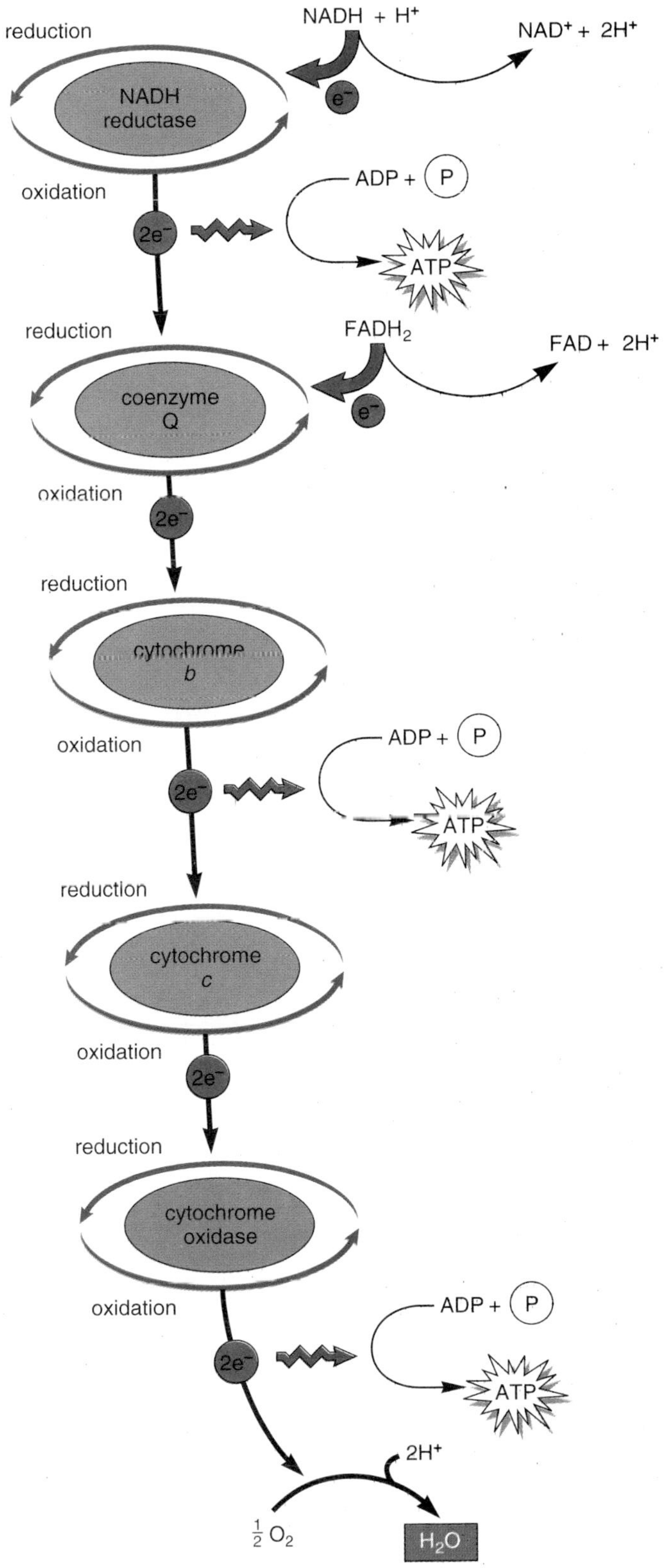
NADH + H⁺
NAD⁺ + 2H⁺
reduction
NADH reductase
e⁻
oxidation
2e⁻
ADP + P
ATP
reduction
FADH₂
FAD + 2H⁺
coenzyme Q
e⁻
oxidation
2e⁻
reduction
cytochrome b
oxidation
2e⁻
ADP + P
ATP
reduction
cytochrome c
oxidation
2e⁻
reduction
cytochrome oxidase
oxidation
2e⁻
ADP + P
ATP
2H⁺
½ O₂
H₂O

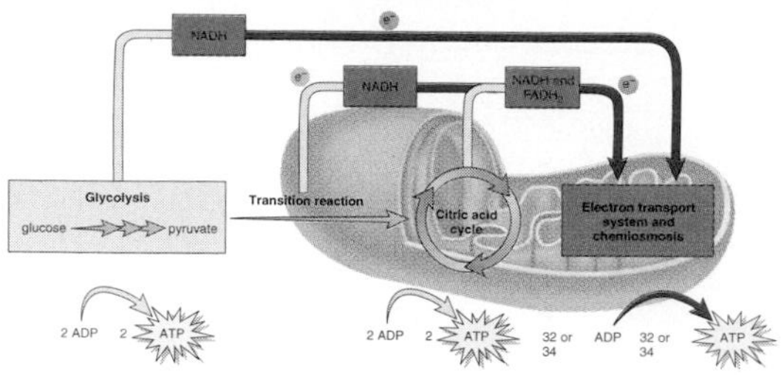

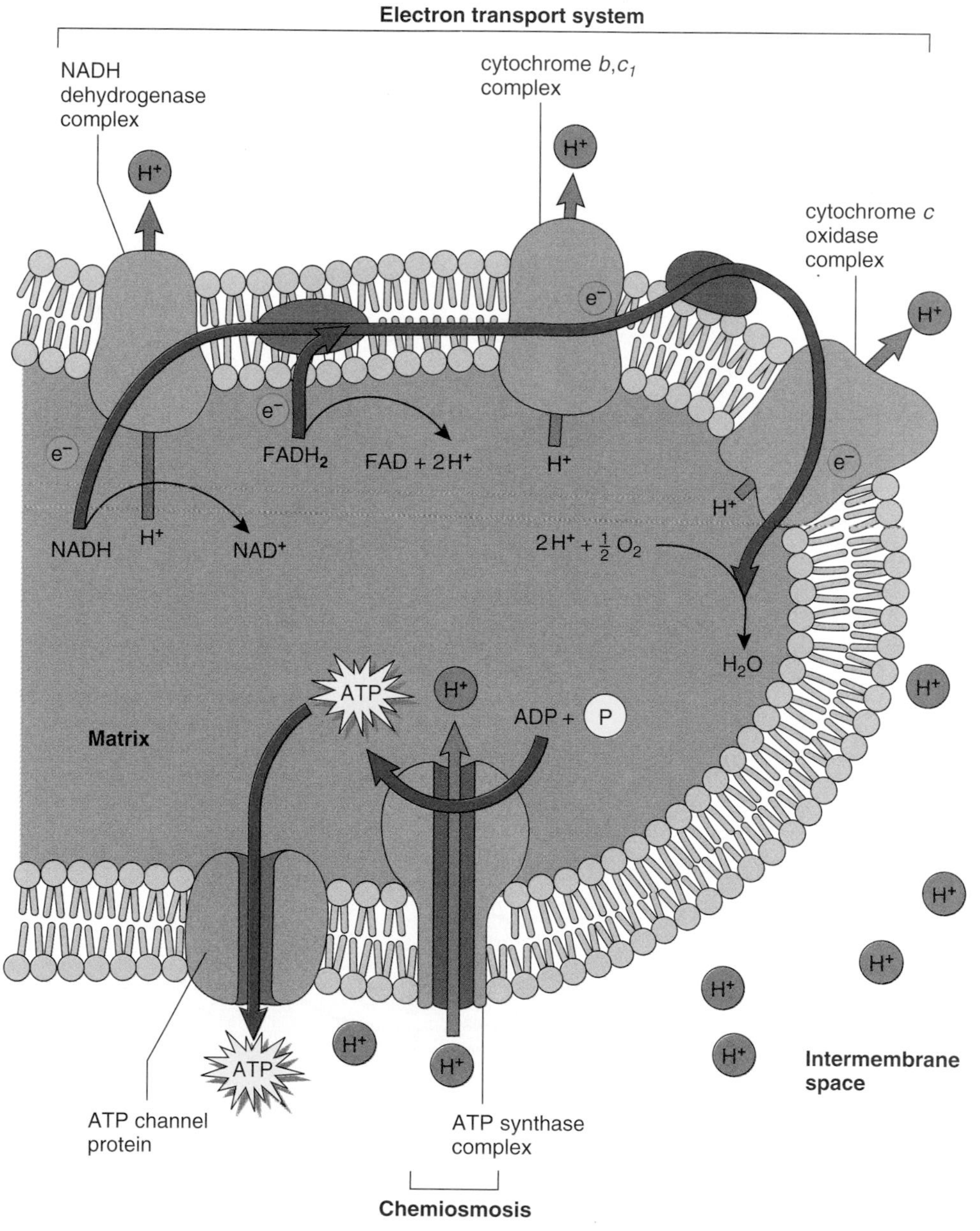

Electron transport system

NADH
dehydrogenase
complex

cytochrome b,c_1
complex

cytochrome c
oxidase
complex

H^+

H^+

H^+

e^-

e^-

e^-

$FADH_2$ $FAD + 2H^+$

H^+

H^+

NADH H^+ NAD^+

$2H^+ + \frac{1}{2}O_2$

H_2O

Matrix

ATP

H^+

ADP + P

H^+

ATP

H^+

H^+

H^+

ATP channel
protein

ATP synthase
complex

H^+

H^+

Intermembrane
space

Chemiosmosis

Organization of cristae
Figure 8.8

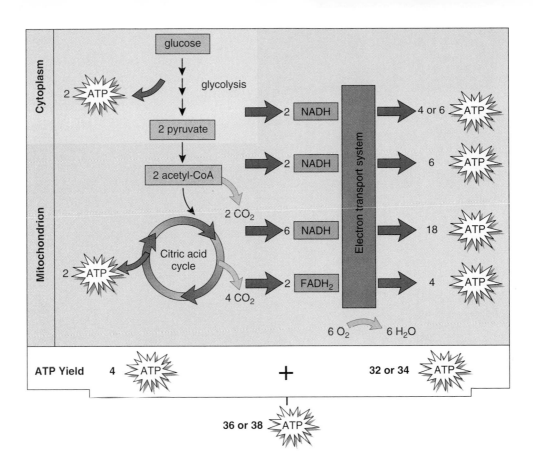

Accounting of energy yield per glucose molecule breakdown
Figure 8.9

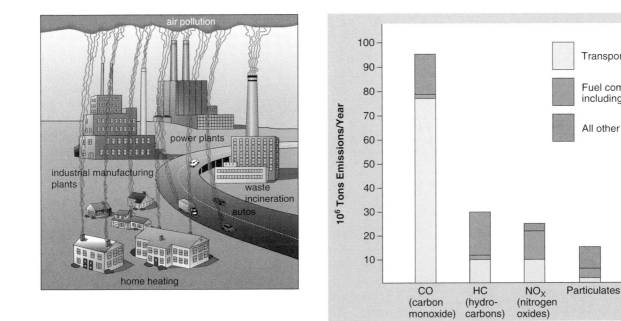

Air pollution
Figure 8A

Fermentation
Figure 8.10

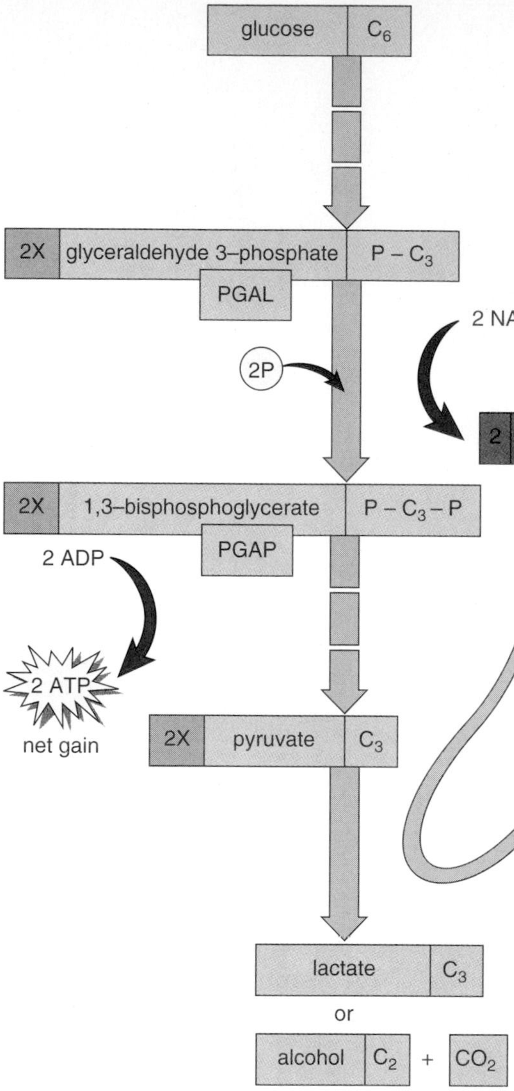

The metabolic pool concept
Figure 8.11

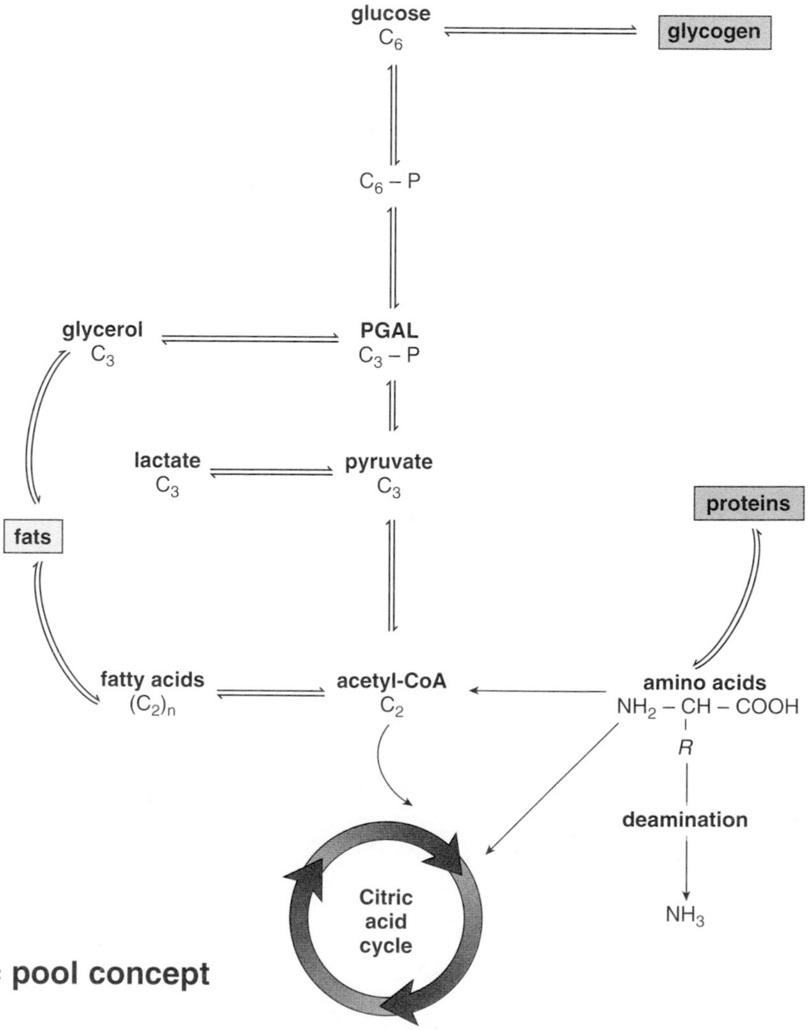

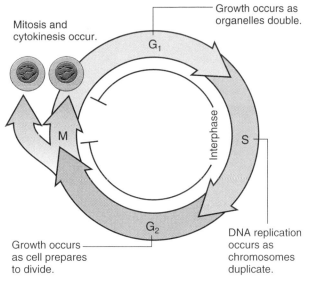

Mitosis and cytokinesis occur.

Growth occurs as organelles double.

G₁

M

Interphase

S

DNA replication occurs as chromosomes duplicate.

G₂

Growth occurs as cell prepares to divide.

The cell cycle
Figure 9.1

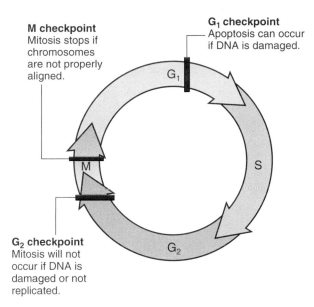

M checkpoint
Mitosis stops if chromosomes are not properly aligned.

G₁ checkpoint
Apoptosis can occur if DNA is damaged.

G₁

M

S

G₂ checkpoint
Mitosis will not occur if DNA is damaged or not replicated.

G₂

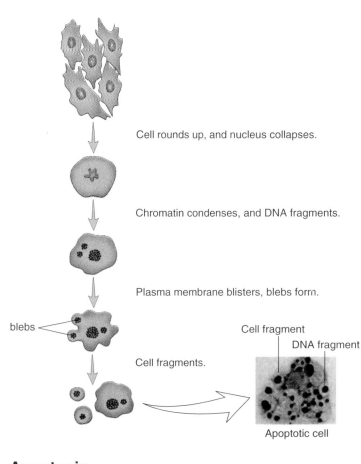

Cell rounds up, and nucleus collapses.

Chromatin condenses, and DNA fragments.

Plasma membrane blisters, blebs form.

blebs

Cell fragments.

Cell fragment

DNA fragment

Apoptotic cell

Apoptosis
Figure 9.2
Courtesy Douglas R. Green/LaJolla Institute for Allergy and Immunology

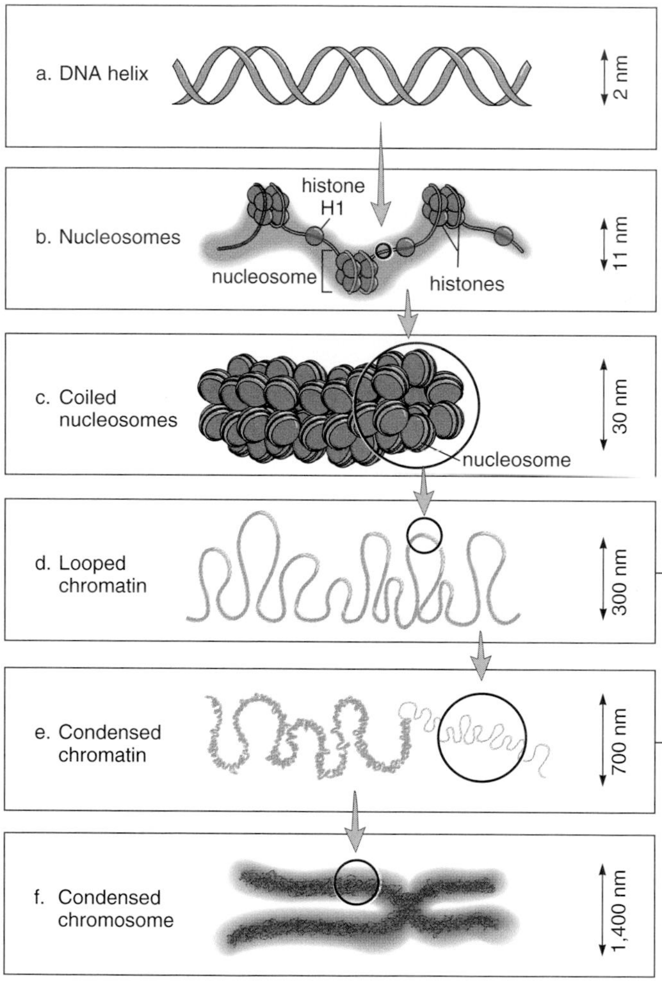

a. DNA helix
2 nm

b. Nucleosomes
histone H1
nucleosome
histones
11 nm

c. Coiled nucleosomes
nucleosome
30 nm

d. Looped chromatin
300 nm

e. Condensed chromatin
700 nm

f. Condensed chromosome
1,400 nm

Levels of chromosome structure
Figure 9A
Courtesy Dr. Stephen Wolfe

euchromatin

nucleolus

heterochromatin

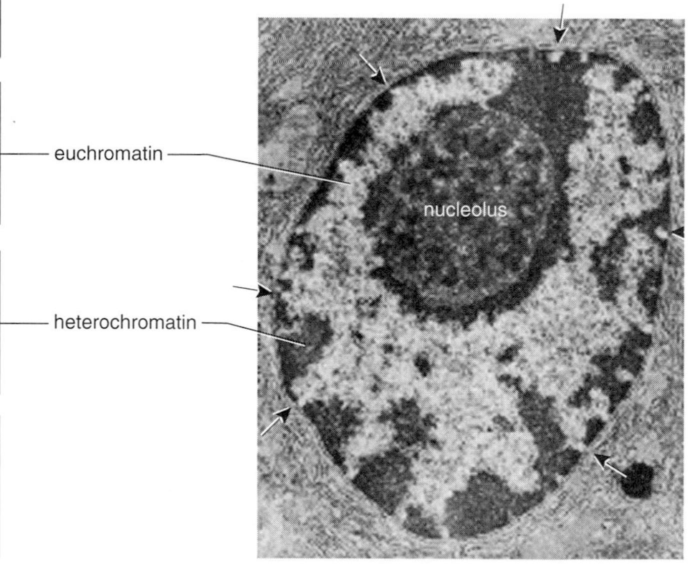

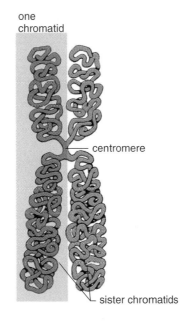

one
chromatid

centromere

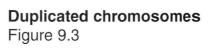

sister chromatids

Duplicated chromosomes
Figure 9.3

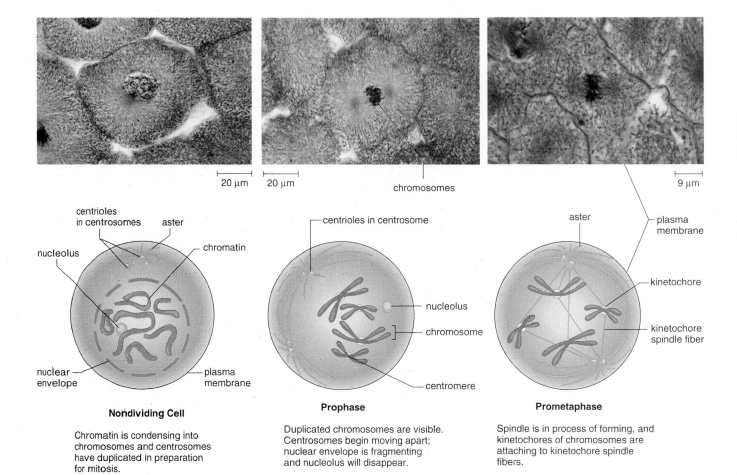

20 μm   **20 μm**   chromosomes   **9 μm**

Nondividing Cell

centrioles in centrosomes   aster

nucleolus   chromatin

nuclear envelope   plasma membrane

Chromatin is condensing into chromosomes and centrosomes have duplicated in preparation for mitosis.

Prophase

centrioles in centrosome

nucleolus

chromosome

centromere

Duplicated chromosomes are visible. Centrosomes begin moving apart; nuclear envelope is fragmenting and nucleolus will disappear.

Prometaphase

aster   plasma membrane

kinetochore

kinetochore spindle fiber

Spindle is in process of forming, and kinetochores of chromosomes are attaching to kinetochore spindle fibers.

Phases of mitosis in animal cells
Figure 9.4

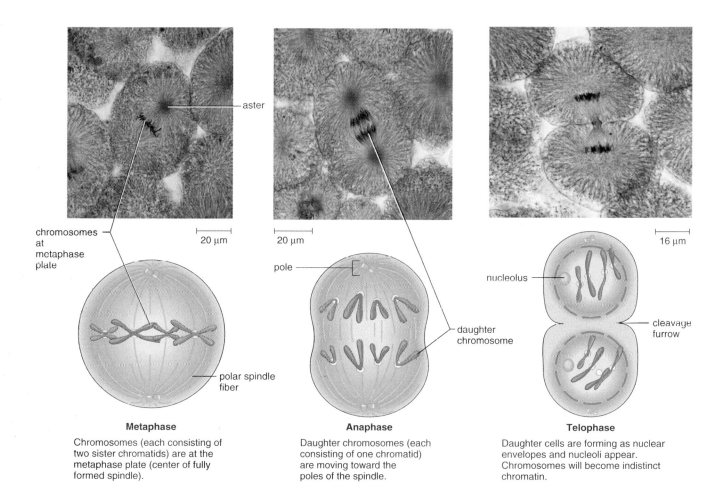

aster

chromosomes
at
metaphase
plate

20 μm

20 μm

pole

16 μm

nucleolus

polar spindle
fiber

daughter
chromosome

cleavage
furrow

Metaphase

Chromosomes (each consisting of
two sister chromatids) are at the
metaphase plate (center of fully
formed spindle).

Anaphase

Daughter chromosomes (each
consisting of one chromatid)
are moving toward the
poles of the spindle.

Telophase

Daughter cells are forming as nuclear
envelopes and nucleoli appear.
Chromosomes will become indistinct
chromatin.

Phases of mitosis in animal cells (continued)

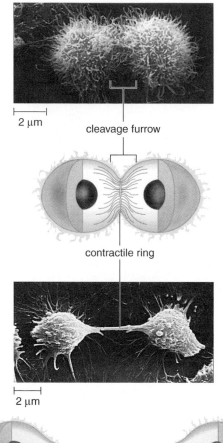

2 µm

cleavage furrow

contractile ring

2 µm

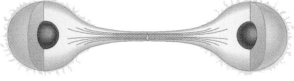

Cytokinesis in animal cells
Figure 9.6

a, b: © R.G. Kessel and C.Y. Shih, "Scanning Electron Microscopy in Biology: A Students' Atlas on Biological Organization," 1974 Springer-Verlag, New York

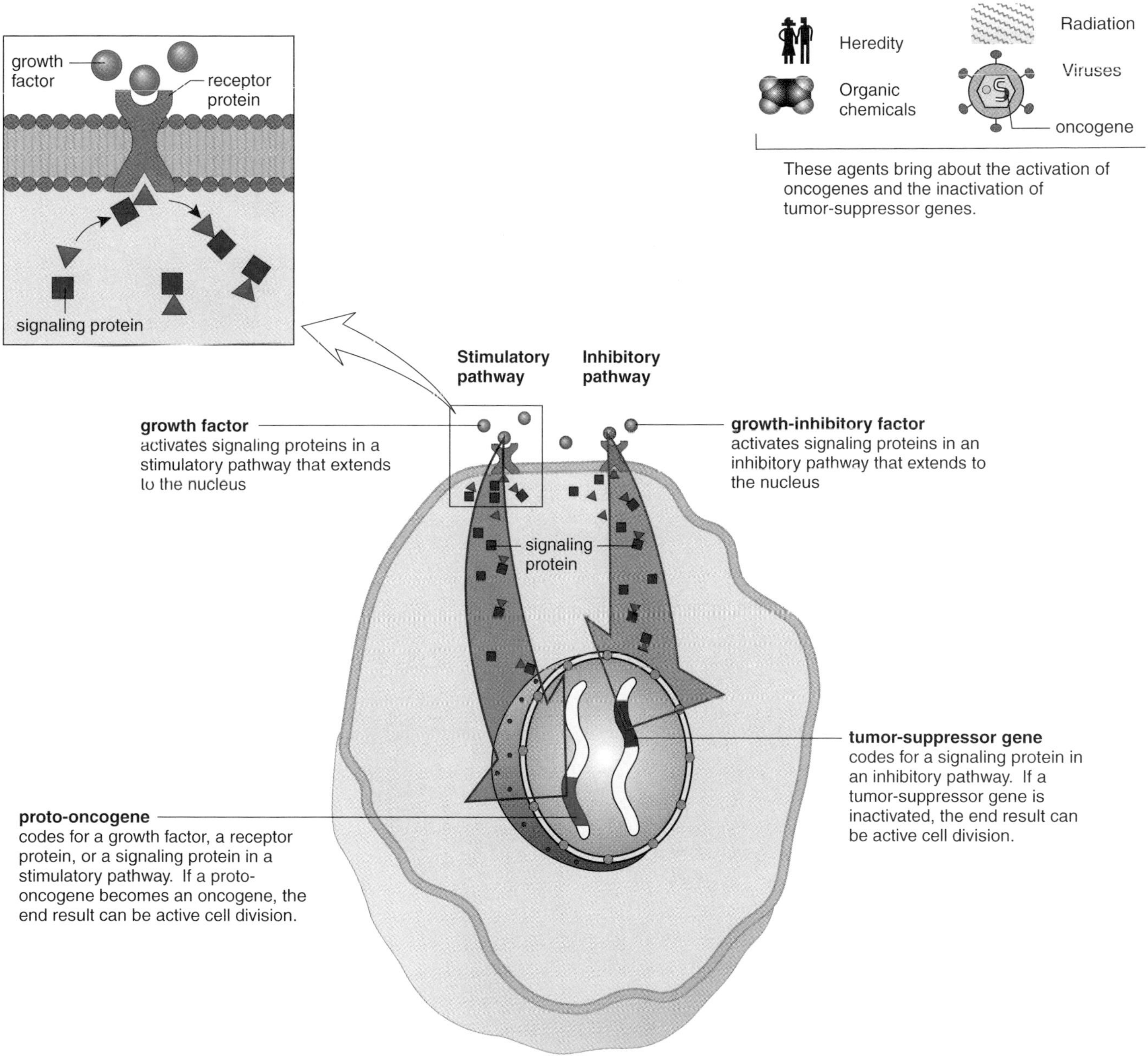

growth factor

receptor protein

signaling protein

Heredity

Organic chemicals

Radiation

Viruses

oncogene

These agents bring about the activation of oncogenes and the inactivation of tumor-suppressor genes.

Stimulatory pathway

Inhibitory pathway

growth factor activates signaling proteins in a stimulatory pathway that extends to the nucleus

growth-inhibitory factor activates signaling proteins in an inhibitory pathway that extends to the nucleus

signaling protein

tumor-suppressor gene codes for a signaling protein in an inhibitory pathway. If a tumor-suppressor gene is inactivated, the end result can be active cell division.

proto-oncogene codes for a growth factor, a receptor protein, or a signaling protein in a stimulatory pathway. If a proto-oncogene becomes an oncogene, the end result can be active cell division.

Causes of cancer
Figure 9.9

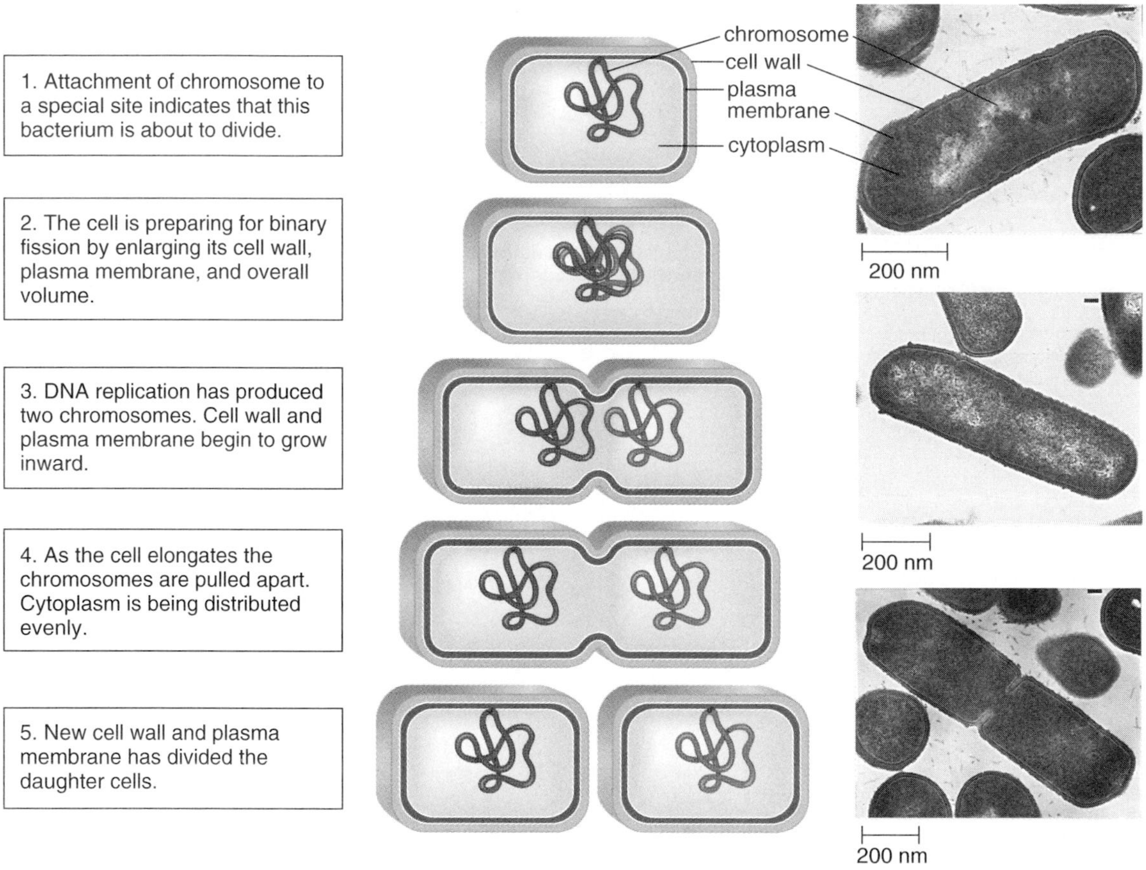

1. Attachment of chromosome to a special site indicates that this bacterium is about to divide.

2. The cell is preparing for binary fission by enlarging its cell wall, plasma membrane, and overall volume.

3. DNA replication has produced two chromosomes. Cell wall and plasma membrane begin to grow inward.

4. As the cell elongates the chromosomes are pulled apart. Cytoplasm is being distributed evenly.

5. New cell wall and plasma membrane has divided the daughter cells.

chromosome
cell wall
plasma membrane
cytoplasm

200 nm

200 nm

200 nm

Binary fission
Figure 9.10

a-c: © Stanley C. Holt/Biological Photo Service

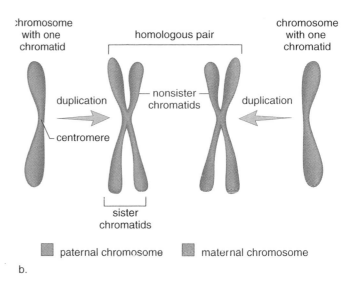

Homologous chromosomes
Figure 10.1

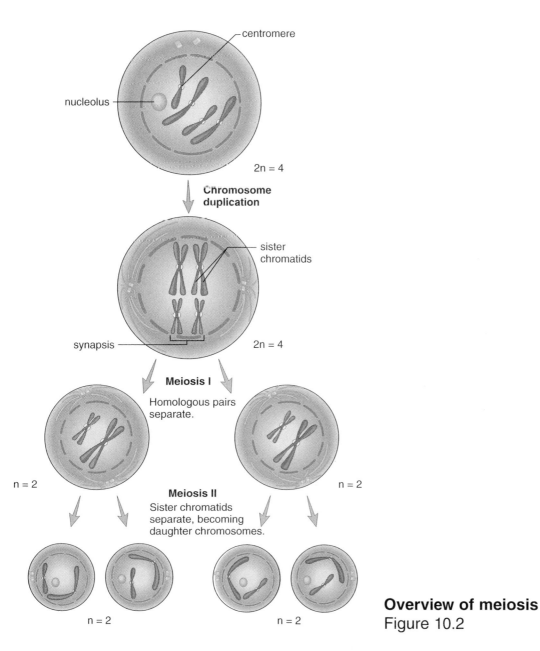

Overview of meiosis
Figure 10.2

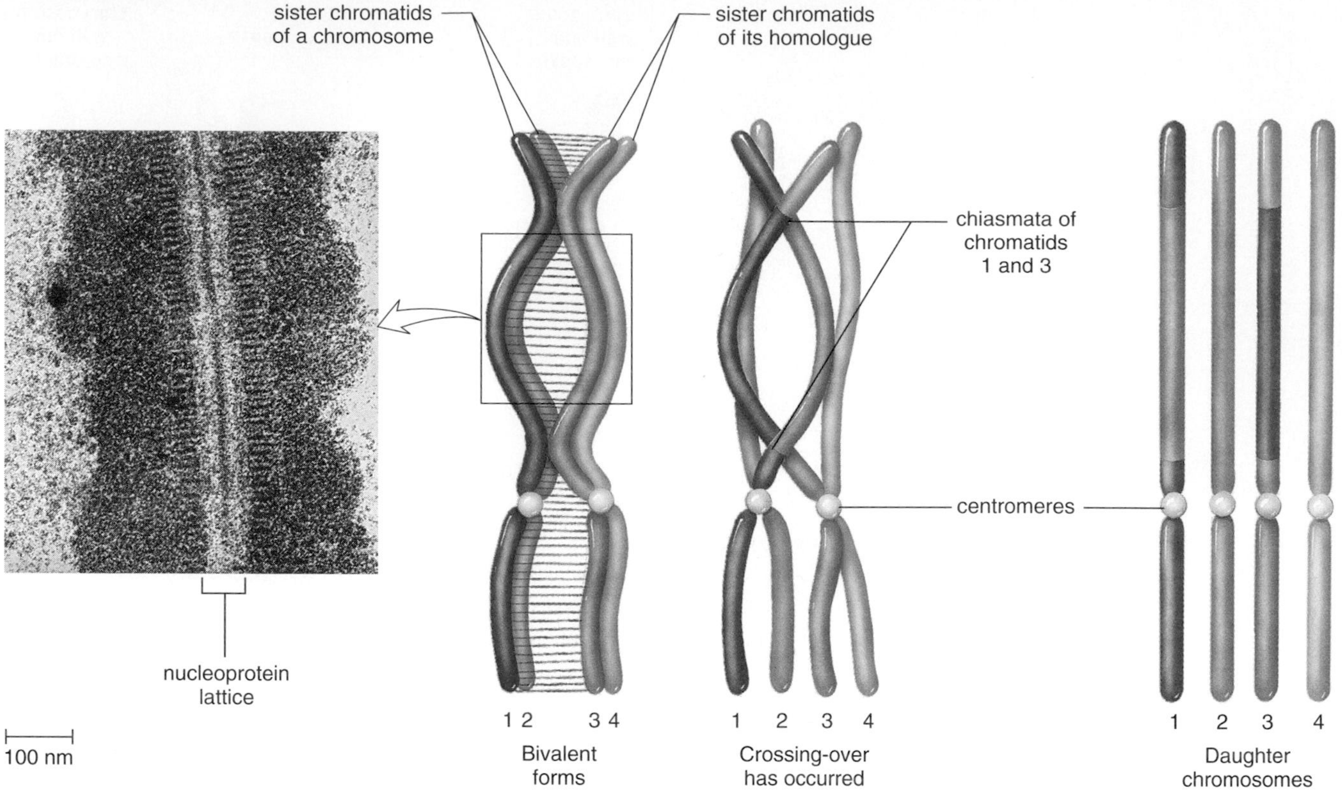

sister chromatids of a chromosome

sister chromatids of its homologue

chiasmata of chromatids 1 and 3

centromeres

nucleoprotein lattice

100 nm

1 2   3 4
Bivalent forms

1   2   3   4
Crossing-over has occurred

1   2   3   4
Daughter chromosomes

Crossing-over occurs during meiosis I
Figure 10.3

a: Courtesy of Dr. D. Von Wettstein

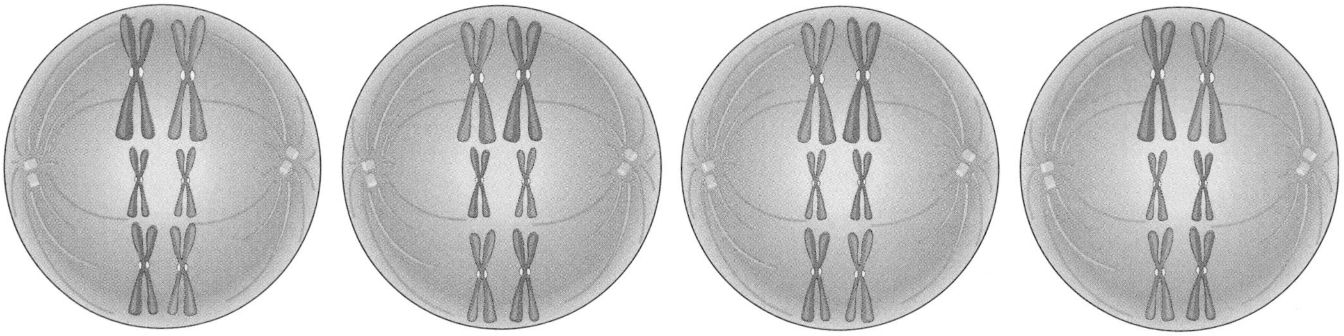

Independent assortment
Figure 10.4

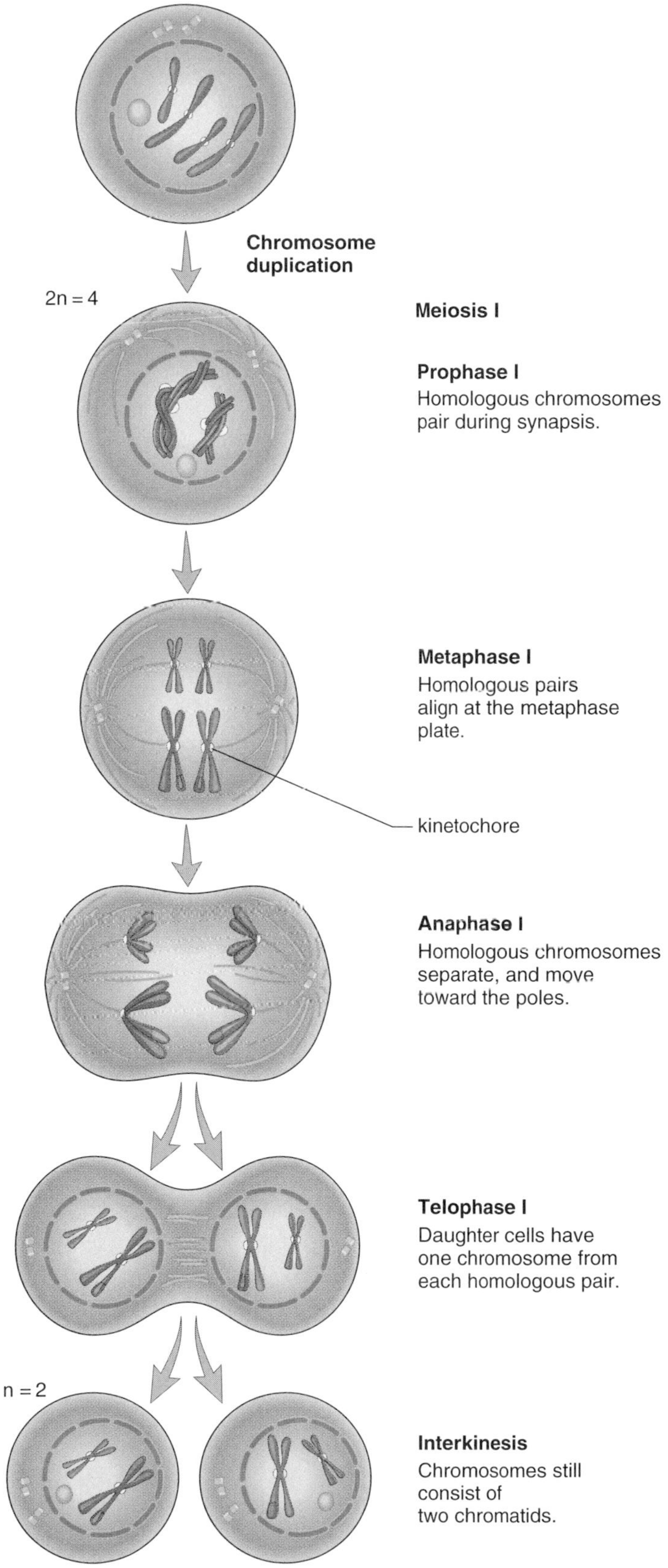

Chromosome duplication

2n = 4

Meiosis I

Prophase I
Homologous chromosomes pair during synapsis.

Metaphase I
Homologous pairs align at the metaphase plate.

—— kinetochore

Anaphase I
Homologous chromosomes separate, and move toward the poles.

Telophase I
Daughter cells have one chromosome from each homologous pair.

n = 2

Interkinesis
Chromosomes still consist of two chromatids.

Meiosis I
Figure 10.6

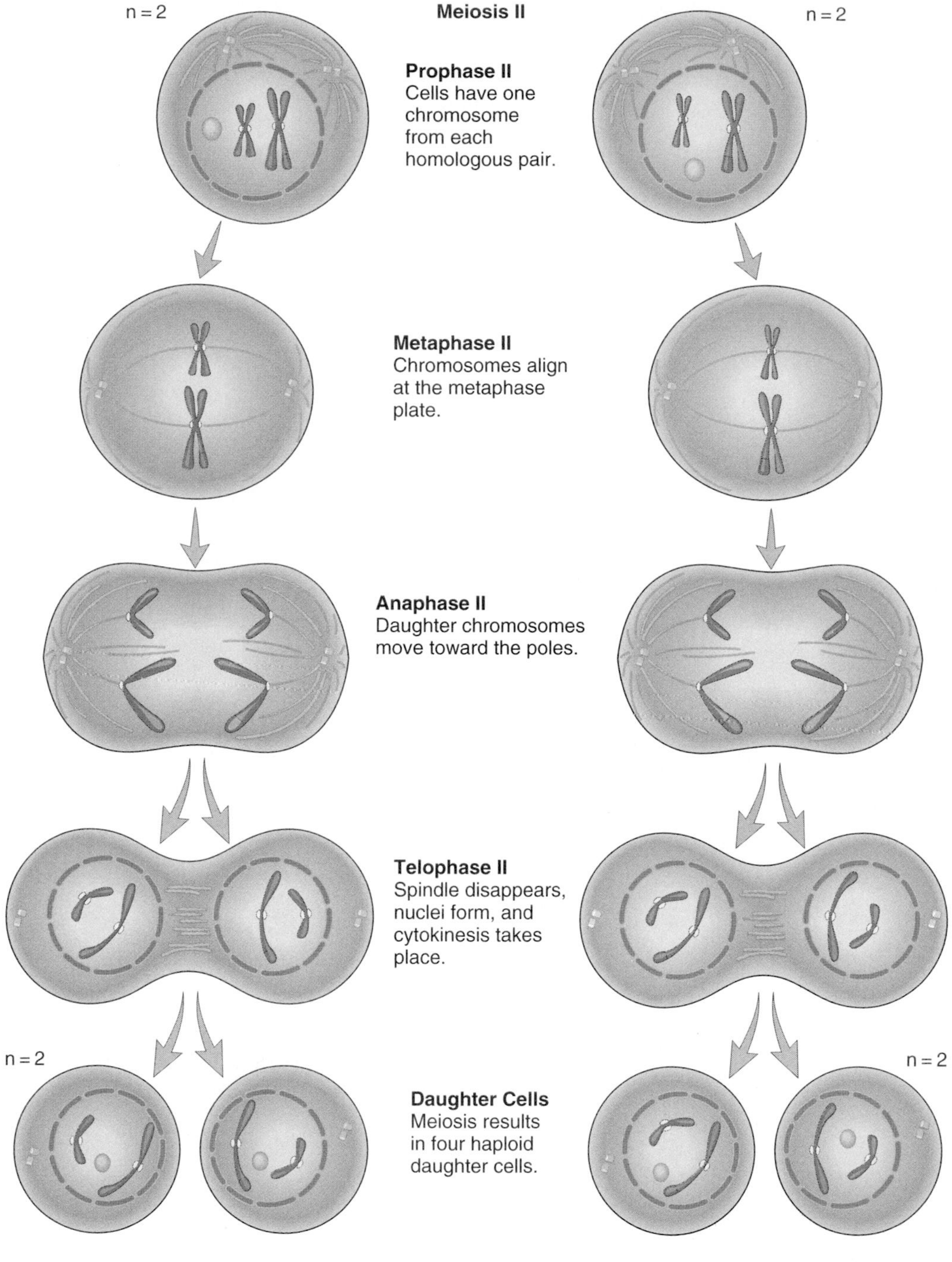

n = 2

Meiosis II

Prophase II
Cells have one chromosome from each homologous pair.

n = 2

Metaphase II
Chromosomes align at the metaphase plate.

Anaphase II
Daughter chromosomes move toward the poles.

Telophase II
Spindle disappears, nuclei form, and cytokinesis takes place.

n = 2

Daughter Cells
Meiosis results in four haploid daughter cells.

n = 2

Meiosis II
Figure 10.7

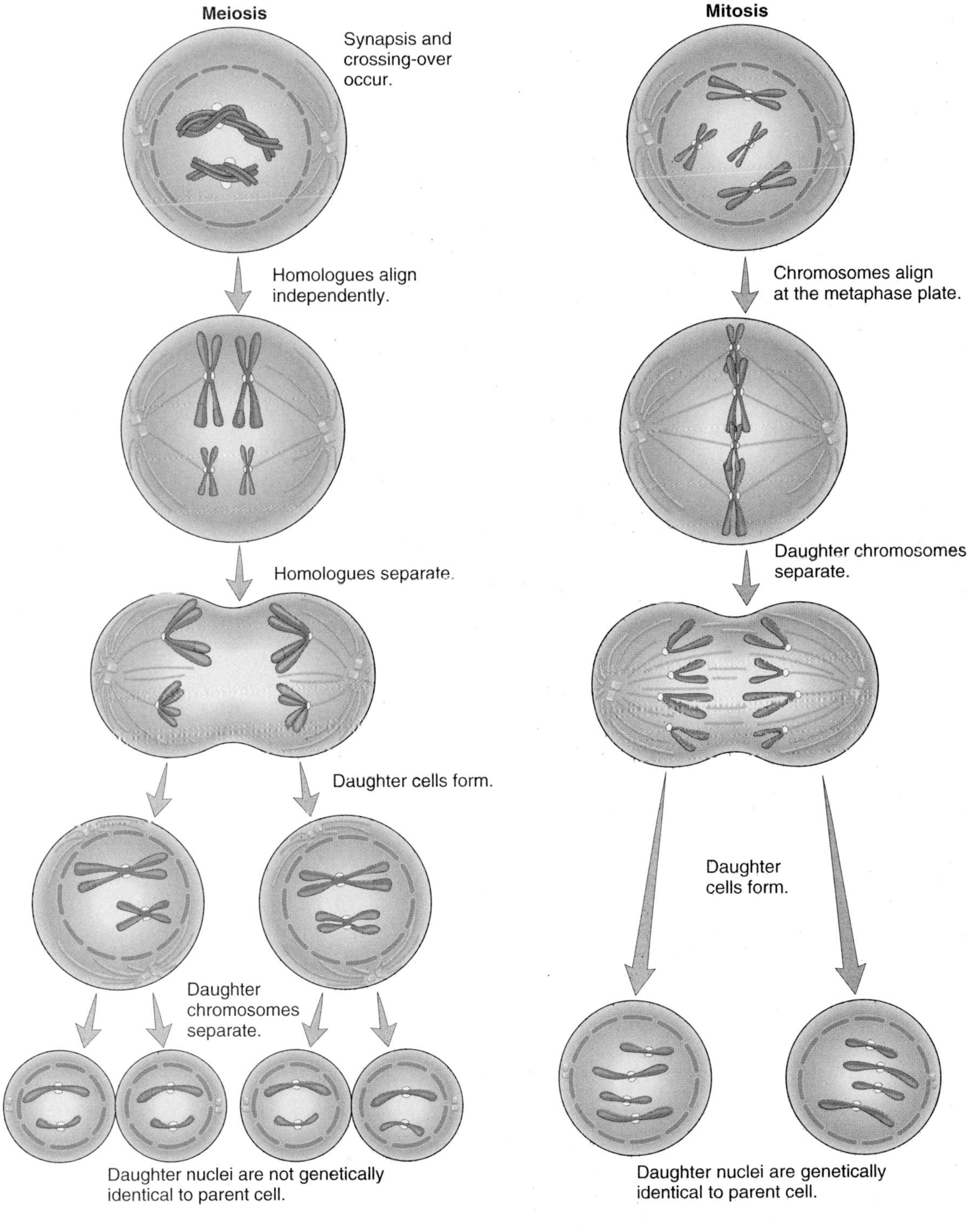

Meiosis

Synapsis and crossing-over occur.

Homologues align independently.

Homologues separate.

Daughter cells form.

Daughter chromosomes separate.

Daughter nuclei are not genetically identical to parent cell.

Mitosis

Chromosomes align at the metaphase plate.

Daughter chromosomes separate.

Daughter cells form.

Daughter nuclei are genetically identical to parent cell.

Meiosis compared to mitosis
Figure 10.8

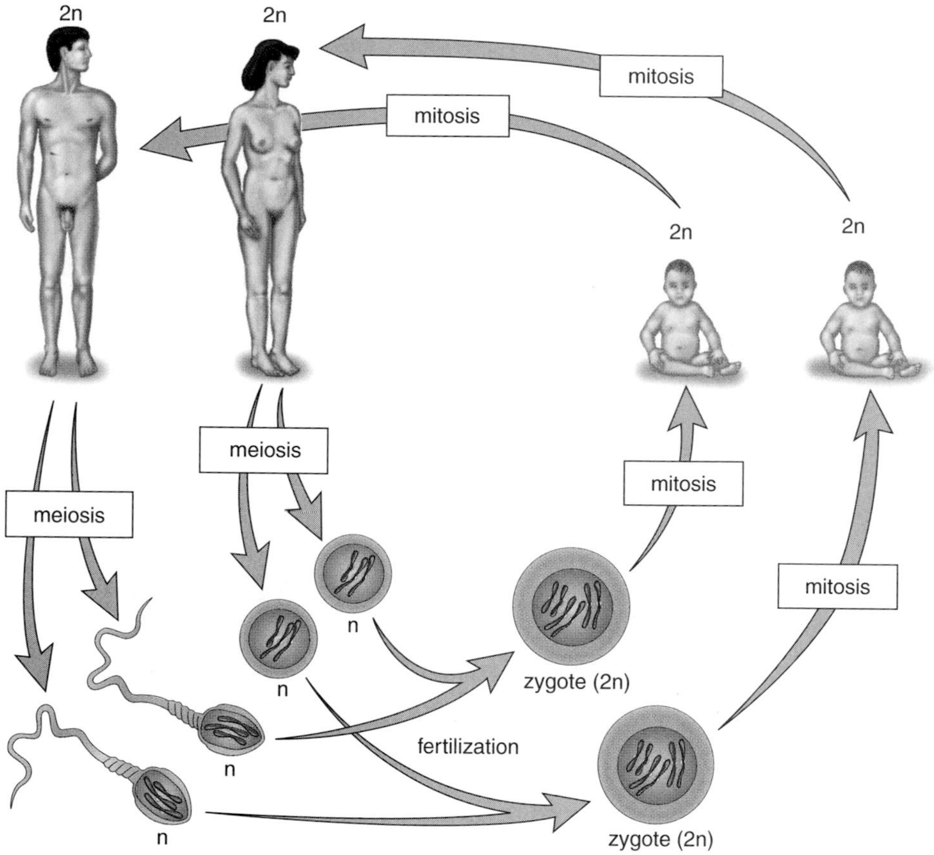
2n
2n
mitosis
mitosis
2n
2n
meiosis
meiosis
mitosis
mitosis
n
n
zygote (2n)
n
fertilization
n
zygote (2n)

Life cycle of humans
Figure 10.9

Spermatogenesis

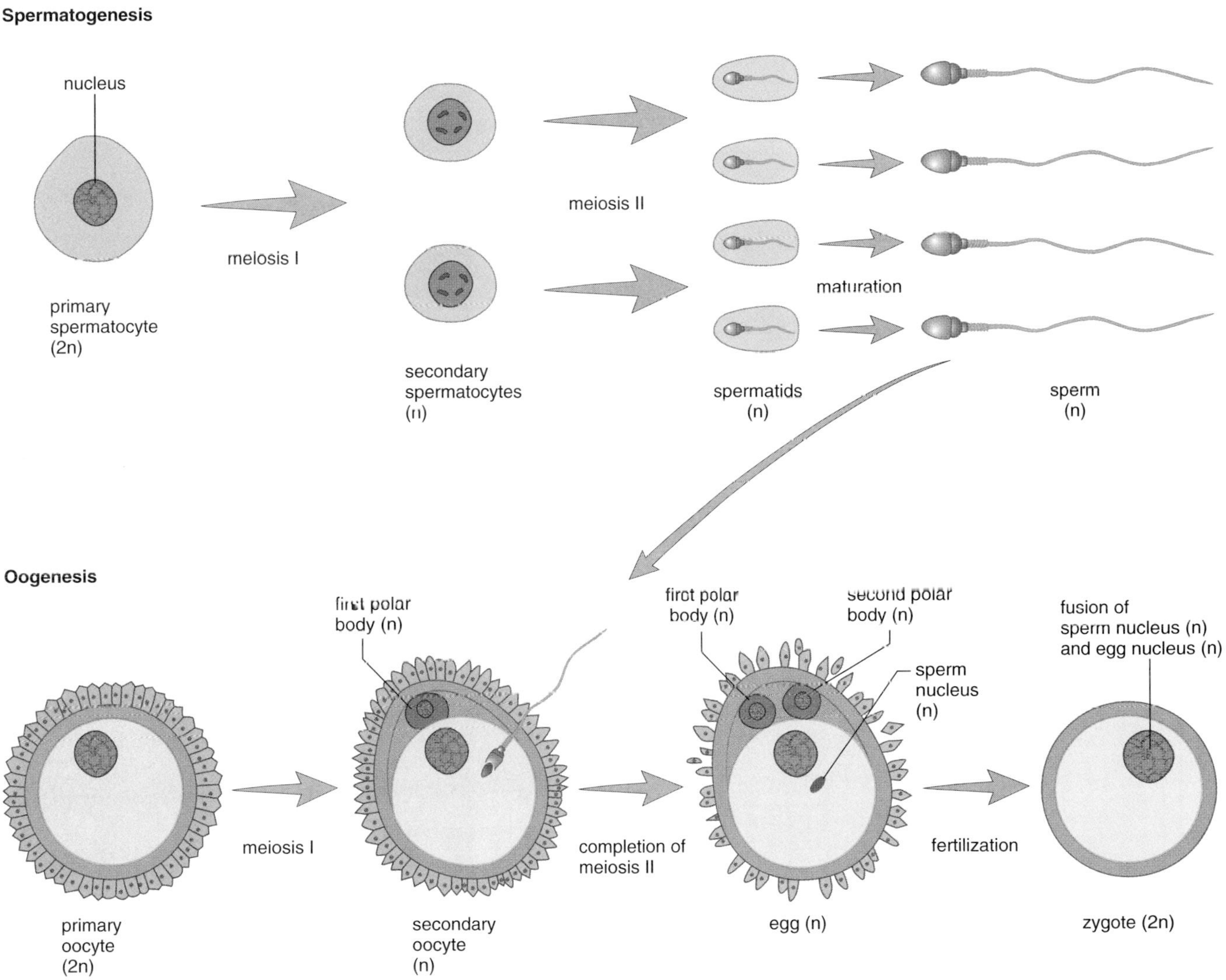

Spermatogenesis and oogenesis in mammals
Figure 10.10

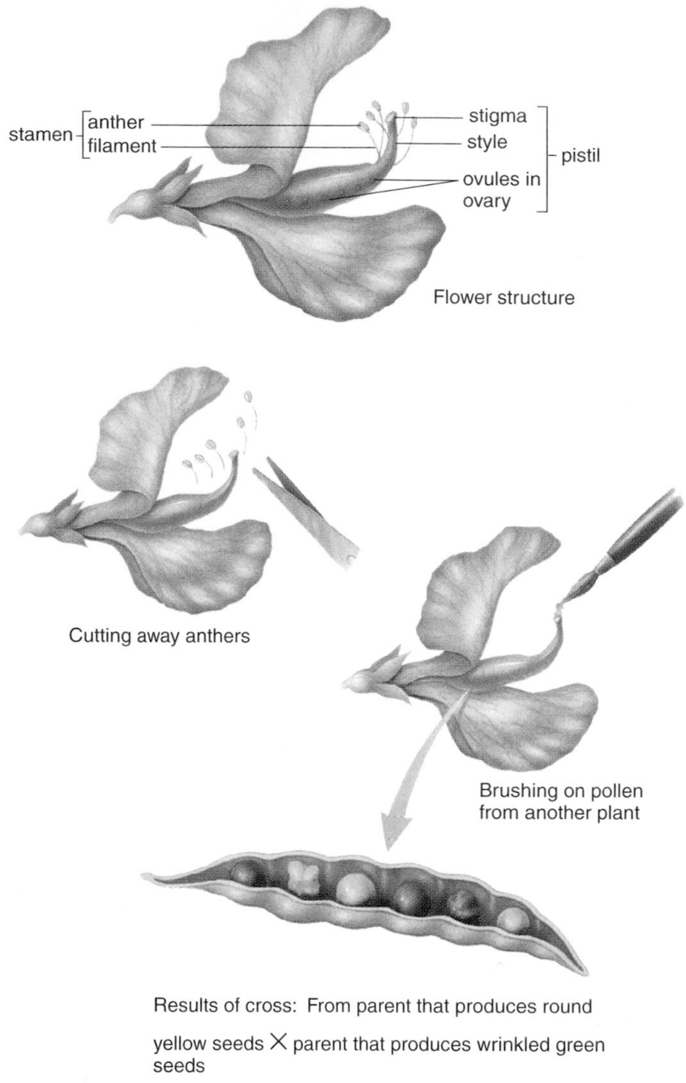

stamen { anther
         filament

stigma
style
ovules in
ovary
} pistil

Flower structure

Cutting away anthers

Brushing on pollen
from another plant

Results of cross: From parent that produces round
yellow seeds X parent that produces wrinkled green
seeds

Trait	Characteristics		F₂ Results*	
			Dominant	Recessive
Stem length	Tall	Short	787	277
Pod shape	Inflated	Constricted	882	299
Seed shape	Round	Wrinkled	5,474	1,850
Seed color	Yellow	Green	7,022	2,001
Flower position	Axial	Terminal	651	207
Flower color	Purple	White	705	224
Pod color	Green	Yellow	428	152

*All of these produce approximately a 3:1 ratio. For example,
$$\frac{787}{277} \approx \frac{3}{1}$$

Garden pea anatomy and traits
Figure 11.2

Monohybrid cross done by Mendel
Figure 11.3

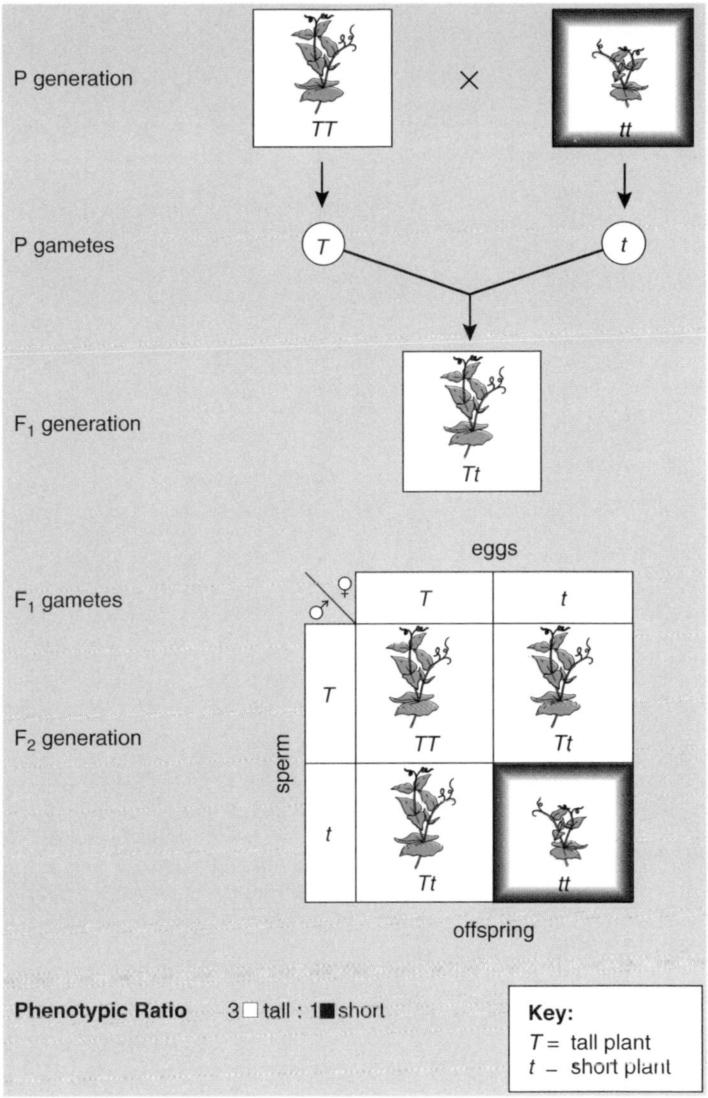

P generation   TT   ×   tt

P gametes   T   t

F₁ generation   Tt

F₁ gametes

F₂ generation

eggs

sperm

T   t

T   TT   Tt

t   Tt   tt

offspring

Phenotypic Ratio   3 ☐ tall : 1 ■ short

Key:
T = tall plant
t − short plant

sister
chromatids

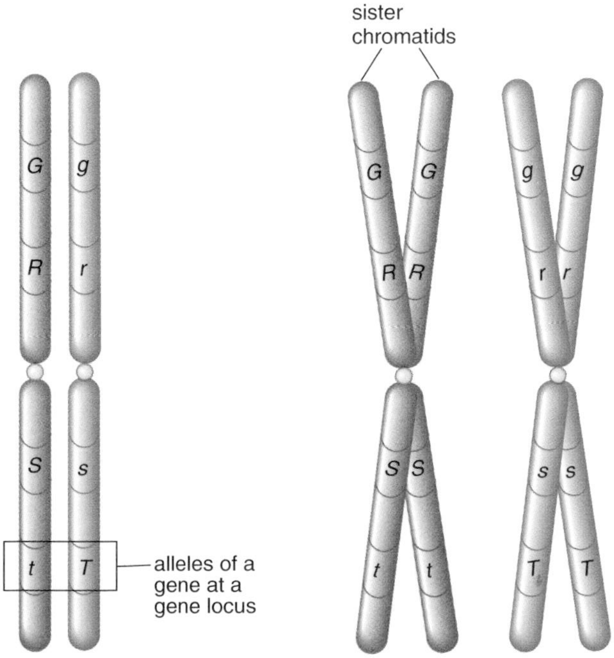

alleles of a
gene at a
gene locus

a. Diagram of homologous
chromosomes shows that
they have various alleles
at specific loci.

b. Diagram of duplicated
chromosomes shows that
sister chromatids have
identical alleles.

Homologous chromosomes
Figure 11.4

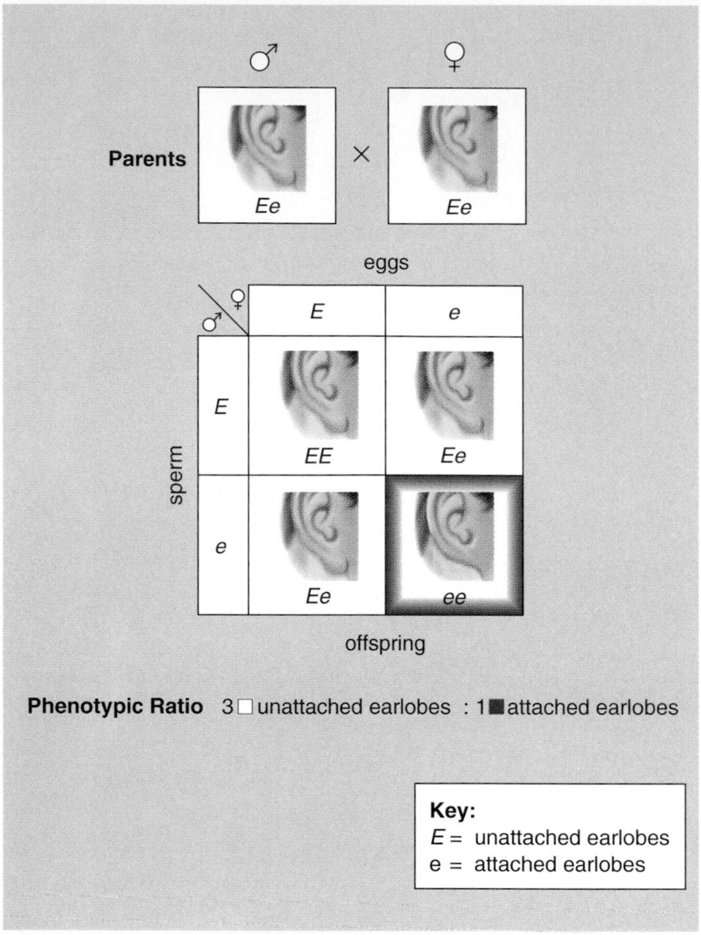

Genetic inheritance in humans
Figure 11.5

Parents

eggs

sperm

offspring

Phenotypic Ratio 3 □ unattached earlobes : 1 ■ attached earlobes

Key:
E = unattached earlobes
e = attached earlobes

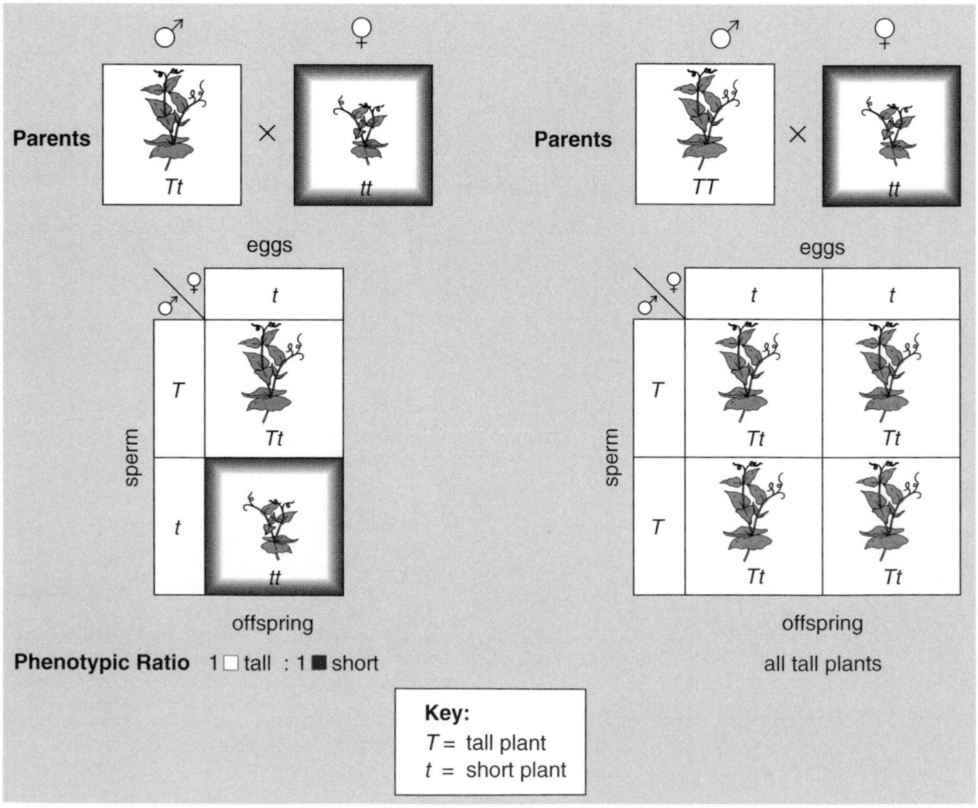

One-trait testcross
Figure 11.6

Parents   ×   Parents   ×

eggs

sperm

offspring

Phenotypic Ratio 1 □ tall : 1 ■ short

eggs

sperm

offspring

all tall plants

Key:
T = tall plant
t = short plant

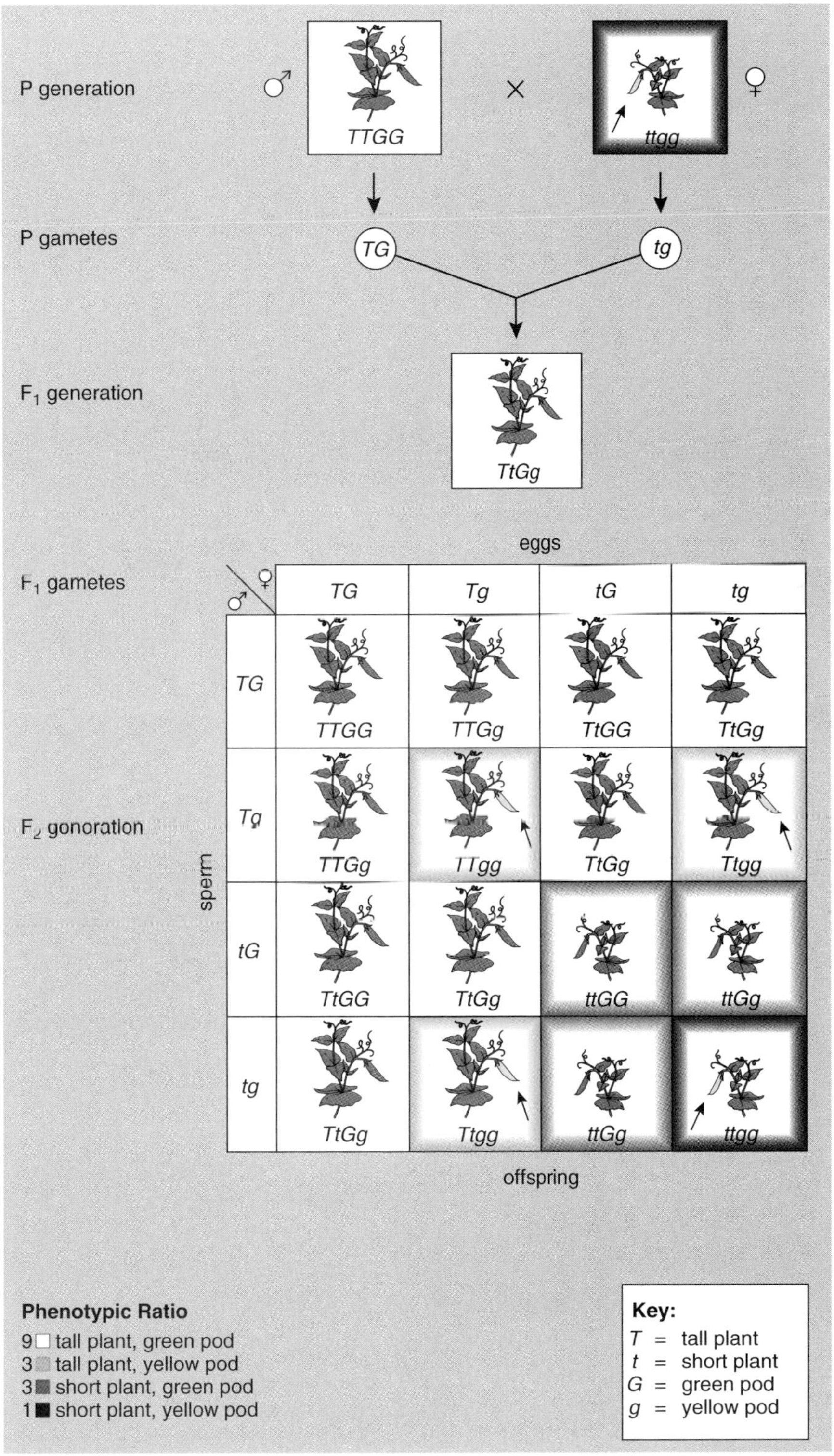
P generation
TTGG
×
ttgg
P gametes
TG
tg
F1 generation
TtGg
eggs
F1 gametes
TG
Tg
tG
tg
TTGG
TTGg
TtGG
TtGg
F2 generation
sperm
TTGg
TTgg
TtGg
Ttgg
TtGG
TtGg
ttGG
ttGg
TtGg
Ttgg
ttGg
ttgg
offspring
Phenotypic Ratio
9 tall plant, green pod
3 tall plant, yellow pod
3 short plant, green pod
1 short plant, yellow pod
Key:
T = tall plant
t = short plant
G = green pod
g = yellow pod

Dihybrid cross done by Mendel
Figure 11.7

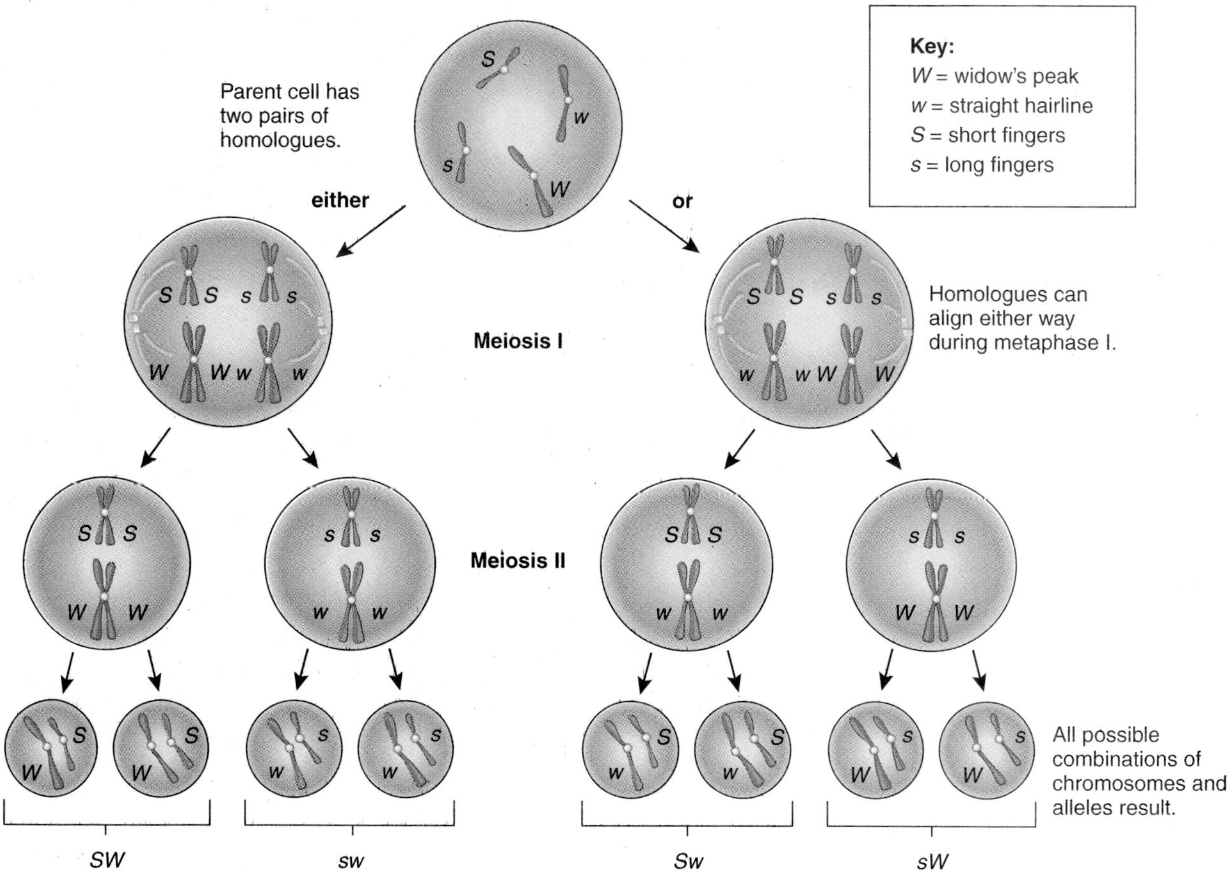

Parent cell has two pairs of homologues.

either

or

Key:
W = widow's peak
w = straight hairline
S = short fingers
s = long fingers

Meiosis I

Homologues can align either way during metaphase I.

Meiosis II

All possible combinations of chromosomes and alleles result.

SW   *sw*   *Sw*   *sW*

Segregation and independent assortment
Figure 11A

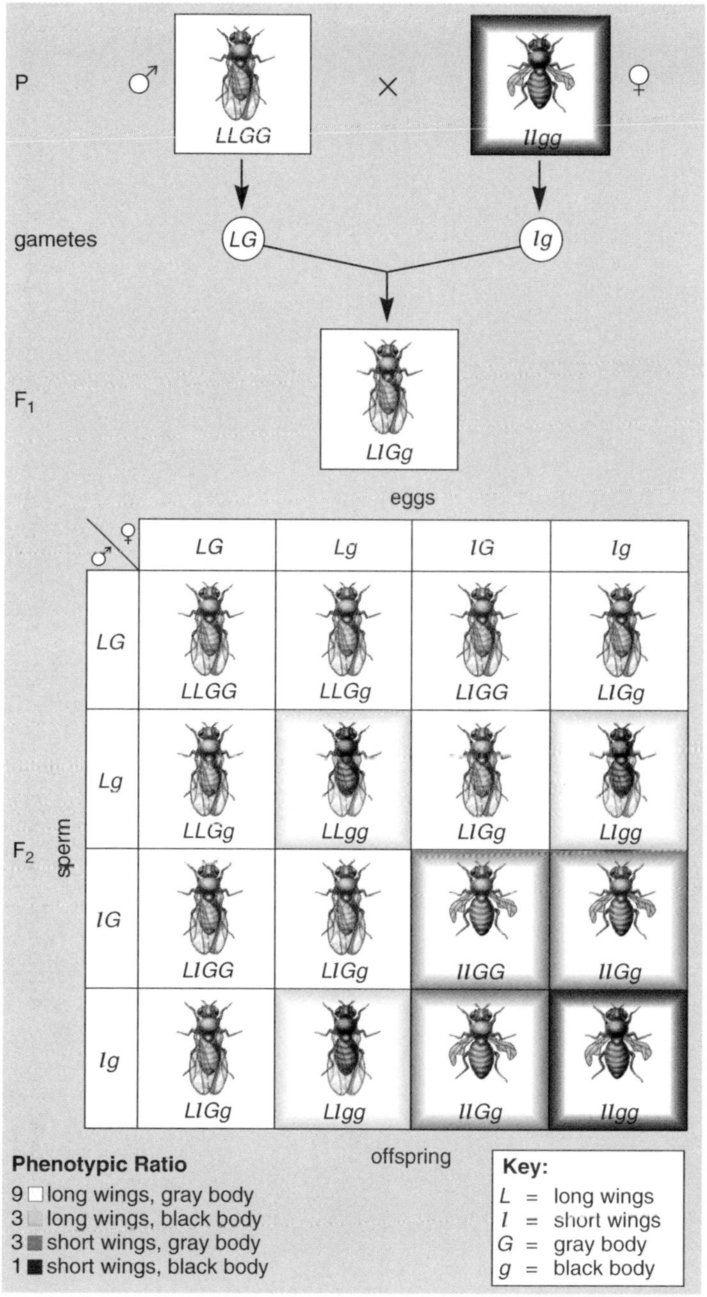

Phenotypic Ratio

9 ☐ long wings, gray body
3 ▨ long wings, black body
3 ▦ short wings, gray body
1 ■ short wings, black body

Key:

L	=	long wings
l	=	short wings
G	=	gray body
g	=	black body

Inheritance in fruit flies

Figure 11.8

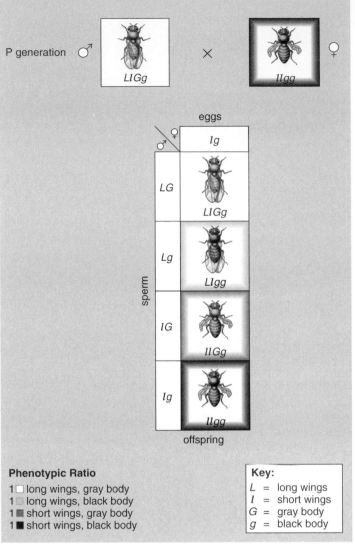

Phenotypic Ratio

1 ☐ long wings, gray body
1 ☐ long wings, black body
1 ■ short wings, gray body
1 ■ short wings, black body

Key:

L = long wings
l = short wings
G = gray body
g = black body

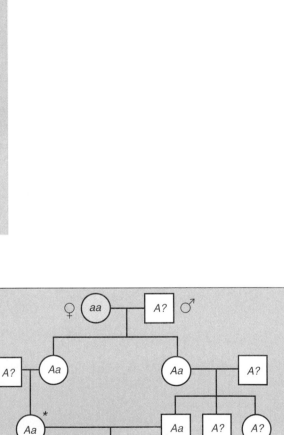

Autosomal recessive disorders

- Affected children can have unaffected parents.
- Heterozygotes (*Aa*) have a normal phenotype.
- Two affected parents will always have affected children.
- Affected individuals with homozygous dominant mates will have unaffected children.
- Close unaffected relatives who reproduce are more likely to have affected children if they have joint affected relatives.
- Both males and females are affected with equal frequency.

Key:

aa = affected
Aa = carrier (unaffected)
AA = normal
A? = unaffected (one allele unknown)

Autosomal recessive pedigree chart
Figure 11.10

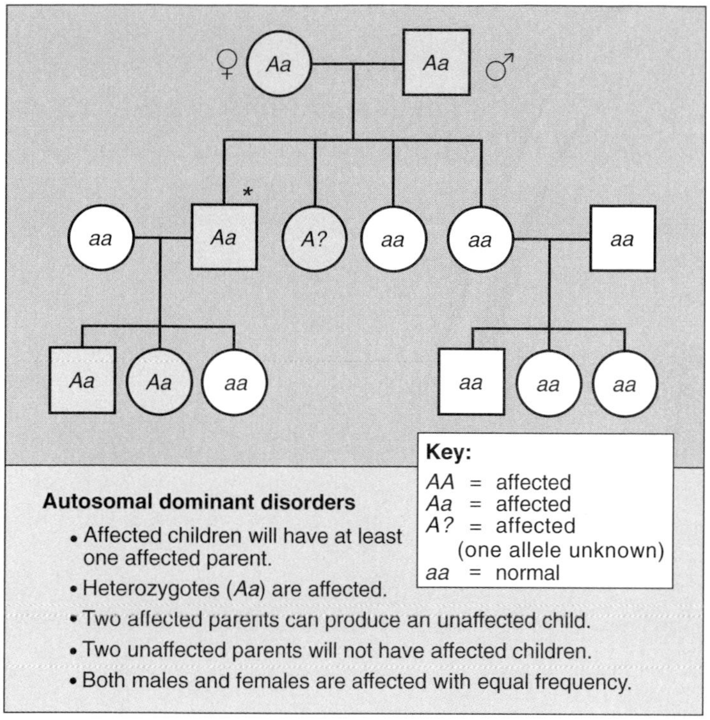

Autosomal dominant disorders

- Affected children will have at least one affected parent.
- Heterozygotes (*Aa*) are affected.
- Two affected parents can produce an unaffected child.
- Two unaffected parents will not have affected children.
- Both males and females are affected with equal frequency.

Key:
AA = affected
Aa = affected
A? = affected
    (one allele unknown)
aa = normal

Autosomal dominant pedigree chart
Figure 11.11

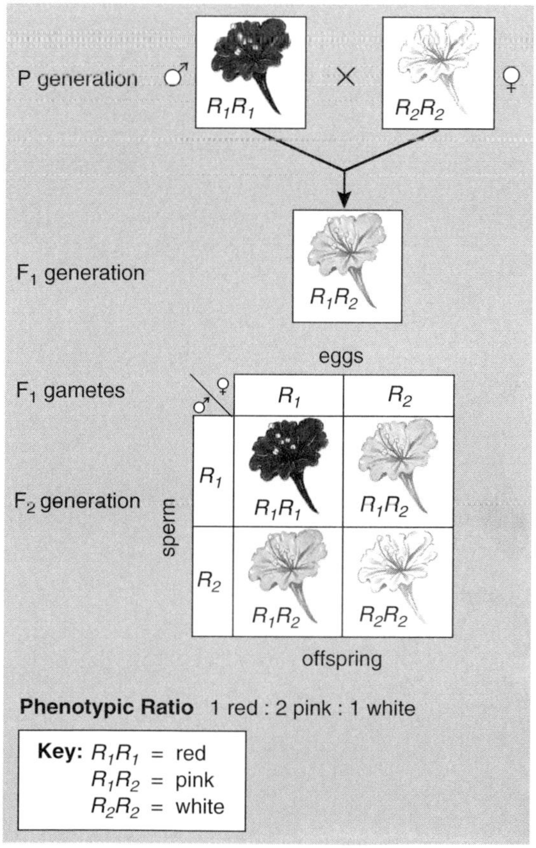

Phenotypic Ratio 1 red : 2 pink : 1 white

Key: R_1R_1 = red
R_1R_2 = pink
R_2R_2 = white

Incomplete dominance
Figure 11.14

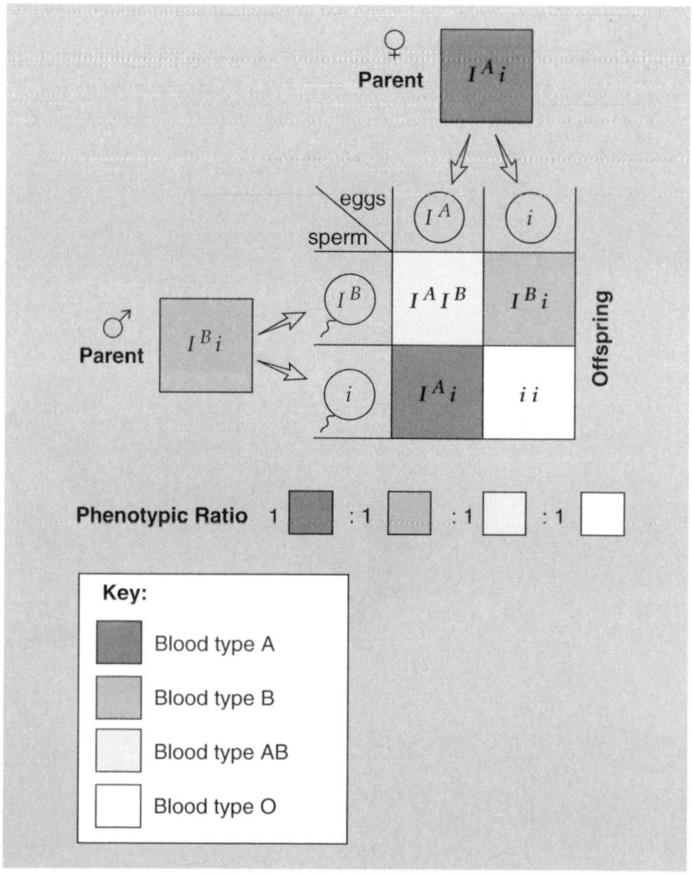

Phenotypic Ratio 1 ▢ : 1 ▢ : 1 ▢ : 1 ▢

Key:
▢ Blood type A
▢ Blood type B
▢ Blood type AB
▢ Blood type O

Inheritance of blood type
Figure 11.15

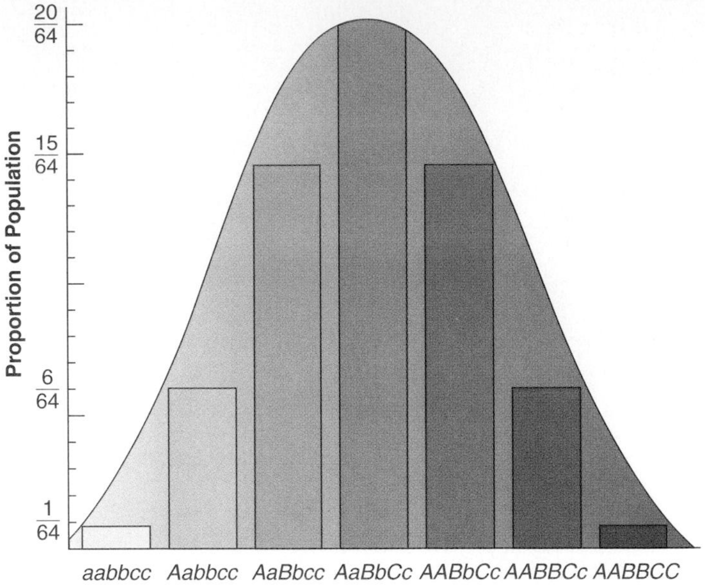

Polygenic inheritance
Figure 11.17

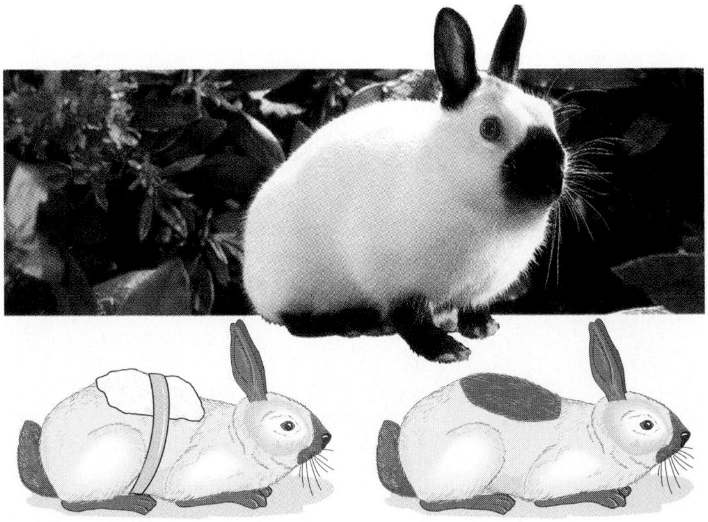

Coat color in Himalayan rabbits
Figure 11.18
© Jane Burton/Bruce Coleman, Inc.

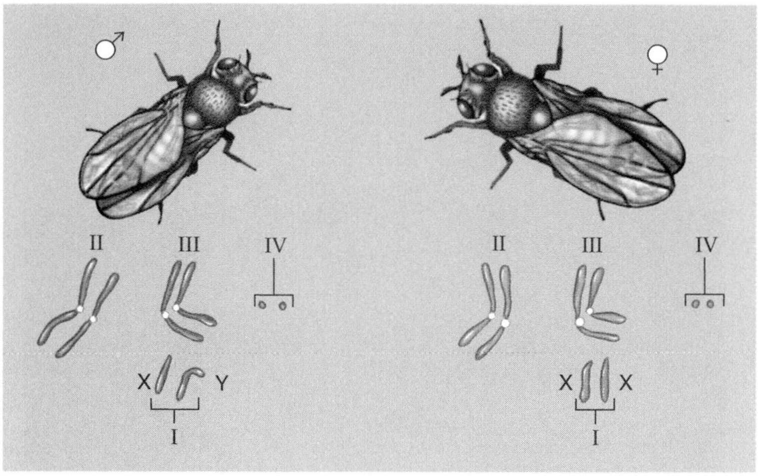

Drosophila

Figure 12.1

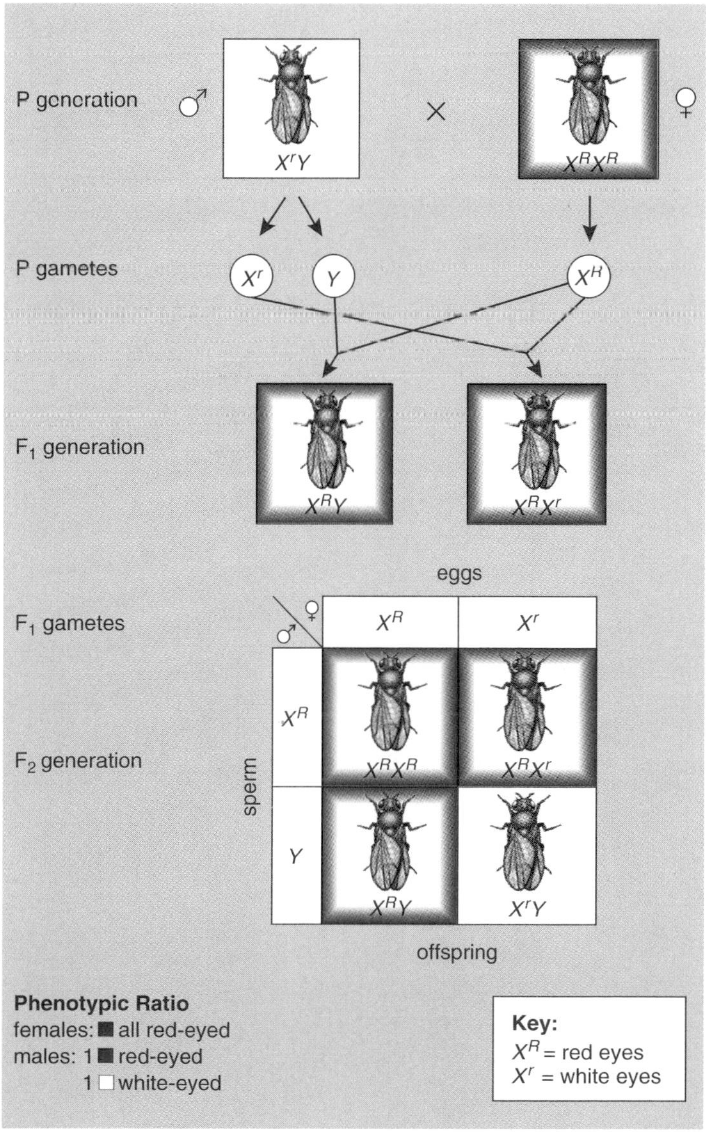

Phenotypic Ratio
females: ■ all red-eyed
males: 1 ■ red-eyed
     1 □ white-eyed

Key:
X^R = red eyes
X^r = white eyes

X-linked inheritance

Figure 12.2

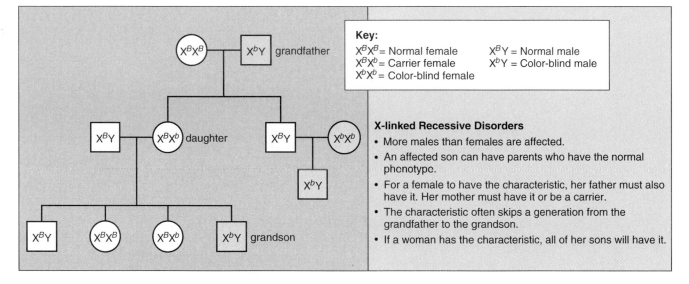

Key:

X^BX^B = Normal female X^BY = Normal male
X^BX^b = Carrier female X^bY = Color-blind male
X^bX^b = Color-blind female

X-linked Recessive Disorders

- More males than females are affected.
- An affected son can have parents who have the normal phenotype.
- For a female to have the characteristic, her father must also have it. Her mother must have it or be a carrier.
- The characteristic often skips a generation from the grandfather to the grandson.
- If a woman has the characteristic, all of her sons will have it.

X-linked recessive pedigree chart
Figure 12.3

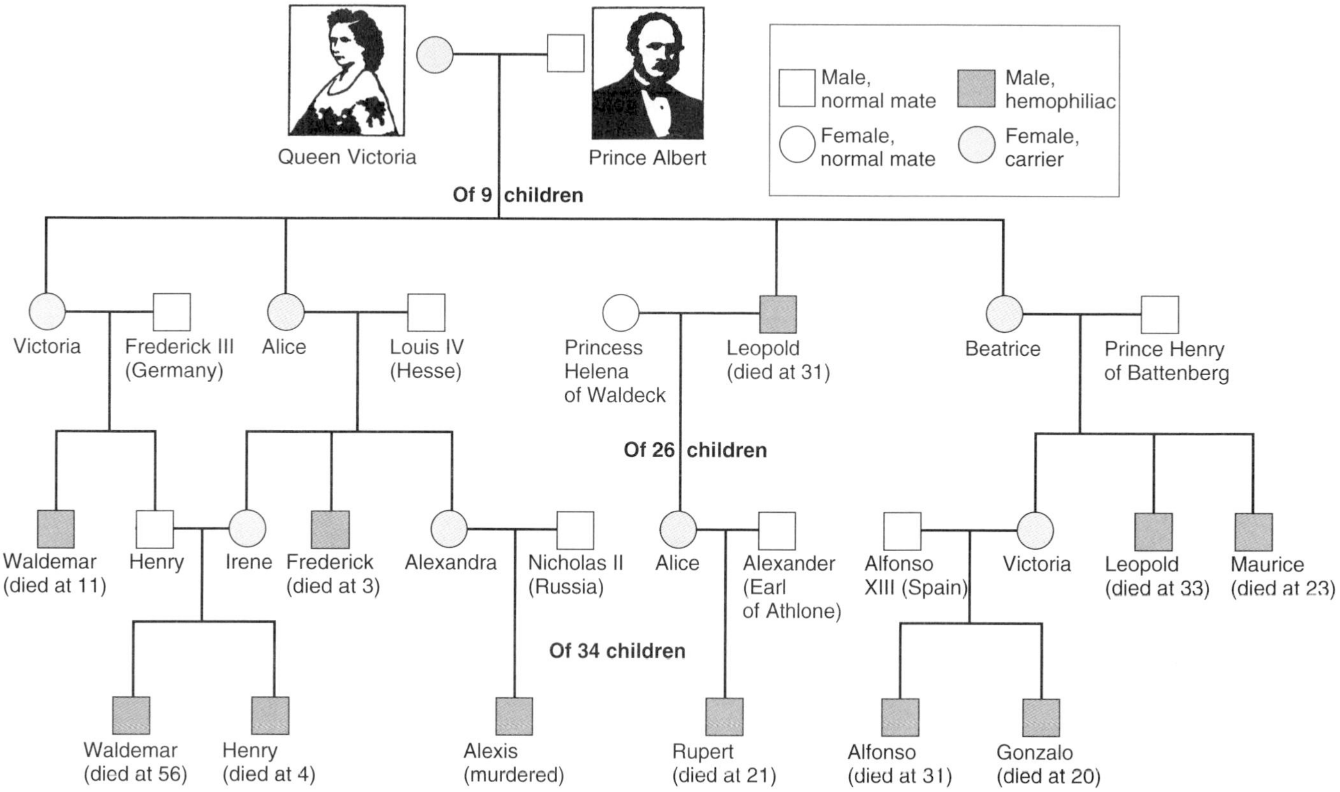

A simplified pedigree showing the X-linked inheritance of hemophilia in European royal families

Figure 12.4

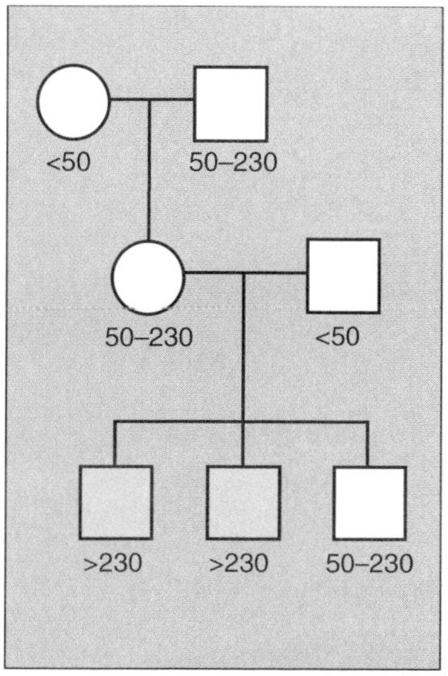

c. Inheritance pattern for fragile X syndrome

Fragile X syndrome

Figure 12A

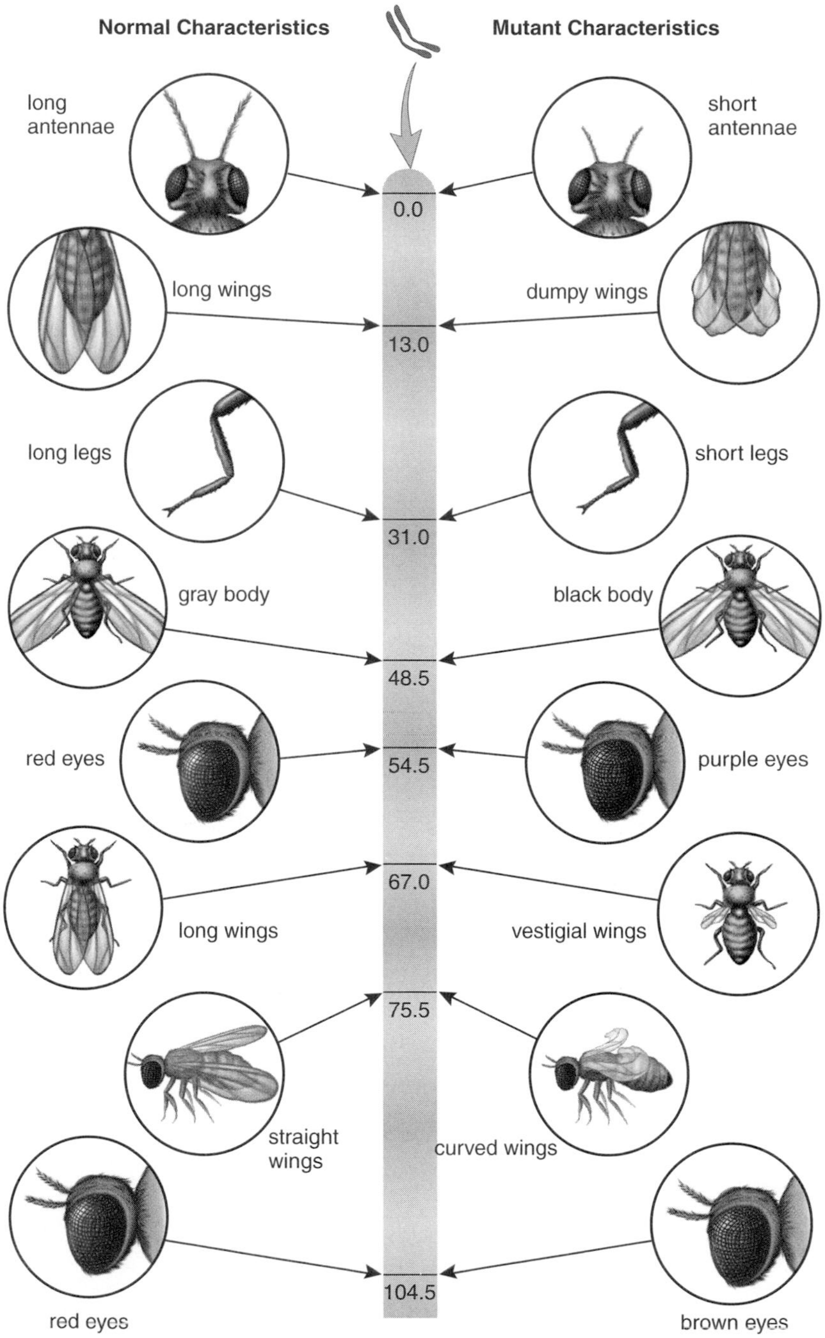

Normal Characteristics   Mutant Characteristics

long antennae — 0.0 — short antennae

long wings — 13.0 — dumpy wings

long legs — 31.0 — short legs

gray body — 48.5 — black body

red eyes — 54.5 — purple eyes

long wings — 67.0 — vestigial wings

straight wings — 75.5 — curved wings

red eyes — 104.5 — brown eyes

Linkage group
Figure 12.5

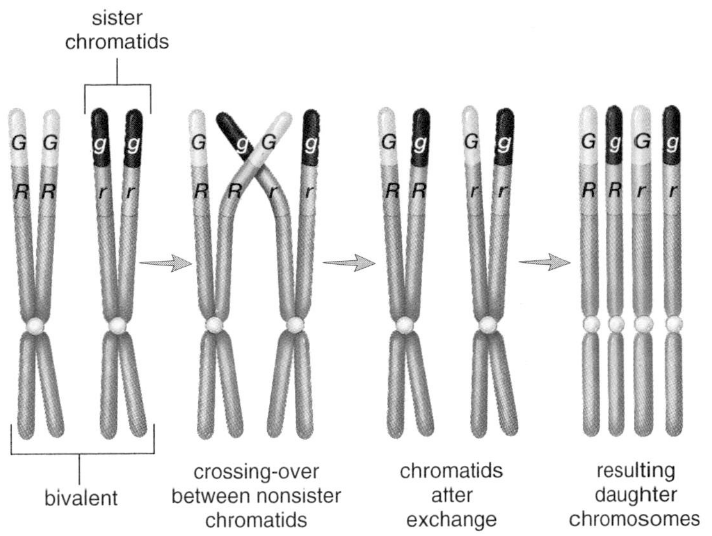

Crossing-over
Figure 12.6

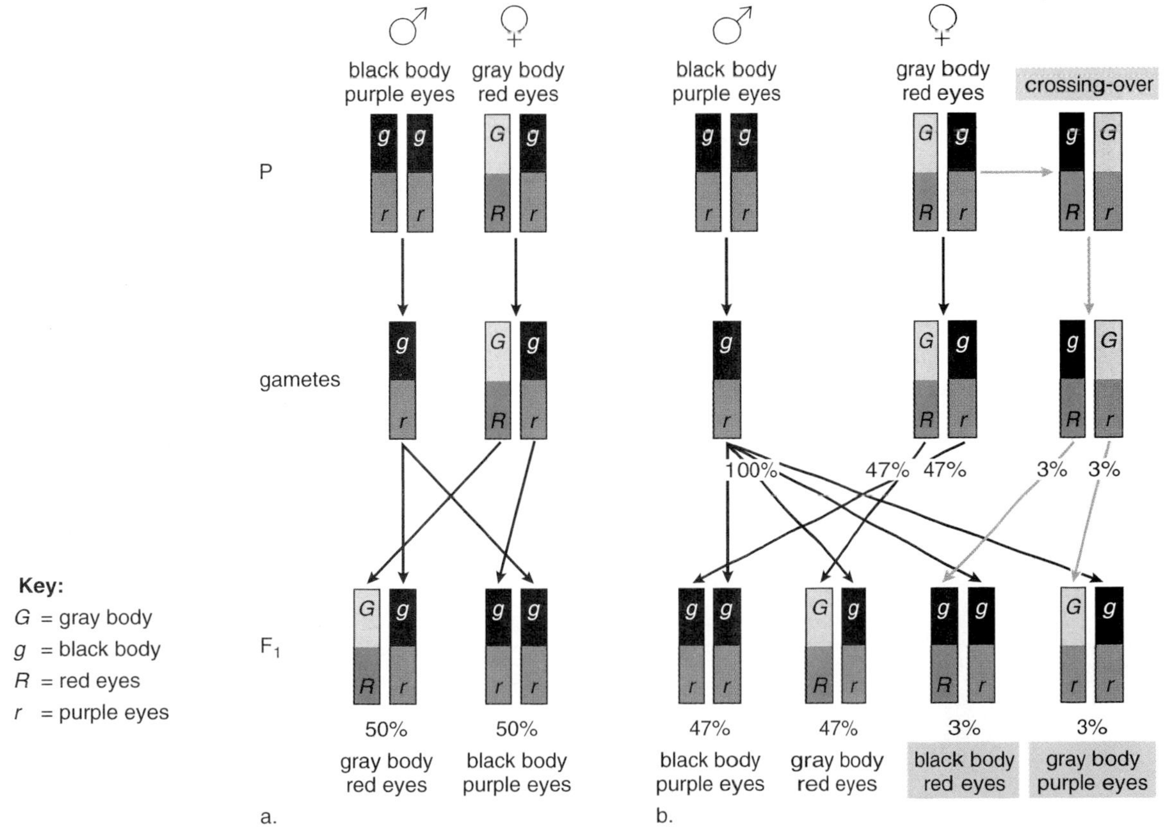

Key:
G = gray body
g = black body
R = red eyes
r = purple eyes

Complete linkage versus incomplete linkage
Figure 12.7

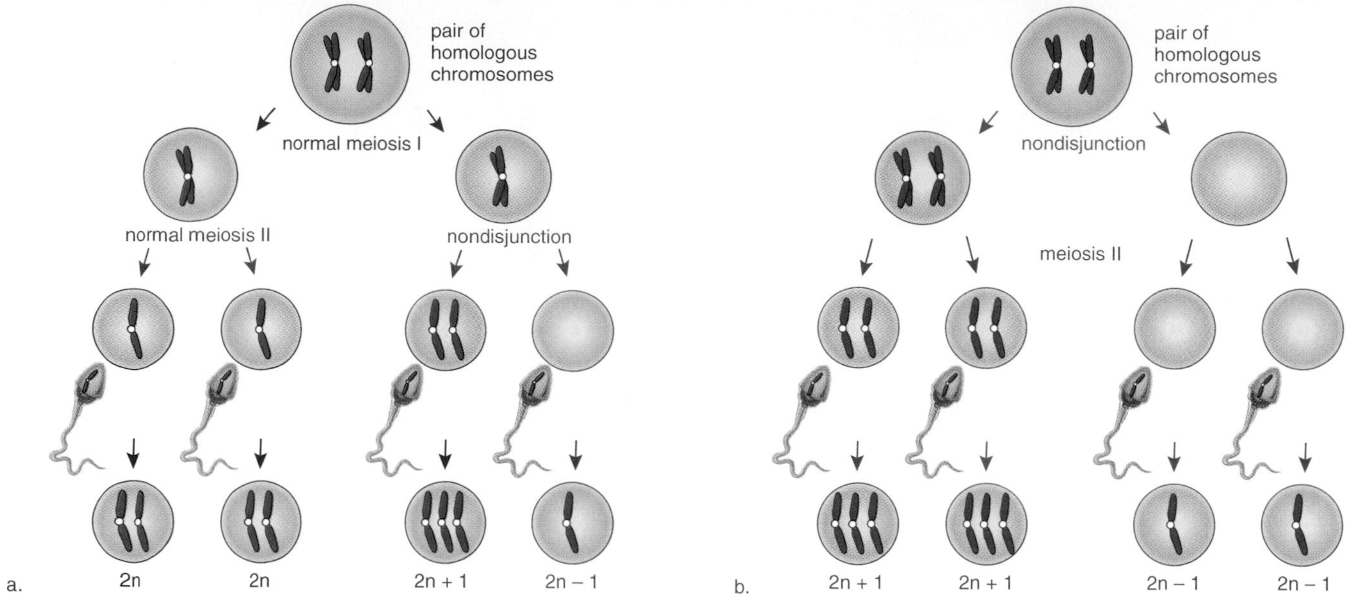

a.   2n   2n   2n + 1   2n − 1

b.   2n + 1   2n + 1   2n − 1   2n − 1

Nondisjunction of chromosomes during oogenesis, followed by fertilization with normal sperm
Figure 12.8

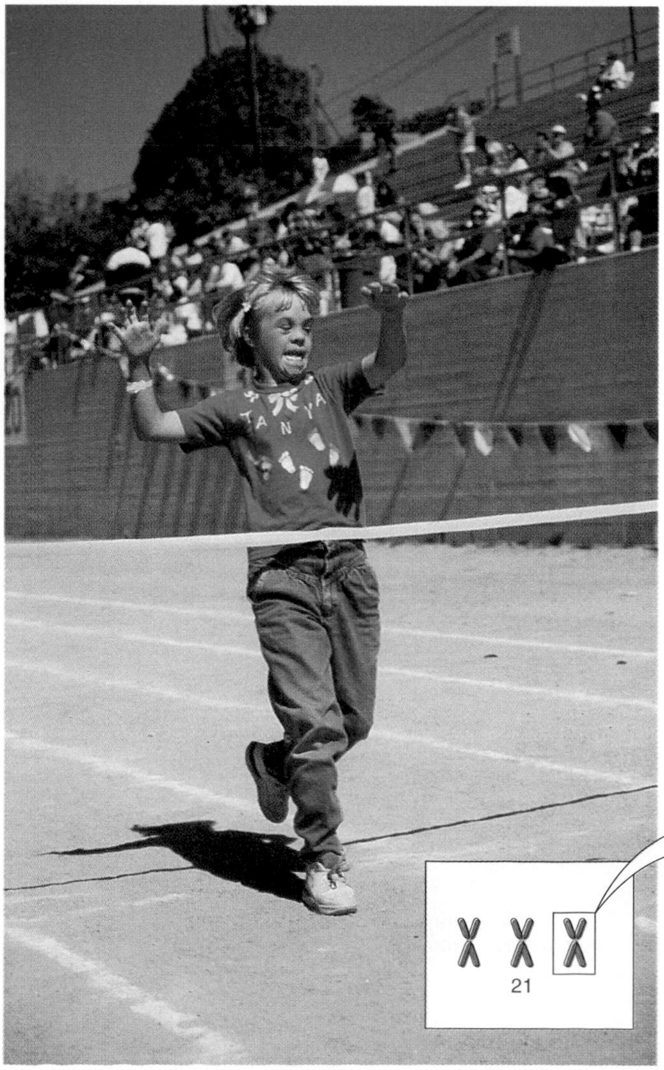

Trisomy 21
Figure 12.9
© Jose Carrilo/PhotoEdit

a.

b.

Gart gene

21

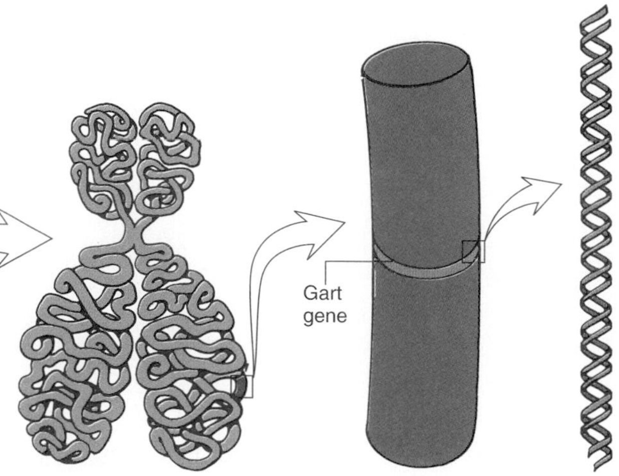

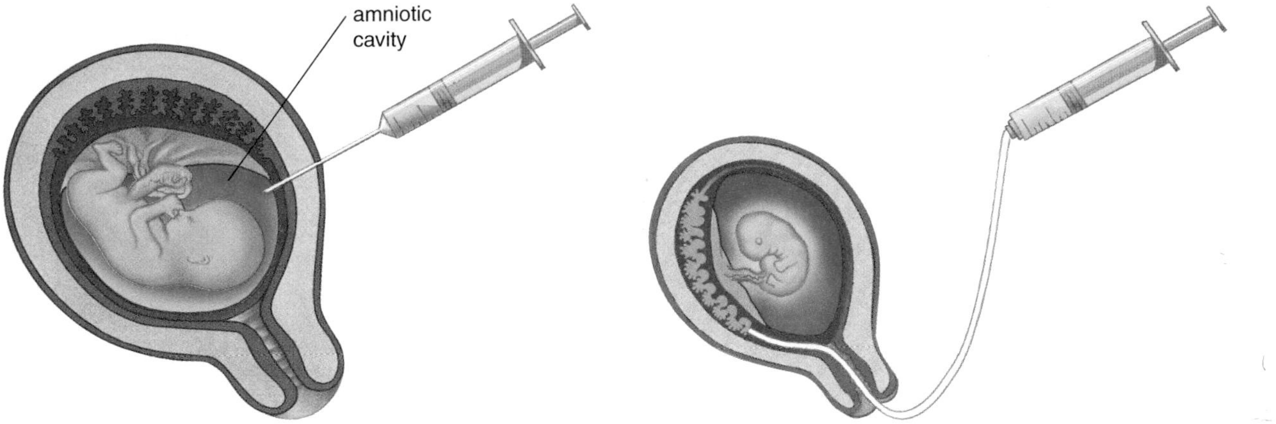

a. During amniocentesis, a long needle is used to withdraw amniotic fluid containing fetal cells.

b. During chorionic villi sampling, a suction tube is used to remove cells from the chorion, where the placenta will develop.

Human karyotype preparation
Figure 12B

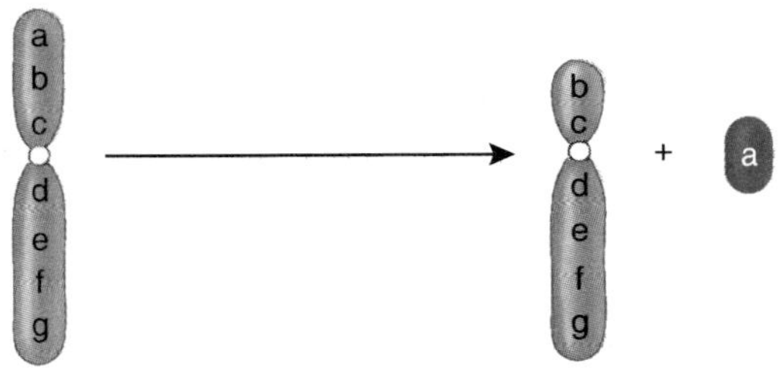

a. Deletion

b. Translocation

c. Duplication

d. Inversion

Types of chromosomal mutations
Figure 12.11

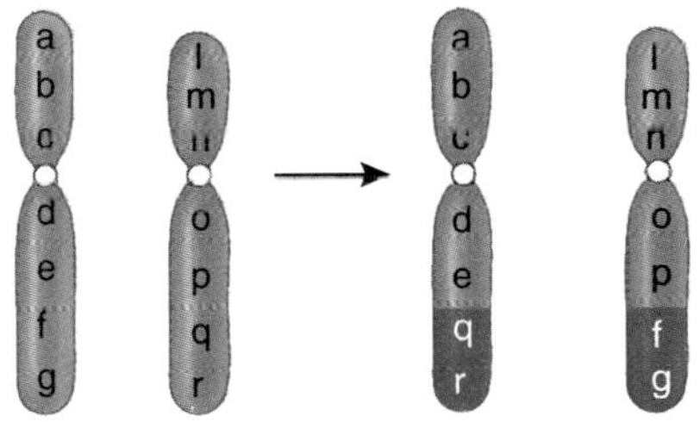

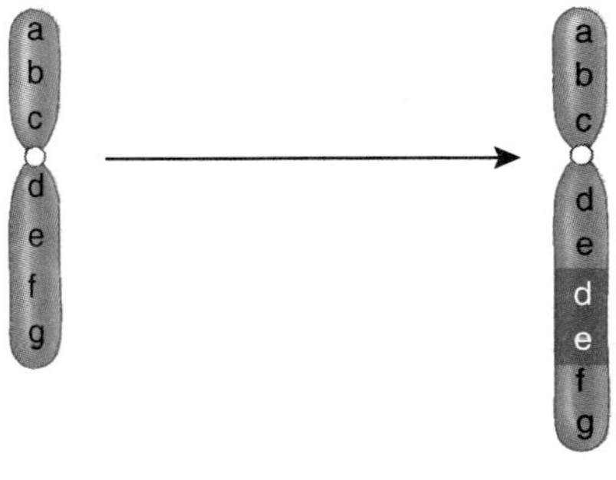

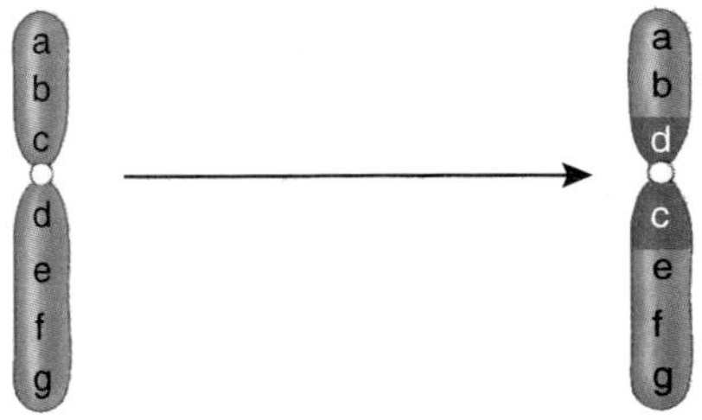

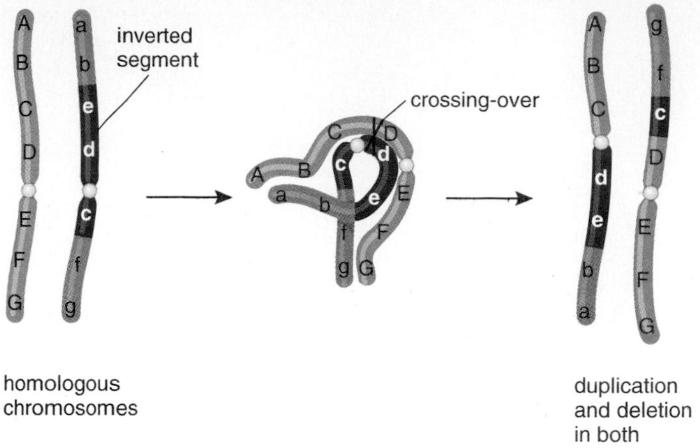

Inversion
Figure 12.12

homologous
chromosomes

inverted
segment

crossing-over

duplication
and deletion
in both

Deletion
Figure 12.13

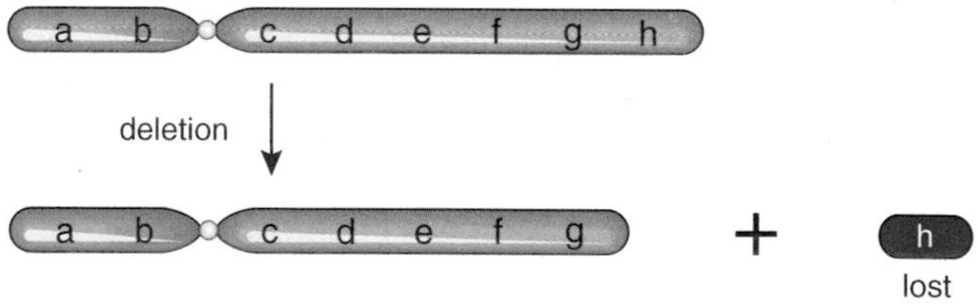

deletion

lost

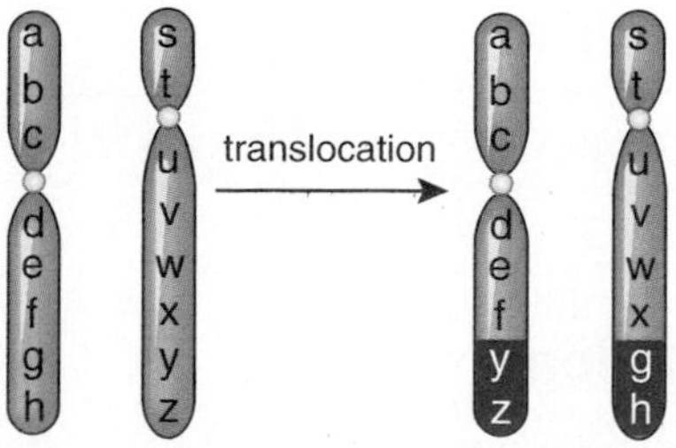

translocation

Translocation
Figure 12.14

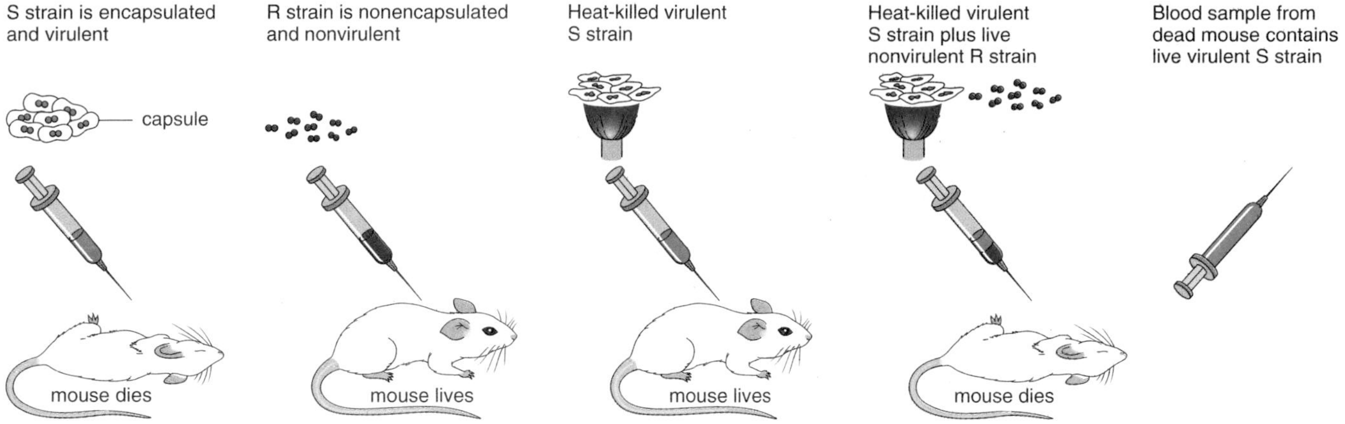

S strain is encapsulated and virulent

capsule

mouse dies

R strain is nonencapsulated and nonvirulent

mouse lives

Heat-killed virulent S strain

mouse lives

Heat-killed virulent S strain plus live nonvirulent R strain

mouse dies

Blood sample from dead mouse contains live virulent S strain

Griffith's transformation experiment
Figure 13.1

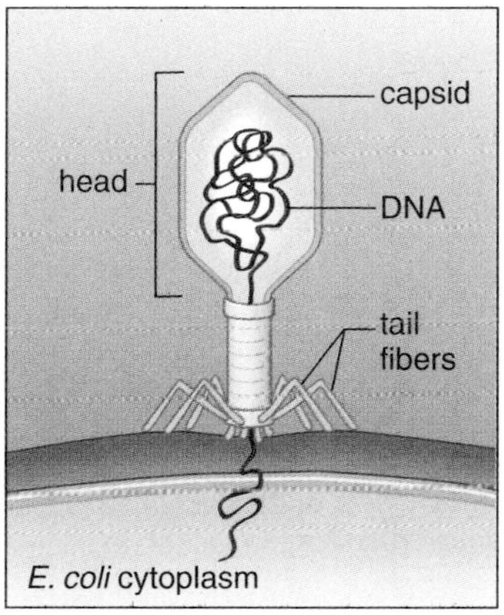

capsid

head

DNA

tail fibers

E. coli cytoplasm

Bacteria and bacteriophages
Figure 13.2

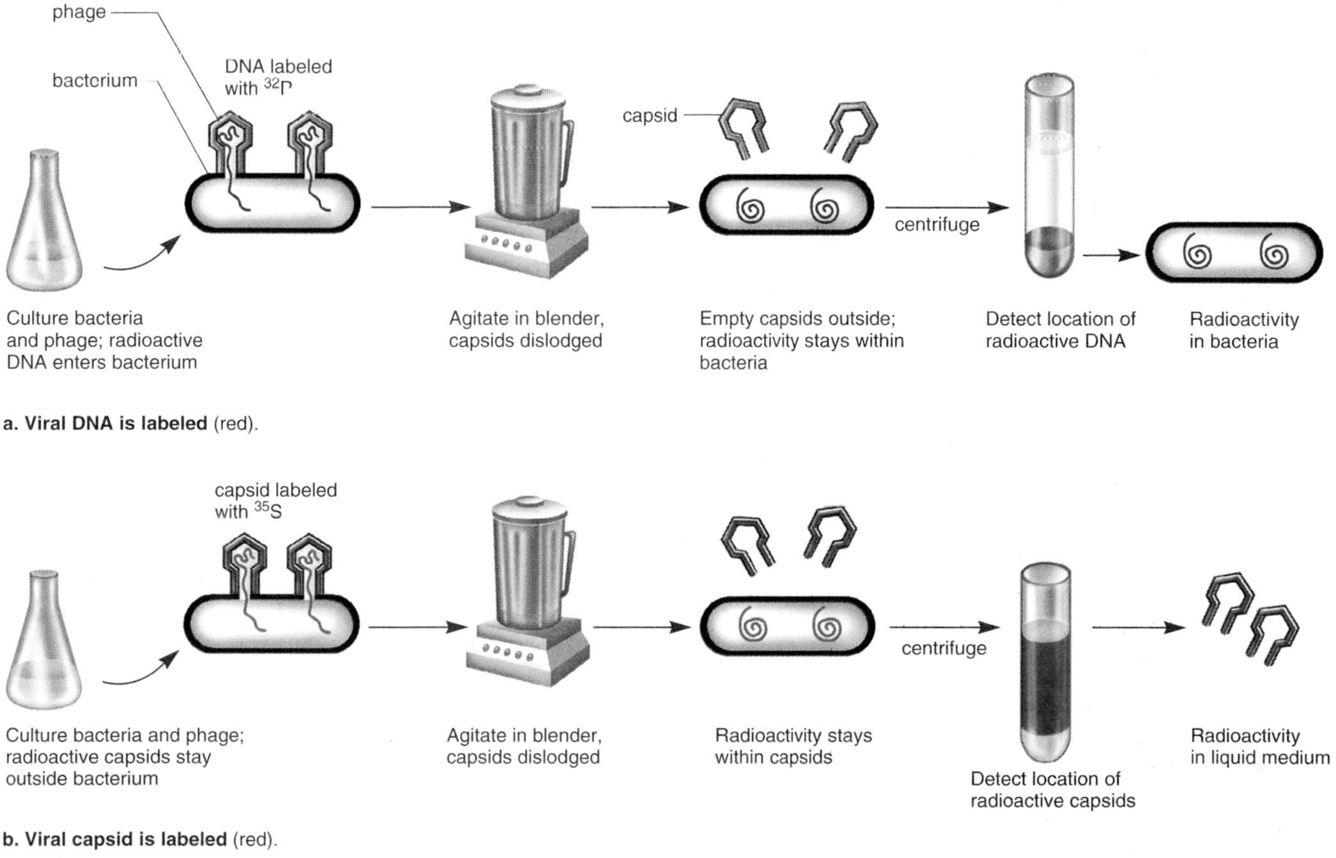

phage

bacterium

DNA labeled with ^{32}P

capsid

centrifuge

Culture bacteria and phage; radioactive DNA enters bacterium

Agitate in blender, capsids dislodged

Empty capsids outside; radioactivity stays within bacteria

Detect location of radioactive DNA

Radioactivity in bacteria

a. Viral DNA is labeled (red).

capsid labeled with ^{35}S

centrifuge

Culture bacteria and phage; radioactive capsids stay outside bacterium

Agitate in blender, capsids dislodged

Radioactivity stays within capsids

Detect location of radioactive capsids

Radioactivity in liquid medium

b. Viral capsid is labeled (red).

Hershey and Chase experiments
Figure 13.3

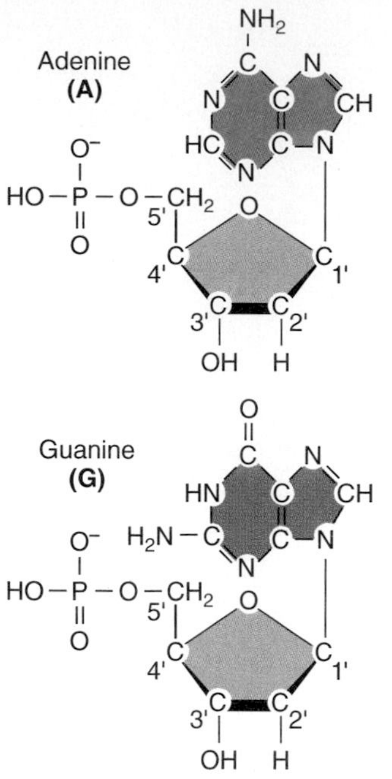

a. **Purine nucleotides**

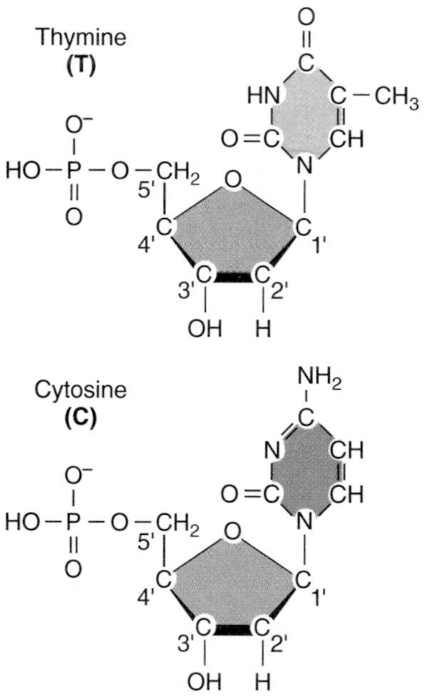

b. **Pyrimidine nucleotides**

Chargaff's DNA Database Composition in Various Species (%)				
Species	**A**	**T**	**G**	**C**
Homo sapiens	31.0	31.5	19.1	18.4
Drosophila melanogaster	27.3	27.6	22.5	22.5
Zea mays	25.6	25.3	24.5	24.6
Neurospora crassa	23.0	23.3	27.1	26.6
Escherichia coli	24.6	24.3	25.5	25.6
Bacillus subtilis	28.4	29.0	21.0	21.6

c.

Nucleotide composition of DNA

Figure 13.4

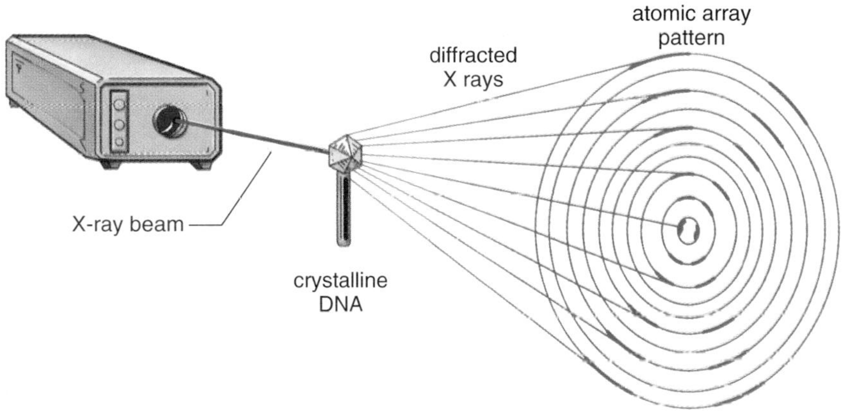
atomic array
pattern
diffracted
X rays
X-ray beam
crystalline
DNA

a.

X-ray diffraction of DNA
Figure 13.5

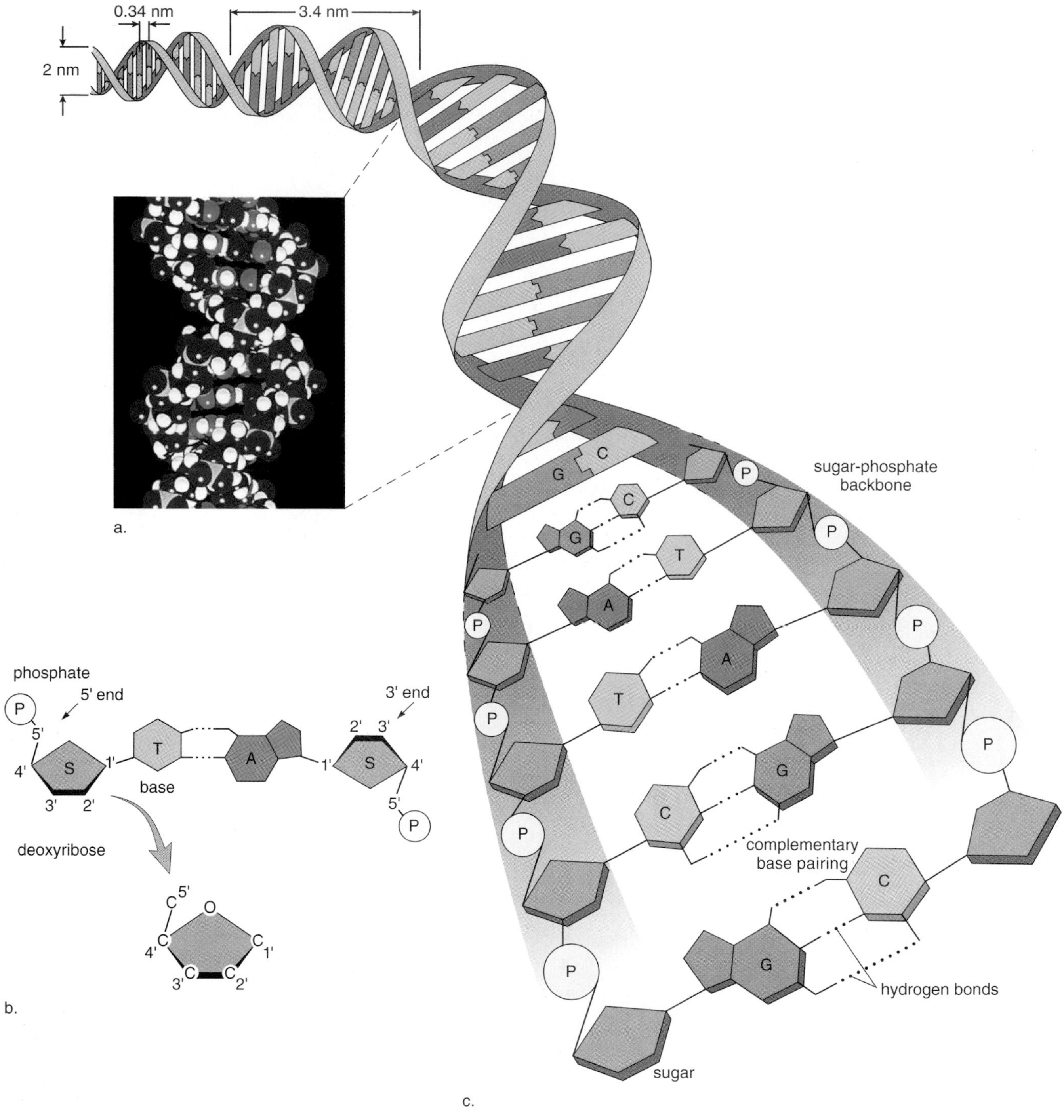

0.34 nm

3.4 nm

2 nm

a.

phosphate

5' end

3' end

P

5'

4'   1'

S

3'   2'

base

T ┊┊ A

1'

2' 3'

S

4'

5'

P

deoxyribose

C 5'

O

C   C 1'

4'

3'   2'

C   C

b.

sugar-phosphate
backbone

G   C

C

P

G

P

T

A

P

A

P

T

P

G

P

C

complementary
base pairing

C

P

G

sugar

hydrogen bonds

c.

Watson and Crick model of DNA
Figure 13.6

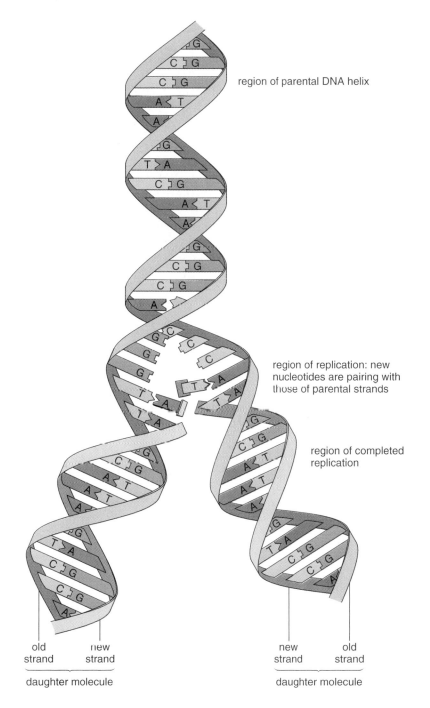

region of parental DNA helix

region of replication: new nucleotides are pairing with those of parental strands

region of completed replication

old strand | new strand
new strand | old strand
daughter molecule | daughter molecule

Semiconservative replication (simplified)
Figure 13.7

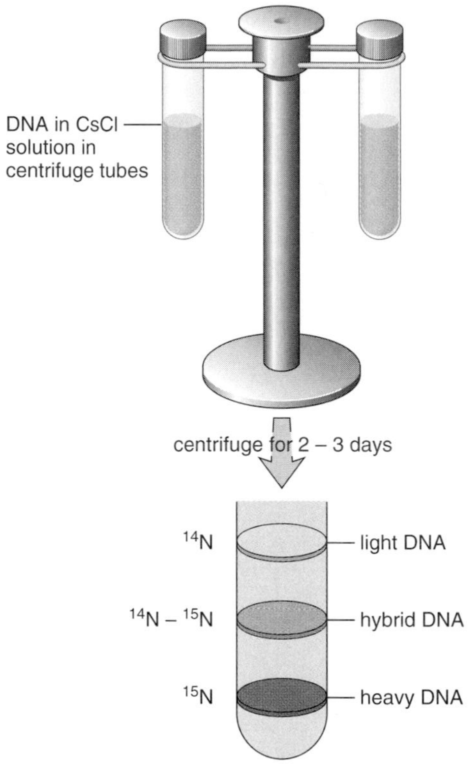

a. Possible results when DNA is centrifuged in CsCl

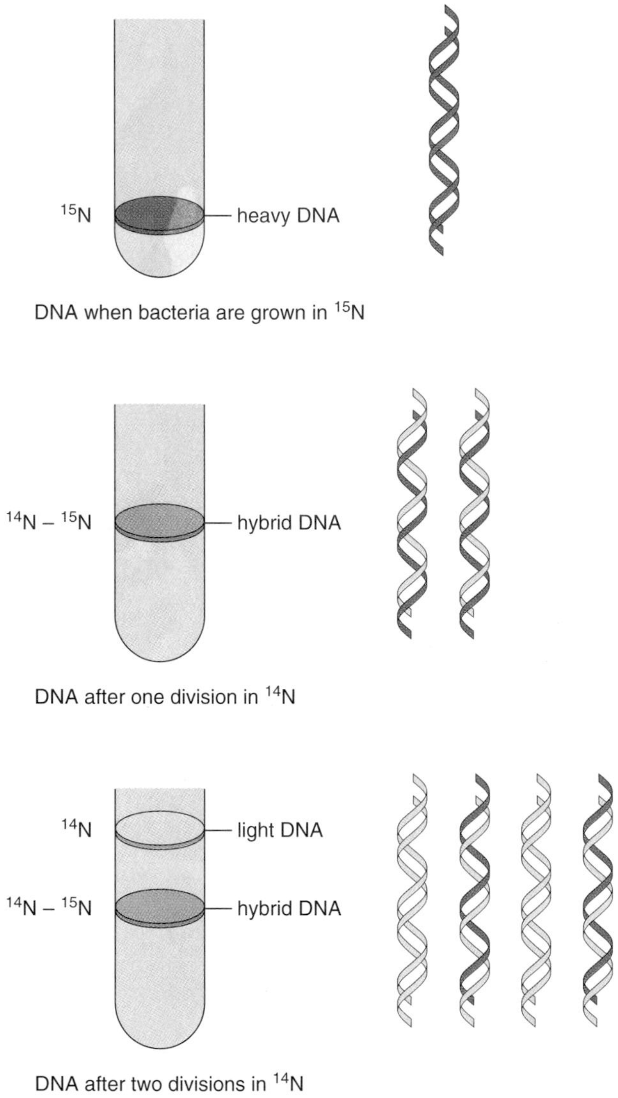

DNA when bacteria are grown in ^{15}N

DNA after one division in ^{14}N

DNA after two divisions in ^{14}N

b. Steps in Meselson and Stahl experiment

Meselson and Stahl's DNA replication experiment
Figure 13.8

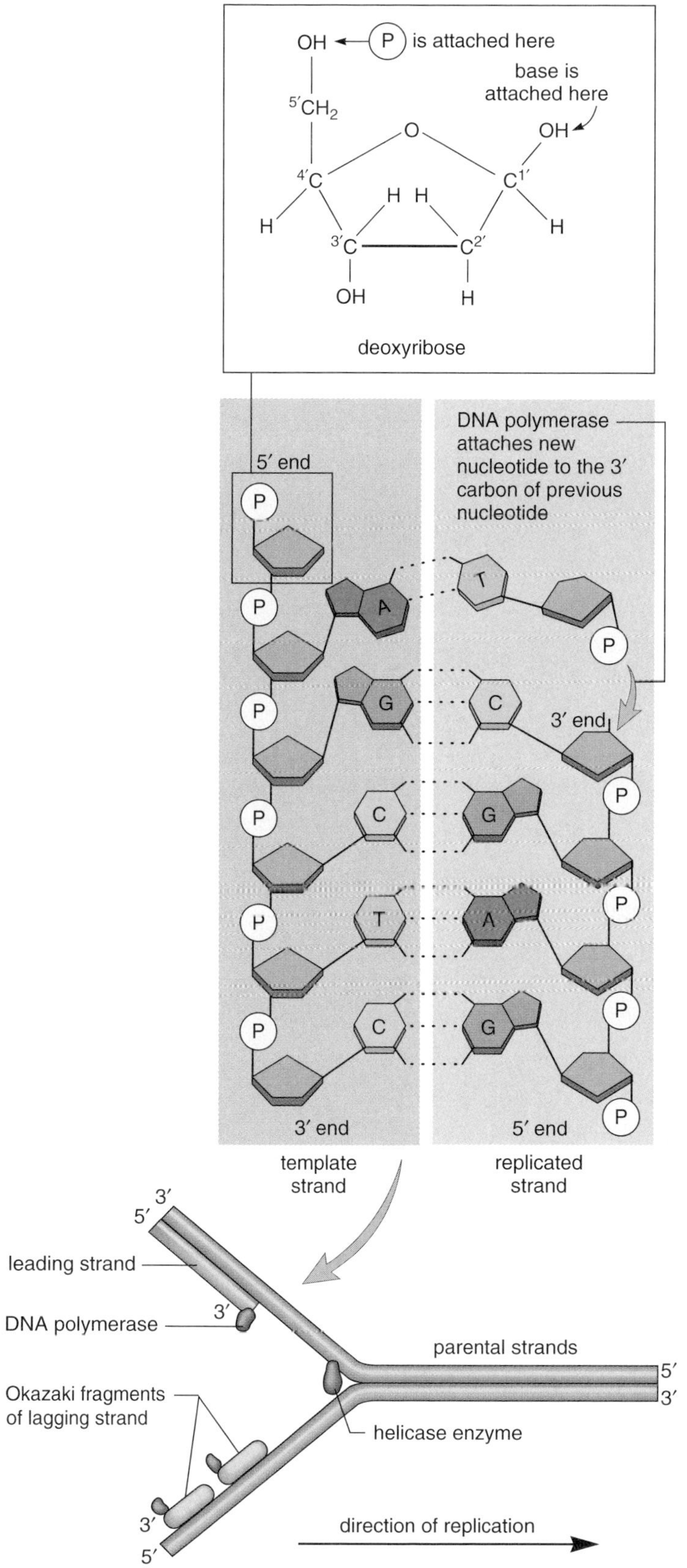
OH ← (P) is attached here
base is attached here
$^{5'}CH_2$
O
OH
$^{4'}C$
$C^{1'}$
H H
H
H
$^{3'}C$
$C^{2'}$
OH
H
deoxyribose
DNA polymerase attaches new nucleotide to the 3′ carbon of previous nucleotide
5′ end
P
A
T
G
C
C
G
T
A
C
G
3′ end
3′ end
5′ end
template strand
replicated strand
3′
5′
leading strand
DNA polymerase
3′
parental strands
5′
3′
Okazaki fragments of lagging strand
helicase enzyme
3′
5′
direction of replication

DNA replications (in depth)
Figure 13A

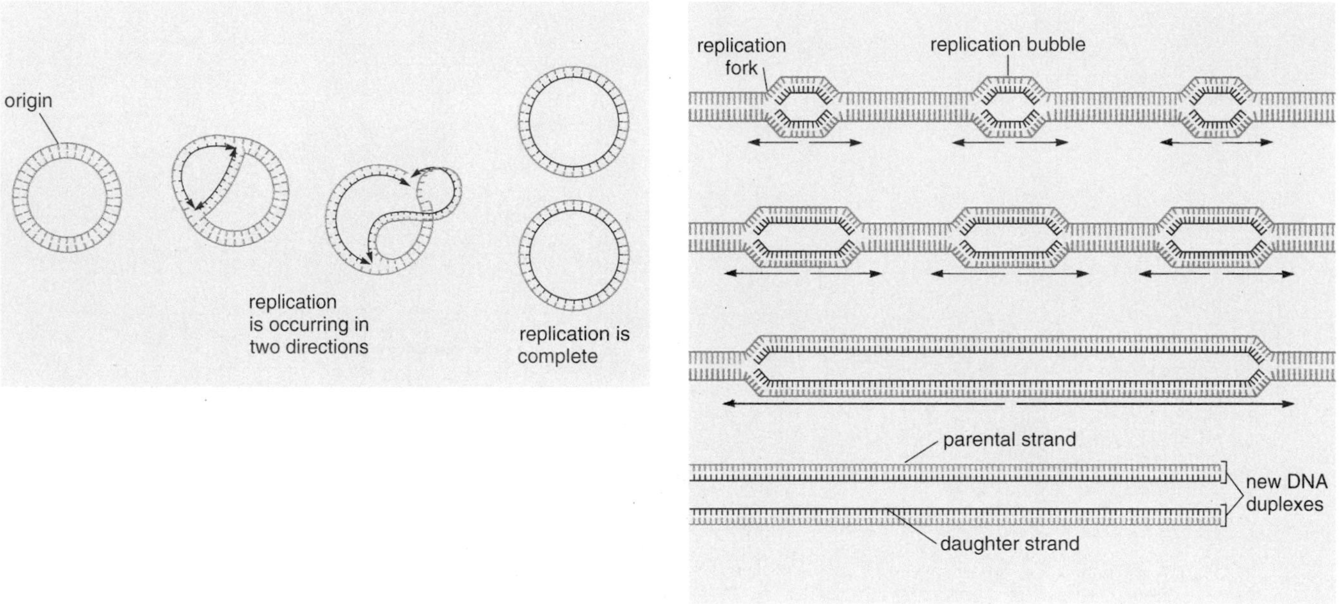

origin

replication
is occurring in
two directions

replication is
complete

replication
fork

replication bubble

parental strand

new DNA
duplexes

daughter strand

Prokaryotic versus eukaryotic replication
Figure 13.9

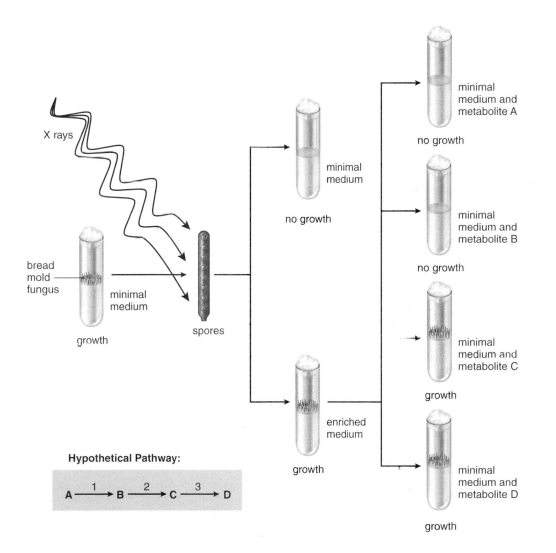

Hypothetical Pathway:

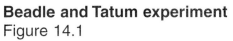

$$A \xrightarrow{\ \ 1\ \ } B \xrightarrow{\ \ 2\ \ } C \xrightarrow{\ \ 3\ \ } D$$

X rays

bread
mold
fungus

minimal
medium

growth

spores

minimal
medium

no growth

enriched
medium

growth

minimal
medium and
metabolite A

no growth

minimal
medium and
metabolite B

no growth

minimal
medium and
metabolite C

growth

minimal
medium and
metabolite D

growth

Beadle and Tatum experiment
Figure 14.1

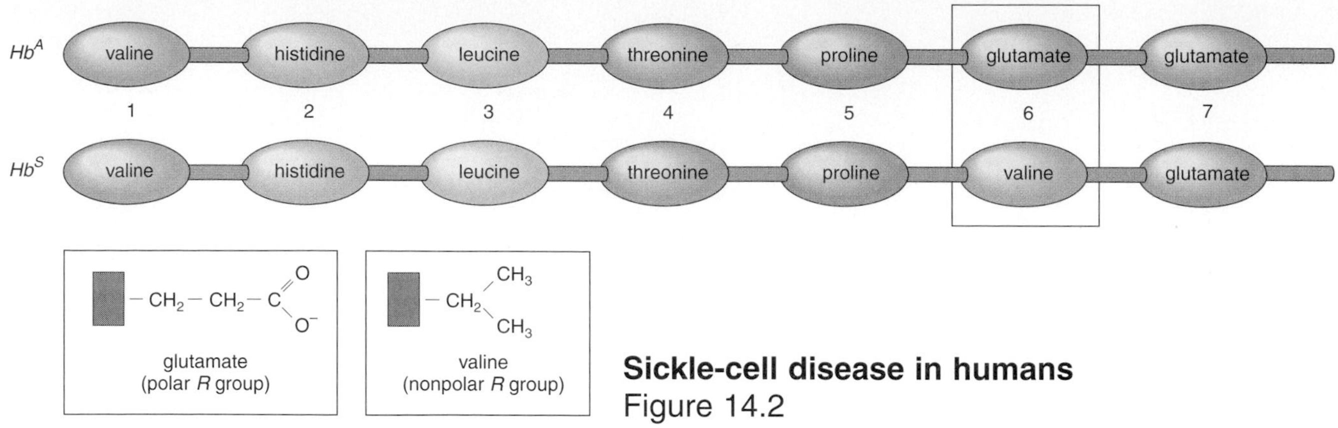

glutamate
(polar *R* group)

valine
(nonpolar *R* group)

Sickle-cell disease in humans
Figure 14.2

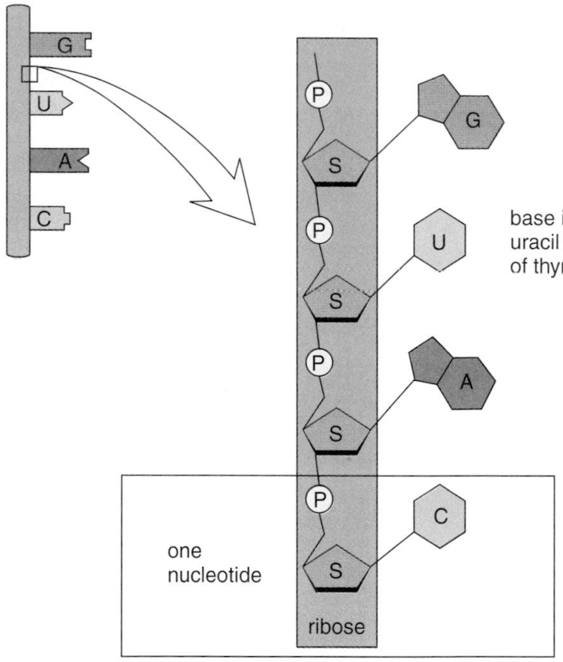

base is
uracil instead
of thymine

one
nucleotide

ribose

Structure of RNA
Figure 14.3

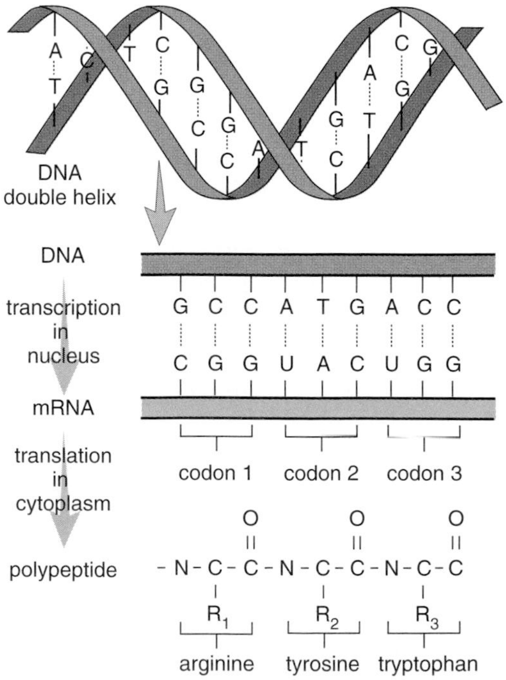

Overview of gene expression
Figure 14.4

First Base	Second Base				Third Base
	U	C	A	G	
U	UUU phenylalanine	UCU serine	UAU tyrosine	UGU cysteine	U
	UUC phenylalanine	UCC serine	UAC tyrosine	UGC cysteine	C
	UUA leucine	UCA serine	UAA *stop*	UGA *stop*	A
	UUG leucine	UCG serine	UAG *stop*	UGG tryptophan	G
C	CUU leucine	CCU proline	CAU histidine	CGU arginine	U
	CUC leucine	CCC proline	CAC histidine	CGC arginine	C
	CUA leucine	CCA proline	CAA glutamine	CGA arginine	A
	CUG leucine	CCG proline	CAG glutamine	CGG arginine	G
A	AUU isoleucine	ACU threonine	AAU asparagine	AGU serine	U
	AUC isoleucine	ACC threonine	AAC asparagine	AGC serine	C
	AUA isoleucine	ACA threonine	AAA lysine	AGA arginine	A
	AUG (*start*) methionine	ACG threonine	AAG lysine	AGG arginine	G
G	GUU valine	GCU alanine	GAU aspartate	GGU glycine	U
	GUC valine	GCC alanine	GAC aspartate	GGC glycine	C
	GUA valine	GCA alanine	GAA glutamate	GGA glycine	A
	GUG valine	GCG alanine	GAG glutamate	GGG glycine	G

Messenger RNA codons
Figure 14.5

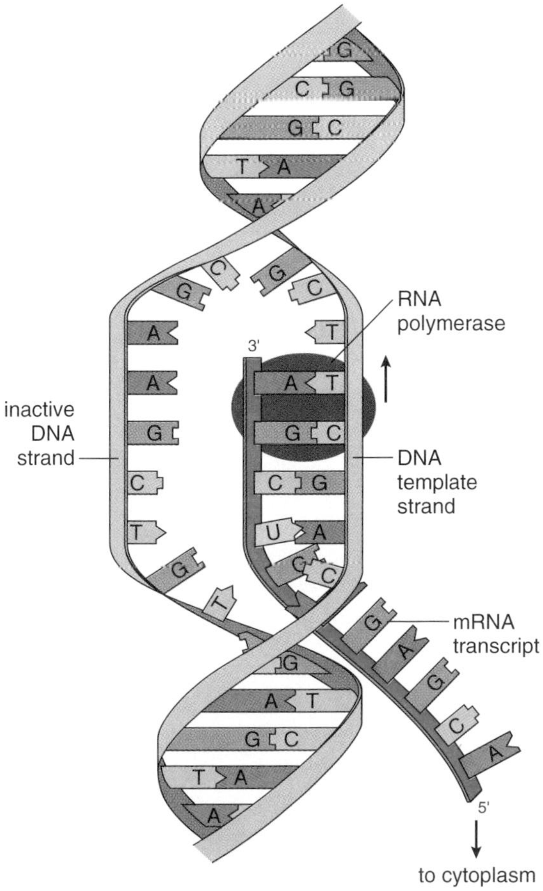

Transcription
Figure 14.6

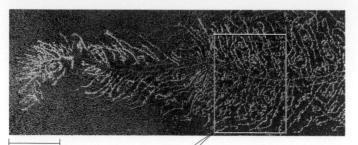

200 nm

RNA polymerase
Figure 14.7

a: © Oscar L. Miller/Photo Researchers, Inc.

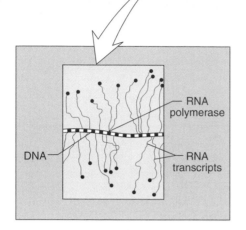

RNA
polymerase

DNA

RNA
transcripts

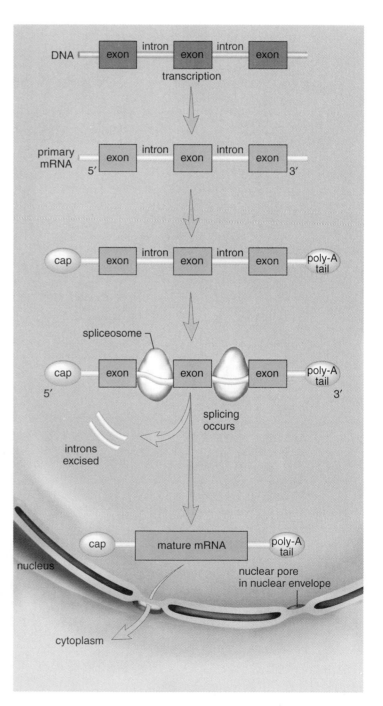

DNA   exon   intron   exon   intron   exon

transcription

primary
mRNA   exon   intron   exon   intron   exon
5'   3'

cap   exon   intron   exon   intron   exon   poly-A
tail

spliceosome

cap   exon   exon   exon   poly-A
tail
5'   3'

splicing
occurs

introns
excised

cap   mature mRNA   poly-A
tail

nucleus   nuclear pore
in nuclear envelope

cytoplasm

Messenger RNA (mRNA) processing
in eukaryotes
Figure 14.8

Structure of a transfer RNA (tRNA) molecule
Figure 14.9

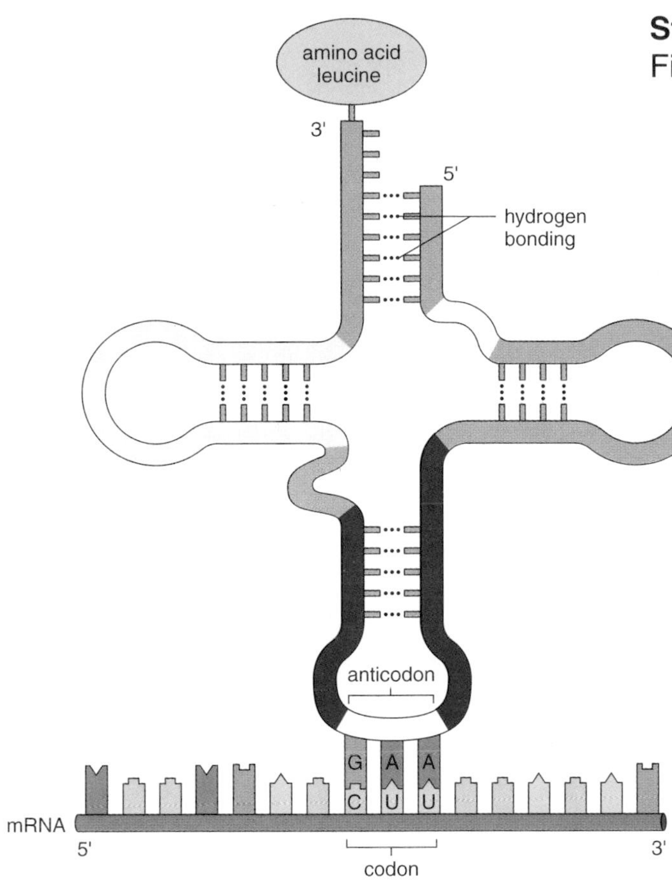

amino acid
leucine

3'

5'

hydrogen
bonding

anticodon

G A A
C U U

mRNA
5' 3'

codon

Ribosome structure and function
Figure 14.10

d: Courtesy Alexander Rich

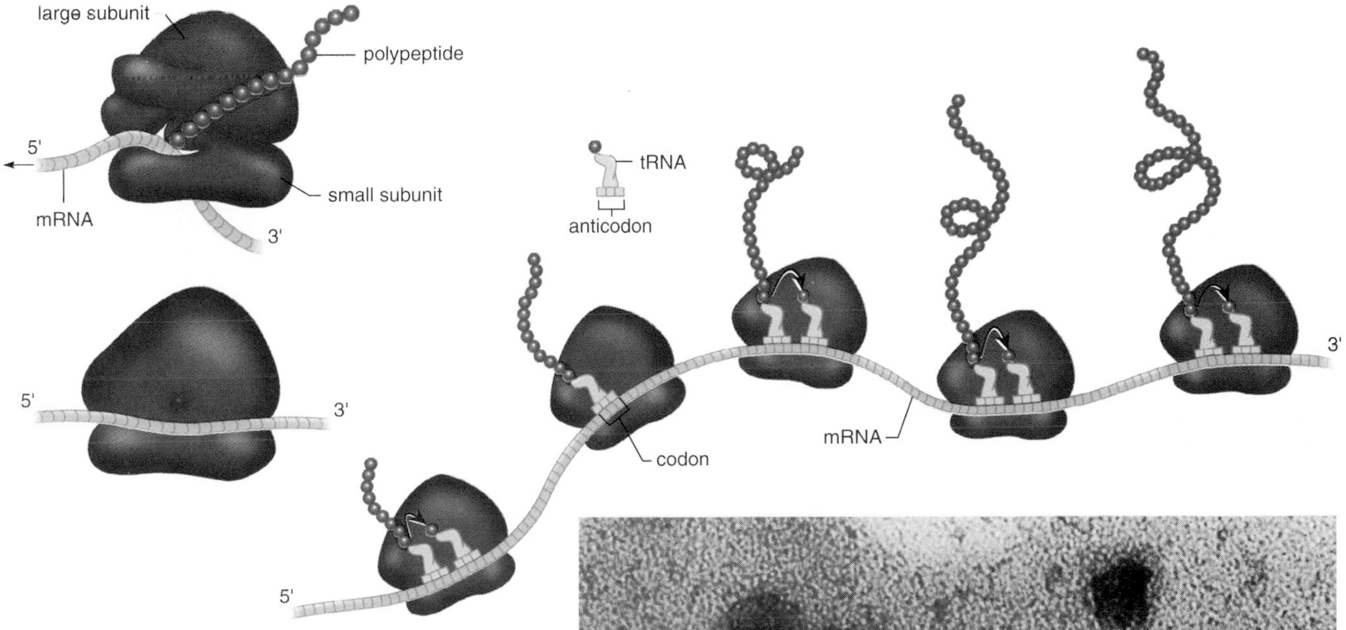

large subunit

polypeptide

5'

mRNA 3'

small subunit

tRNA

anticodon

codon

mRNA

5' 3'

5' 3'

5'

3'

20 nm

111

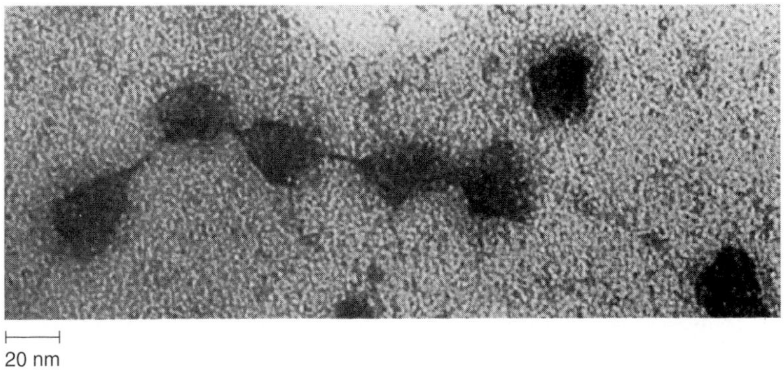

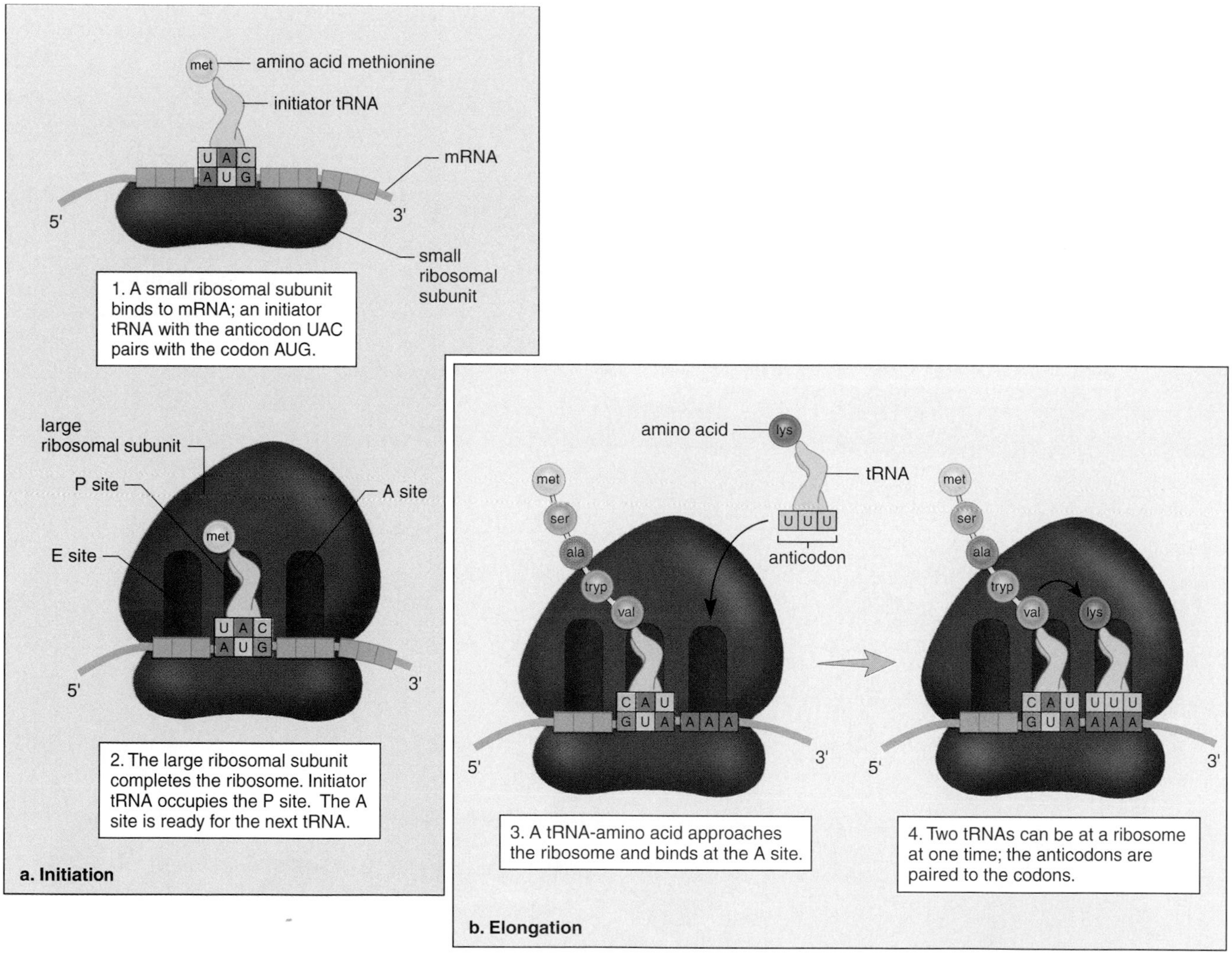
met
amino acid methionine
initiator tRNA
U A C
A U G
mRNA
5'
3'
small ribosomal subunit
1. A small ribosomal subunit binds to mRNA; an initiator tRNA with the anticodon UAC pairs with the codon AUG.
large ribosomal subunit
P site
A site
E site
met
U A C
A U G
5'
3'
2. The large ribosomal subunit completes the ribosome. Initiator tRNA occupies the P site. The A site is ready for the next tRNA.
a. Initiation
amino acid
lys
tRNA
U U U
anticodon
met
ser
ala
tryp
val
C A U
G U A A A A
5'
3'
3. A tRNA-amino acid approaches the ribosome and binds at the A site.
met
ser
ala
tryp
val
lys
C A U U U U
G U A A A A
5'
3'
4. Two tRNAs can be at a ribosome at one time; the anticodons are paired to the codons.
b. Elongation

Protein synthesis
Figure 14.11

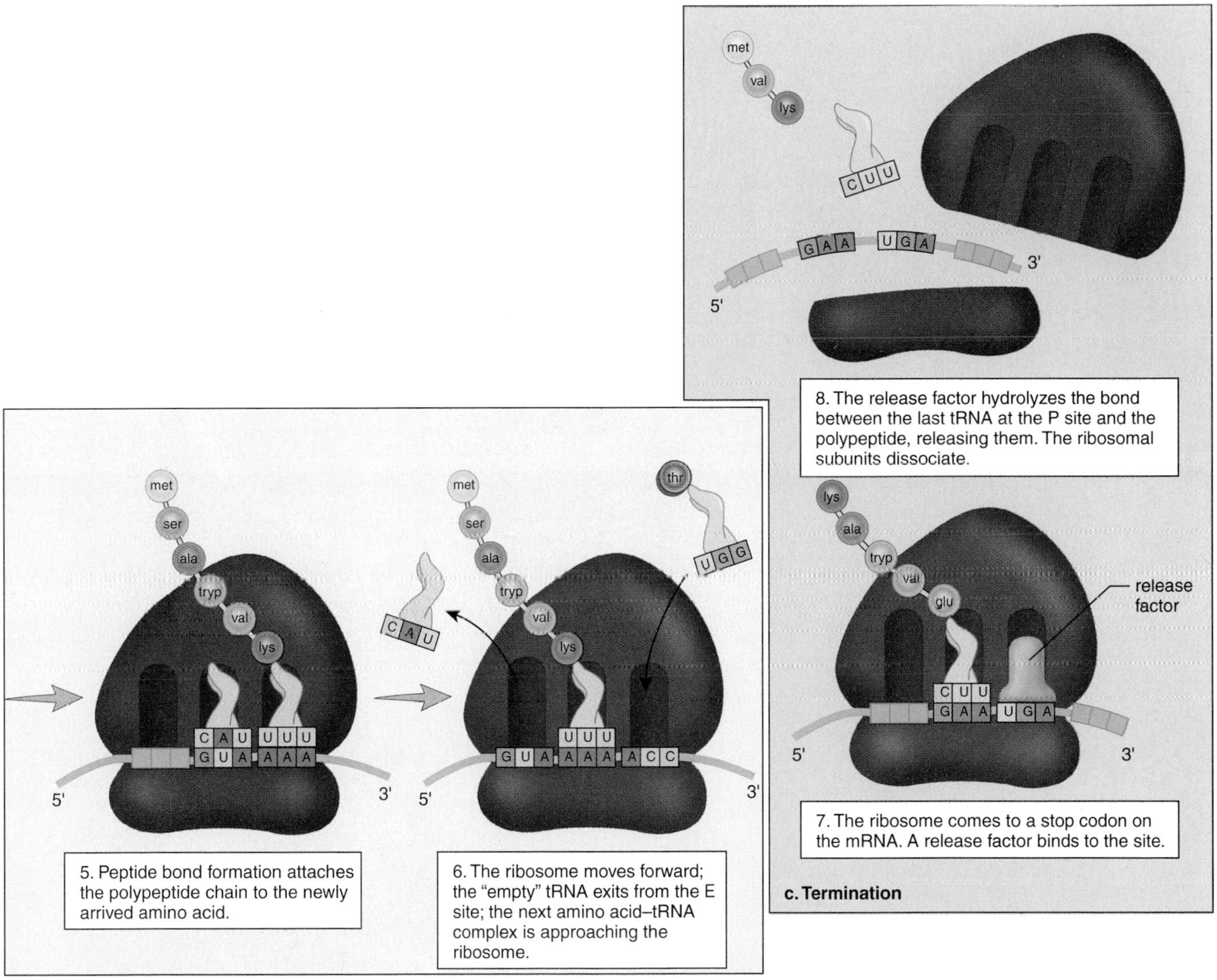

8. The release factor hydrolyzes the bond between the last tRNA at the P site and the polypeptide, releasing them. The ribosomal subunits dissociate.

release factor

7. The ribosome comes to a stop codon on the mRNA. A release factor binds to the site.

c. Termination

5. Peptide bond formation attaches the polypeptide chain to the newly arrived amino acid.

6. The ribosome moves forward; the "empty" tRNA exits from the E site; the next amino acid–tRNA complex is approaching the ribosome.

Protein synthesis (continued)

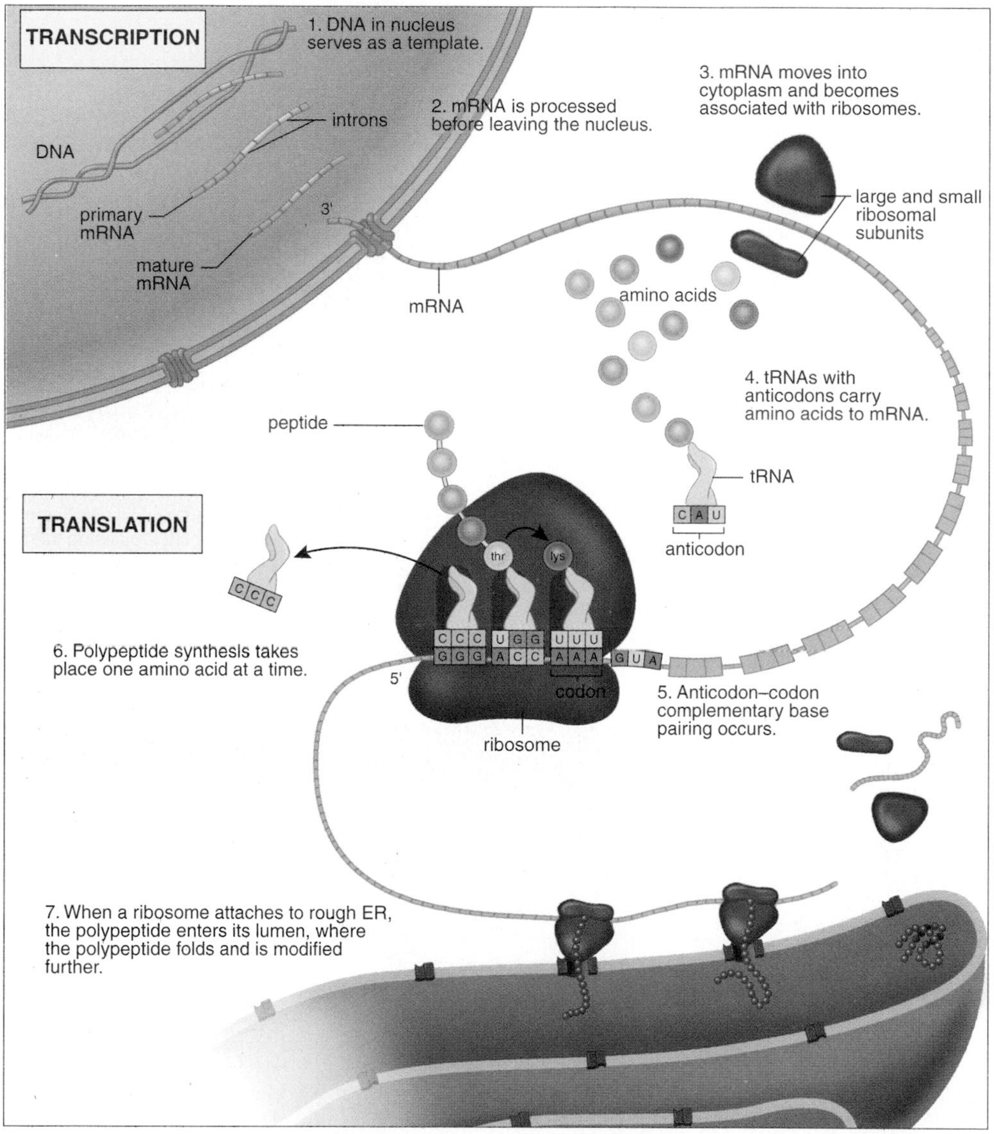

Summary of gene expression in eukaryotes
Figure 14.12

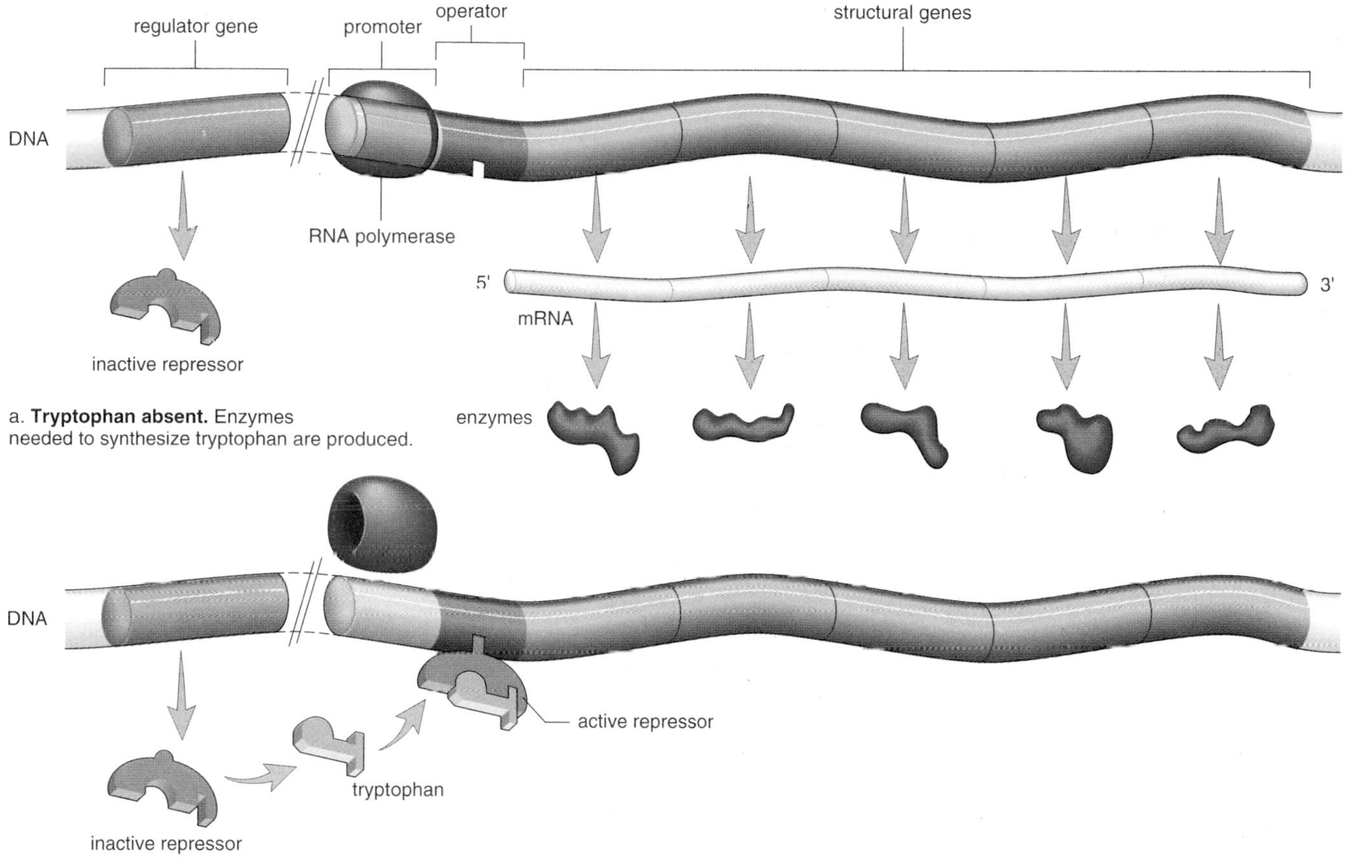

regulator gene   promoter   operator   structural genes

DNA

RNA polymerase

inactive repressor

5'   mRNA   3'

a. **Tryptophan absent.** Enzymes needed to synthesize tryptophan are produced.

enzymes

DNA

active repressor

tryptophan

inactive repressor

b. **Tryptophan present.** Presence of tryptophan blocks production of enzymes.

The *trp* operon
Figure 15.1

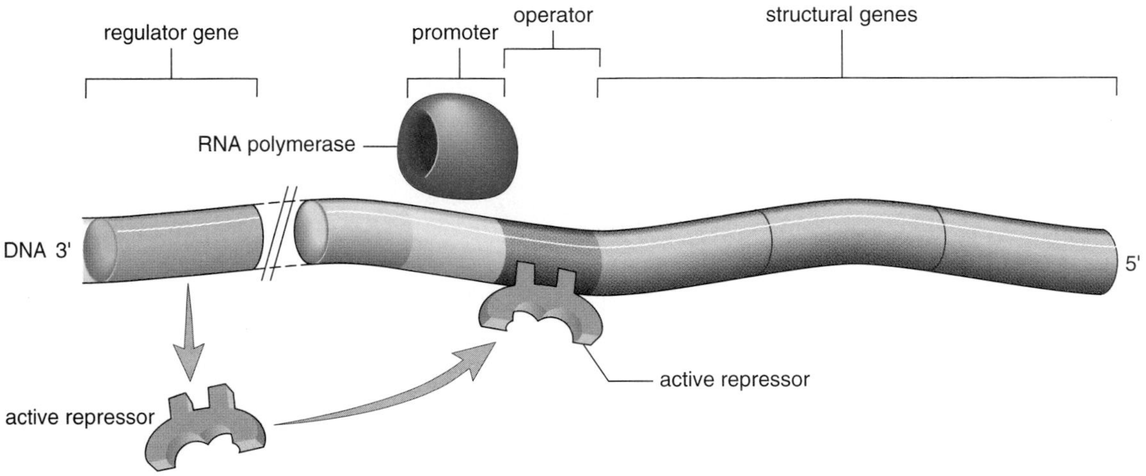

regulator gene   promoter   operator   structural genes

RNA polymerase

DNA 3'   5'

active repressor   active repressor

a. **Lactose absent.** Enzymes needed to take up and utilize lactose are not produced.

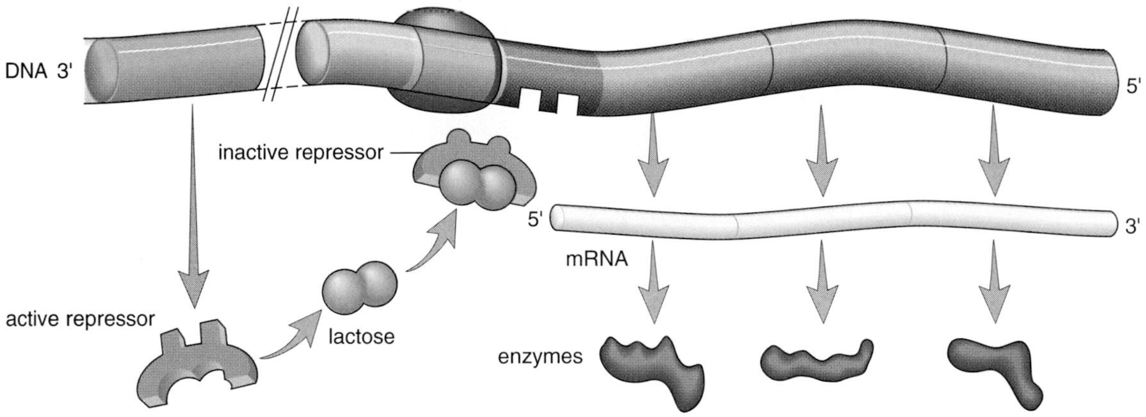

DNA 3'   5'

inactive repressor

5'   3'
mRNA

active repressor   lactose   enzymes

b. **Lactose present.** Enzymes needed to take up and utilize lactose are produced.

The *lac* operon
Figure 15.2

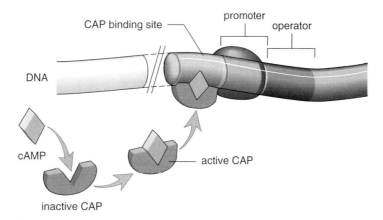

a. **Lactose present, glucose absent (cAMP level high)**

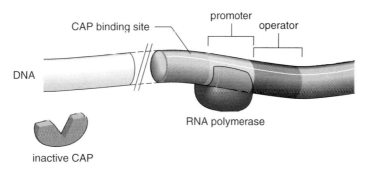

b. **Lactose present, glucose present (cAMP level low)**

Action of CAP
Figure 15.3

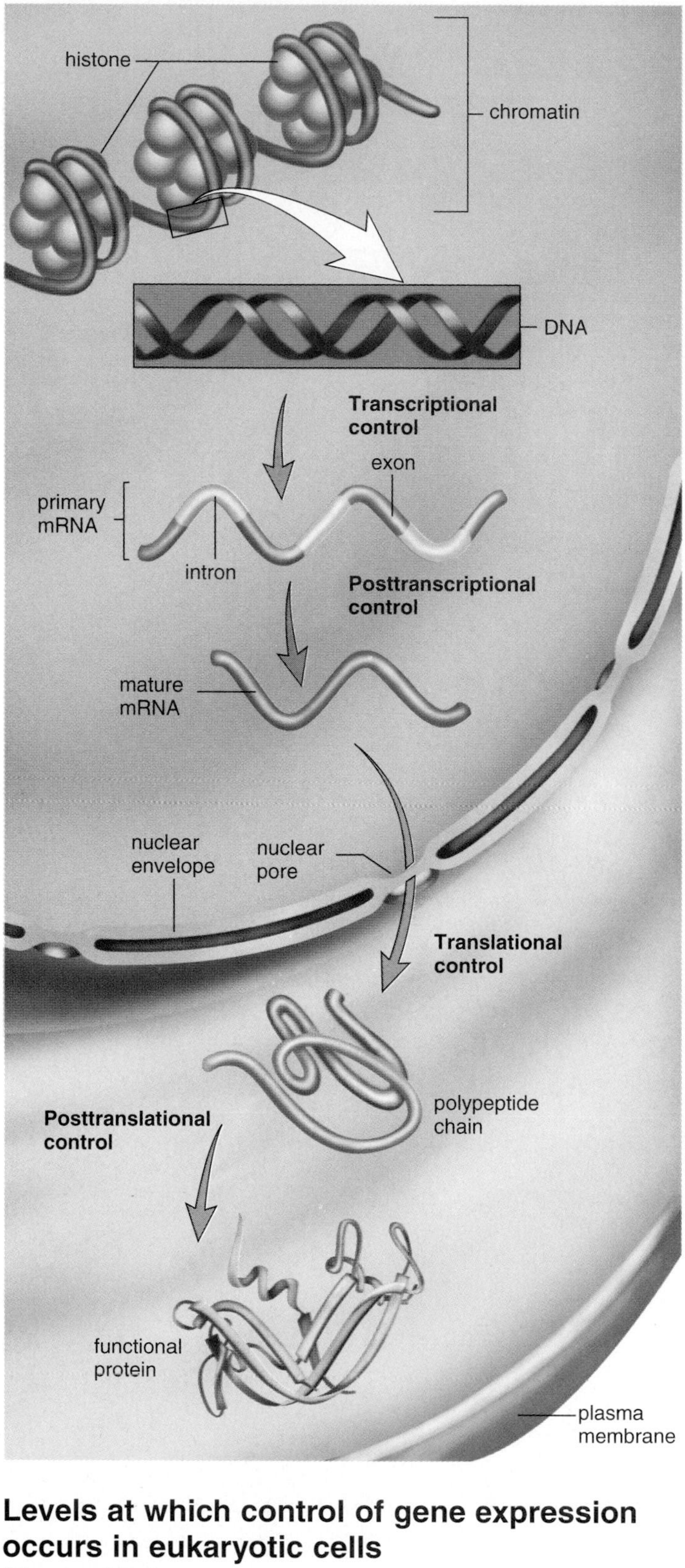

histone

chromatin

DNA

Transcriptional control

exon

primary mRNA

intron

Posttranscriptional control

mature mRNA

nuclear envelope

nuclear pore

Translational control

Posttranslational control

polypeptide chain

functional protein

plasma membrane

Levels at which control of gene expression occurs in eukaryotic cells

Figure 15.4

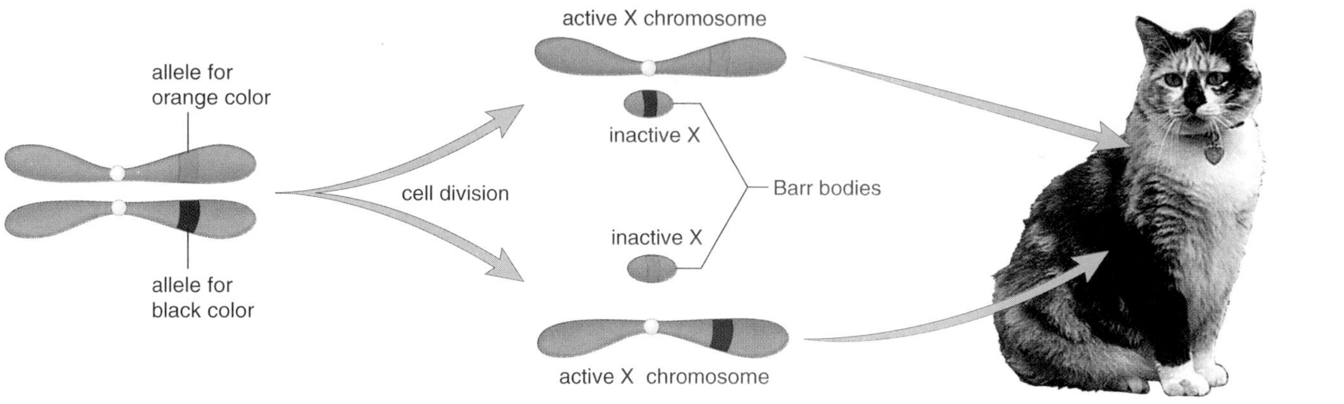

Females have two X chromosomes.

One X chromosome is inactivated in each cell. Which one is by chance.

Coats of tortoiseshell cats have patches of orange and black.

X-inactivation in mammalian females
Figure 15.5

Levels of chromatin structure
Figure 15.6

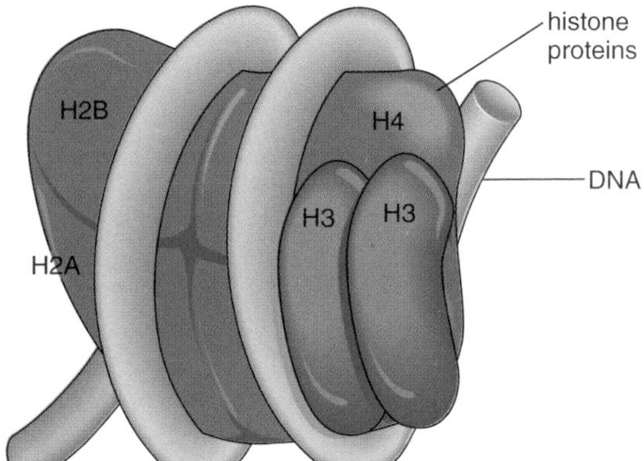

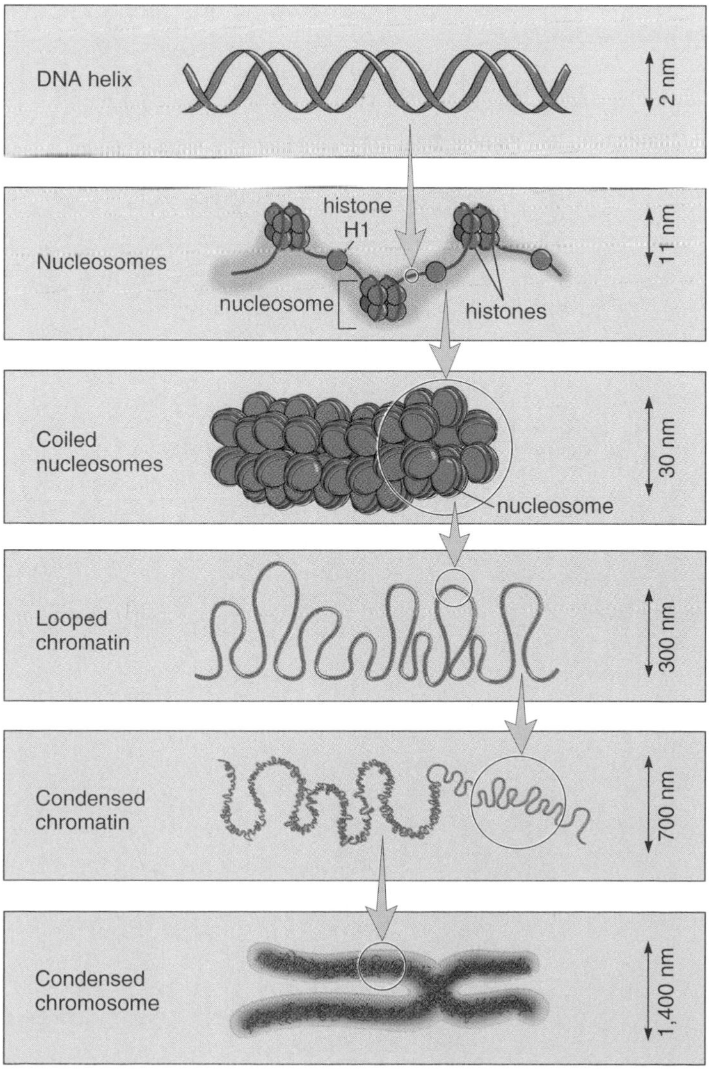

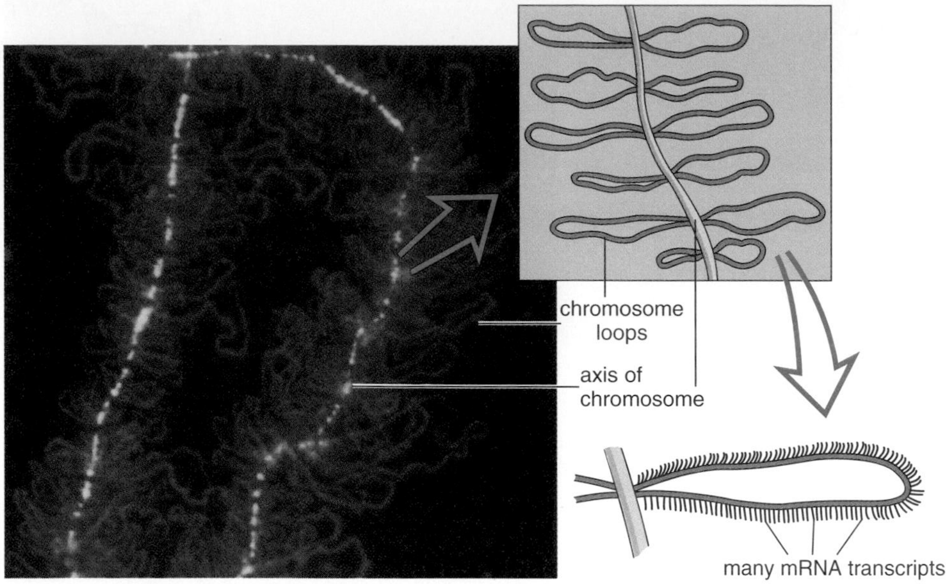

chromosome
loops

axis of
chromosome

many mRNA transcripts

Lampbrush chromosomes
Figure 15.7

From M.B. Roth and J.G. Gall, *Journal of Cell Biology*, 105:1047-1054, 1987.
Reproduced by copyright permission of The Rockefeller University Press

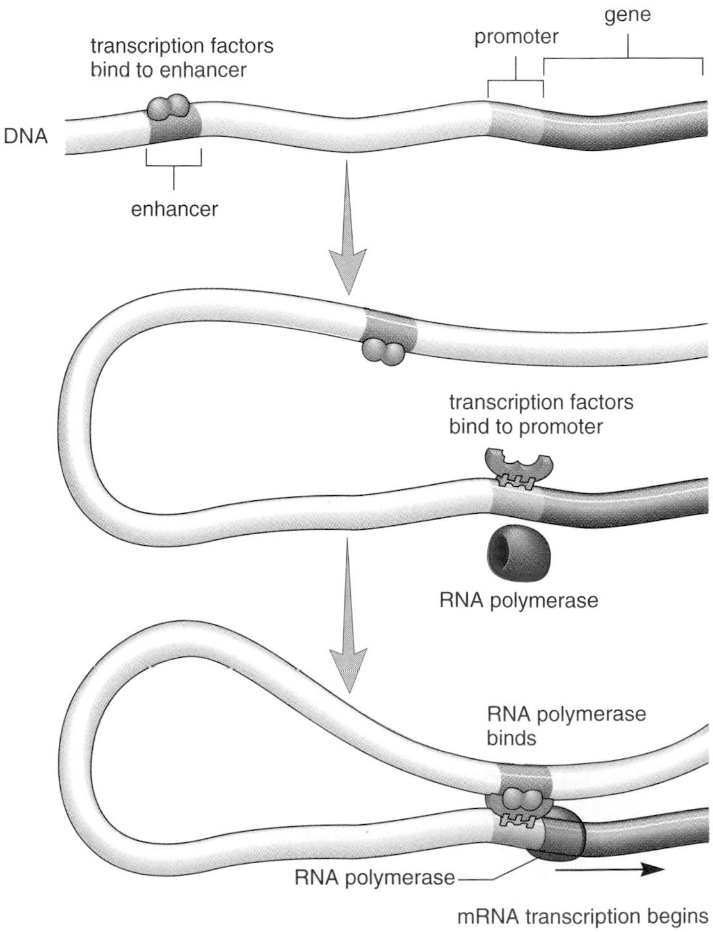

transcription factors
bind to enhancer

gene

promoter

DNA

enhancer

transcription factors
bind to promoter

RNA polymerase

RNA polymerase
binds

RNA polymerase

mRNA transcription begins

Transcription factors
Figure 15.9

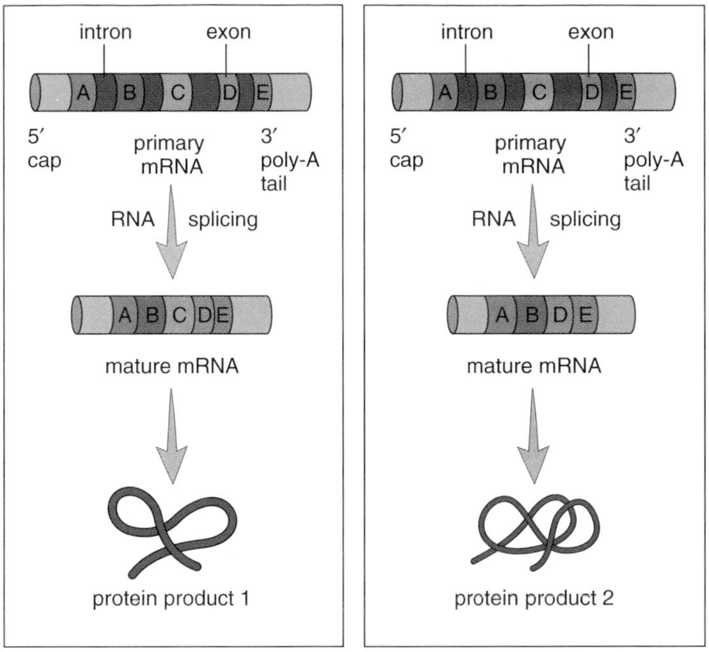

Processing of mRNA transcripts
Figure 15.10

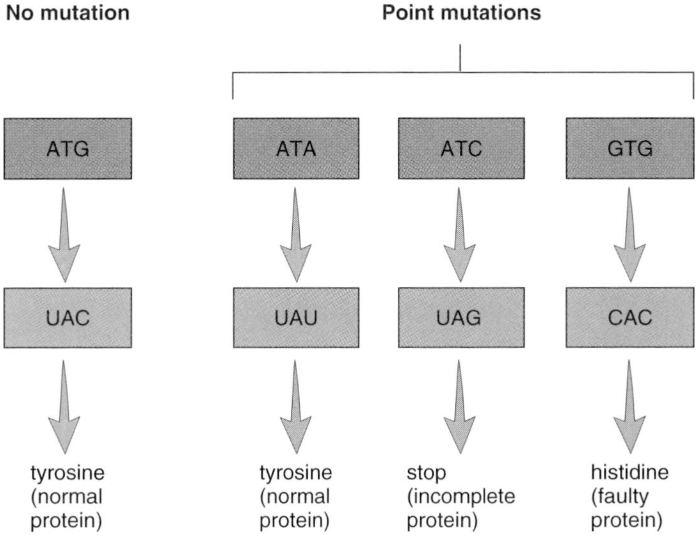

Point mutation
Figure 15.11

New mutations arise, and one cell (green) has the ability to start a tumor.

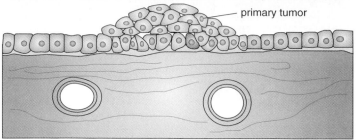

primary tumor

Cancer in situ. The tumor is at its place of origin. One cell (purple) mutates further.

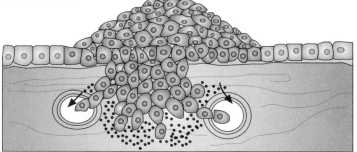

Cancer cells now have the ability to invade lymphatic and blood vessels.

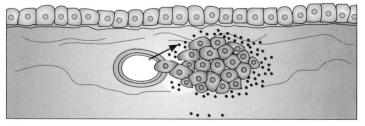

New metastatic tumors are found some distance from the primary tumor.

Carcinogenesis
Figure 15.13

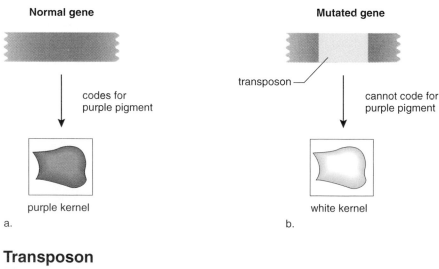

Normal gene

codes for purple pigment

purple kernel

a.

Mutated gene

transposon

cannot code for purple pigment

white kernel

b.

Transposon
Figure 15C

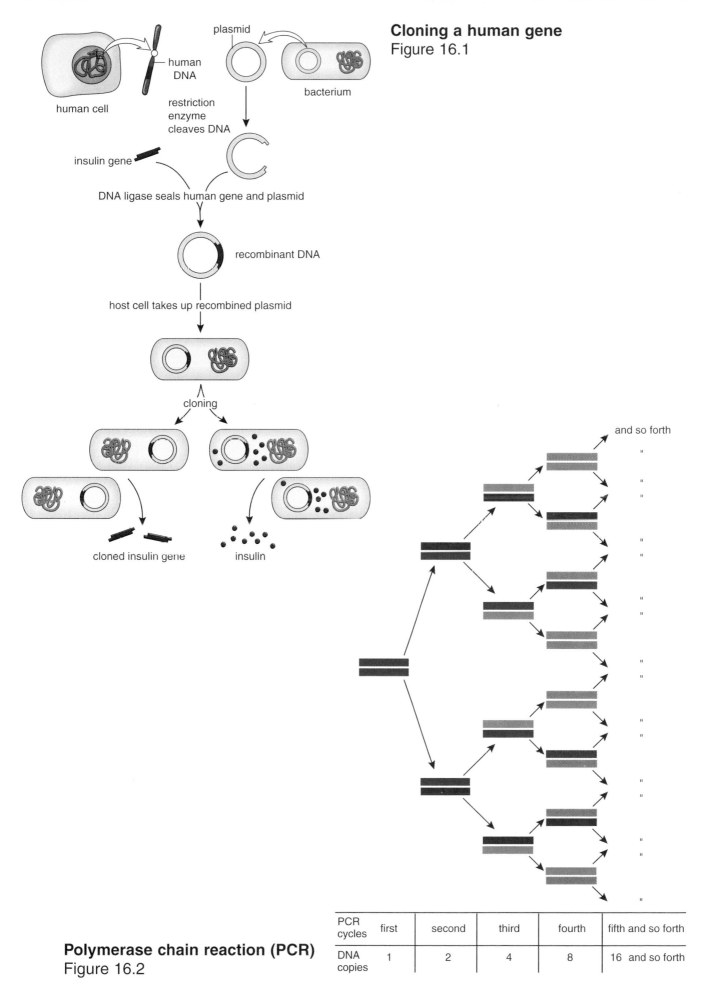

Cloning a human gene
Figure 16.1

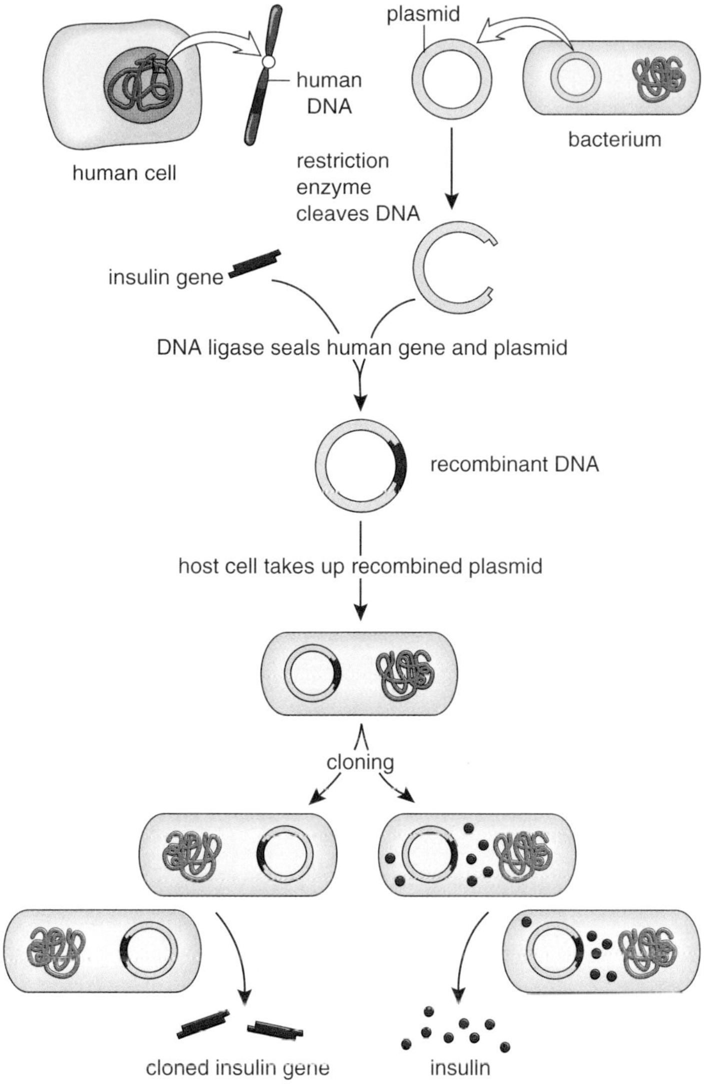

plasmid

bacterium

human cell

human DNA

restriction enzyme cleaves DNA

insulin gene

DNA ligase seals human gene and plasmid

recombinant DNA

host cell takes up recombined plasmid

cloning

cloned insulin gene

insulin

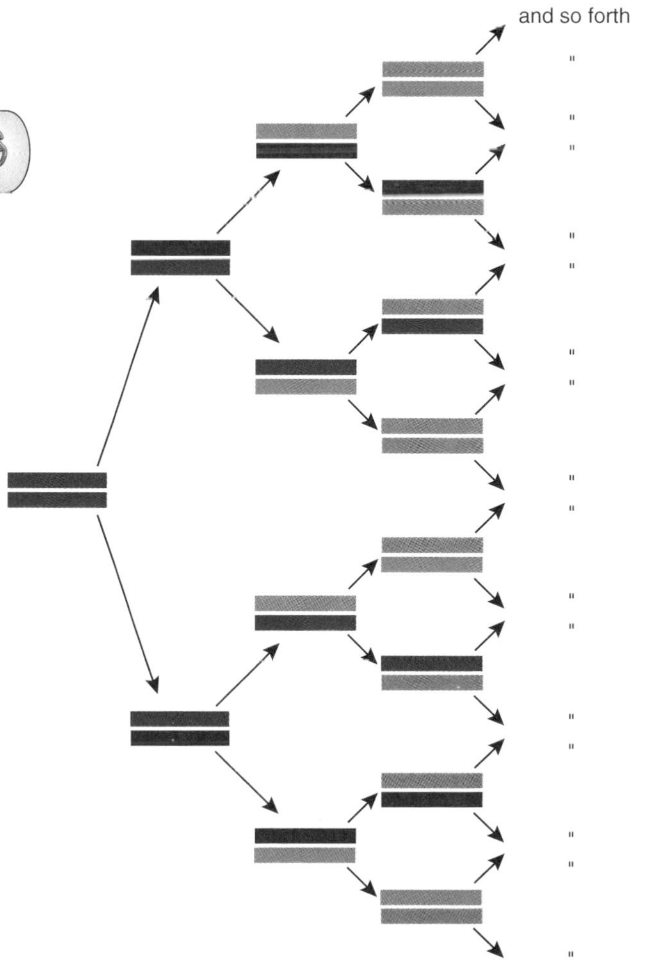

and so forth

Polymerase chain reaction (PCR)
Figure 16.2

PCR cycles	first	second	third	fourth	fifth and so forth
DNA copies	1	2	4	8	16 and so forth

123

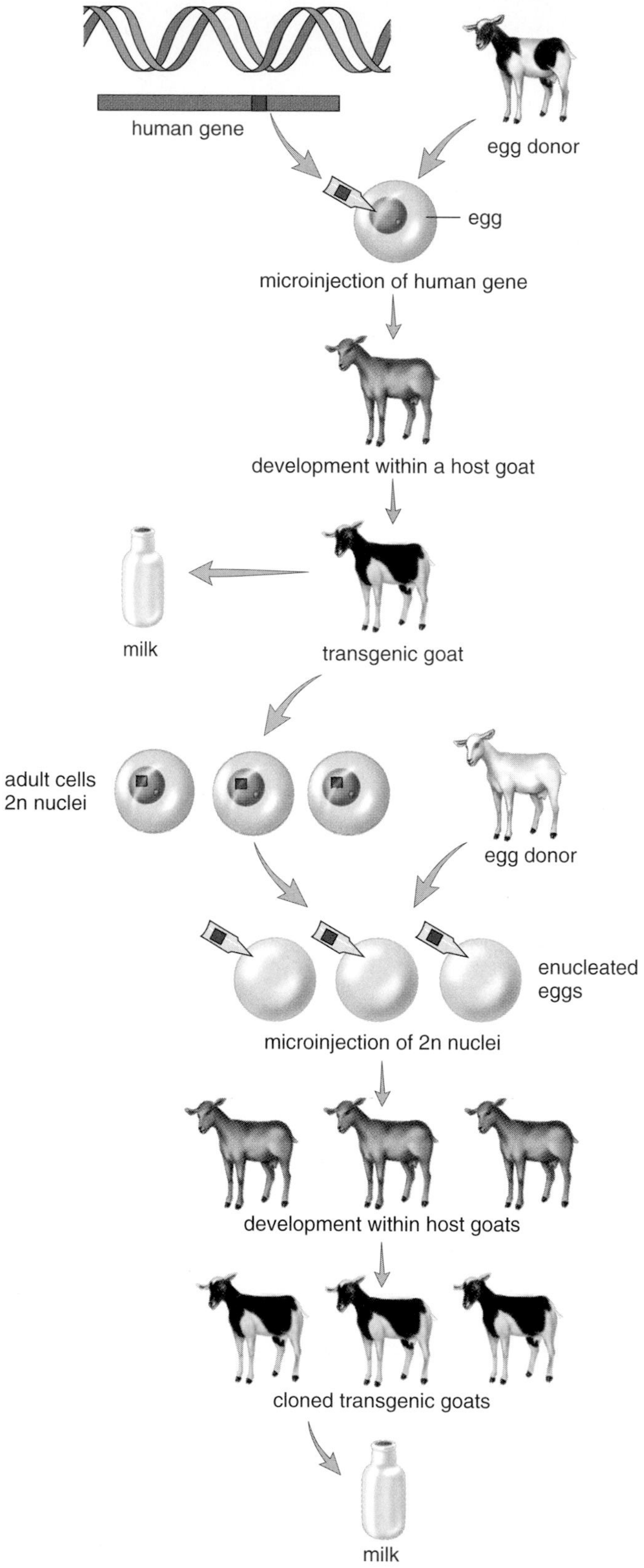

Transgenic mammals
Figure 16.4

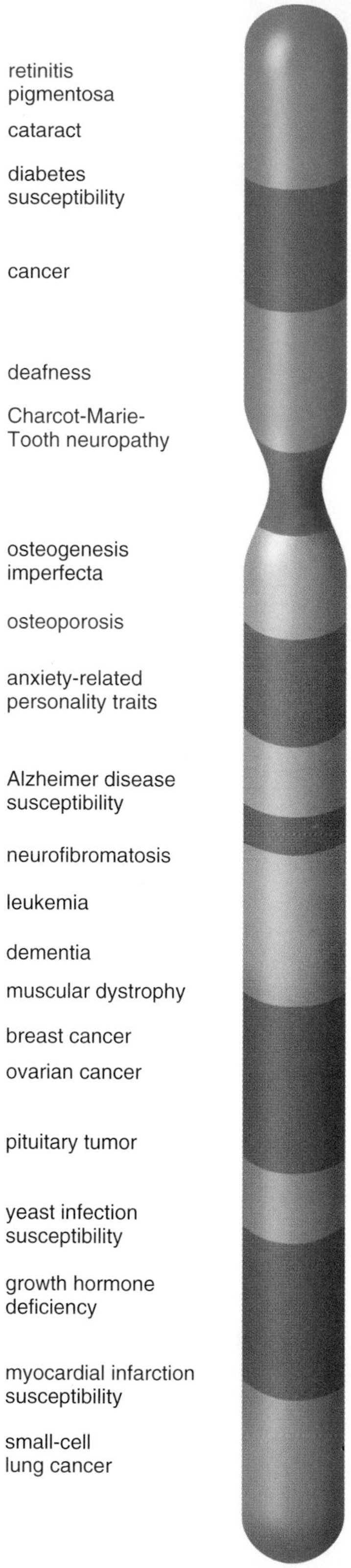

retinitis
pigmentosa

cataract

diabetes
susceptibility

cancer

deafness

Charcot-Marie-
Tooth neuropathy

osteogenesis
imperfecta

osteoporosis

anxiety-related
personality traits

Alzheimer disease
susceptibility

neurofibromatosis

leukemia

dementia

muscular dystrophy

breast cancer

ovarian cancer

pituitary tumor

yeast infection
susceptibility

growth hormone
deficiency

myocardial infarction
susceptibility

small-cell
lung cancer

Genetic map of chromosome 17
Figure 16.5

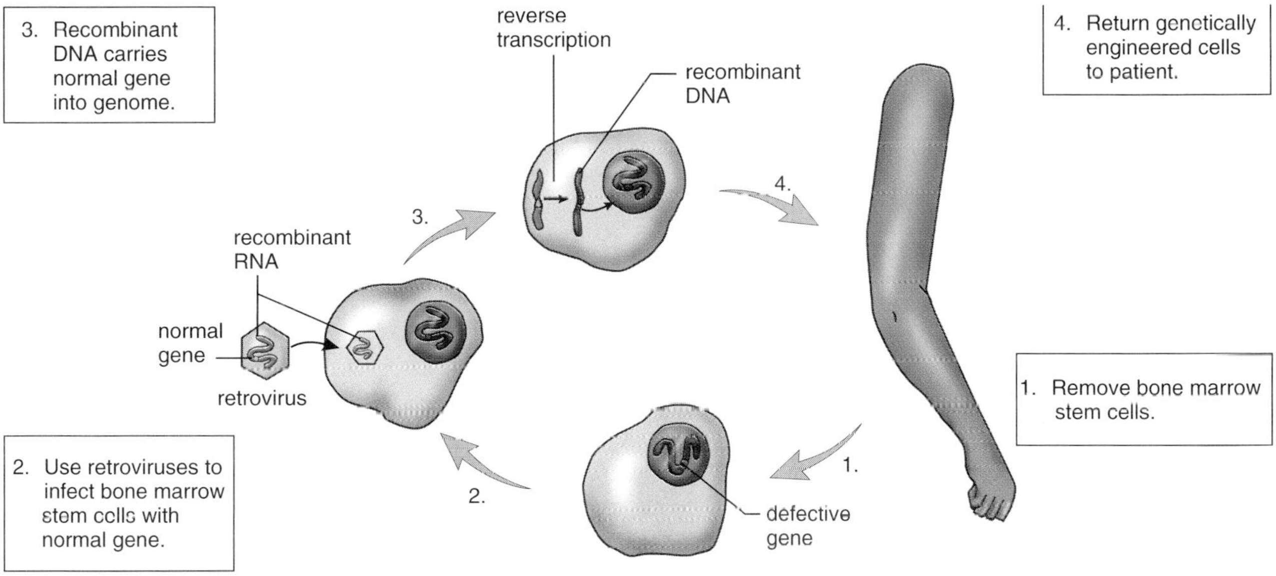

3. Recombinant DNA carries normal gene into genome.

reverse transcription

recombinant DNA

4. Return genetically engineered cells to patient.

recombinant RNA

normal gene

retrovirus

1. Remove bone marrow stem cells.

2. Use retroviruses to infect bone marrow stem cells with normal gene.

defective gene

Ex vivo gene therapy in humans
Figure 16.6

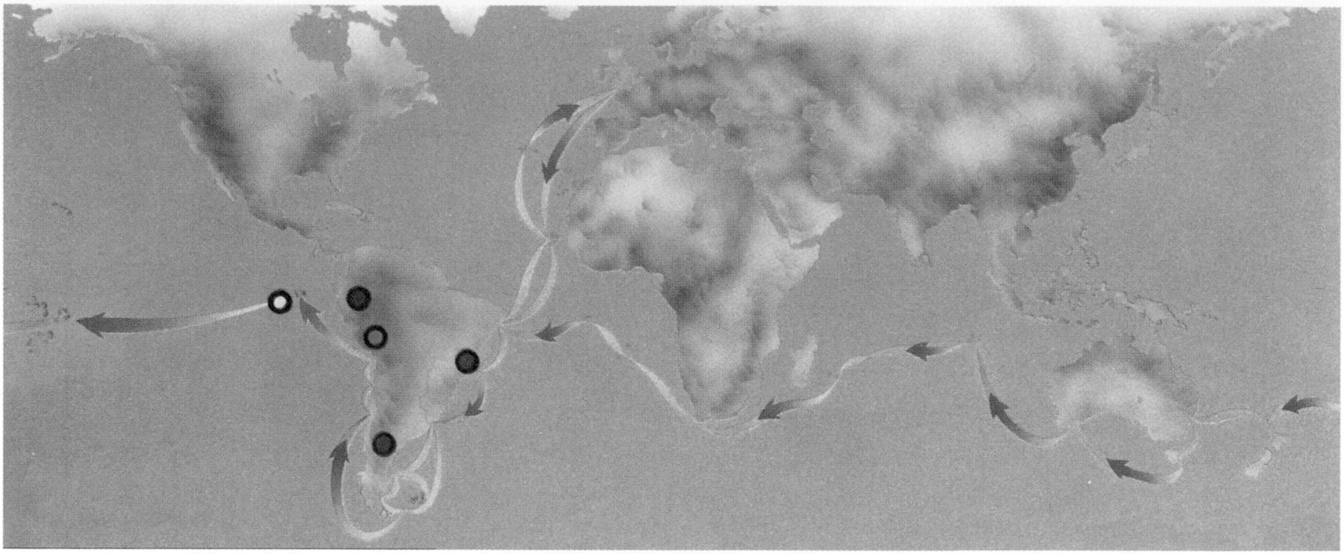

Voyage of the HMS *Beagle*
Figure 17.1

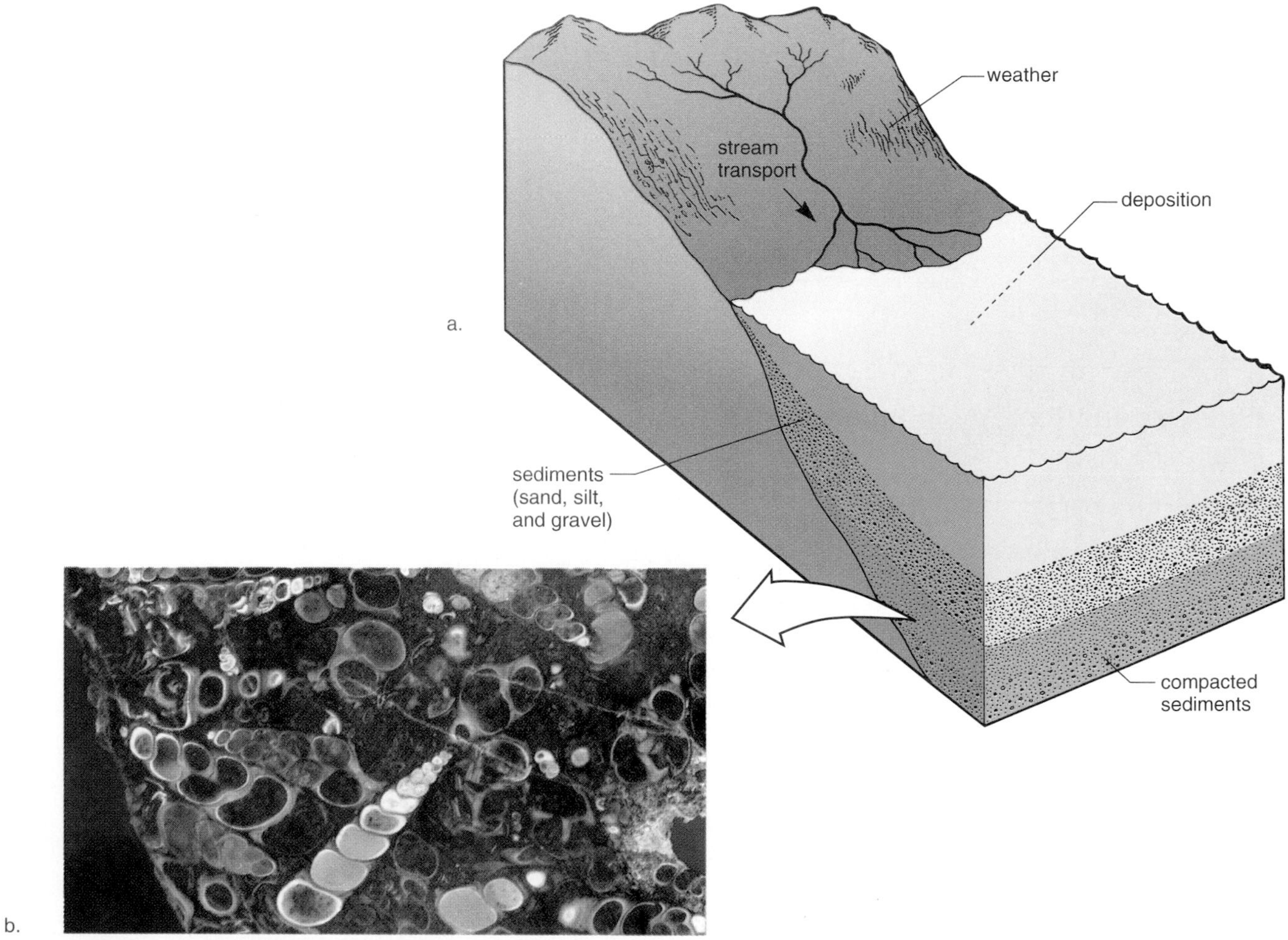

a.

weather

deposition

stream transport

sediments (sand, silt, and gravel)

compacted sediments

b.

Formation of sedimentary rock
Figure 17.4

b: © J. & L. Weber/Peter Arnold, Inc.

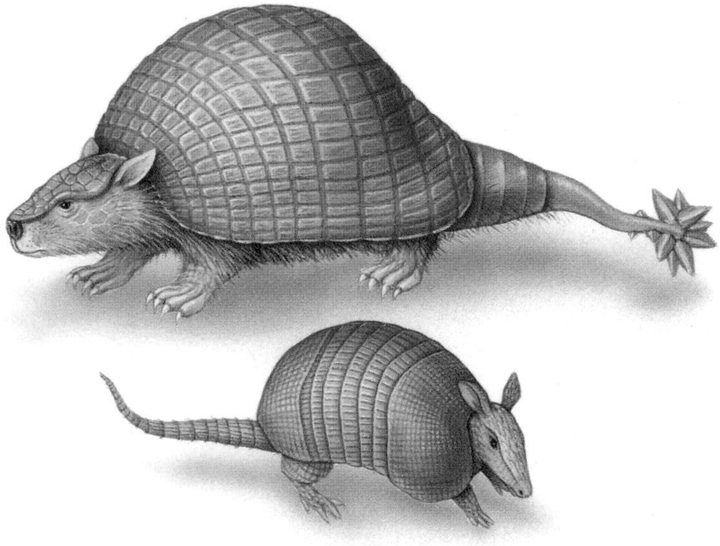

A glyptodont compared to an armadillo
Figure 17.5

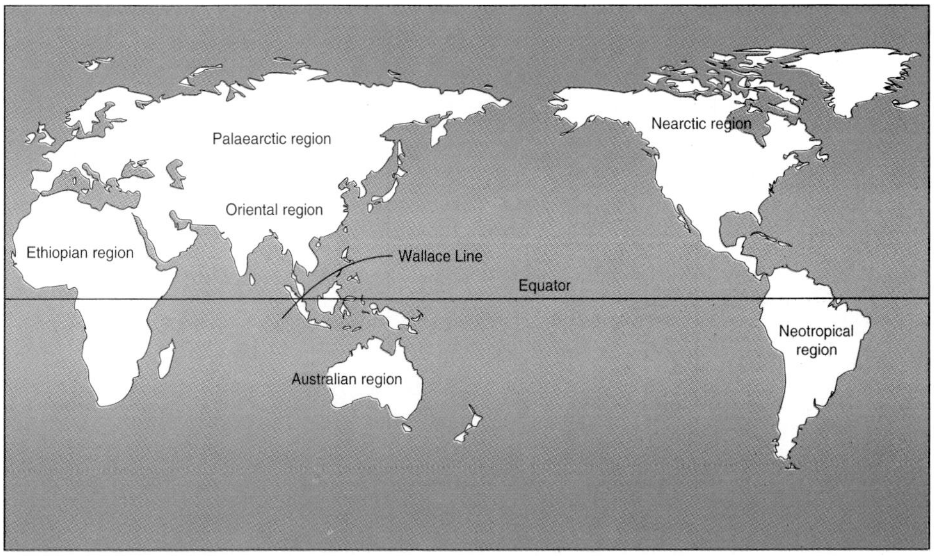

Biogeographical regions
Figure 17B

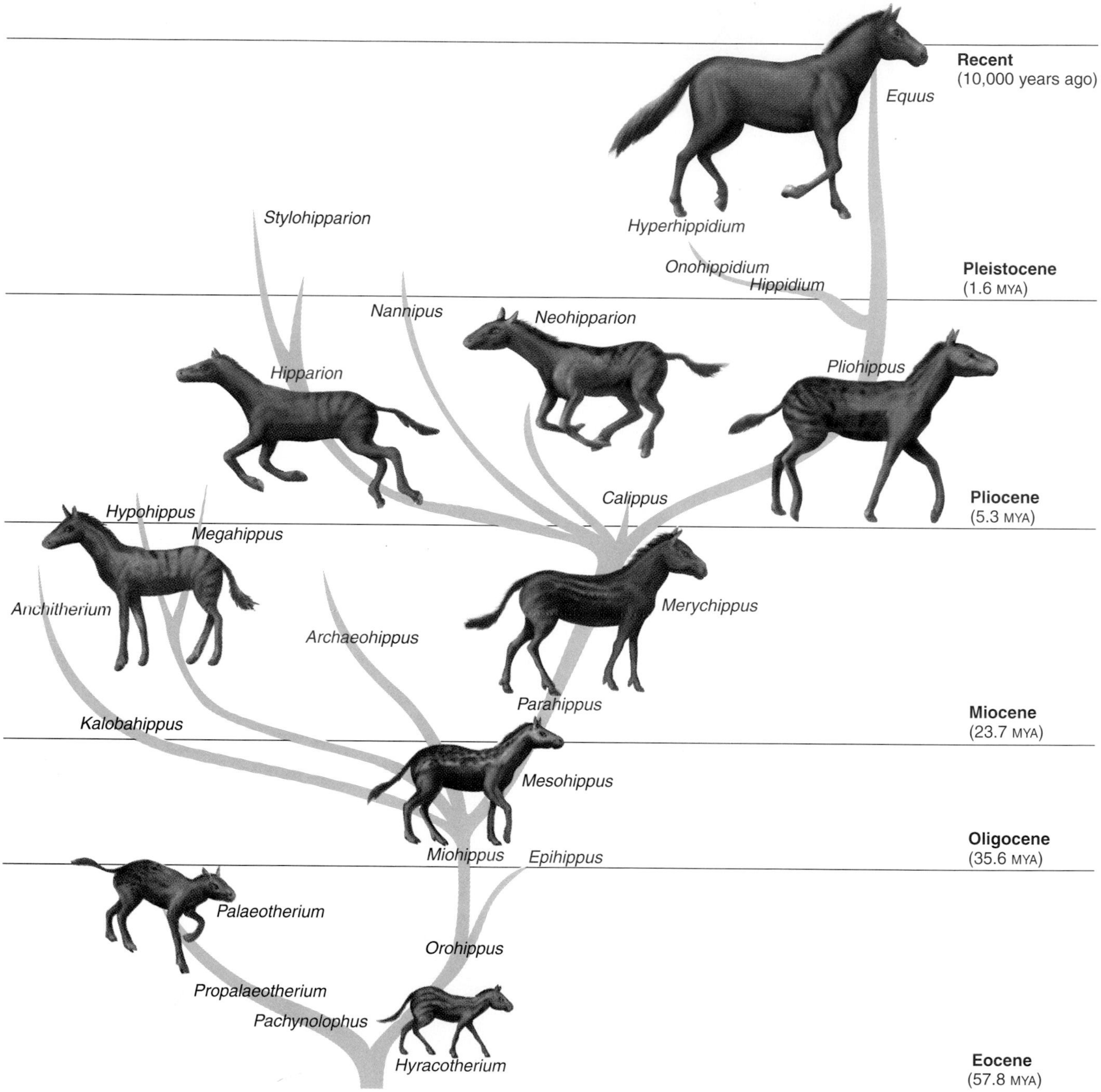

Recent (10,000 years ago)

Equus

Hyperhippidium

Onohippidium

Hippidium

Pleistocene (1.6 MYA)

Stylohipparion

Nannipus

Neohipparion

Hipparion

Pliohippus

Pliocene (5.3 MYA)

Hypohippus

Megahippus

Calippus

Anchitherium

Archaeohippus

Merychippus

Kalobahippus

Parahippus

Miocene (23.7 MYA)

Mesohippus

Miohippus

Epihippus

Oligocene (35.6 MYA)

Palaeotherium

Orohippus

Propalaeotherium

Pachynolophus

Hyracotherium

Eocene (57.8 MYA)

Evolutionary history of *Equus*
Figure 17.13

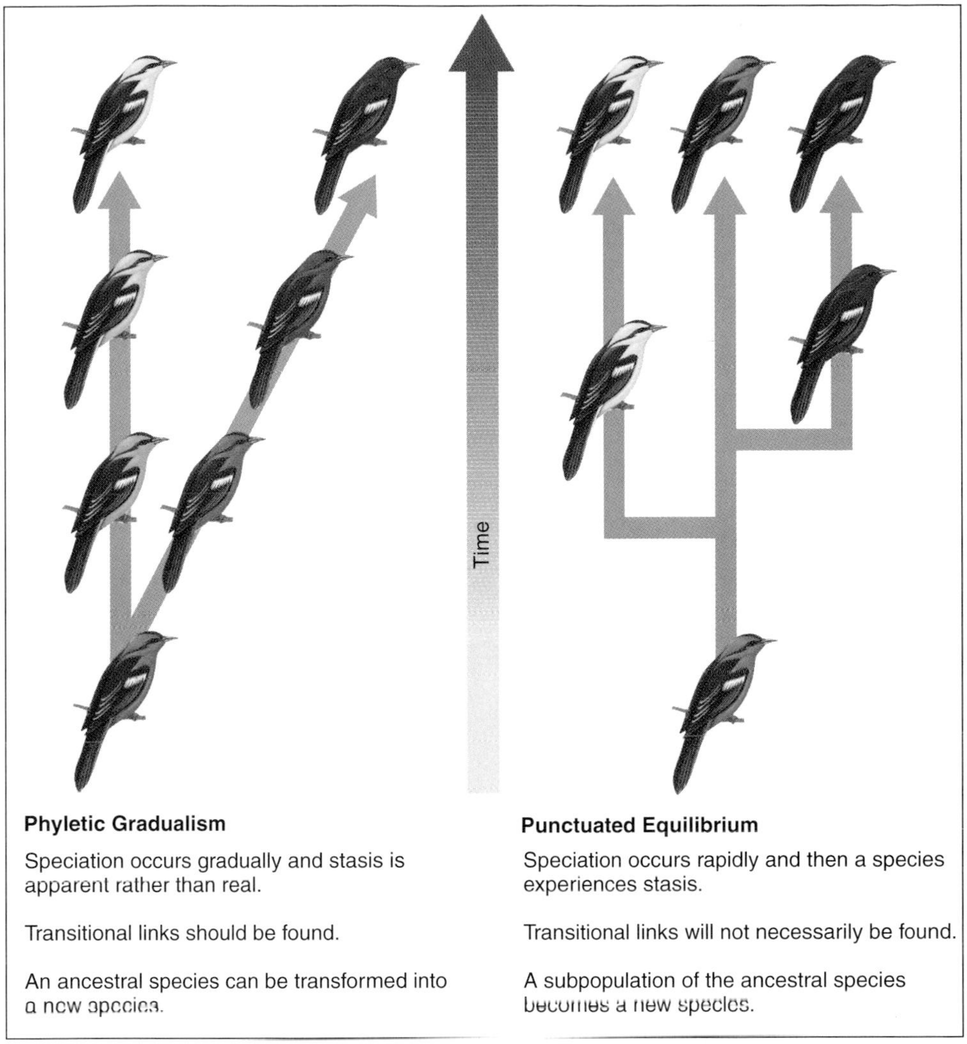

Phyletic Gradualism

Speciation occurs gradually and stasis is apparent rather than real.

Transitional links should be found.

An ancestral species can be transformed into a new species.

Punctuated Equilibrium

Speciation occurs rapidly and then a species experiences stasis.

Transitional links will not necessarily be found.

A subpopulation of the ancestral species becomes a new species.

Phyletic gradualism versus punctuated equilibrium
Figure 17C

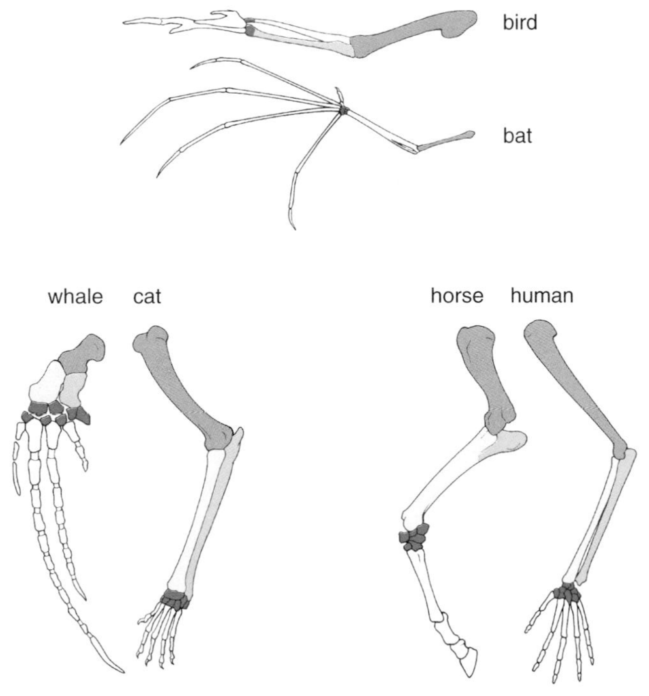

Significance of structural similarities
Figure 17.15

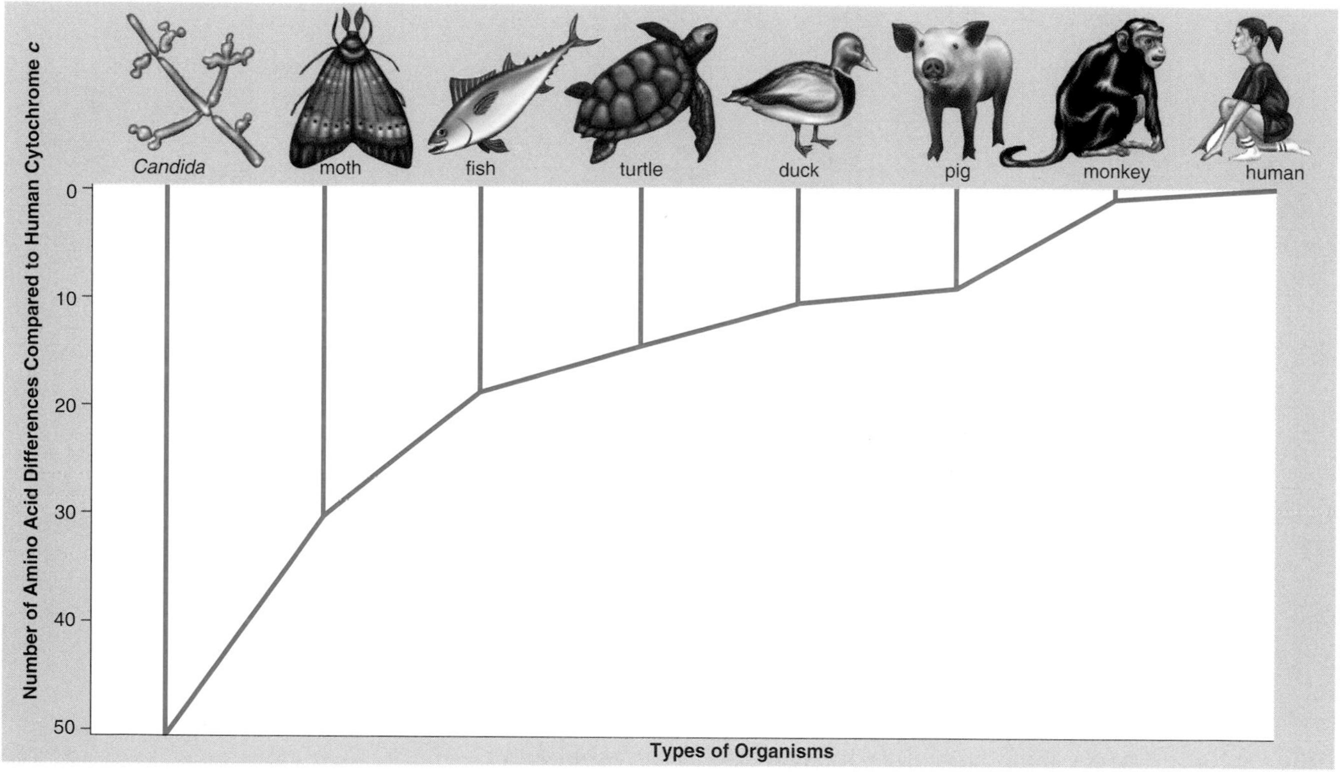

Significance of biochemical differences
Figure 17.17

$$p^2 + 2\,pq + q^2$$

p^2 = frequency of homozygous dominant individuals

p = frequency of dominant allele

q^2 = frequency of homozygous recessive individuals

q = frequency of recessive allele

$2\,pq$ = frequency of heterozygous individuals

Realize that $p + q = 1$ (There are only 2 alleles.)

$p^2 + 2\,pq + q^2 = 1$ (These are the only genotypes.)

Example

An investigator has determined by inspection that 16% of a human population has a recessive trait. Using this information, we can complete all the genotype and allele frequencies for the population, provided the conditions for Hardy-Weinberg equilibrium are met.

Given: $q^2 = 16\% = 0.16$ are homozygous recessive individuals

Therefore, $q = \sqrt{0.16} = 0.4$ = frequency of recessive allele
$p = 1.0 - 0.4 = 0.6$ = frequency of dominant allele
$p^2 = (0.6)(0.6) - 0.36 = 36\%$ are homozygous
                dominant individuals
$2\,pq = 2(0.6)(0.4) = 0.48 = 48\%$ are heterozygous } 84% have the
                individuals   dominant phenotype
                or
                $= 1.00 - 0.52 = 0.48$

Calculating gene pool frequencies using the Hardy-Weinberg equation
Figure 18.1

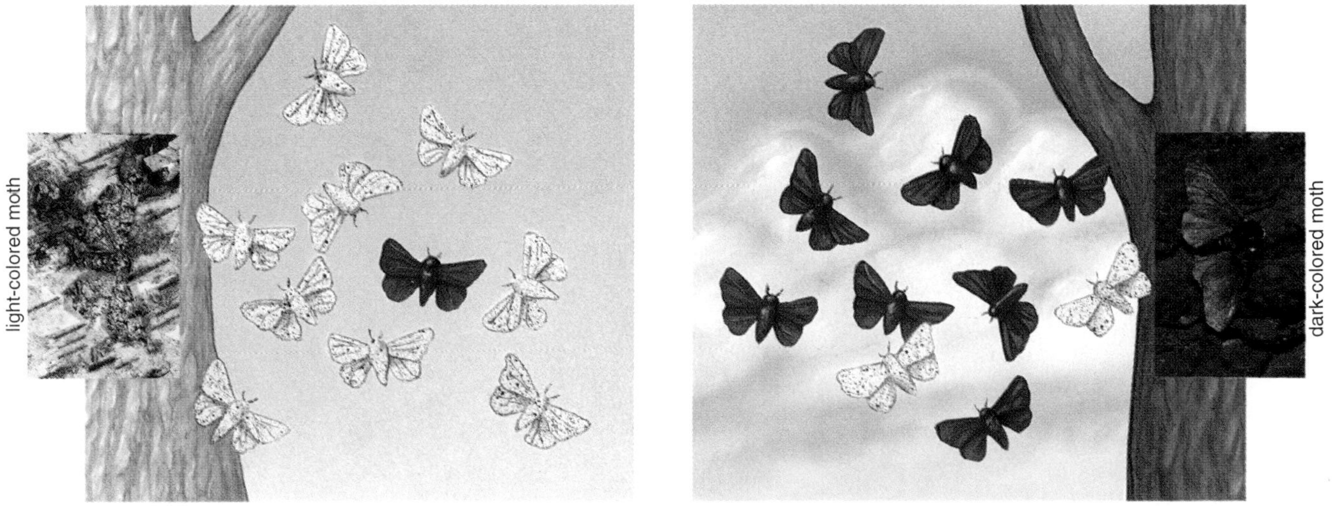

light-colored moth

dark-colored moth

Industrial melanism and microevolution
Figure 18.2

a: © Breck Kent/Animals Animals/Earth Scenes; b: © Michael Tweedie/Photo Researchers, Inc.

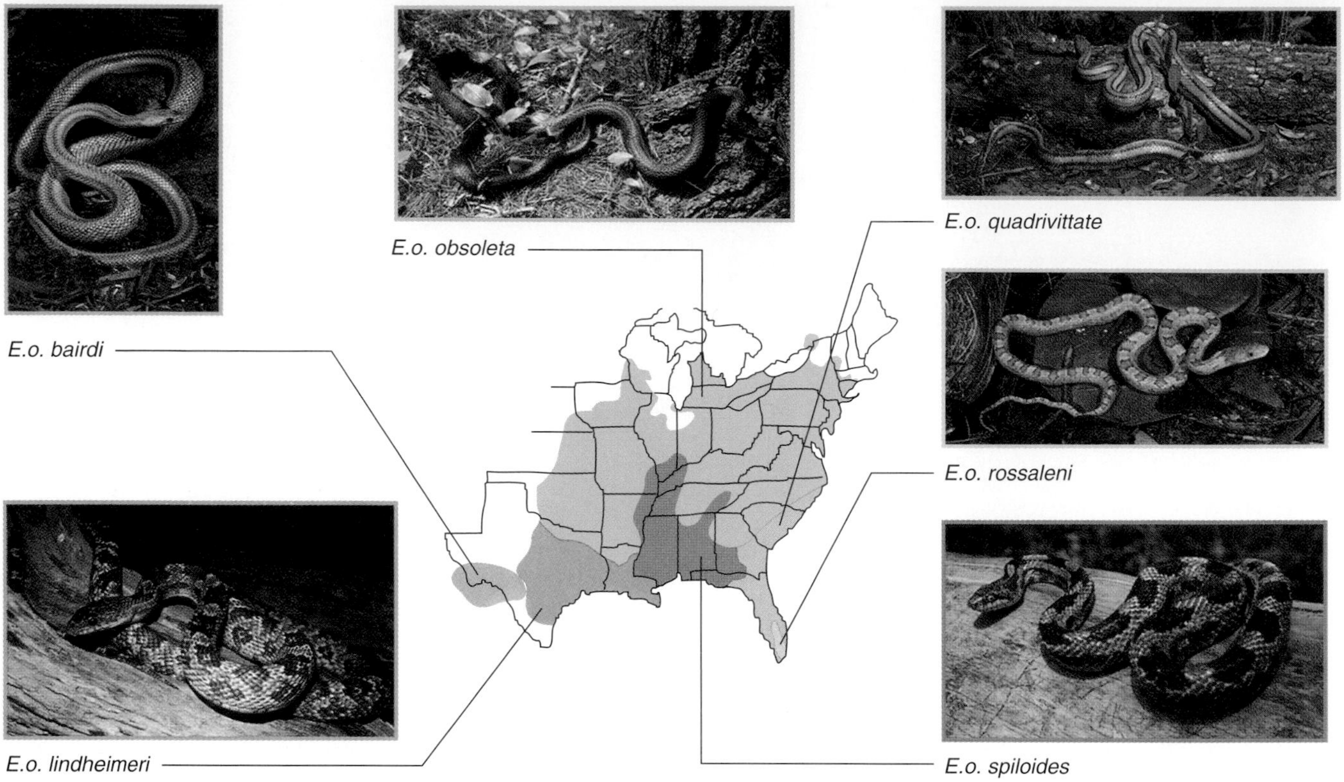
E.o. obsoleta
E.o. quadrivittate
E.o. bairdi
E.o. rossaleni
E.o. lindheimeri
E.o. spiloides

Gene flow
Figure 18.3

b (top left): © Visuals Unlimited; c (bottom left): © Zig Leszczynski/Animals Animals/Earth Scenes; d (bottom right): © Joseph Collins/Photo Researchers, Inc.; e (middle right): © Dale Jackson/Visuals Unlimited; f (top right): © Zig Leszczynski/Animals Animals/Earth Scenes; g (center): © William Weber/Visuals Unlimited

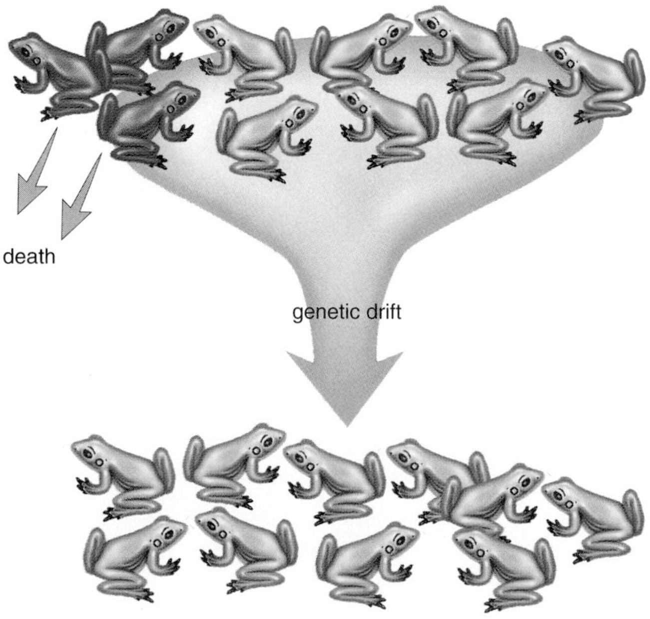
death
genetic drift

Genetic drift
Figure 18.4

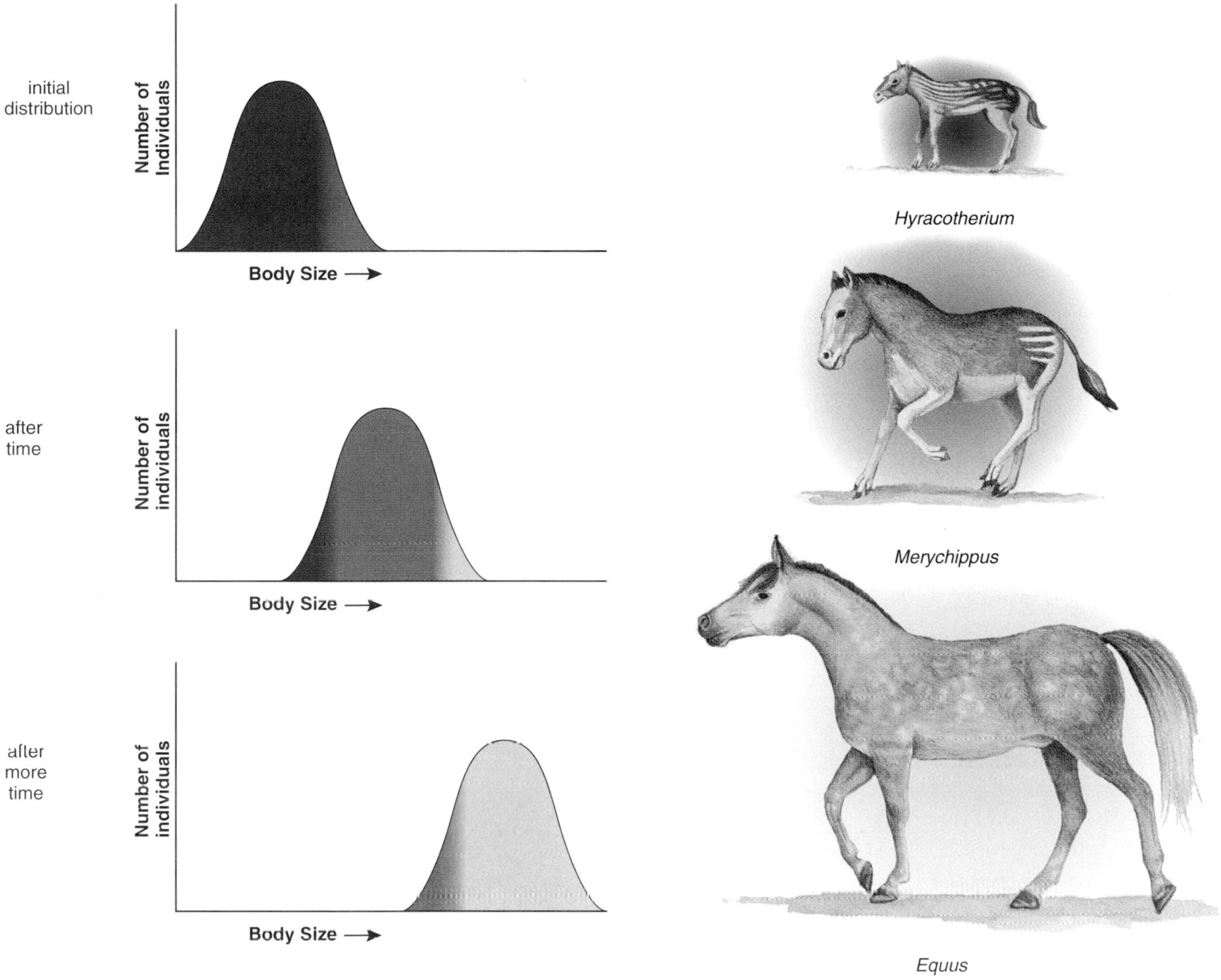

initial distribution

after time

after more time

Hyracotherium

Merychippus

Equus

Directional selection

Figure 18.6

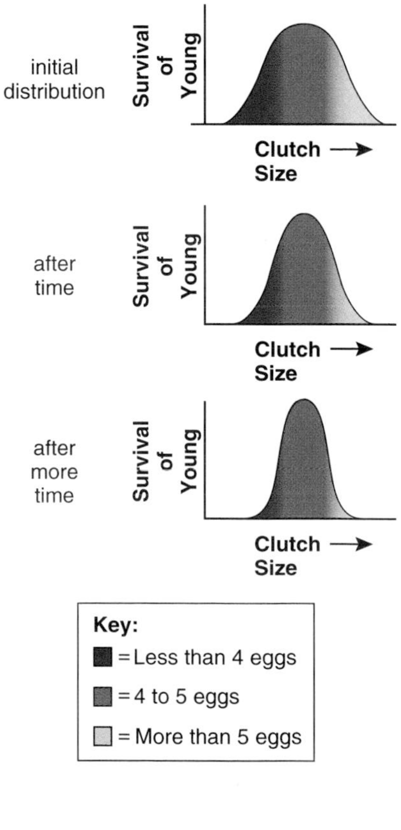

initial distribution

after time

after more time

Key:

■ = Less than 4 eggs

▨ = 4 to 5 eggs

□ = More than 5 eggs

Stabilizing selection

Figure 18.7

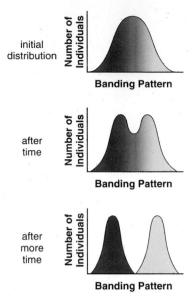

Disruptive selection
Figure 18.8

b: © Bob Evans/Peter Arnold, Inc.

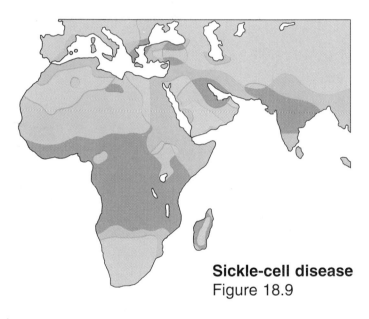

Sickle-cell disease
Figure 18.9

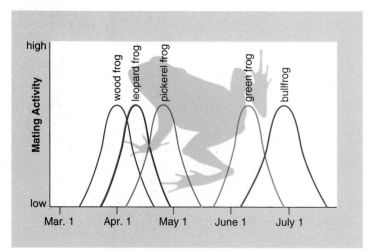

Temporal isolation
Figure 18.11

134

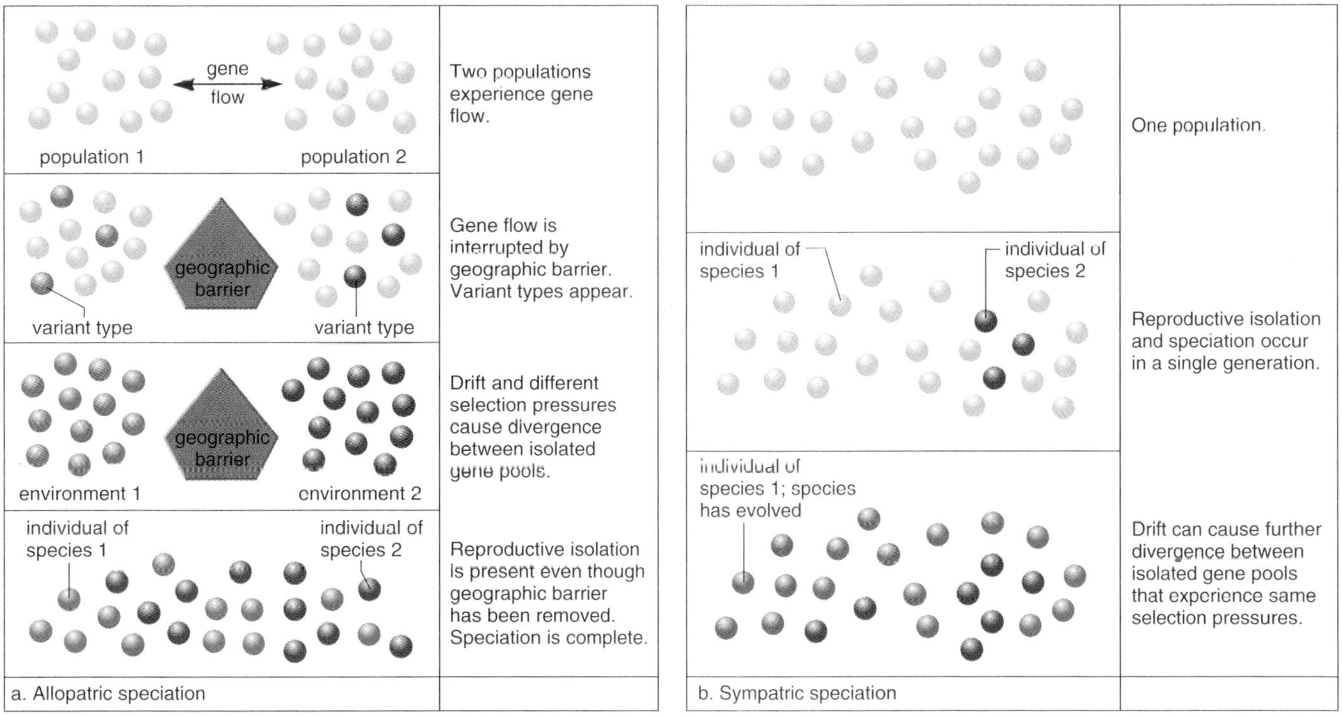

Allopatric versus sympatric speciation
Figure 18.12

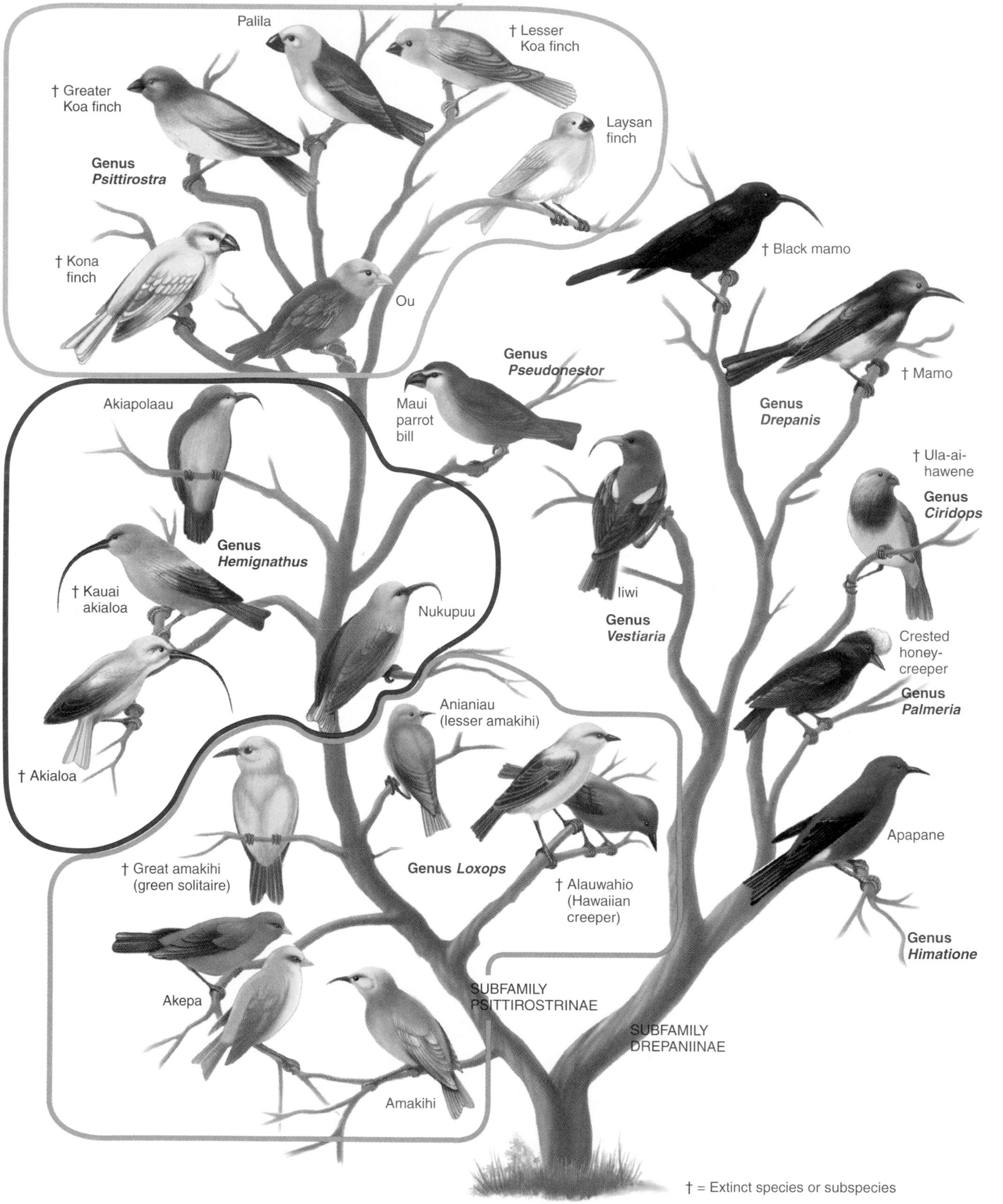

Palila

† Lesser
Koa finch

† Greater
Koa finch

Laysan
finch

Genus
Psittirostra

† Kona
finch

Ou

† Black mamo

† Mamo

Genus
Drepanis

Akiapolaau

Genus
Pseudonestor

Maui
parrot
bill

† Ula-ai-
hawene

Genus
Ciridops

Genus
Hemignathus

† Kauai
akialoa

Nukupuu

Iiwi

Genus
Vestiaria

Crested
honey-
creeper

Genus
Palmeria

† Akialoa

Anianiau
(lesser amakihi)

† Great amakihi
(green solitaire)

Genus *Loxops*

† Alauwahio
(Hawaiian
creeper)

Apapane

Akepa

Amakihi

SUBFAMILY
PSITTIROSTRINAE

Genus
Himatione

SUBFAMILY
DREPANIINAE

† = Extinct species or subspecies

Adaptive radiation in Hawaiian honeycreepers
Figure 18.13

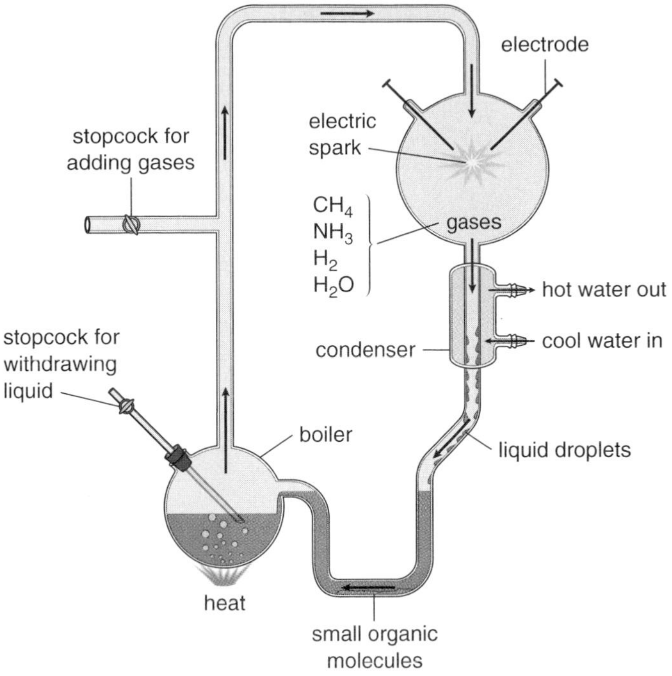

Stanley Miller's apparatus and experiment
Figure 19.1

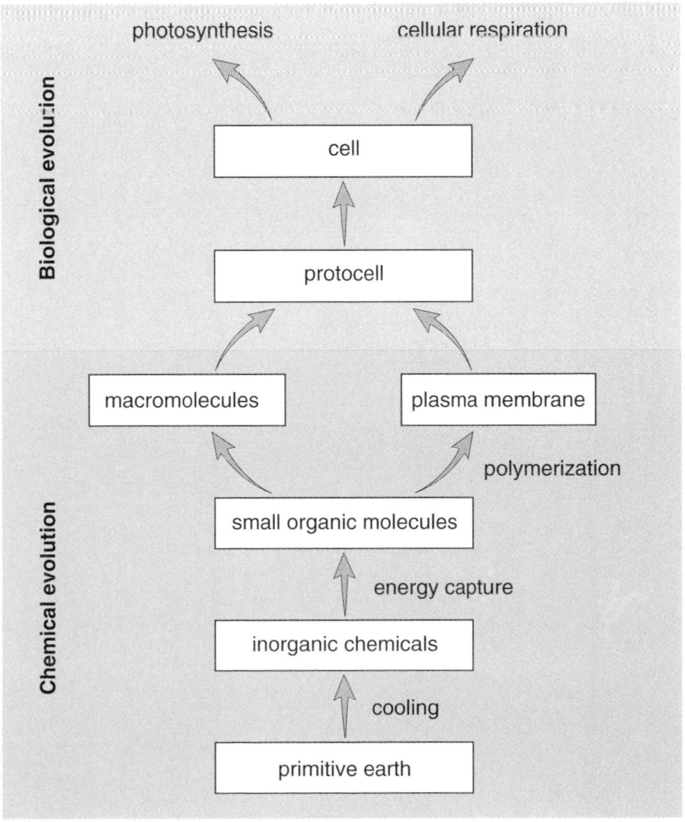

Origin of the first cell(s)
Figure 19.4

Prokaryote fossil of the Precambrian
Figure 19.7

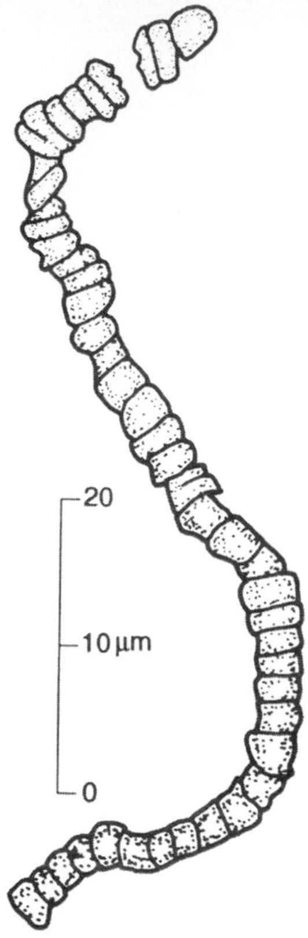

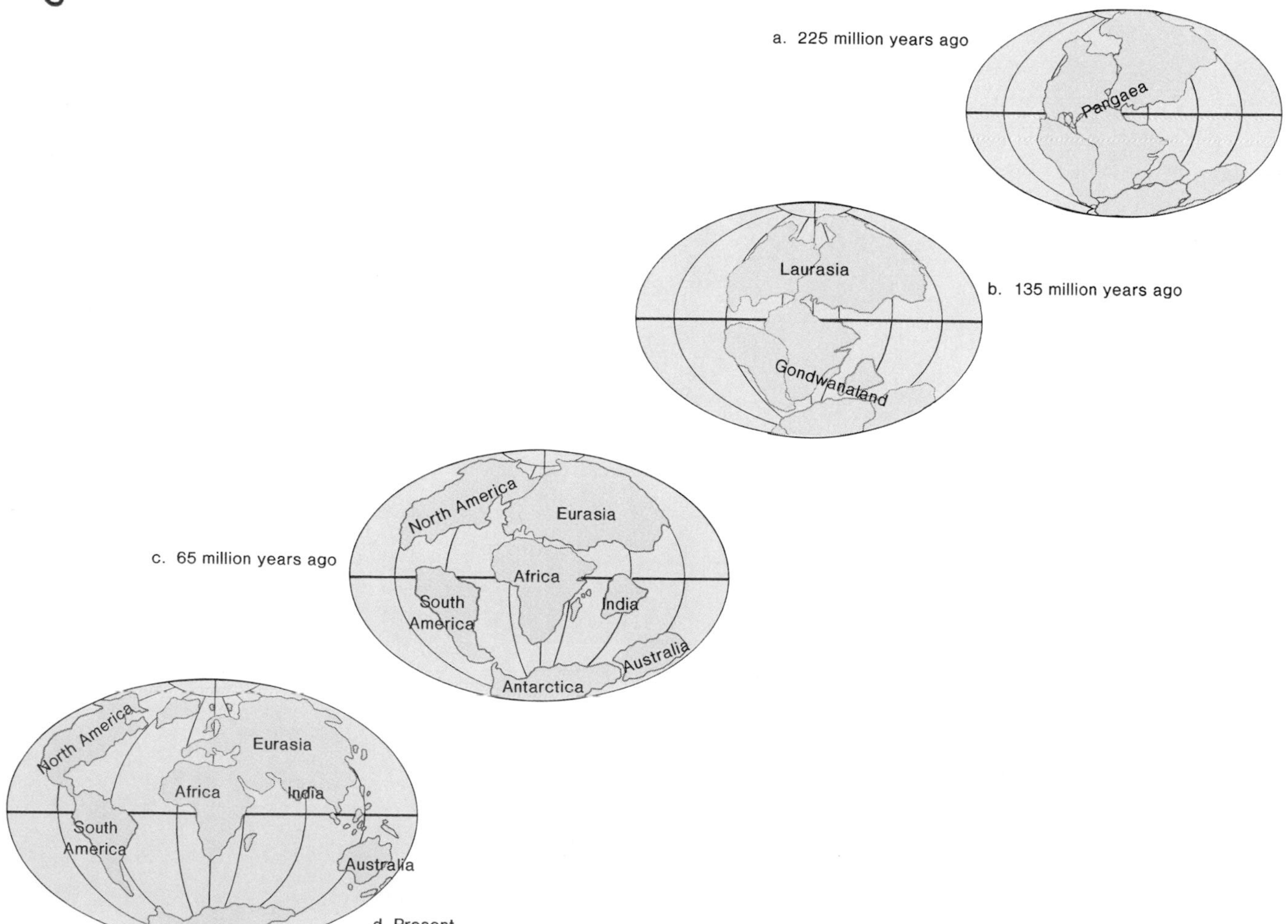

a. 225 million years ago

Pangaea

Laurasia

b. 135 million years ago

Gondwanaland

c. 65 million years ago

North America

Eurasia

Africa

South America

India

Australia

Antarctica

North America

Eurasia

Africa

India

South America

Australia

Antarctica

d. Present

Continental drift
Figure 19.14

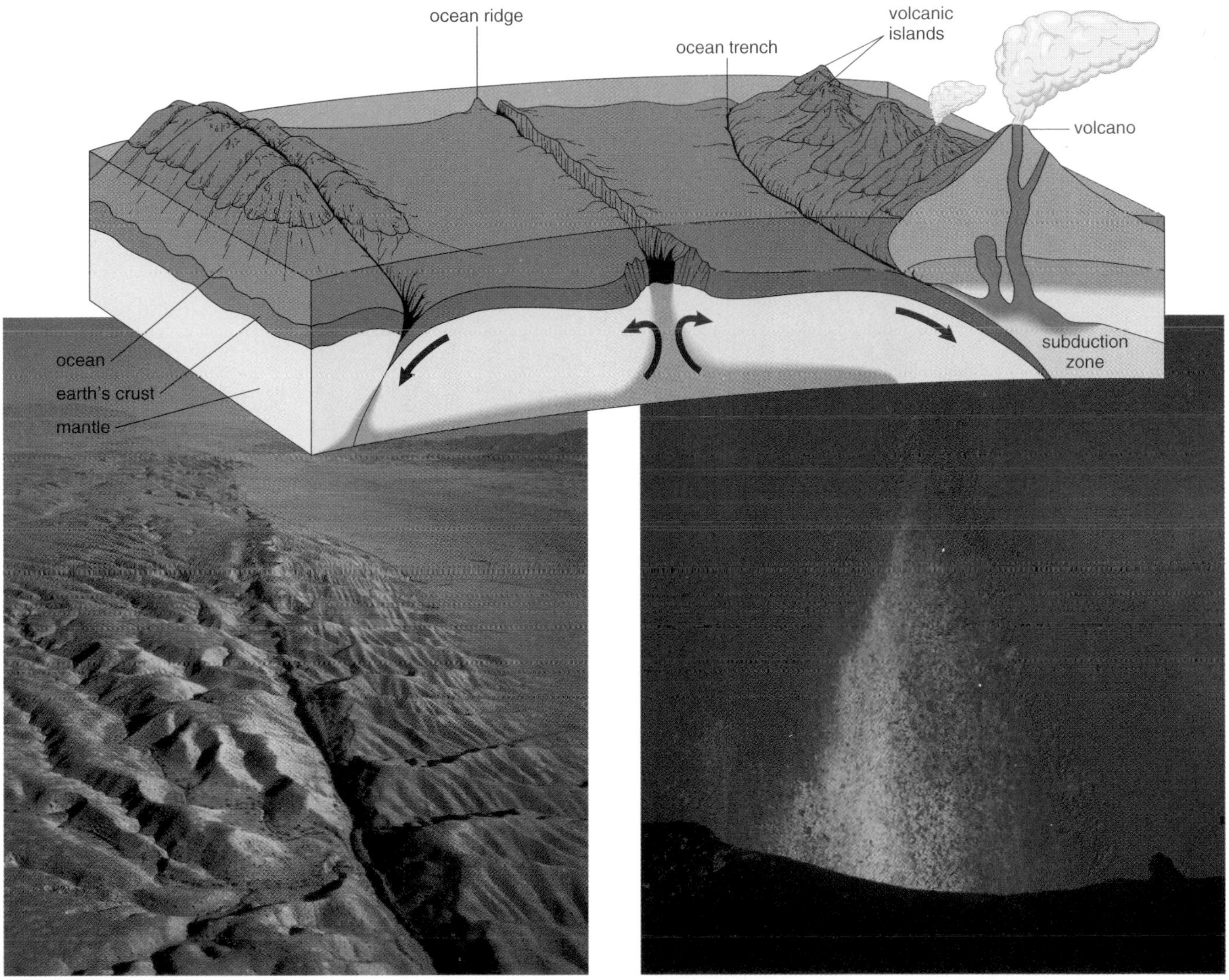

Plate tectonics
Figure 19.15

b: © David Parker/Photo Researchers, Inc.; c: © Matthew Shipp/Photo Researchers, Inc.

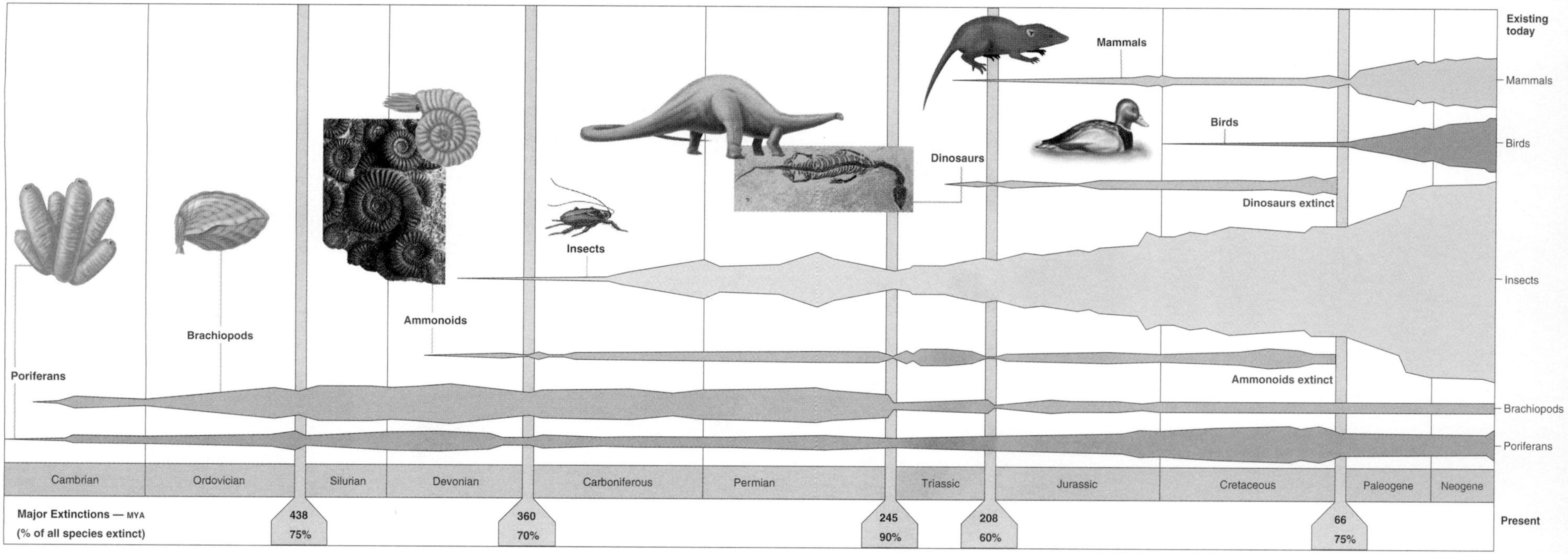

Mass extinctions
Figure 19.16

© The McGraw-Hill Companies, Inc./Carlyn Iverson, photographer

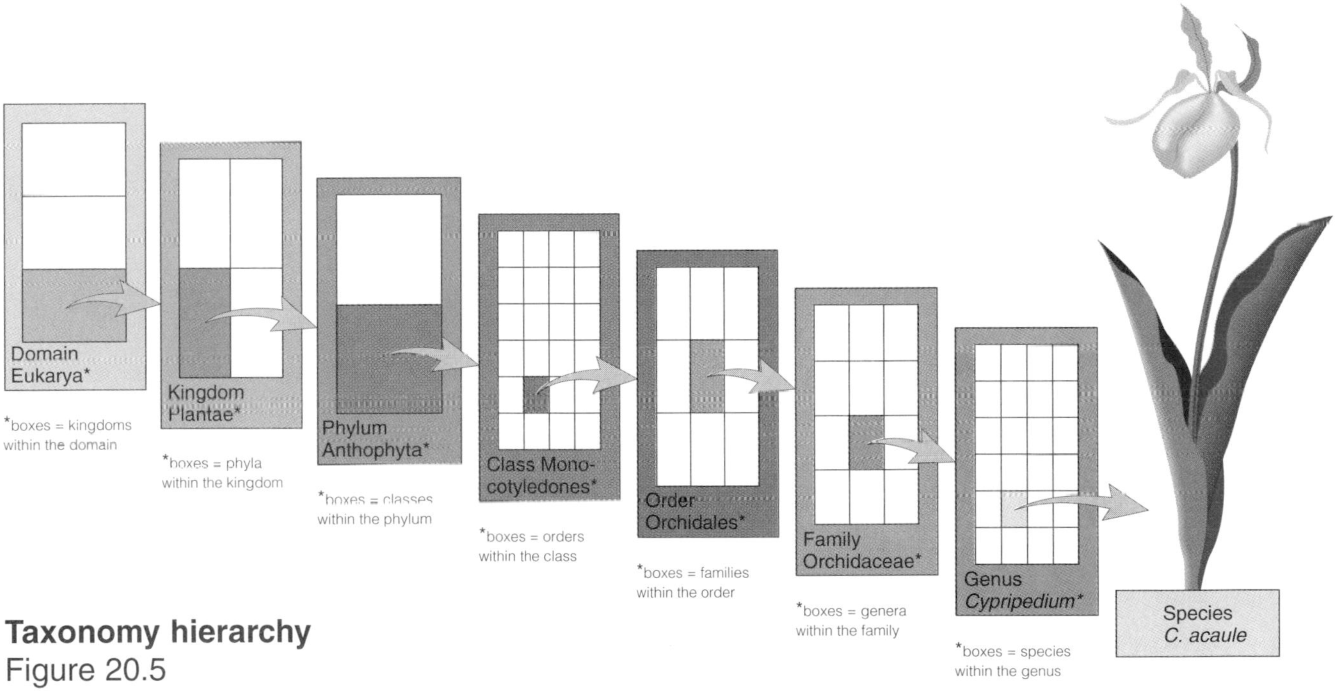

Taxonomy hierarchy
Figure 20.5

Domain
Eukarya*

*boxes = kingdoms
within the domain

Kingdom
Plantae*

*boxes = phyla
within the kingdom

Phylum
Anthophyta*

*boxes = classes
within the phylum

Class Mono-
cotyledones*

*boxes = orders
within the class

Order
Orchidales*

*boxes = families
within the order

Family
Orchidaceae*

*boxes = genera
within the family

Genus
*Cypripedium**

*boxes = species
within the genus

Species
C. acaule

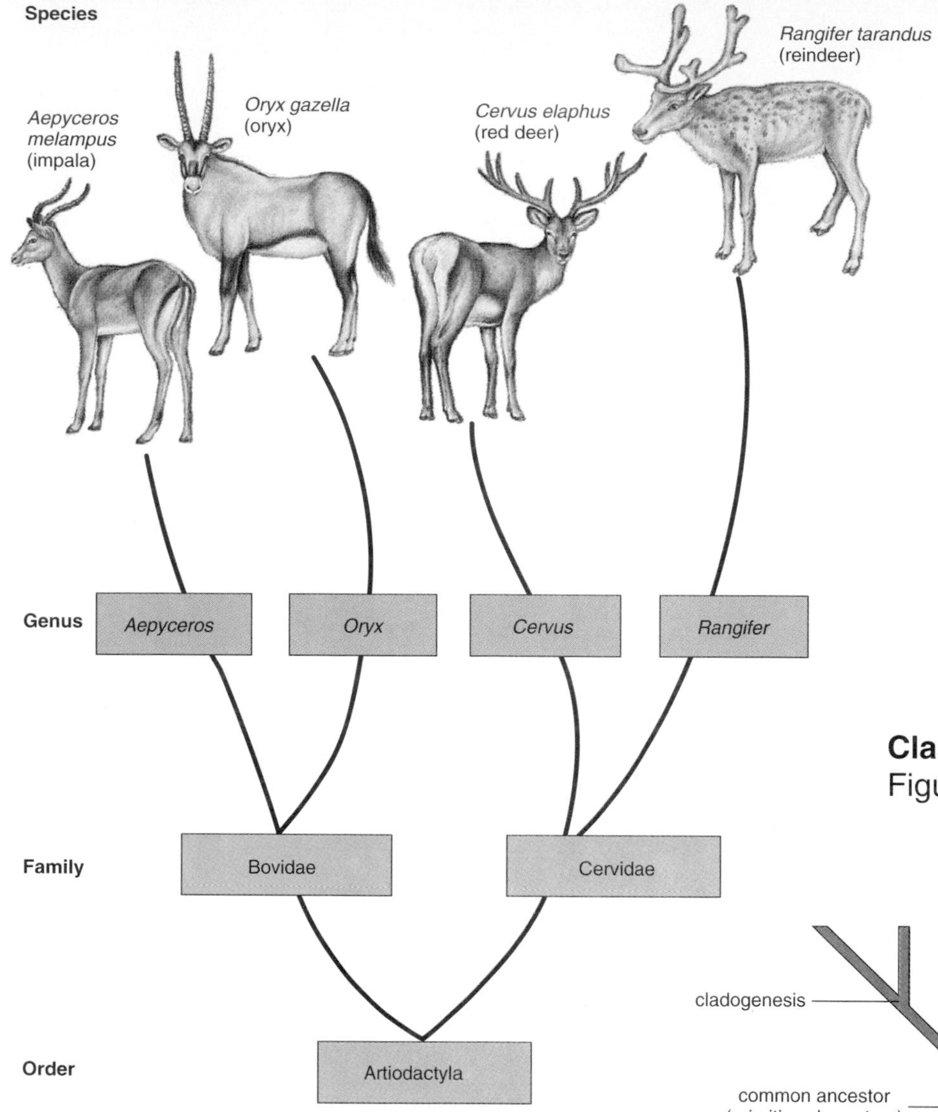

Species

Aepyceros melampus (impala)

Oryx gazella (oryx)

Cervus elaphus (red deer)

Rangifer tarandus (reindeer)

Genus | Aepyceros | Oryx | Cervus | Rangifer

Family | Bovidae | Cervidae

Order | Artiodactyla

Classification and phylogeny
Figure 20.6

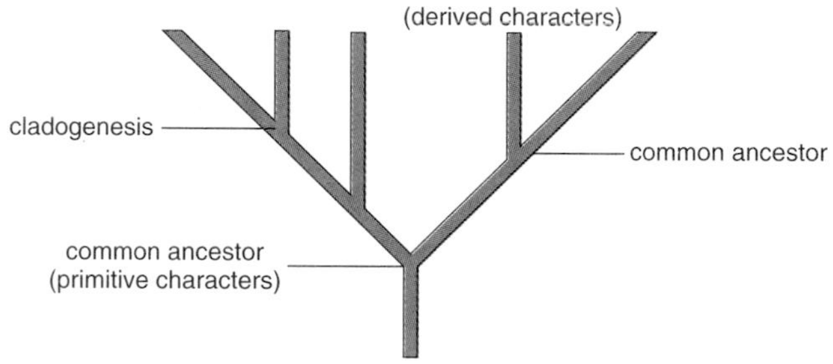

(derived characters)

cladogenesis

common ancestor

common ancestor (primitive characters)

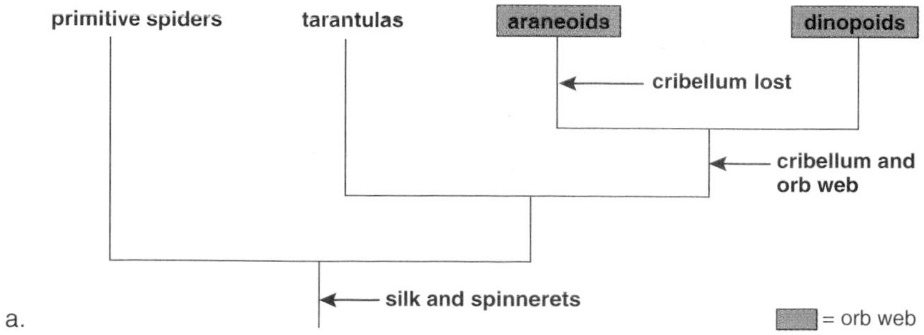

primitive spiders   tarantulas   araneoids   dinopoids

cribellum lost

cribellum and orb web

silk and spinnerets

a.

⬛ = orb web

Evolution of orb web
Figure 20A

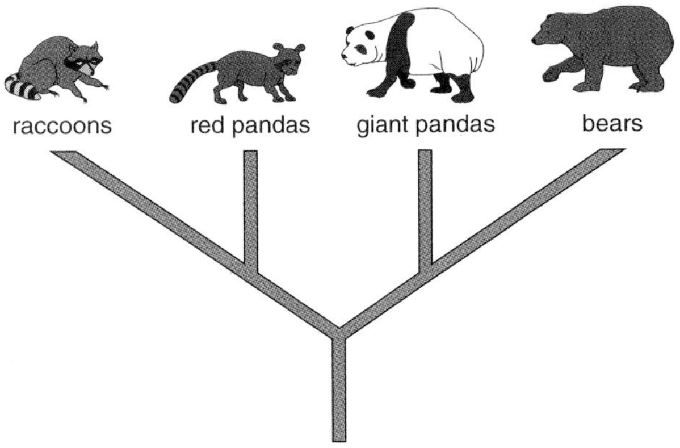

Ancestry of giant pandas
Figure 20.9

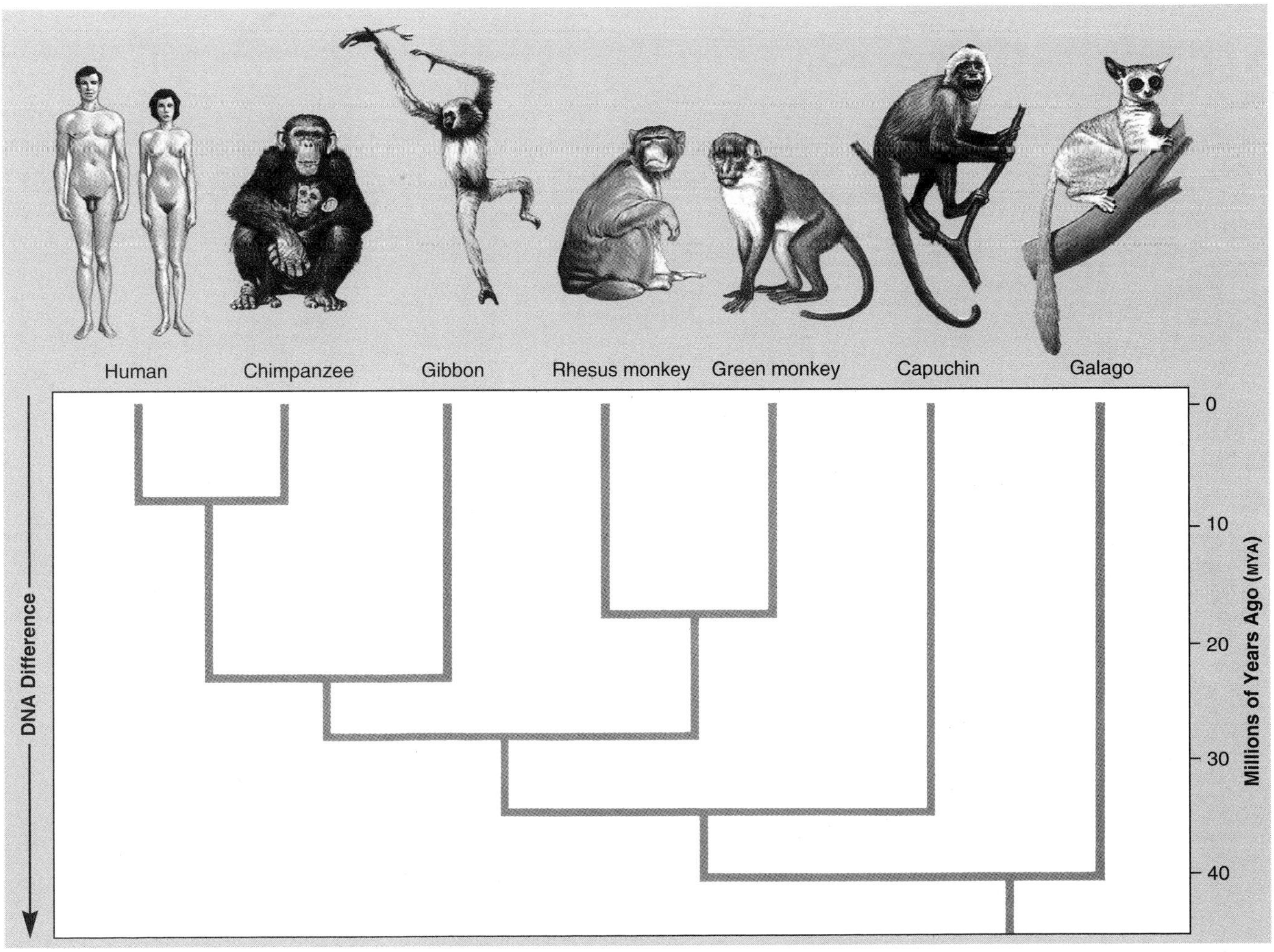

Genetic data
Figure 20.10

	lancelet	eel	newt	snake	lizard
Notochord in embryo					
Vertebrae					
Lungs					
Three-chambered heart					
Internal fertilization					
Amniotic membrane in egg					
Four bony limbs					
Long cylindrical body					

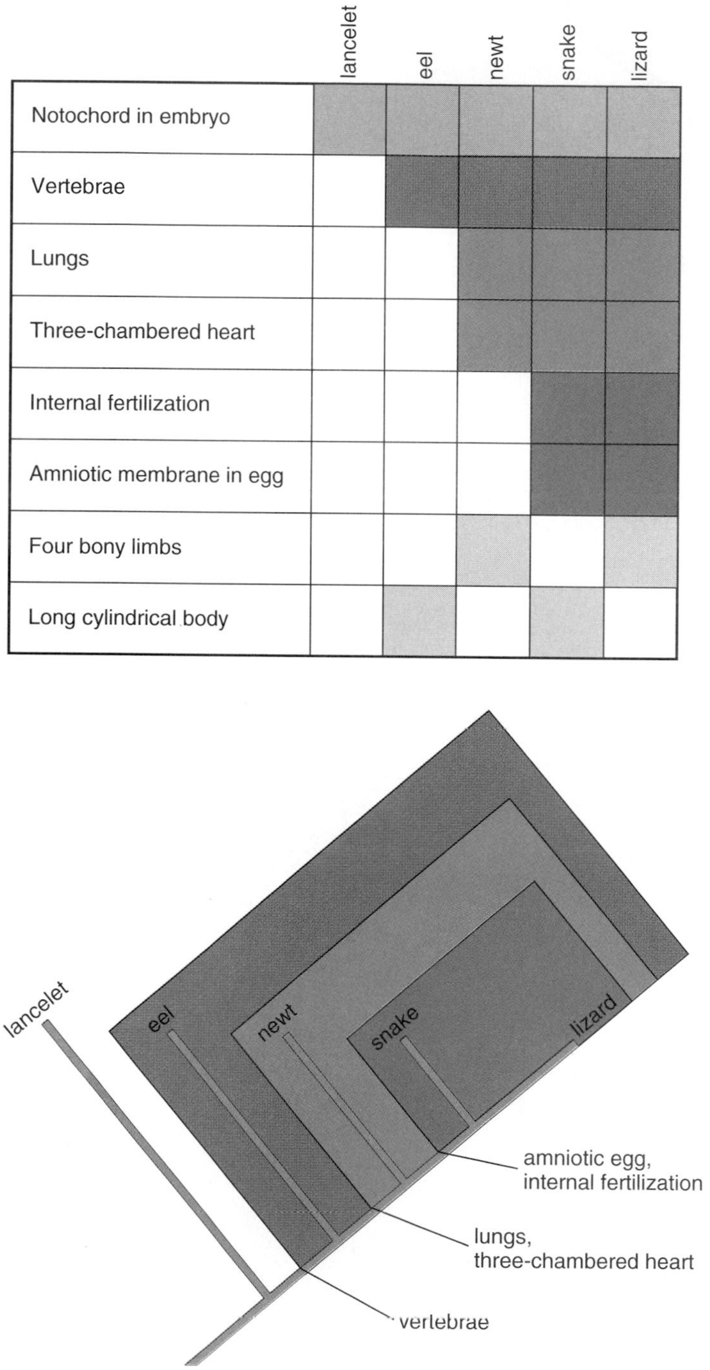

Constructing a cladogram
Figure 20.11

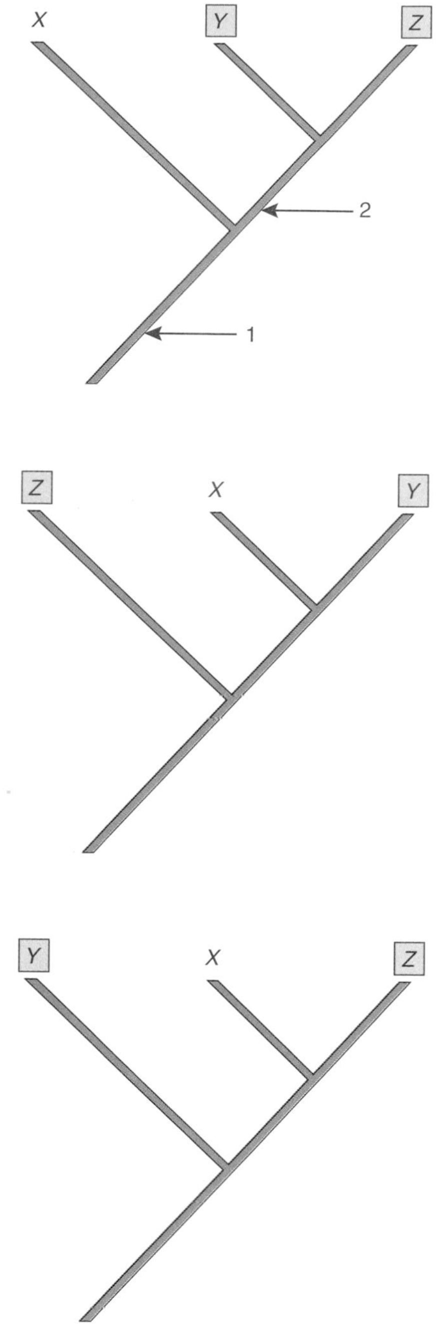

Alternate, simplified cladograms
Figure 20.12

144

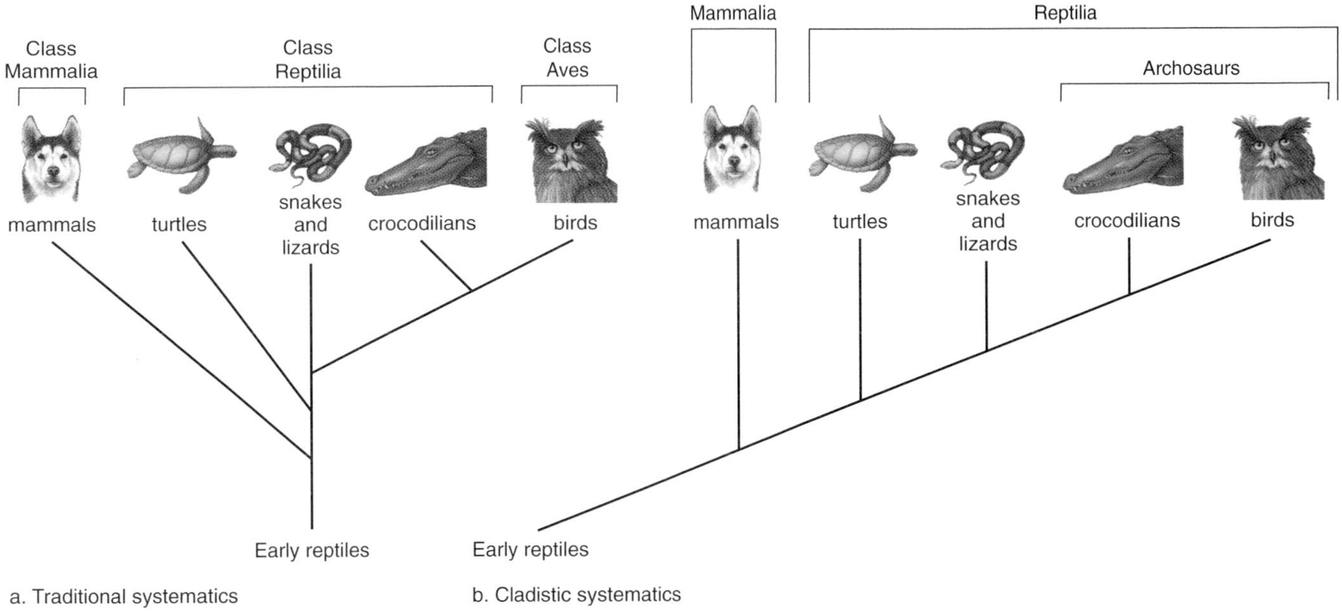

a. Traditional systematics

b. Cladistic systematics

Traditional versus cladistic view of reptilian phylogeny
Figure 20.13

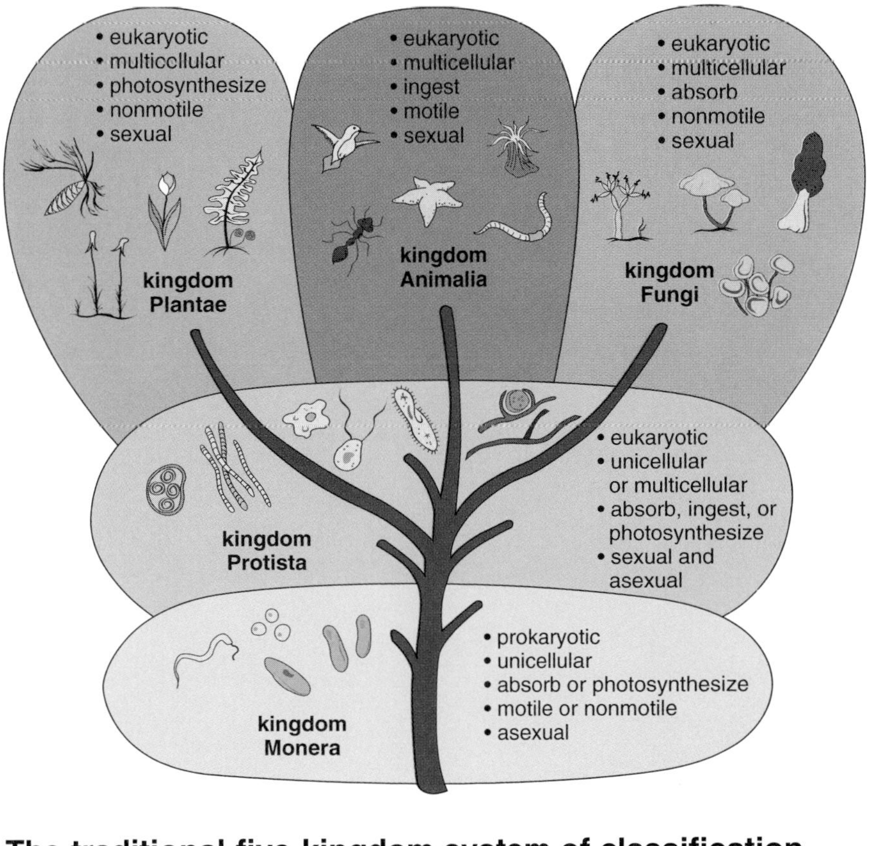

The traditional five-kingdom system of classification
Figure 20.14

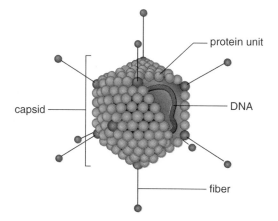
protein unit

capsid

DNA

fiber

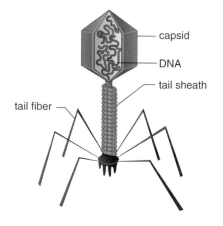
capsid

DNA

tail sheath

tail fiber

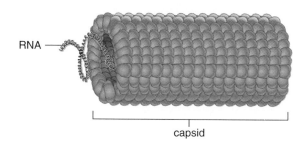
RNA

capsid

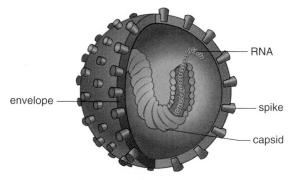
RNA

envelope

spike

capsid

Viruses
Figure 21.1

Lytic and lysogenic cycles in prokaryotes
Figure 21.3

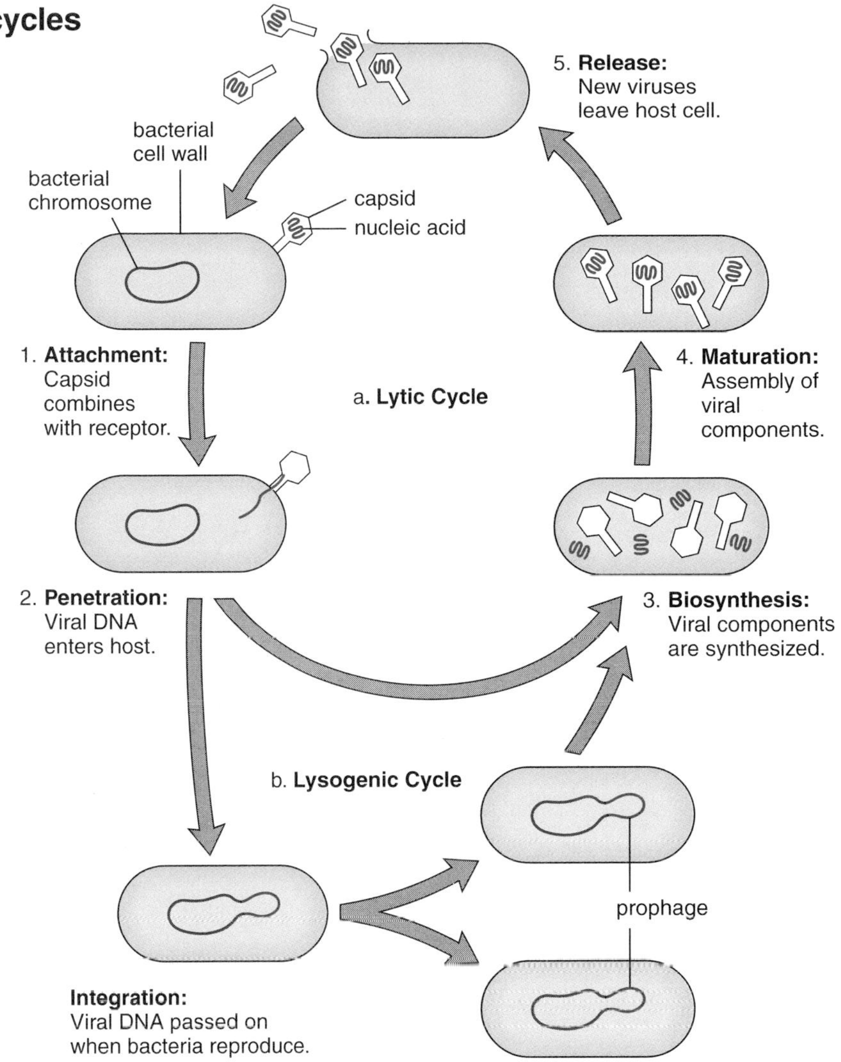

5. **Release:** New viruses leave host cell.

bacterial cell wall
bacterial chromosome
capsid
nucleic acid

1. **Attachment:** Capsid combines with receptor.

a. **Lytic Cycle**

4. **Maturation:** Assembly of viral components.

2. **Penetration:** Viral DNA enters host.

3. **Biosynthesis:** Viral components are synthesized.

b. **Lysogenic Cycle**

prophage

Integration: Viral DNA passed on when bacteria reproduce.

Reproduction of the retrovirus HIV-1
Figure 21.4

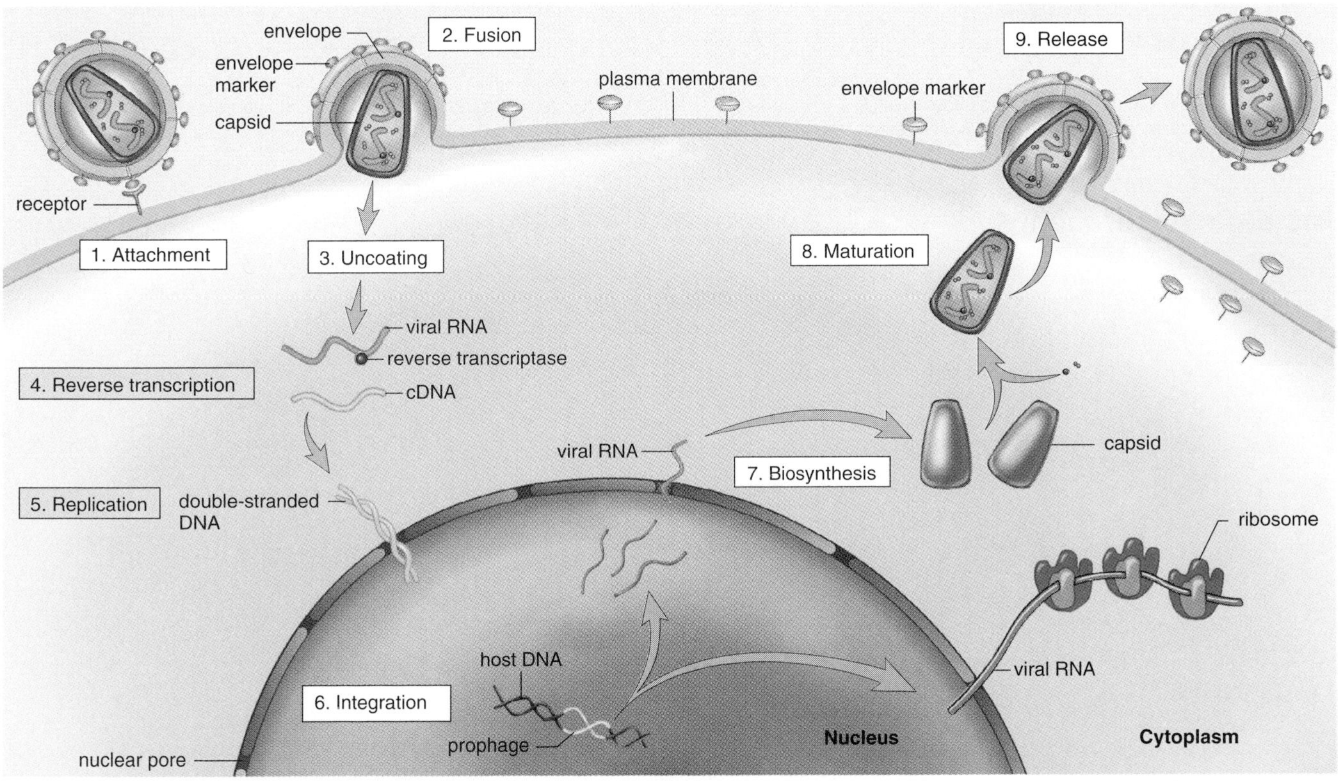

envelope
envelope marker
capsid

2. Fusion

plasma membrane
envelope marker

9. Release

receptor

1. Attachment

3. Uncoating

8. Maturation

viral RNA
reverse transcriptase
cDNA

4. Reverse transcription

viral RNA

7. Biosynthesis

capsid

5. Replication

double-stranded DNA

ribosome

host DNA

viral RNA

6. Integration

prophage

nuclear pore

Nucleus

Cytoplasm

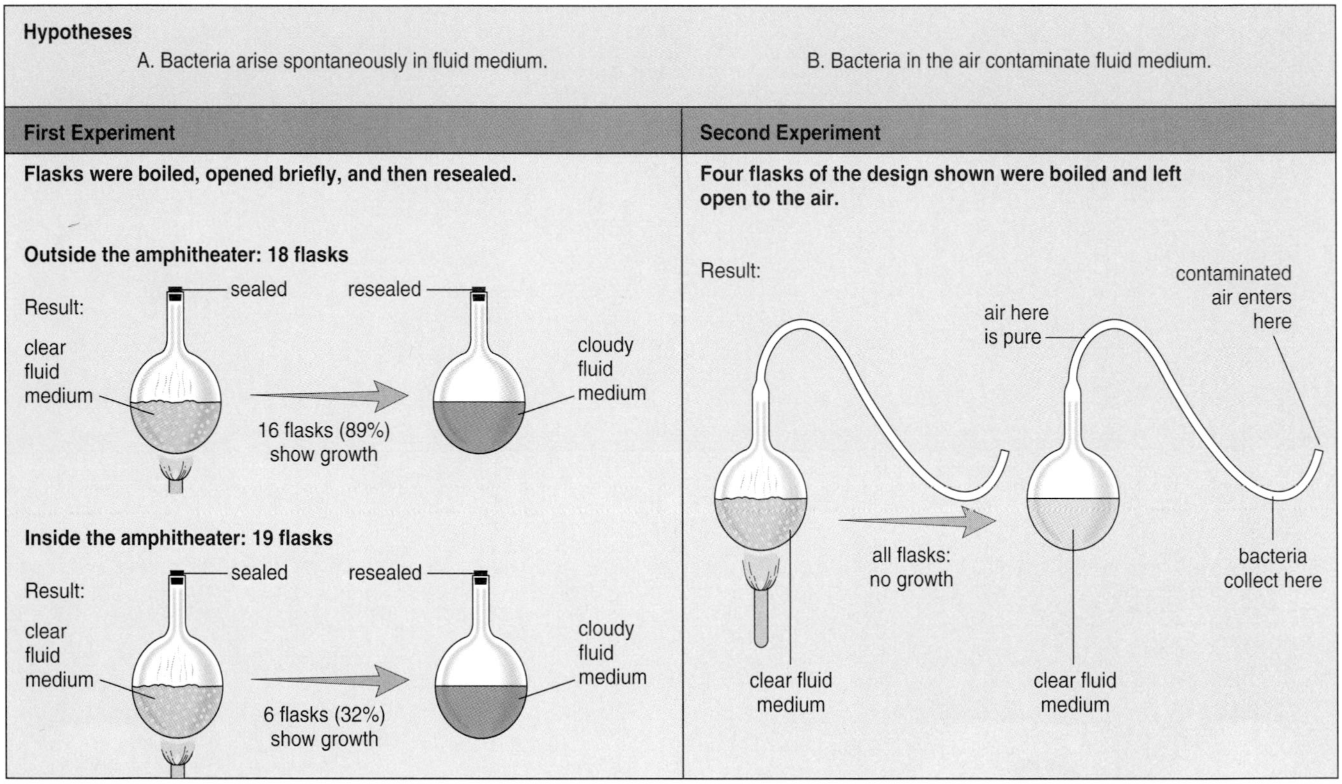

Hypotheses

A. Bacteria arise spontaneously in fluid medium.

B. Bacteria in the air contaminate fluid medium.

First Experiment

Flasks were boiled, opened briefly, and then resealed.

Outside the amphitheater: 18 flasks

Result:

sealed — resealed

clear fluid medium

16 flasks (89%) show growth

cloudy fluid medium

Inside the amphitheater: 19 flasks

Result:

sealed — resealed

clear fluid medium

6 flasks (32%) show growth

cloudy fluid medium

Second Experiment

Four flasks of the design shown were boiled and left open to the air.

Result:

air here is pure

contaminated air enters here

all flasks: no growth

bacteria collect here

clear fluid medium

clear fluid medium

Conclusion:
With hypothesis A, all flasks would be cloudy. Relative concentrations of bacteria in the air explain the results; therefore, hypothesis B is supported.

Conclusion:
When air reaching the medium contains no bacteria, the medium remains free of growth; therefore, hypothesis B is further supported.

Pasteur's experiment
Figure 21.5

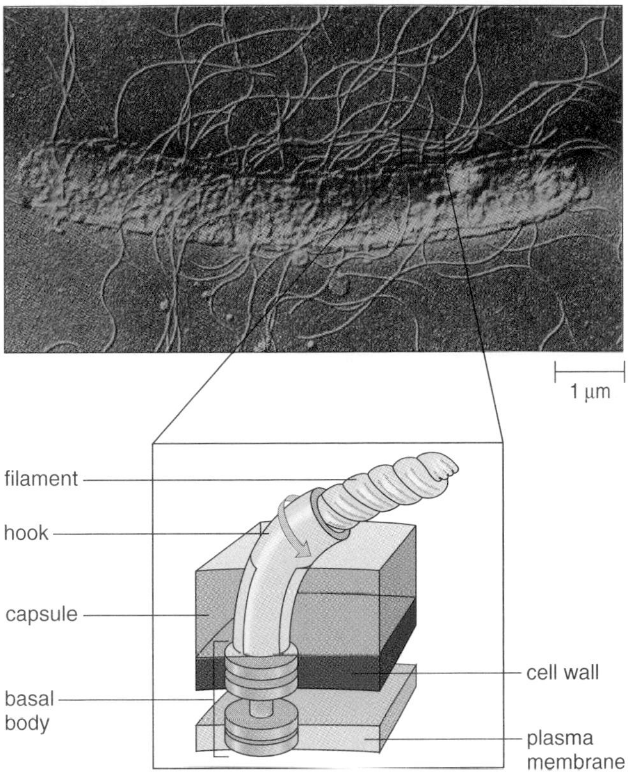

filament

hook

capsule

basal body

cell wall

plasma membrane

1 μm

Flagella

Figure 21.6

© Fred Hossler/Visuals Unlimited

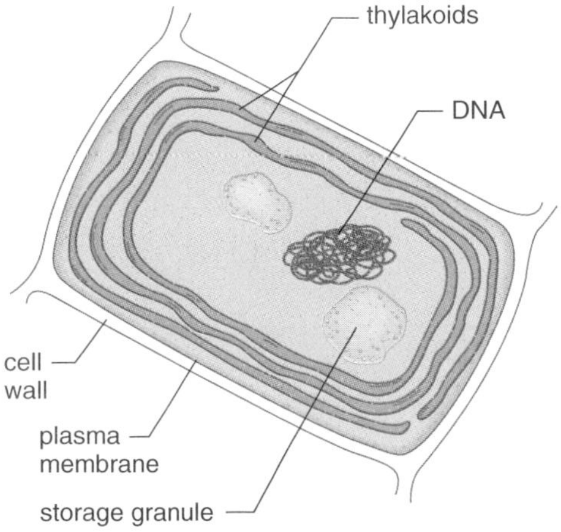

thylakoids

DNA

cell wall

plasma membrane

storage granule

Diversity among the cyanobacteria

Figure 21.12

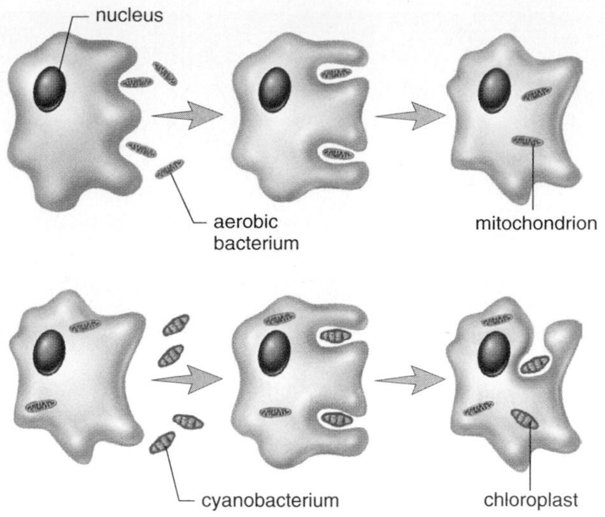

Origin of the eukaryotic cell
Figure 22.1

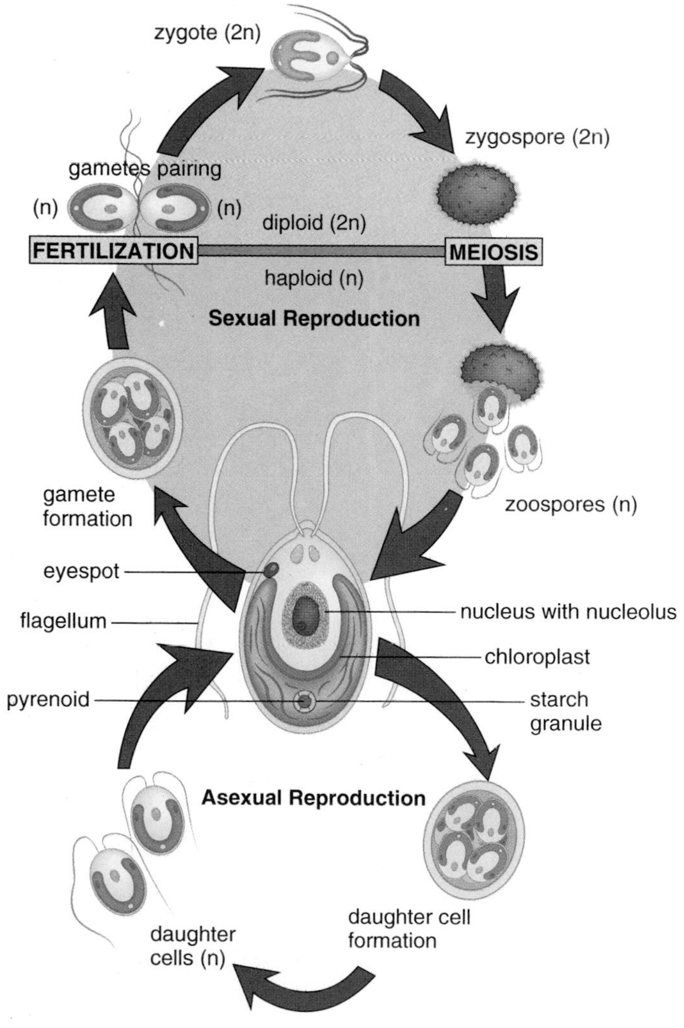

Reproduction in _Chlamydomonas_
Figure 22.4

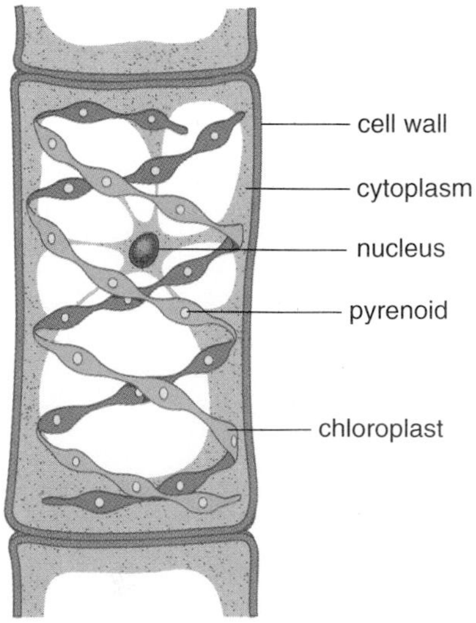

Spirogyra
Figure 22.5

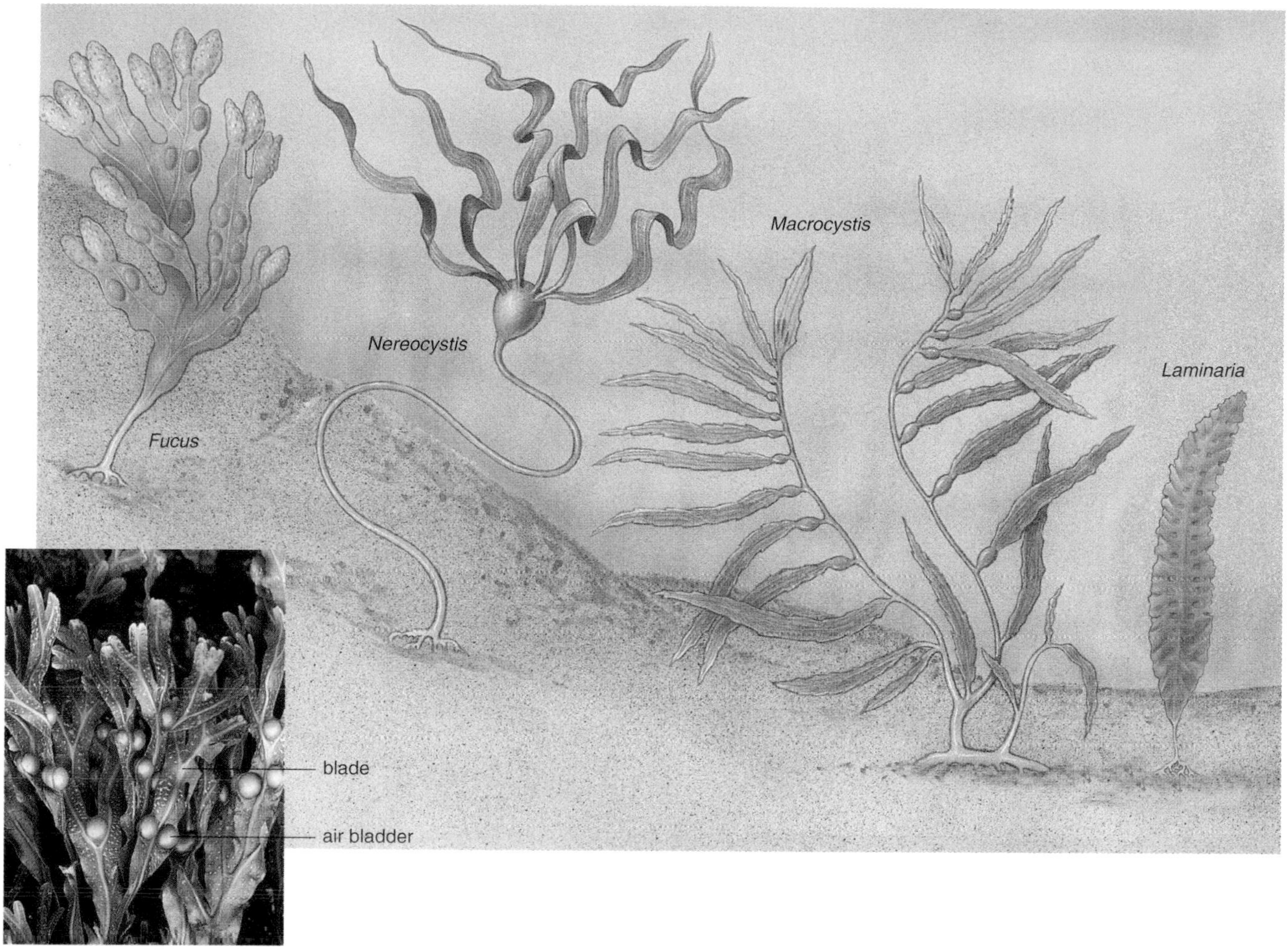

blade

air bladder

Brown algae
Figure 22.9

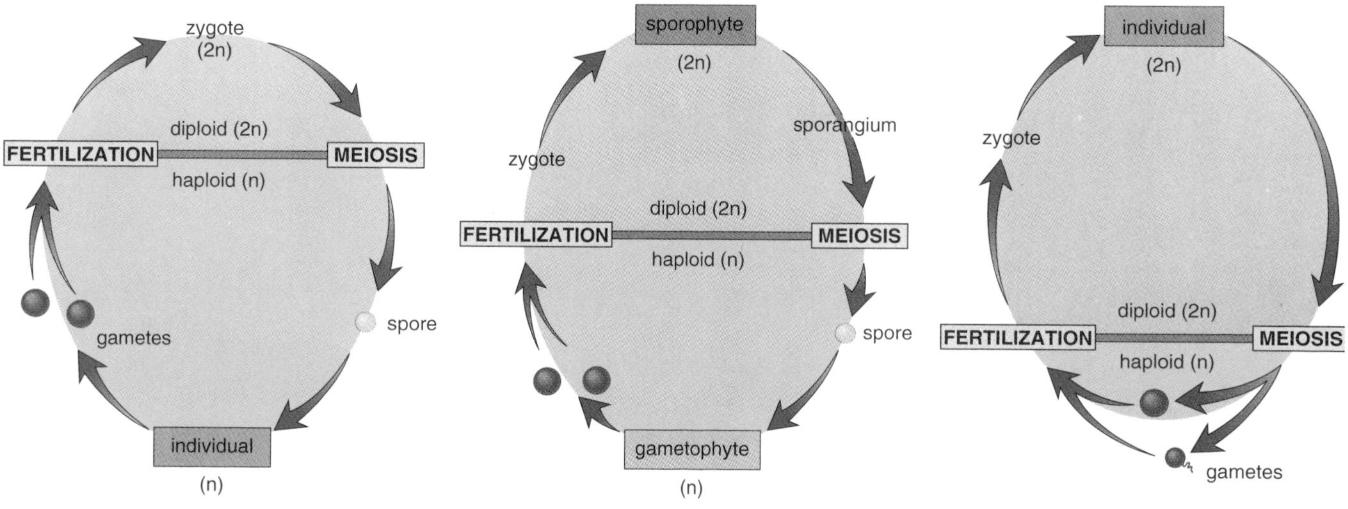

Common life cycles in sexual reproduction
Figure 22A

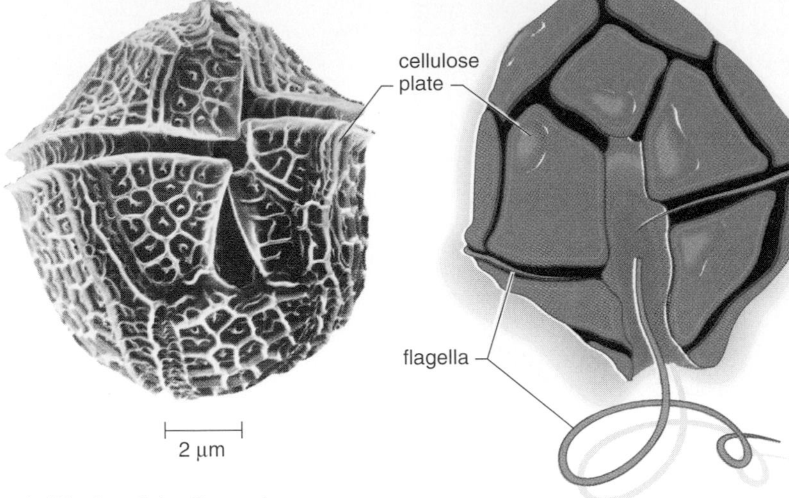

cellulose
plate

flagella

b. Dinoflagellate, *Gonyaulax*

2 µm

Diatoms and dinoflagellates
Figure 22.10

b: © Biophoto Assoc./Photo Researchers, Inc.

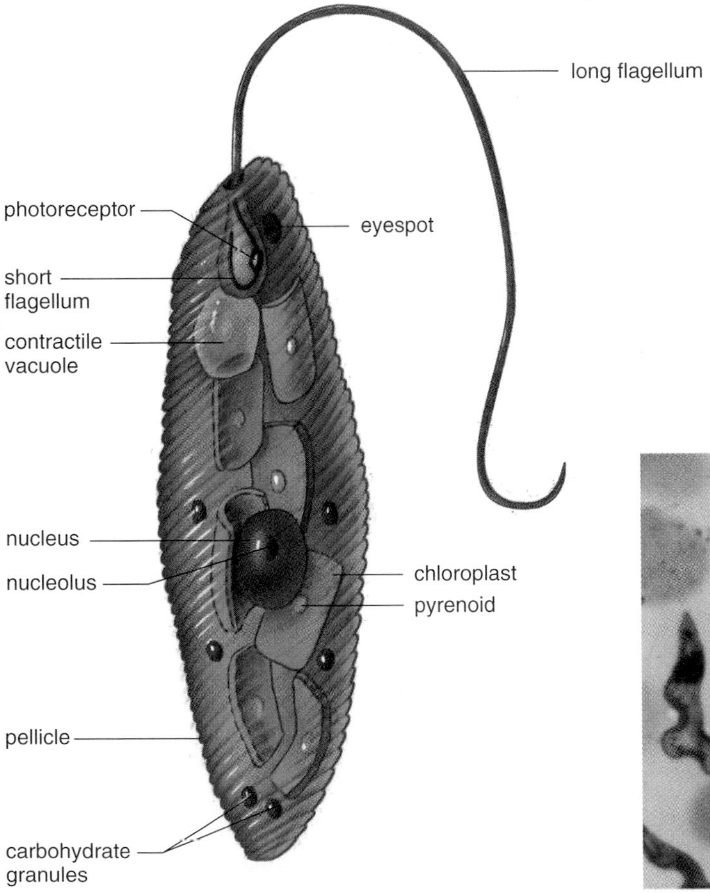

long flagellum

photoreceptor

eyespot

short
flagellum

contractile
vacuole

nucleus

nucleolus

chloroplast

pyrenoid

pellicle

carbohydrate
granules

a.

Euglena
Figure 22.12

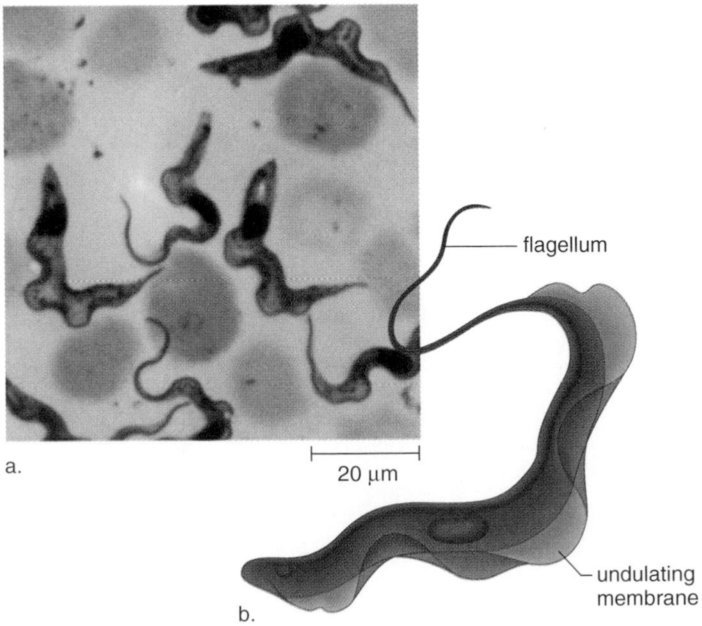

flagellum

a.

20 µm

undulating
membrane

b.

Zooflagellates
Figure 22.13

a: © Ed Reschke/Peter Arnold, Inc.

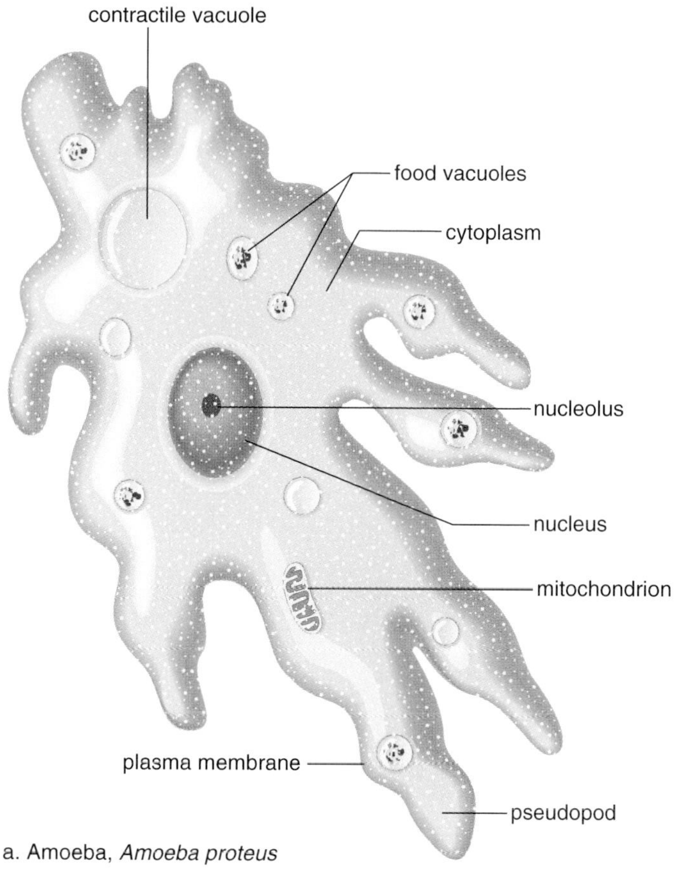

contractile vacuole

food vacuoles

cytoplasm

nucleolus

nucleus

mitochondrion

plasma membrane

pseudopod

a. Amoeba, *Amoeba proteus*

Protists with pseudopods
Figure 22.15

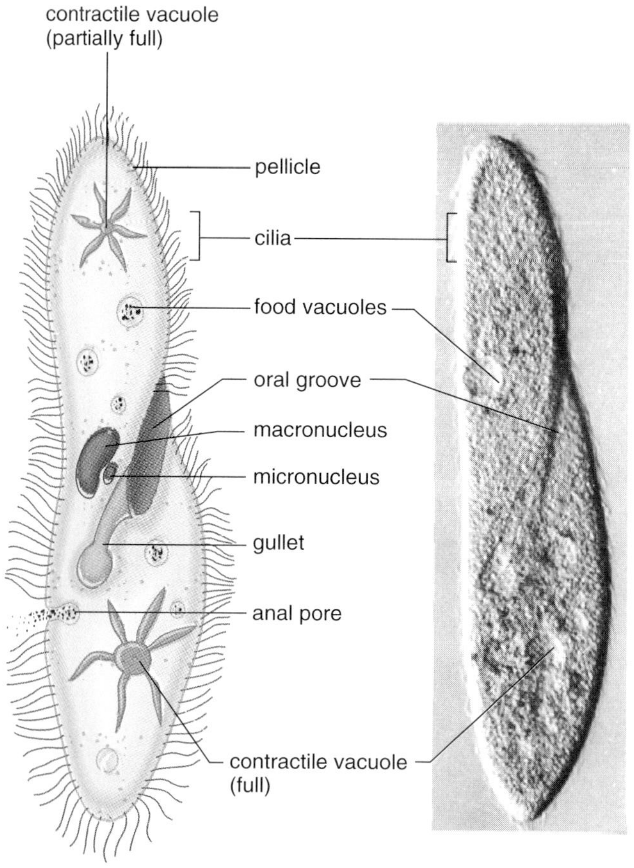

contractile vacuole
(partially full)

pellicle

cilia

food vacuoles

oral groove

macronucleus

micronucleus

gullet

anal pore

contractile vacuole
(full)

b. *Paramecium*

Ciliates
Figure 22.16

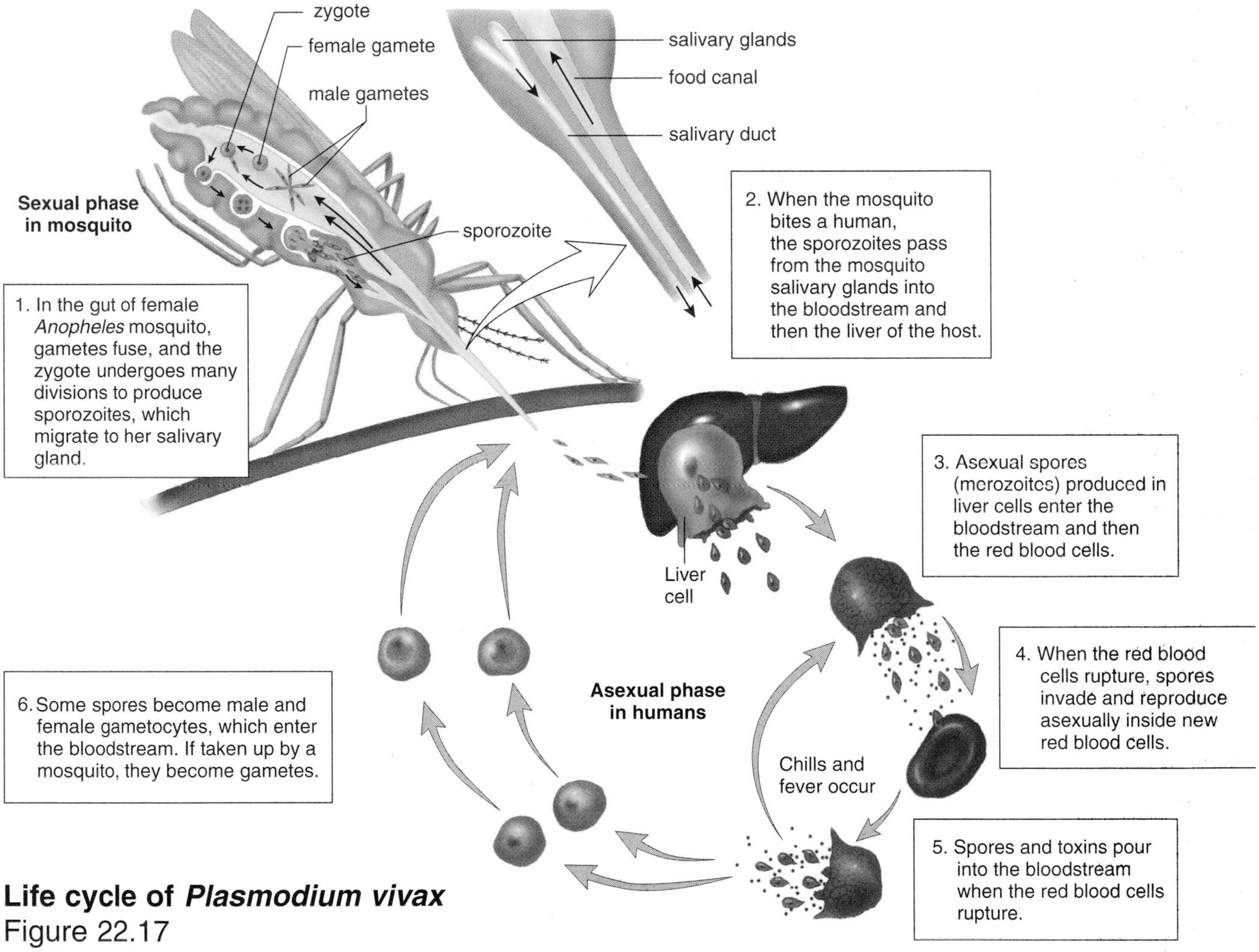

zygote
female gamete
male gametes

salivary glands
food canal
salivary duct

Sexual phase in mosquito

sporozoite

2. When the mosquito bites a human, the sporozoites pass from the mosquito salivary glands into the bloodstream and then the liver of the host.

1. In the gut of female *Anopheles* mosquito, gametes fuse, and the zygote undergoes many divisions to produce sporozoites, which migrate to her salivary gland.

3. Asexual spores (merozoites) produced in liver cells enter the bloodstream and then the red blood cells.

Liver cell

6. Some spores become male and female gametocytes, which enter the bloodstream. If taken up by a mosquito, they become gametes.

Asexual phase in humans

Chills and fever occur

4. When the red blood cells rupture, spores invade and reproduce asexually inside new red blood cells.

5. Spores and toxins pour into the bloodstream when the red blood cells rupture.

Life cycle of *Plasmodium vivax*
Figure 22.17

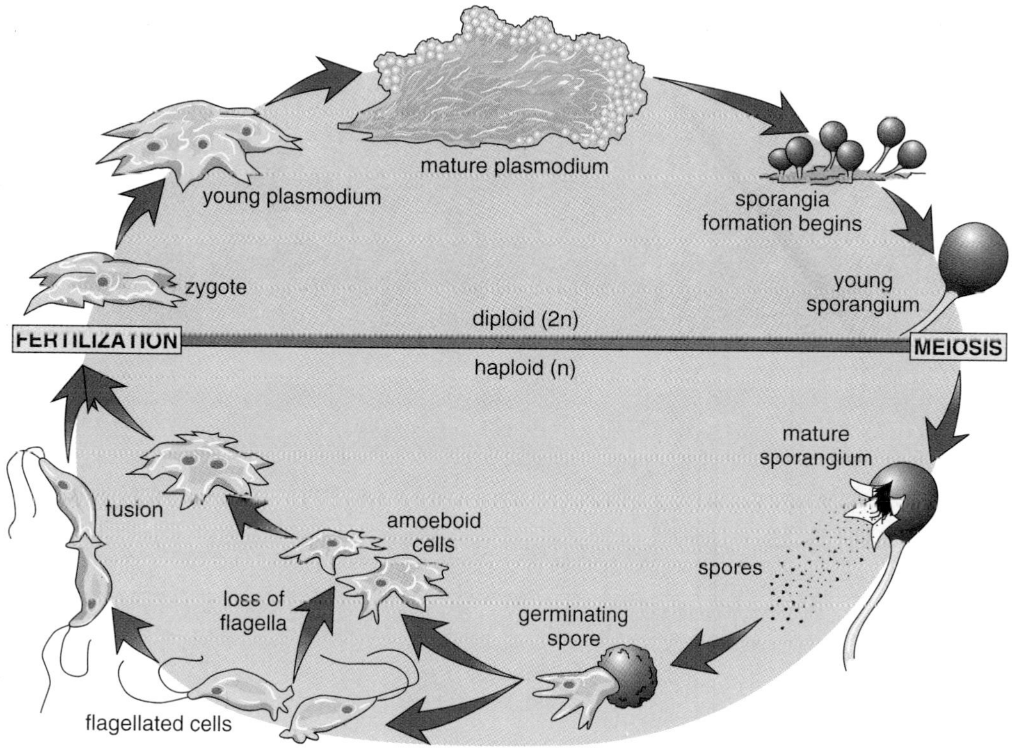
young plasmodium
mature plasmodium
sporangia
formation begins
young
sporangium
zygote
diploid (2n)
FERTILIZATION
MEIOSIS
haploid (n)
mature
sporangium
fusion
amoeboid
cells
spores
loss of
flagella
germinating
spore
flagellated cells

Plasmodial slime molds
Figure 22.18

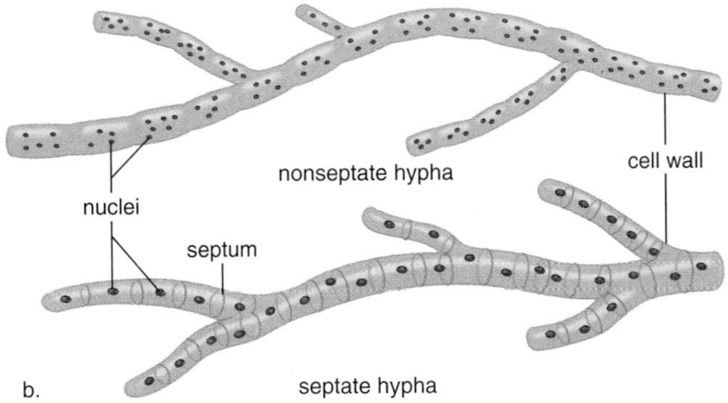

nuclei

nonseptate hypha

cell wall

septum

b.

septate hypha

Mycelium of fungi
Figure 23.1

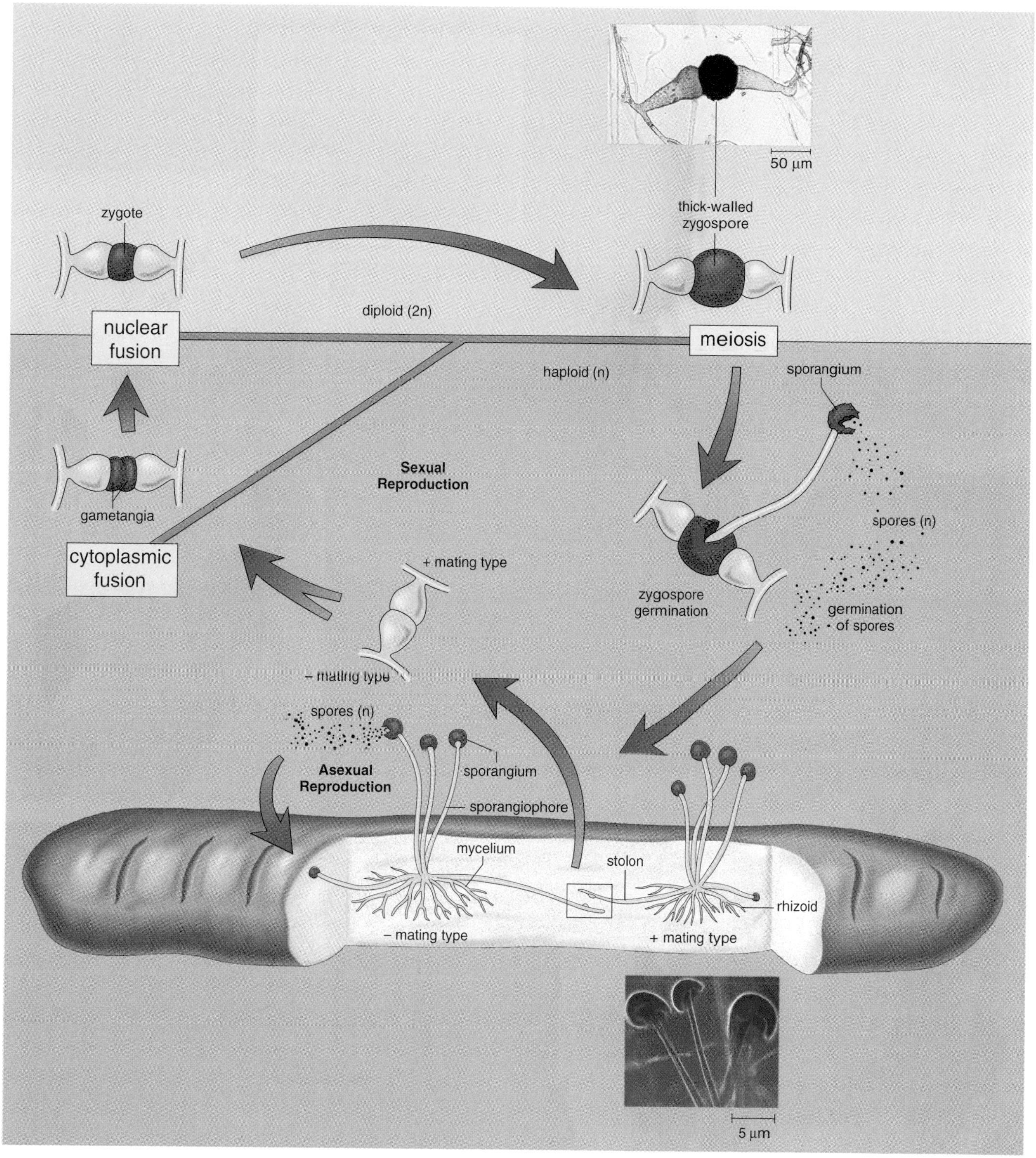

zygote

thick-walled
zygospore

50 μm

nuclear
fusion

diploid (2n)

meiosis

haploid (n)

sporangium

cytoplasmic
fusion

gametangia

Sexual
Reproduction

+ mating type

– mating type

zygospore
germination

spores (n)

germination
of spores

spores (n)

sporangium

sporangiophore

mycelium

stolon

Asexual
Reproduction

– mating type

+ mating type

rhizoid

5 μm

Black bread mold, *Rhizopus stolonifer*

Figure 23.3

top © James W. Richardson/Visuals Unlimited; Bottom: © David M. Phillips/Visuals Unlimited

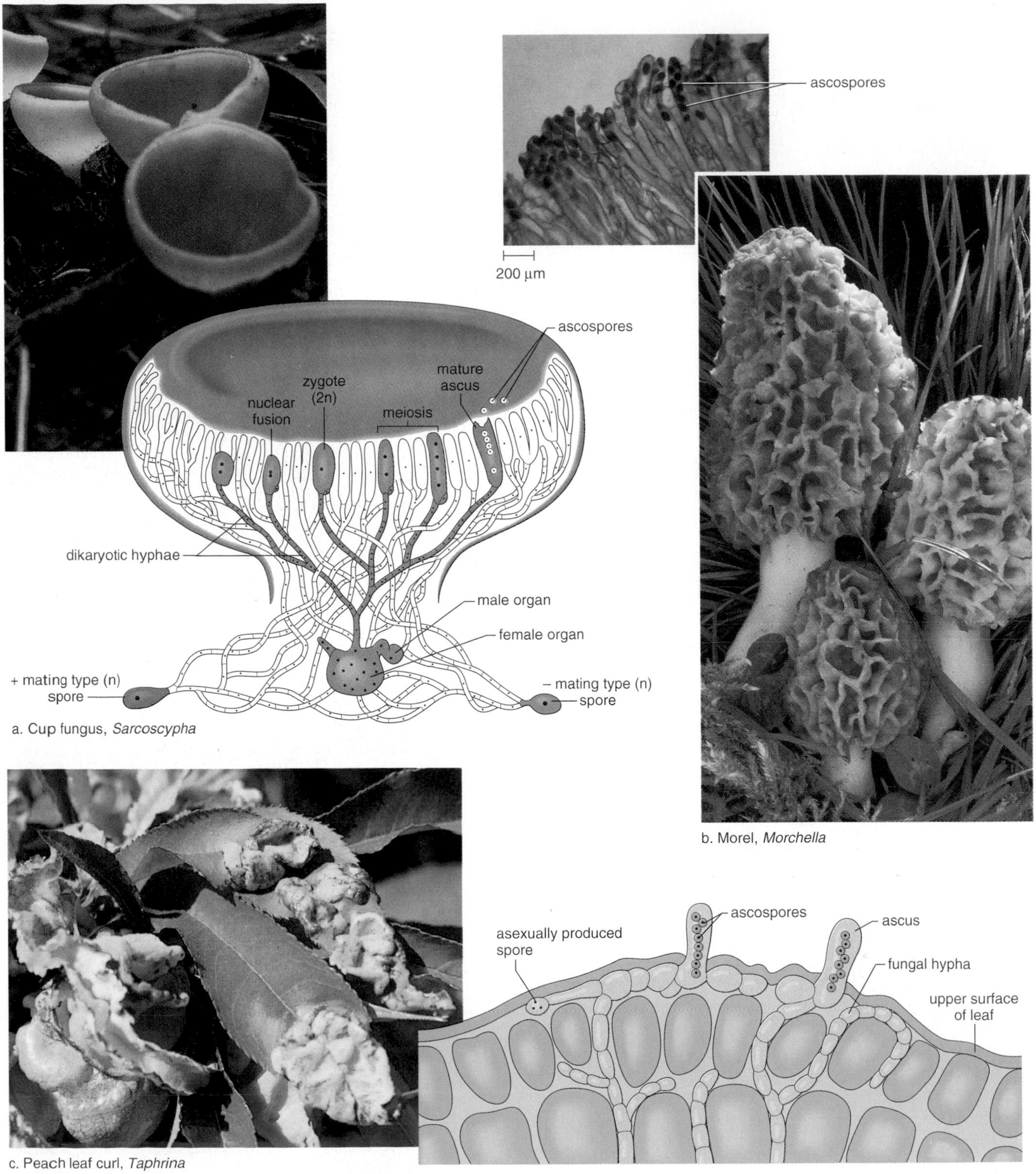

ascospores

200 µm

ascospores

mature
ascus

nuclear
fusion

zygote
(2n)

meiosis

dikaryotic hyphae

male organ

female organ

+ mating type (n)
spore

− mating type (n)
spore

a. Cup fungus, *Sarcoscypha*

b. Morel, *Morchella*

c. Peach leaf curl, *Taphrina*

ascospores

ascus

asexually produced
spore

fungal hypha

upper surface
of leaf

Sac fungi
Figure 23.4

a: © Walter H. Hodge/Peter Arnold, Inc.; b (top): © James Richardson/Visuals Unlimited; b (bottom): © Michael Viard/Peter Arnold, Inc.; c: © Kingsley Stern

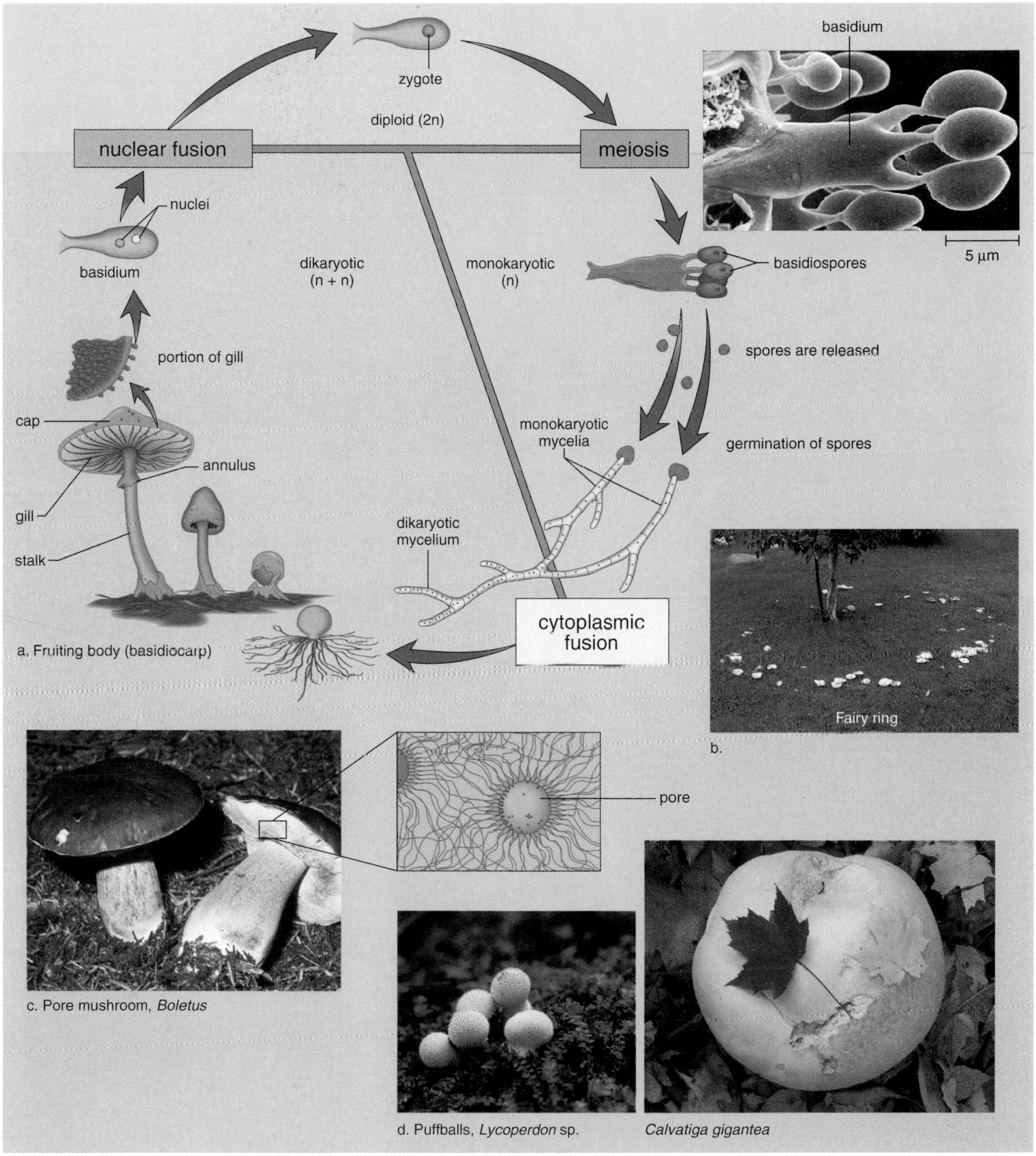

basidium

5 µm

nuclear fusion

zygote

diploid (2n)

meiosis

nuclei

basidium

dikaryotic
(n + n)

monokaryotic
(n)

basidiospores

spores are released

portion of gill

monokaryotic
mycelia

germination of spores

cap

annulus

gill

stalk

dikaryotic
mycelium

cytoplasmic
fusion

a, Fruiting body (basidiocarp)

Fairy ring

b.

pore

c. Pore mushroom, *Boletus*

d. Puffballs, *Lycoperdon* sp.

Calvatiga gigantea

Club fungi

Figure 23.6

a: © Biophoto Associates; b: © Glenn Oliver/Visuals Unlimited; c: © M. Eichelberger/Visuals Unlimited; d (left): © Dick Poe/Visuals Unlimited; d (right): © L. West/Photo Researchers, Inc.

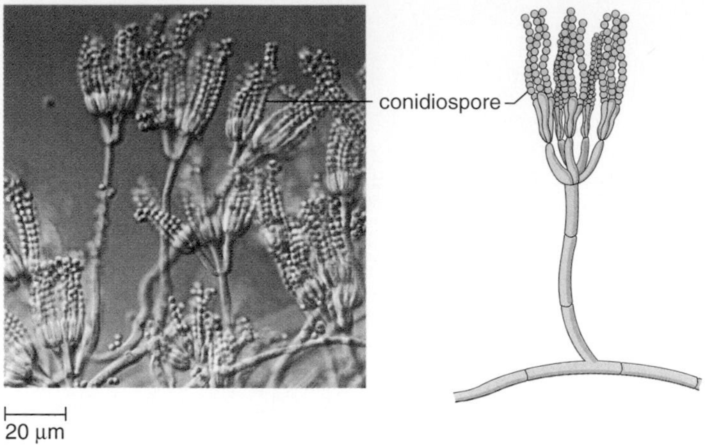

conidiospore

20 µm

b.

Penicillium
Figure 23.8

b: Courtesy G.L. Barron/University of Guelph

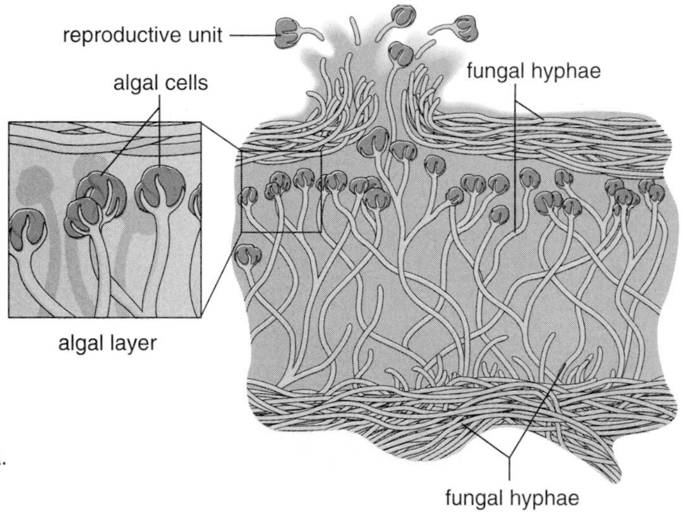

reproductive unit

algal cells

fungal hyphae

algal layer

a.

fungal hyphae

Lichen morphology
Figure 23.10

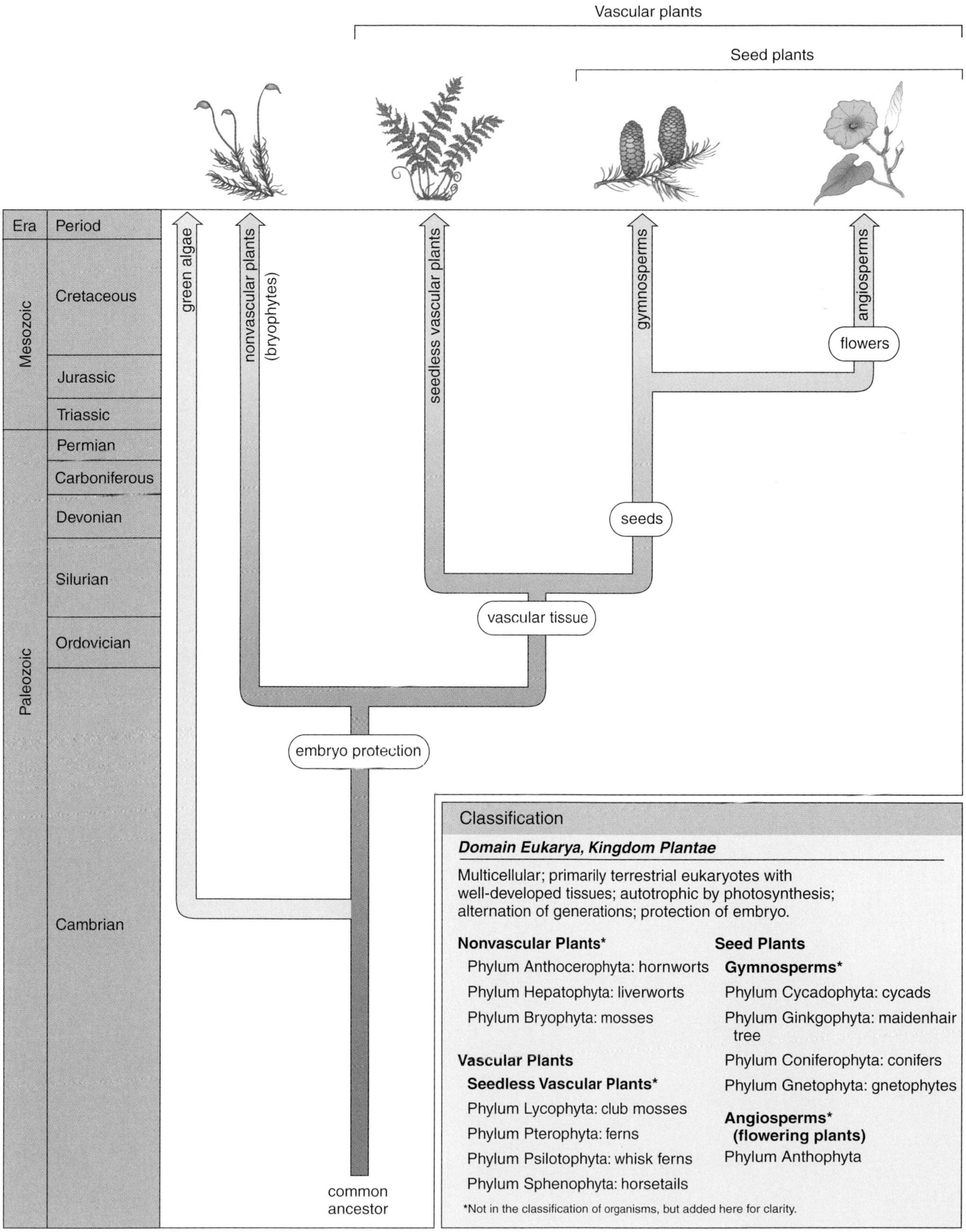

Era	Period									

Vascular plants

Seed plants

green algae

nonvascular plants (bryophytes)

seedless vascular plants

gymnosperms

angiosperms

flowers

seeds

vascular tissue

embryo protection

common ancestor

Era	Period
Mesozoic	Cretaceous
	Jurassic
	Triassic
Paleozoic	Permian
	Carboniferous
	Devonian
	Silurian
	Ordovician
	Cambrian

Classification

Domain Eukarya, Kingdom Plantae

Multicellular; primarily terrestrial eukaryotes with well-developed tissues; autotrophic by photosynthesis; alternation of generations; protection of embryo.

Nonvascular Plants*

  Phylum Anthocerophyta: hornworts

  Phylum Hepatophyta: liverworts

  Phylum Bryophyta: mosses

Vascular Plants

  Seedless Vascular Plants*

  Phylum Lycophyta: club mosses

  Phylum Pterophyta: ferns

  Phylum Psilotophyta: whisk ferns

  Phylum Sphenophyta: horsetails

Seed Plants

  Gymnosperms*

  Phylum Cycadophyta: cycads

  Phylum Ginkgophyta: maidenhair tree

  Phylum Coniferophyta: conifers

  Phylum Gnetophyta: gnetophytes

  Angiosperms*
  (flowering plants)

  Phylum Anthophyta

*Not in the classification of organisms, but added here for clarity.

Evolutionary history of plants
Figure 24.2

Alternation of generations
Figure 24.3

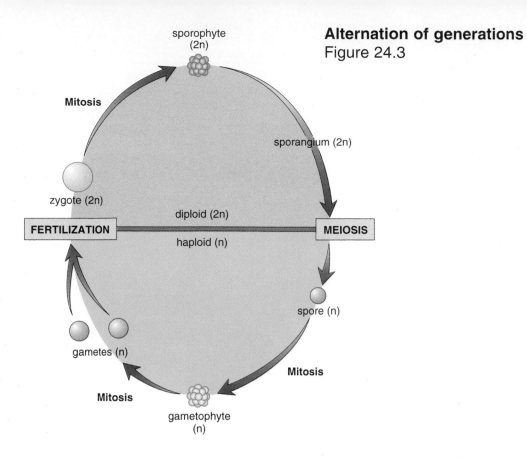

sporophyte
(2n)

Mitosis

sporangium (2n)

zygote (2n)

FERTILIZATION — diploid (2n) — **MEIOSIS**

haploid (n)

spore (n)

gametes (n)

Mitosis

Mitosis

gametophyte
(n)

Reduction in the size of the gametophyte
Figure 24.4

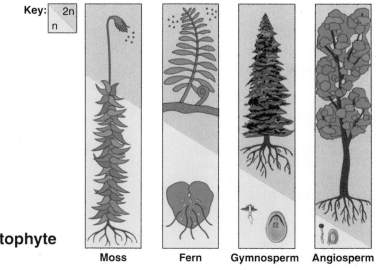

Key: 2n
       n

Moss   Fern   Gymnosperm   Angiosperm

Hornwort (*Anthoceros* sp.)
Figure 24.7

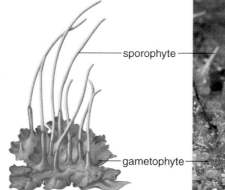

sporophyte

gametophyte

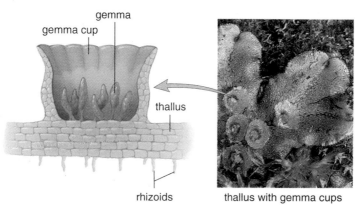

gemma

gemma cup

thallus

rhizoids

a. Gemma cup

thallus with gemma cups

Liverwort, *Marchantia*
Figure 24.8

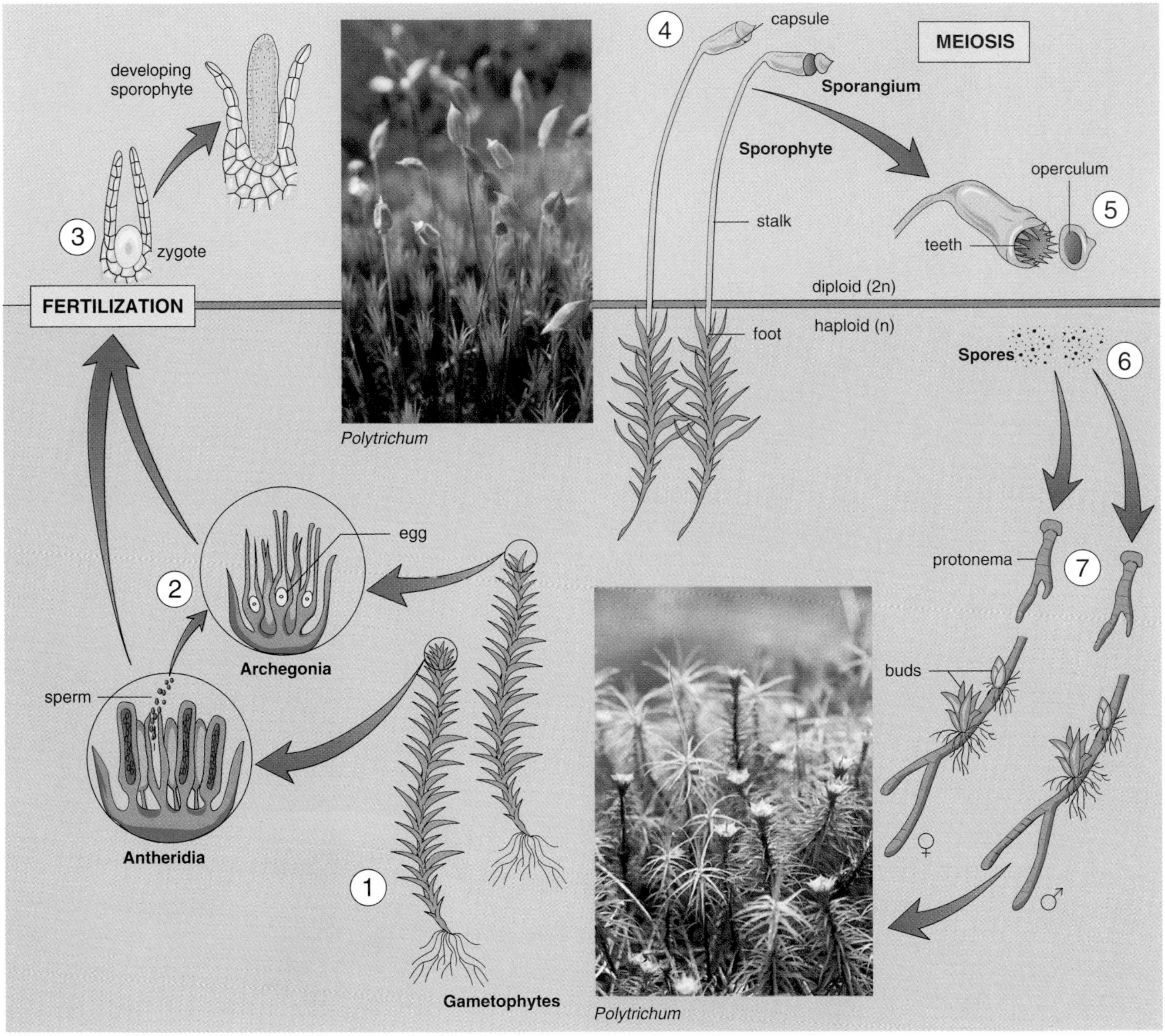

Moss life cycle

Figure 24.9

top: © Heather Angel/Biofotos; 24.9 lower: © Bruce Iverson

Cooksonian fossil
Figure 24.10

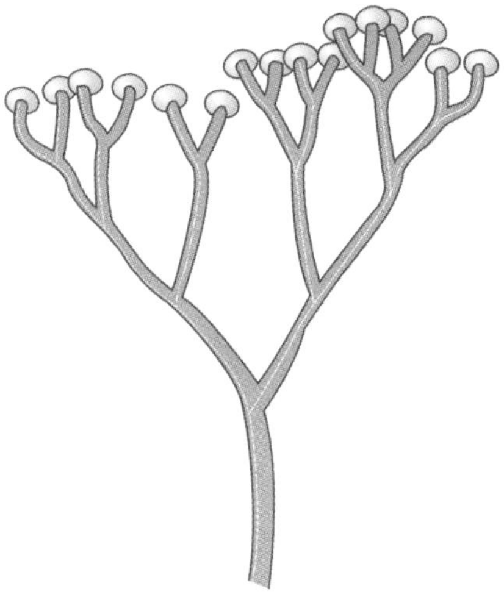

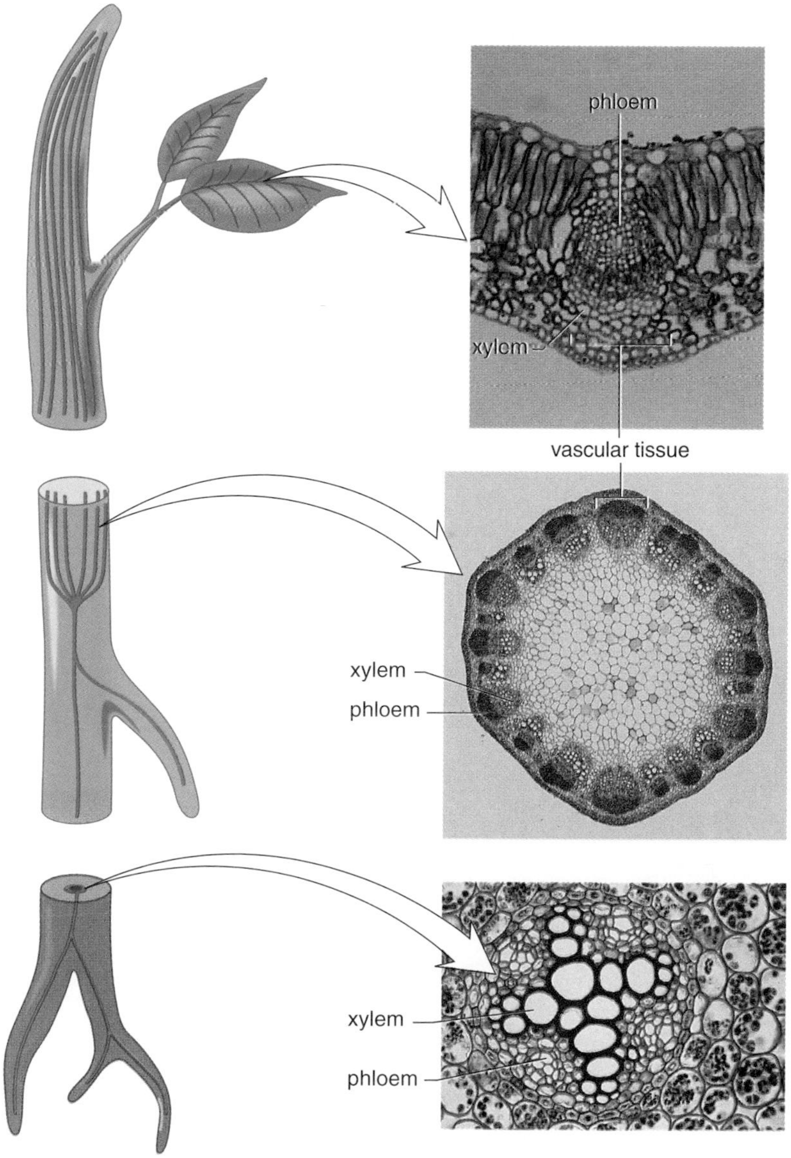

phloem

xylem

vascular tissue

xylem

phloem

xylem

phloem

Vascular tissue
Figure 24.11

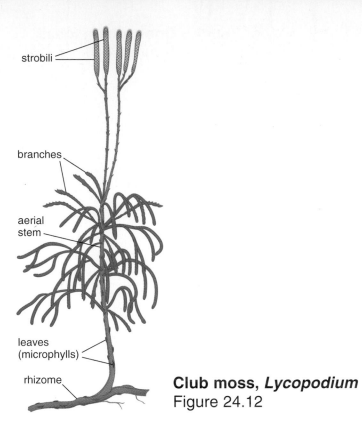

strobili

branches

aerial
stem

leaves
(microphylls)

rhizome

Club moss, *Lycopodium*
Figure 24.12

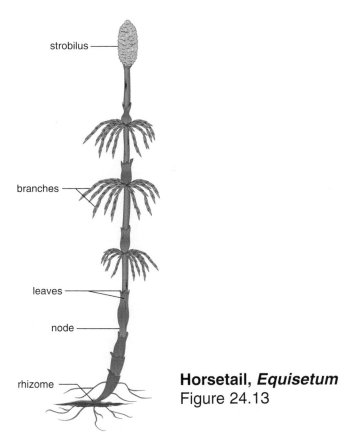

strobilus

branches

leaves

node

rhizome

Horsetail, *Equisetum*
Figure 24.13

Whisk fern, *Psilotum*
Figure 24.14

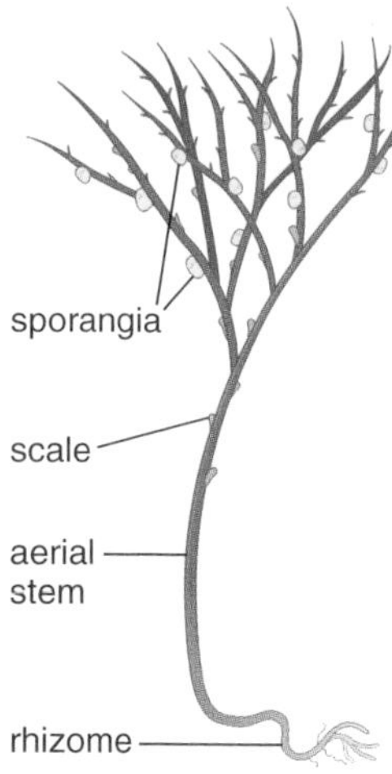

sporangia

scale

aerial
stem

rhizome

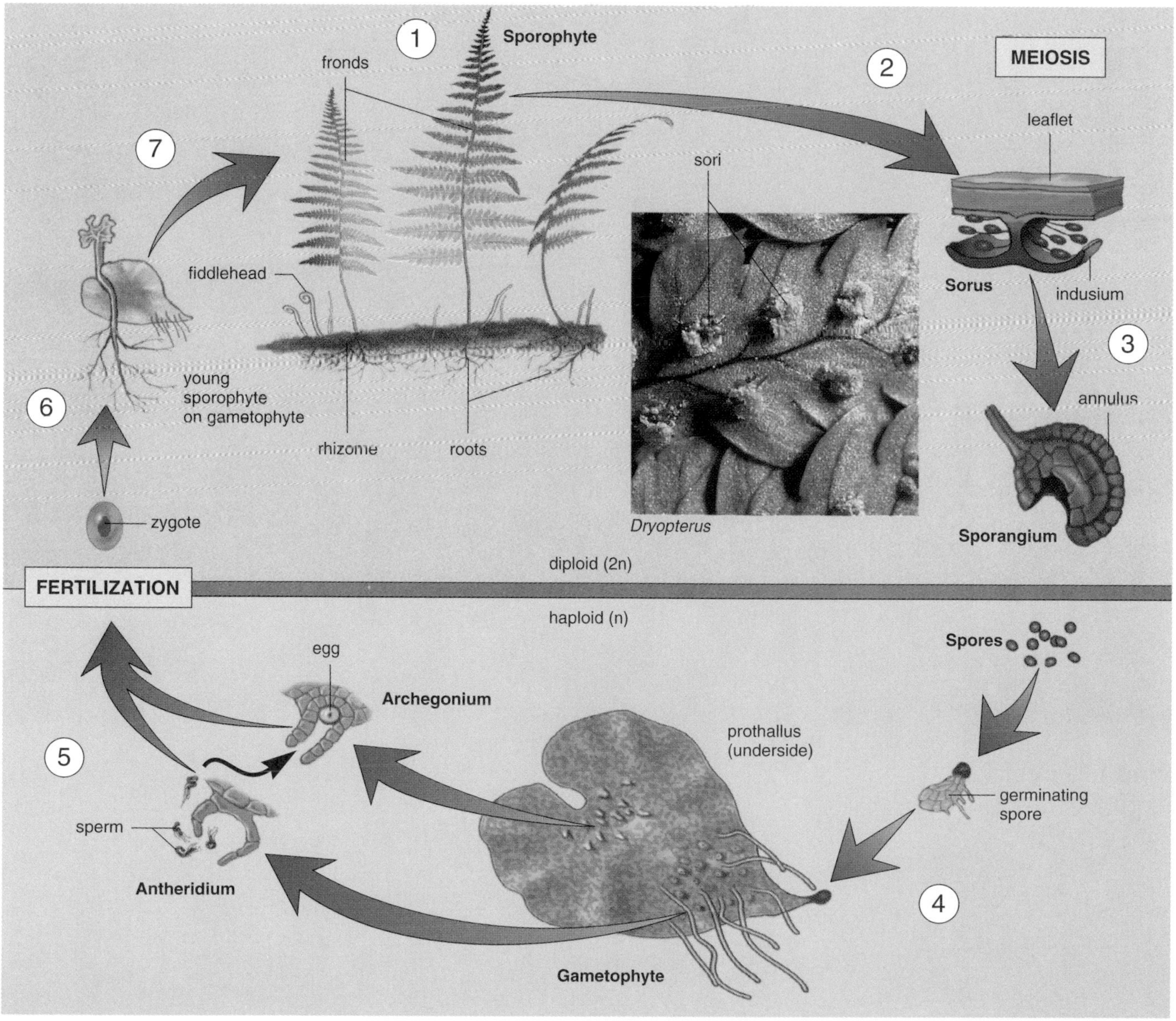

Fern life cycle
Figure 24.16
© Matt Meadows/Peter Arnold, Inc.

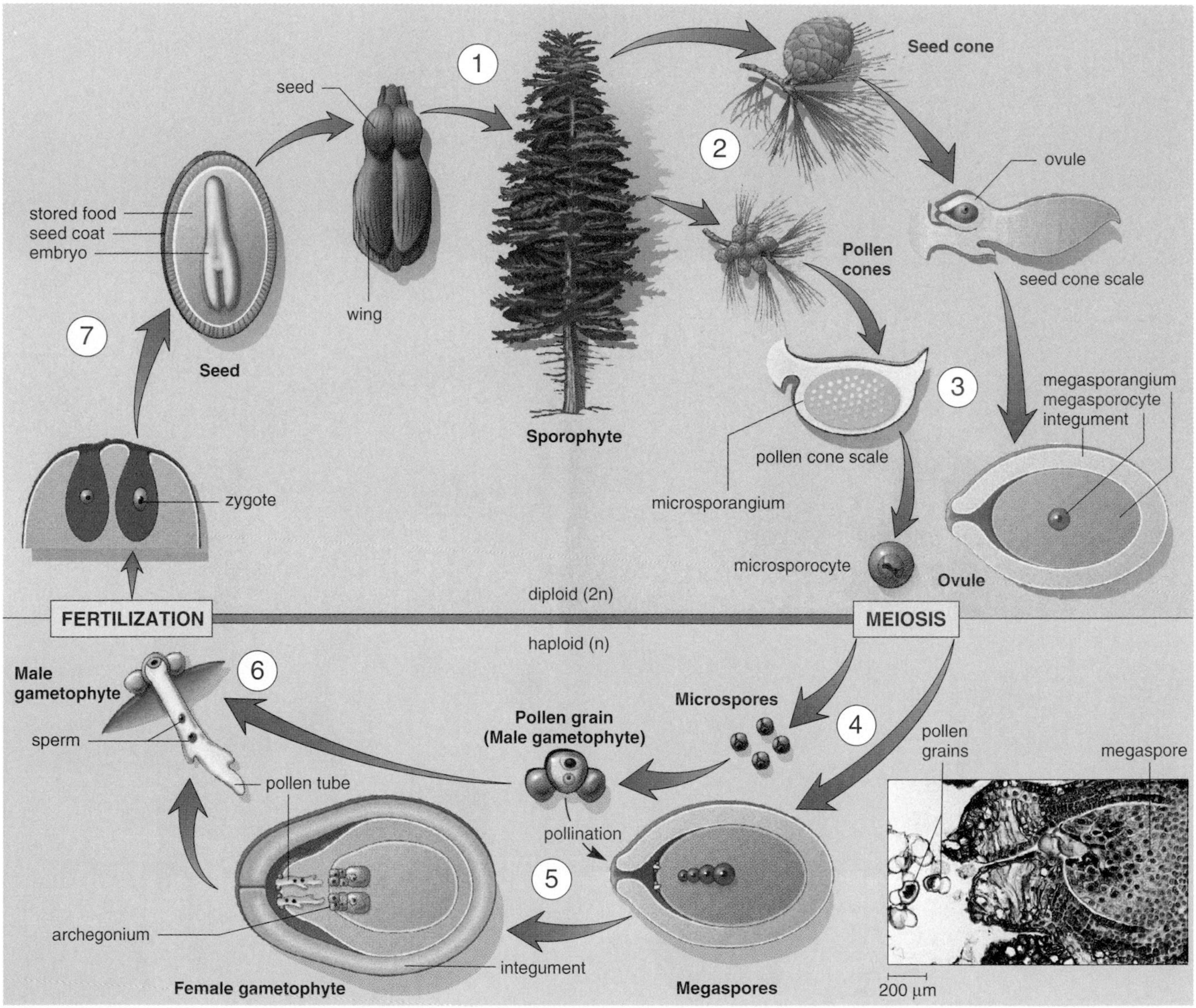

Pine life cycle
Figure 24.18
© Phototake

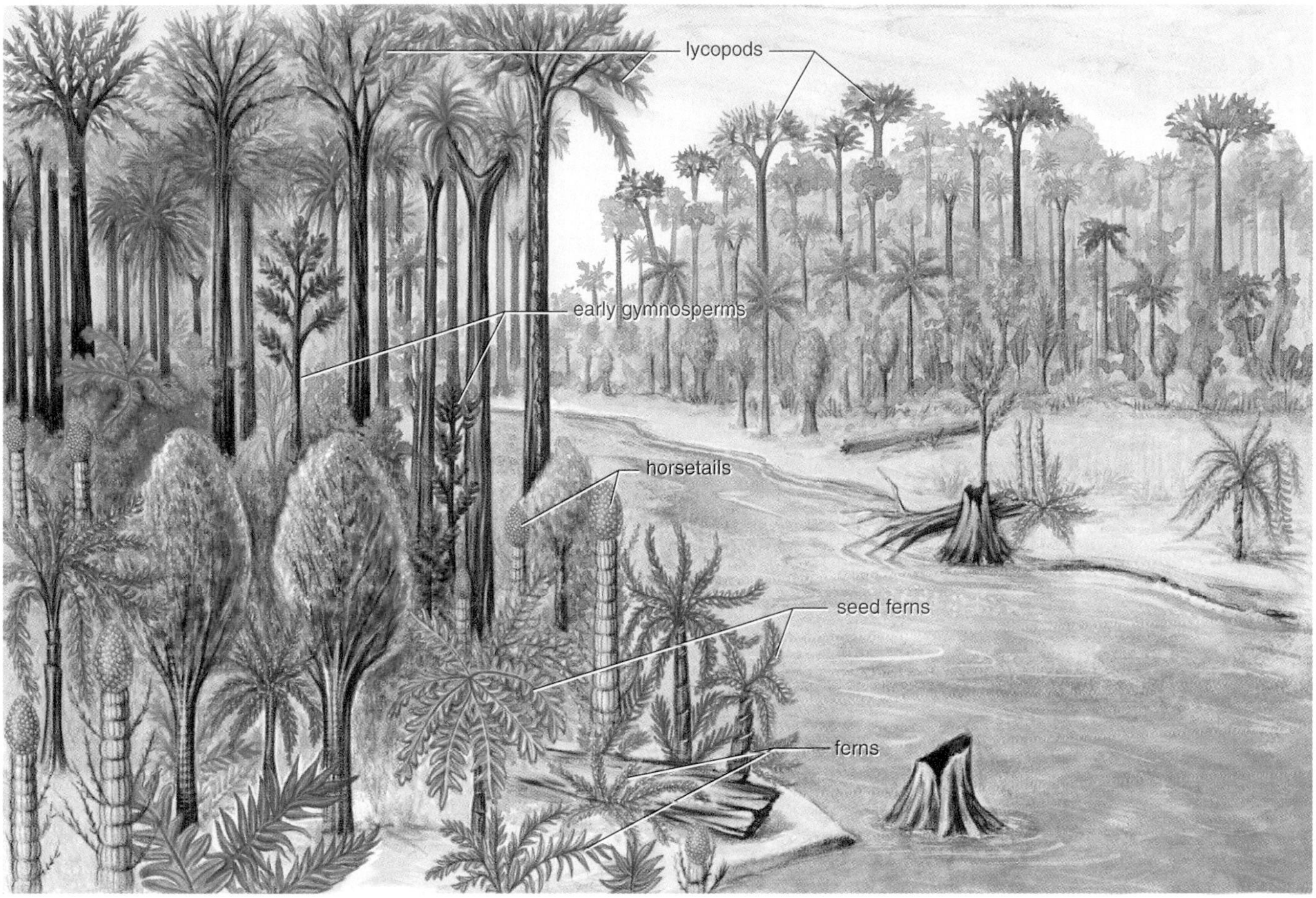

Swamp forest of the Carboniferous period
Figure 24A

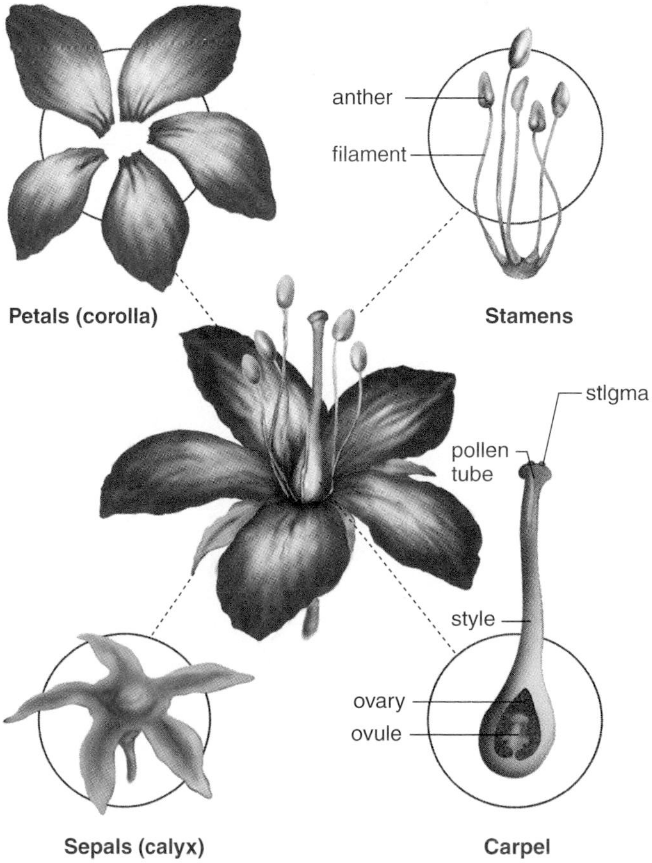

Generalized flower
Figure 24.25

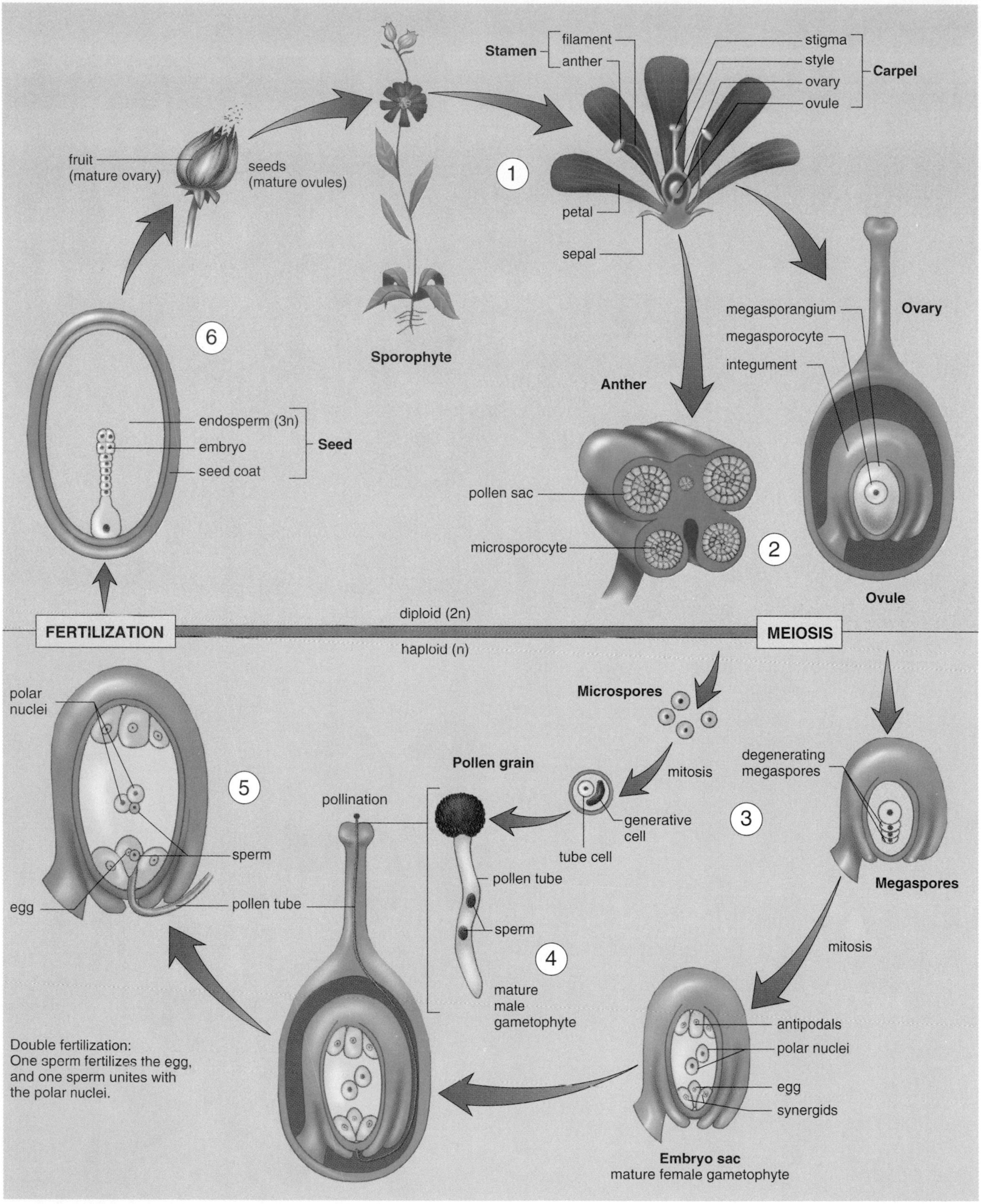

Stamen — filament / anther

Carpel — stigma / style / ovary / ovule

petal

sepal

① **Sporophyte**

fruit (mature ovary)

seeds (mature ovules)

⑥

endosperm (3n) / embryo / seed coat **Seed**

Anther

pollen sac

microsporocyte

②

Ovary

megasporangium / megasporocyte / integument

Ovule

diploid (2n)

FERTILIZATION

MEIOSIS

haploid (n)

Microspores

Pollen grain

mitosis

generative cell

tube cell

degenerating megaspores

③

Megaspores

mitosis

polar nuclei

⑤

pollination

sperm

egg

pollen tube

pollen tube

sperm

④

mature male gametophyte

Double fertilization:
One sperm fertilizes the egg, and one sperm unites with the polar nuclei.

antipodals

polar nuclei

egg

synergids

Embryo sac
mature female gametophyte

Flowering plant life cycle
Figure 24.26

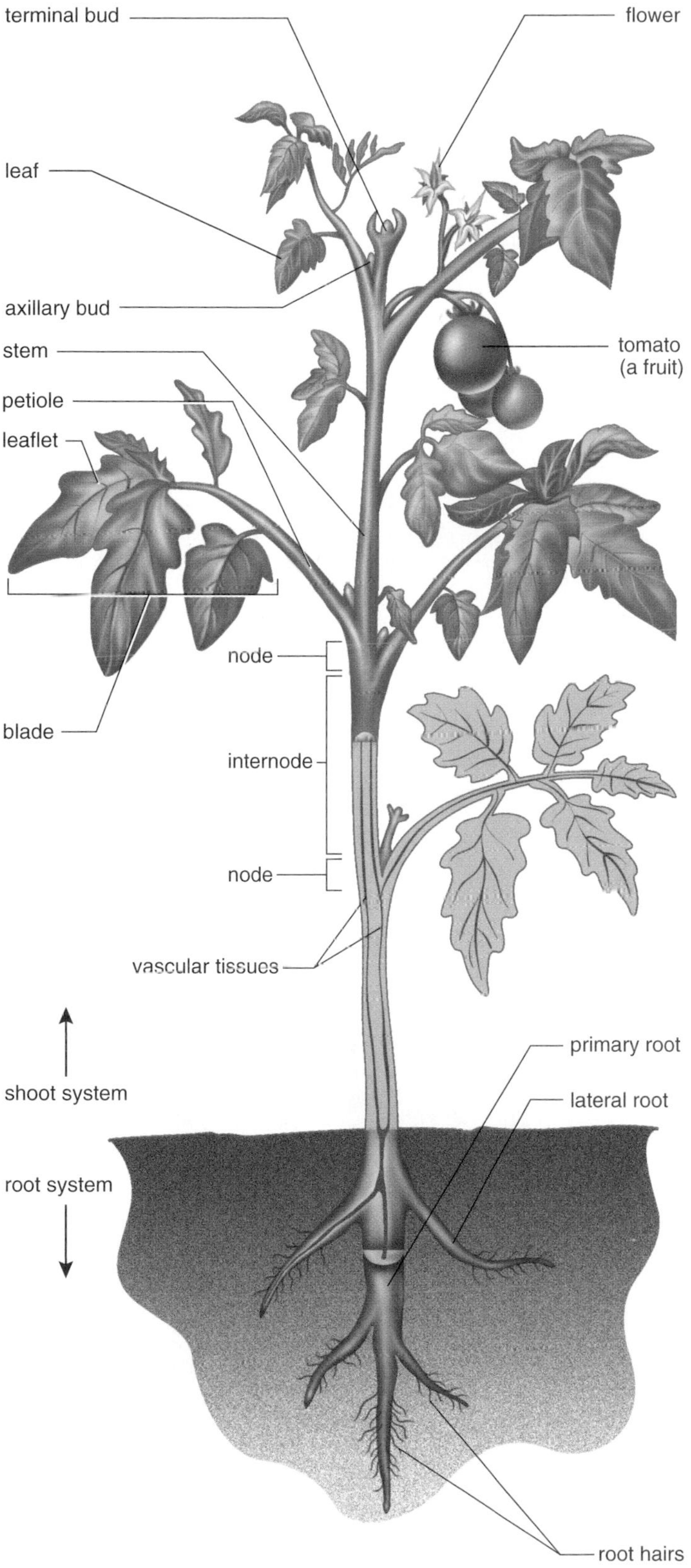

terminal bud

flower

leaf

axillary bud

stem

petiole

leaflet

blade

node

internode

node

vascular tissues

tomato
(a fruit)

primary root

lateral root

root hairs

shoot system

root system

Organization of plant body
Figure 25.1

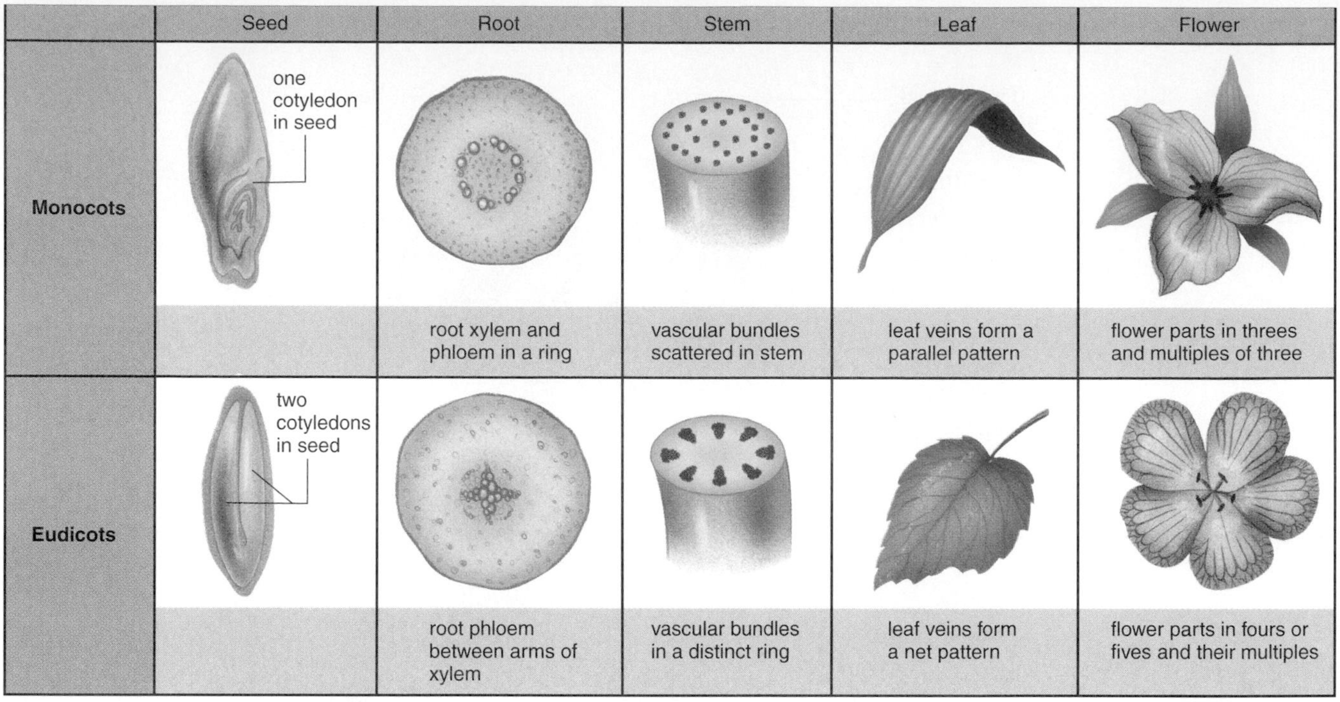

	Seed	Root	Stem	Leaf	Flower
Monocots	one cotyledon in seed	root xylem and phloem in a ring	vascular bundles scattered in stem	leaf veins form a parallel pattern	flower parts in threes and multiples of three
Eudicots	two cotyledons in seed	root phloem between arms of xylem	vascular bundles in a distinct ring	leaf veins form a net pattern	flower parts in fours or fives and their multiples

Flowering plants are either monocots or eudicots
Figure 25.3

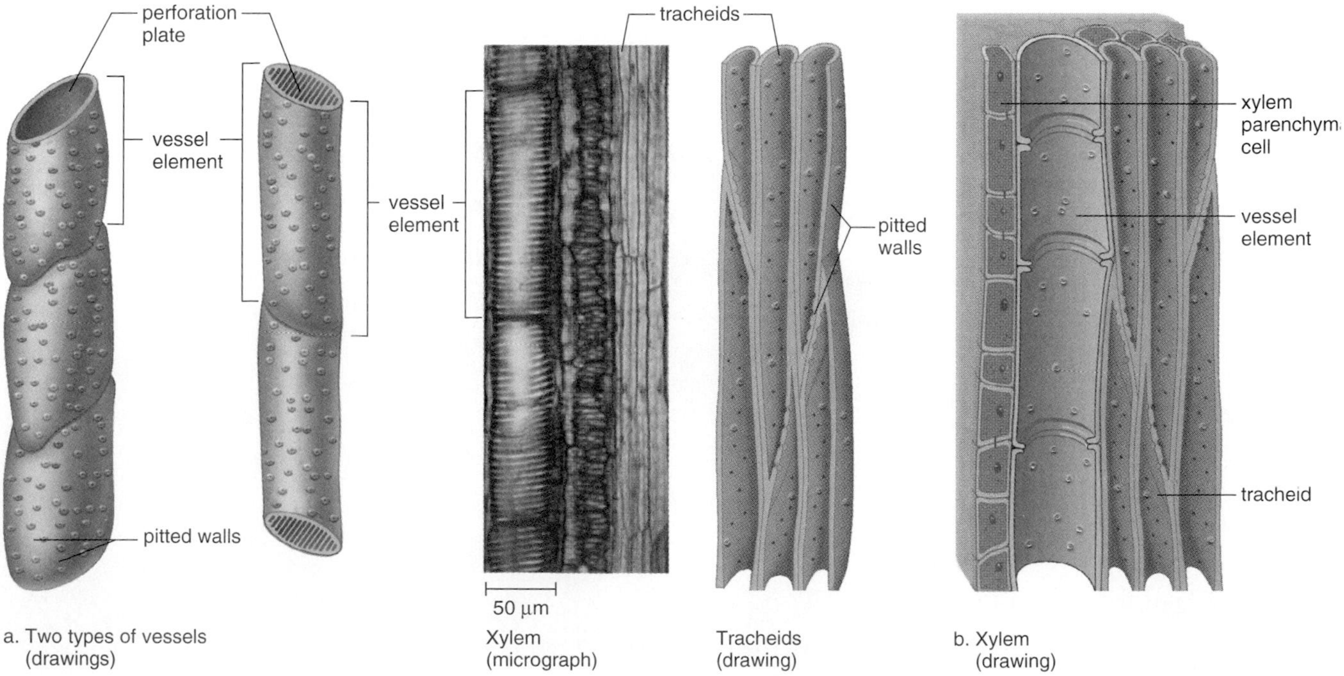

a. Two types of vessels (drawings)

Xylem (micrograph)   50 μm

Tracheids (drawing)

b. Xylem (drawing)

Xylem structure
Figure 25.6
middle: © J. Robert Waaland/Biological Photo Service

172

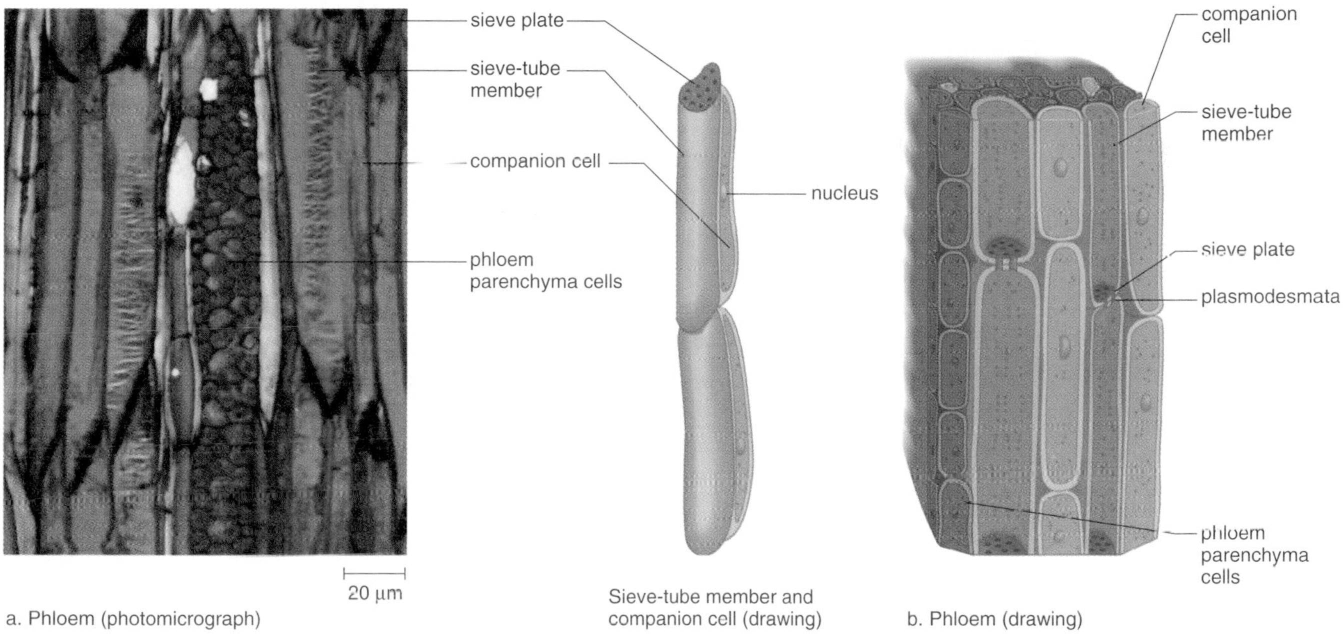

sieve plate

sieve-tube member

companion cell

phloem parenchyma cells

a. Phloem (photomicrograph)

20 µm

nucleus

Sieve-tube member and companion cell (drawing)

companion cell

sieve-tube member

sieve plate

plasmodesmata

phloem parenchyma cells

b. Phloem (drawing)

Pholem structure
Figure 25.7

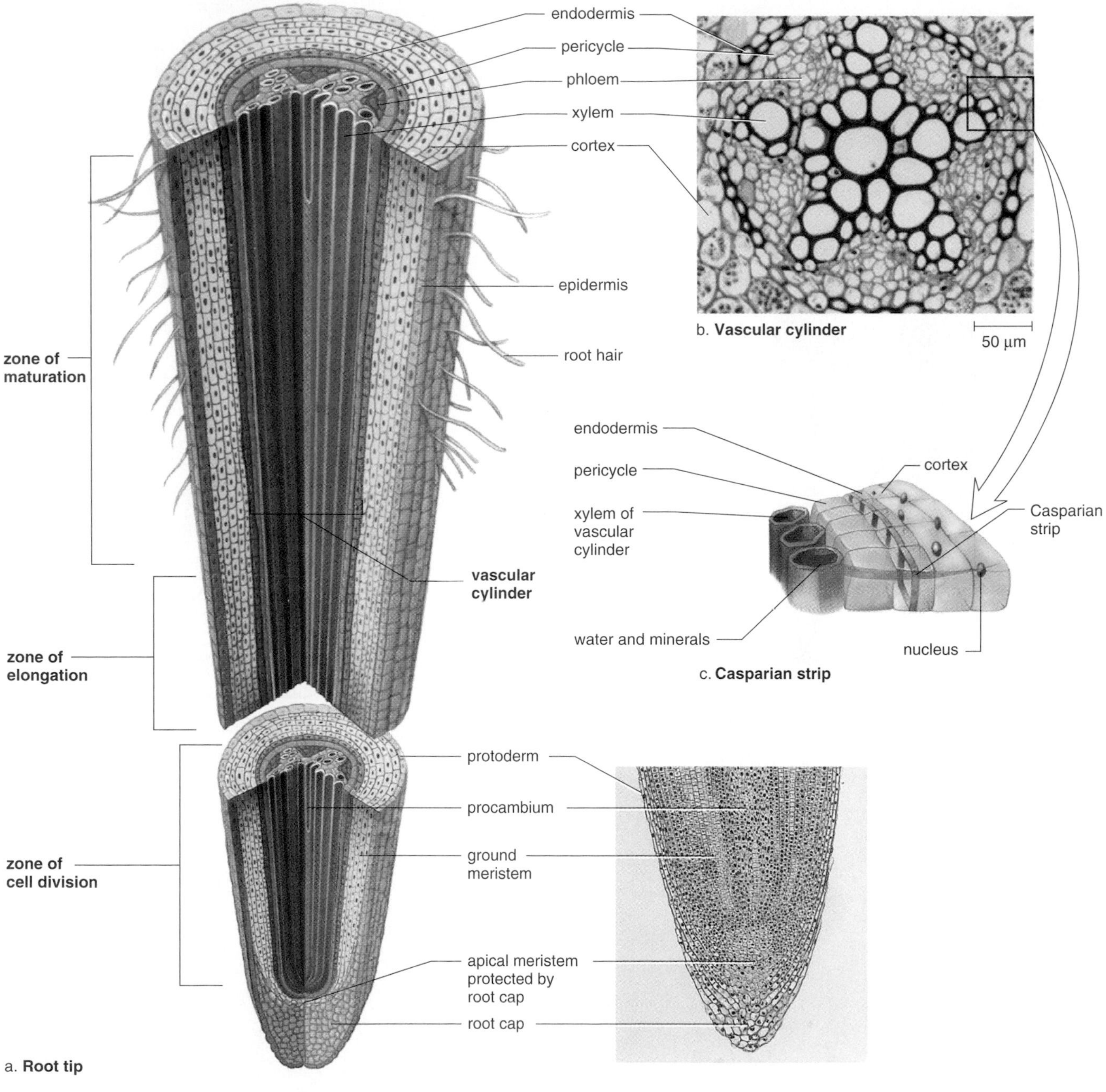

endodermis
pericycle
phloem
xylem
cortex

epidermis

root hair

zone of
maturation

vascular
cylinder

zone of
elongation

zone of
cell division

protoderm

procambium

ground
meristem

apical meristem
protected by
root cap
root cap

a. Root tip

b. **Vascular cylinder**

50 μm

endodermis

pericycle

cortex

xylem of
vascular
cylinder

Casparian
strip

water and minerals

nucleus

c. **Casparian strip**

Eudicot root tip
Figure 25.8
a: Courtesy Ray F. Evert/University of Wisconsin Madison; b: © CABISCO/Phototake

a. Shoot tip

100 µm

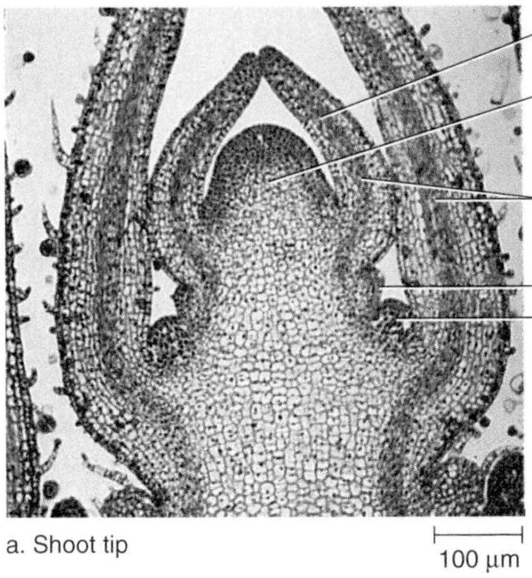

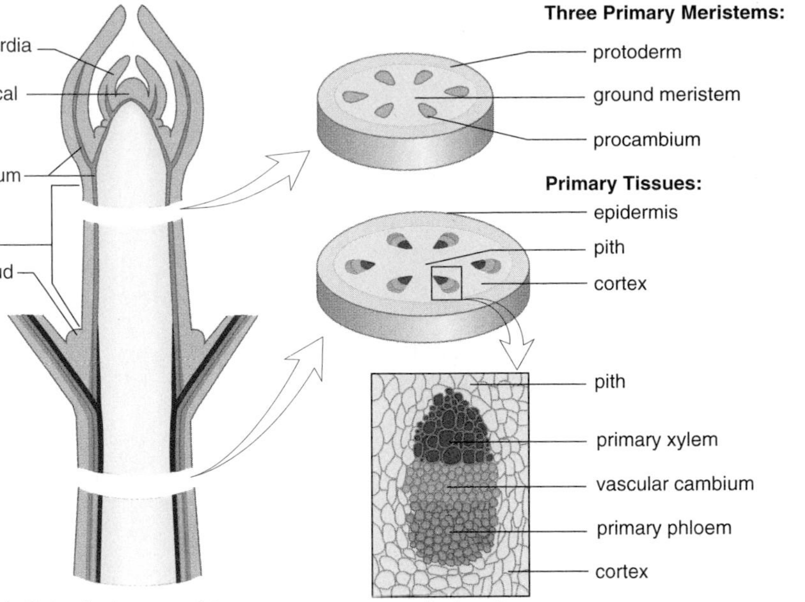

Three Primary Meristems:
- protoderm
- ground meristem
- procambium

leaf primordia

shoot apical meristem

procambium

internode

axillary bud

Primary Tissues:
- epidermis
- pith
- cortex

b. Fate of primary meristems

pith
primary xylem
vascular cambium
primary phloem
cortex

vascular bundle

Shoot tip and primary meristems
Figure 25.12
a: © J. Robert Waaland/Biological Photo Service

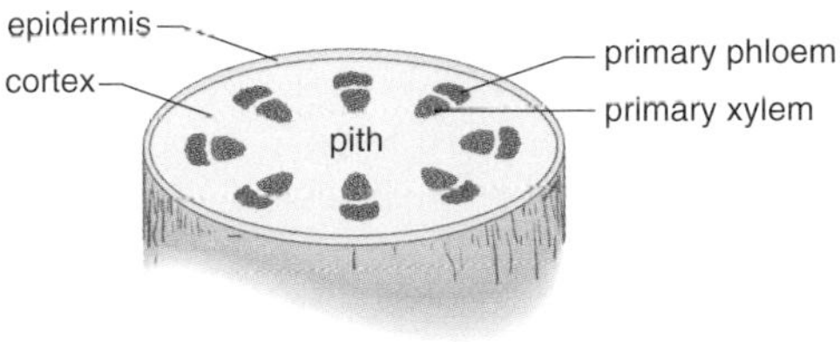

epidermis
cortex
pith
primary phloem
primary xylem

a. Eudicot stem, no secondary growth

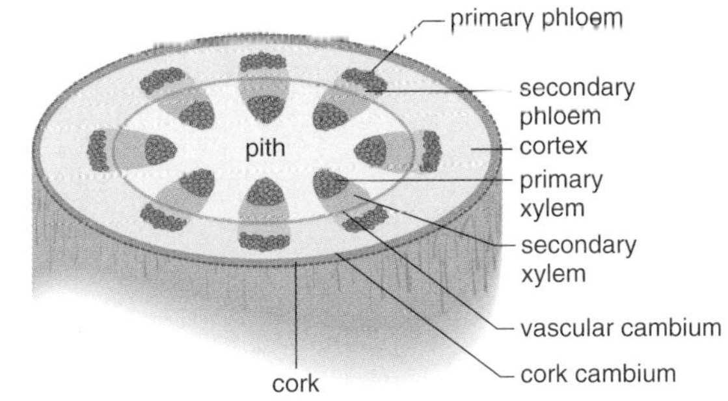

primary phloem
secondary phloem
cortex
primary xylem
secondary xylem
vascular cambium
cork cambium

pith

cork

b. Eudicot stem, some secondary growth

Diagrams of secondary growth of stems
Figure 25.15

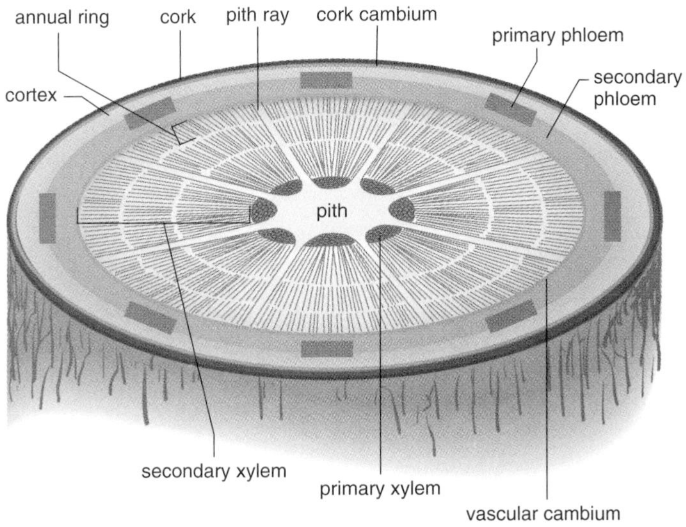

annual ring
cork
pith ray
cork cambium
primary phloem
secondary phloem
cortex
pith
secondary xylem
primary xylem
vascular cambium

c. Eudicot stem, secondary growth well underway

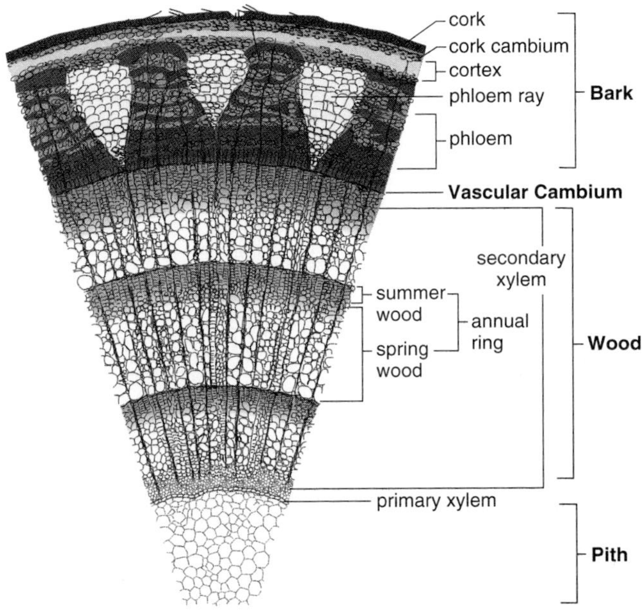

cork
cork cambium
cortex
phloem ray
phloem
} **Bark**

Vascular Cambium

secondary xylem
summer wood
spring wood
} annual ring
} **Wood**

primary xylem

} **Pith**

d. Section of woody stem

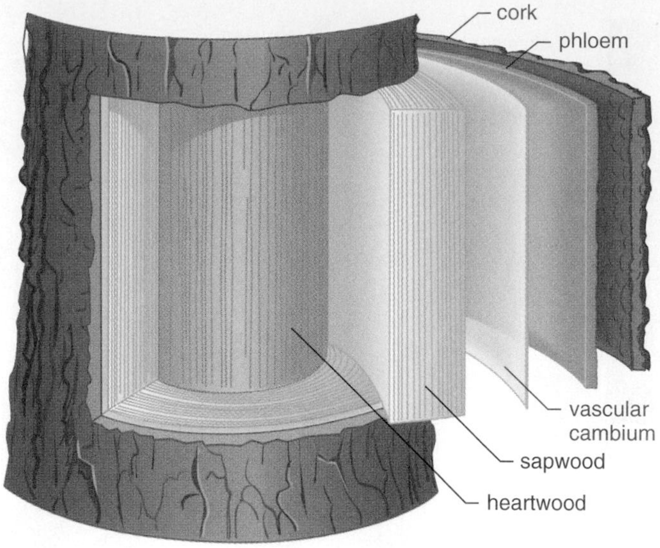

cork
phloem
vascular cambium
sapwood
heartwood

Tree trunk
Figure 25.16

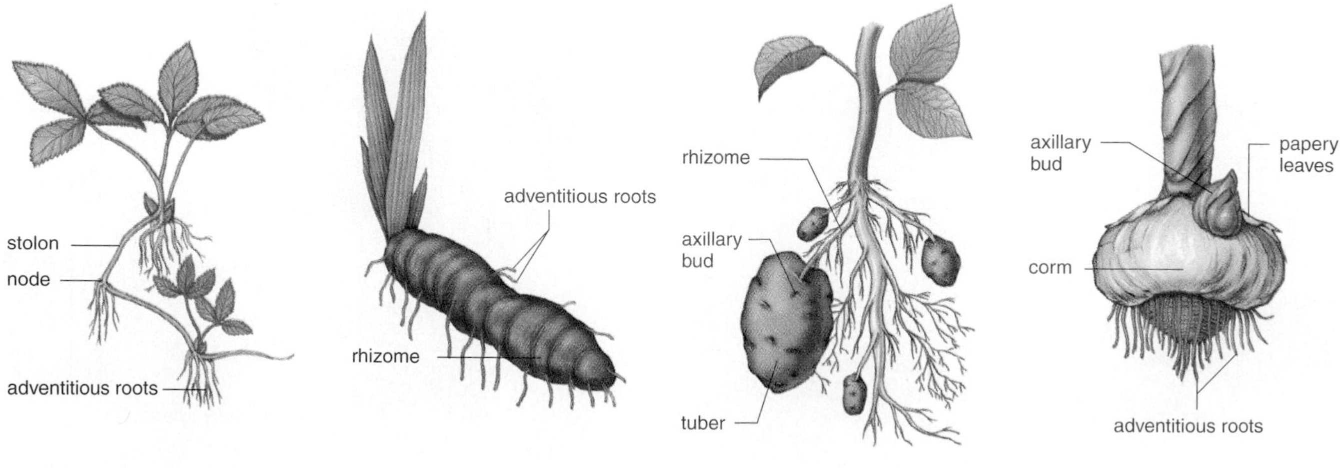

stolon
node
adventitious roots

adventitious roots
rhizome

rhizome
axillary bud
tuber

axillary bud
papery leaves
corm
adventitious roots

Stem diversity
Figure 25.17

Leaf structure
Figure 25.18

right: © Jeremy Burgess/SPL/Photo Researchers, Inc.

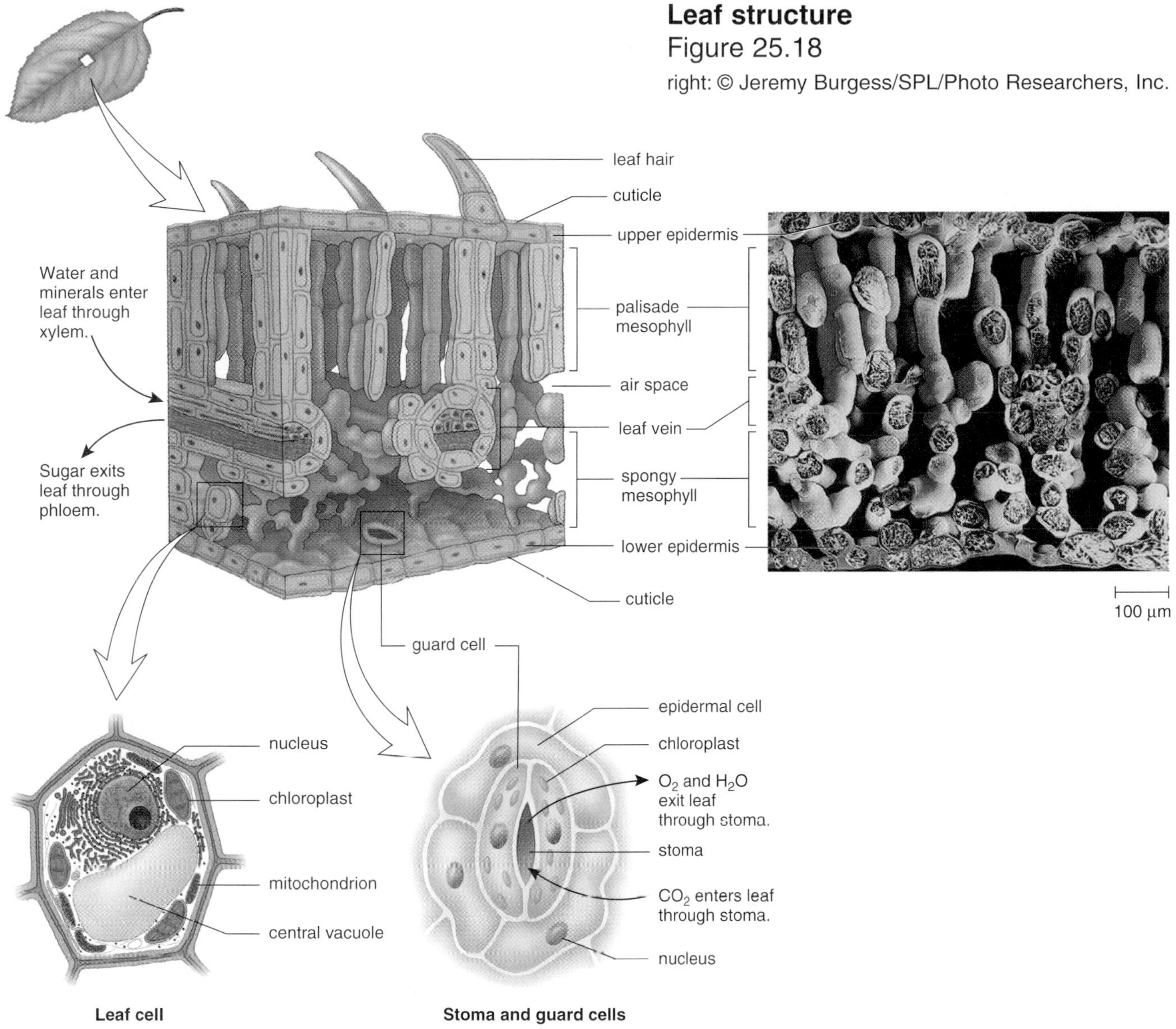

Water and minerals enter leaf through xylem.

Sugar exits leaf through phloem.

leaf hair
cuticle
upper epidermis
palisade mesophyll
air space
leaf vein
spongy mesophyll
lower epidermis
cuticle

100 µm

guard cell

nucleus
chloroplast
mitochondrion
central vacuole

Leaf cell

epidermal cell
chloroplast
O_2 and H_2O exit leaf through stoma.
stoma
CO_2 enters leaf through stoma.
nucleus

Stoma and guard cells

Classification of leaves
Figure 25.19

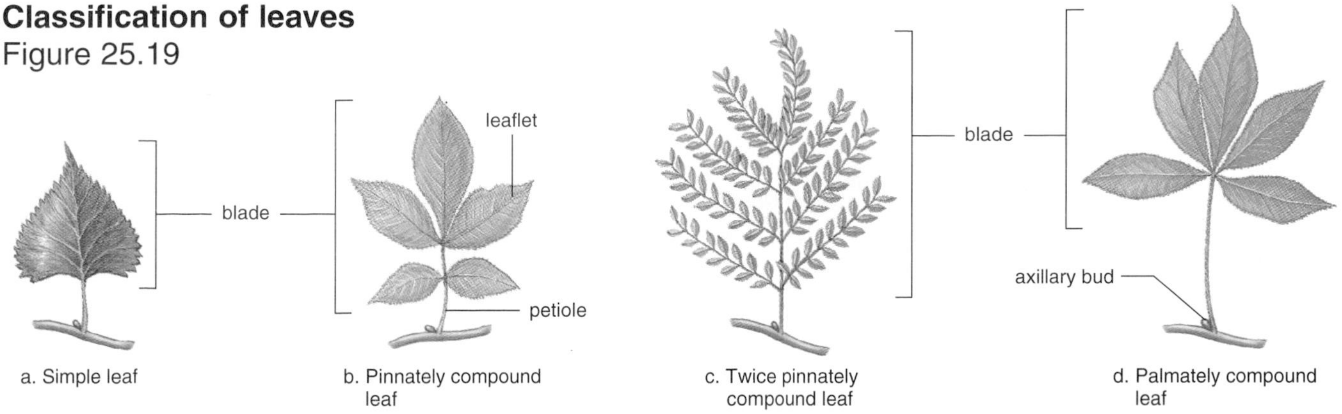

blade

a. Simple leaf

leaflet
blade
petiole

b. Pinnately compound leaf

blade

c. Twice pinnately compound leaf

blade
axillary bud

d. Palmately compound leaf

Absorbing minerals
Figure 26.3

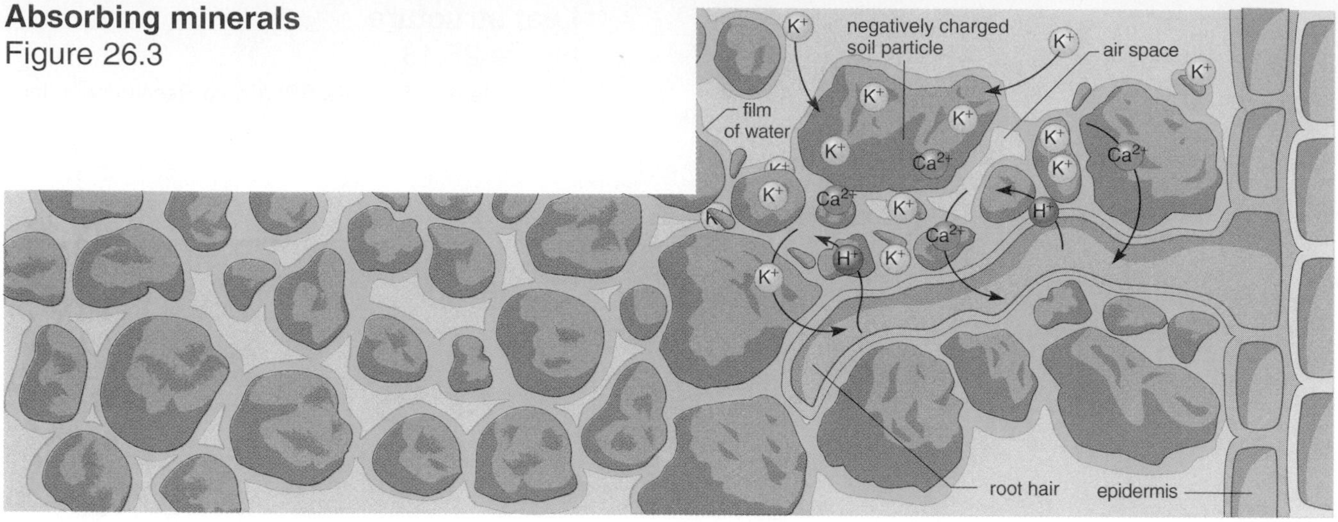

K⁺

K⁺

negatively charged
soil particle

K⁺

air space

K⁺

film
of water

K⁺

K⁺

K⁺

K⁺

Ca²⁺

K⁺

K⁺

Ca²⁺

K⁺

K⁺

Ca²⁺

H⁺

K⁺

H⁺

K⁺

Ca²⁺

H⁺

K⁺

root hair   epidermis

Simplified soil profile
Figure 26.4

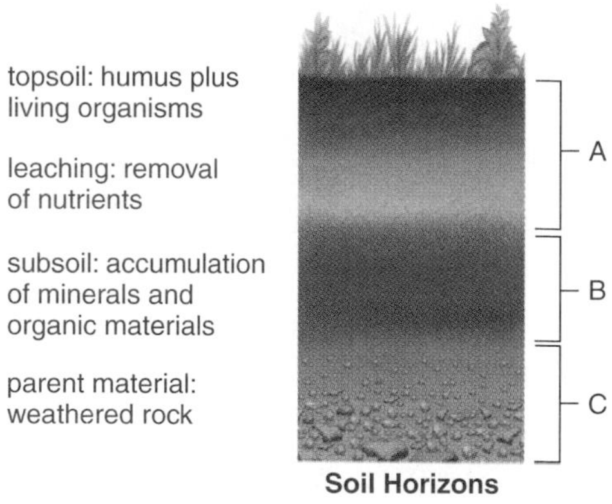

topsoil: humus plus
living organisms

leaching: removal
of nutrients

subsoil: accumulation
of minerals and
organic materials

parent material:
weathered rock

A

B

C

Soil Horizons

Water and mineral uptake
Figure 26.5

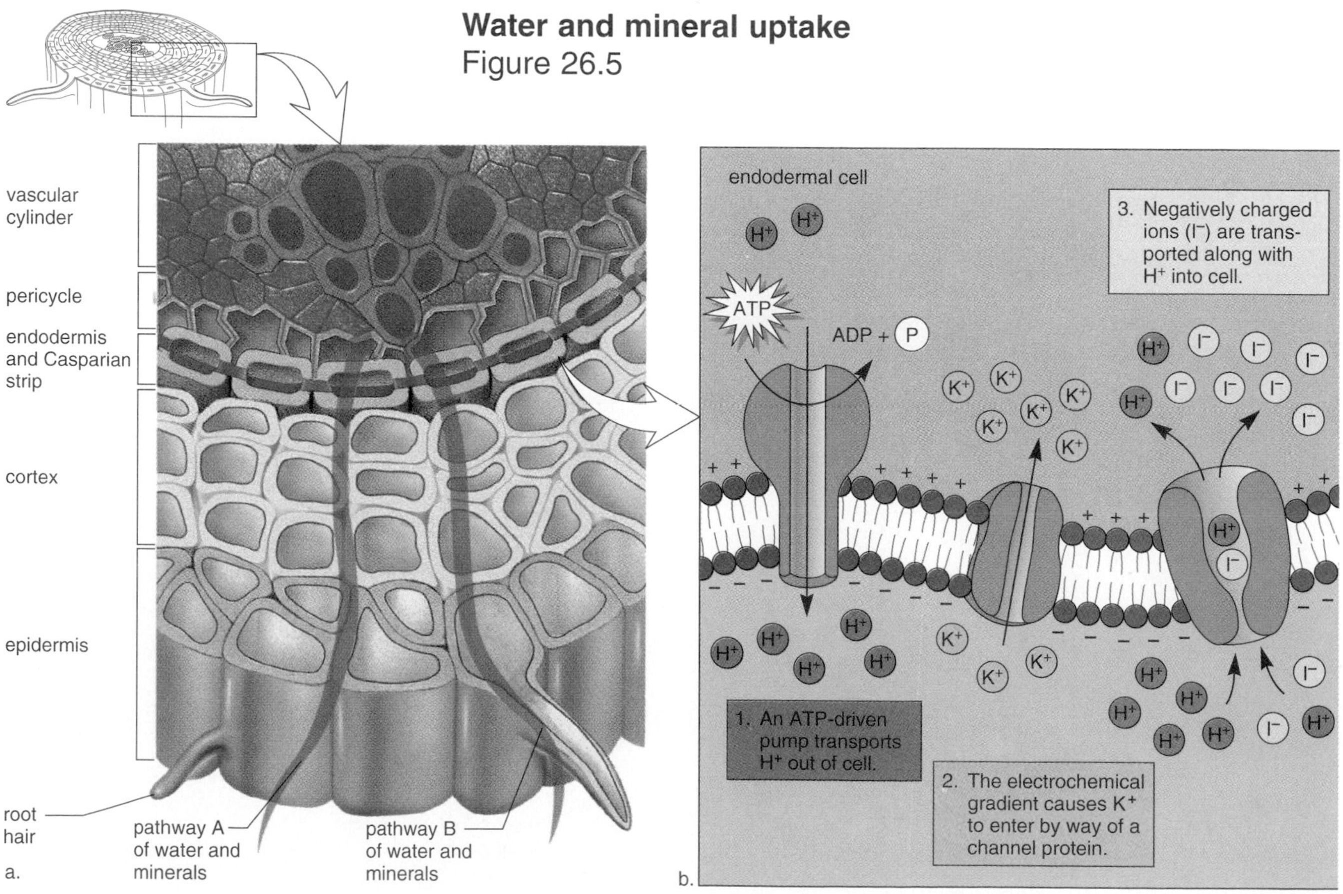

vascular
cylinder

pericycle

endodermis
and Casparian
strip

cortex

epidermis

root
hair

pathway A
of water and
minerals

pathway B
of water and
minerals

a.

endodermal cell

H⁺   H⁺

ATP

ADP + P

3. Negatively charged
ions (I⁻) are trans-
ported along with
H⁺ into cell.

H⁺   I⁻
I⁻   I⁻
H⁺   I⁻   I⁻
I⁻   I⁻
K⁺   K⁺   K⁺   H⁺
K⁺   K⁺   I⁻
K⁺   H⁺
+ +   + + + +   + + +   +   + +
H⁺
I⁻

H⁺   H⁺   K⁺   H⁺
H⁺   H⁺   K⁺   H⁺   I⁻   H⁺
K⁺   K⁺   H⁺   I⁻
H⁺
H⁺   I⁻   H⁺

1. An ATP-driven
pump transports
H⁺ out of cell.

2. The electrochemical
gradient causes K⁺
to enter by way of a
channel protein.

b.

178

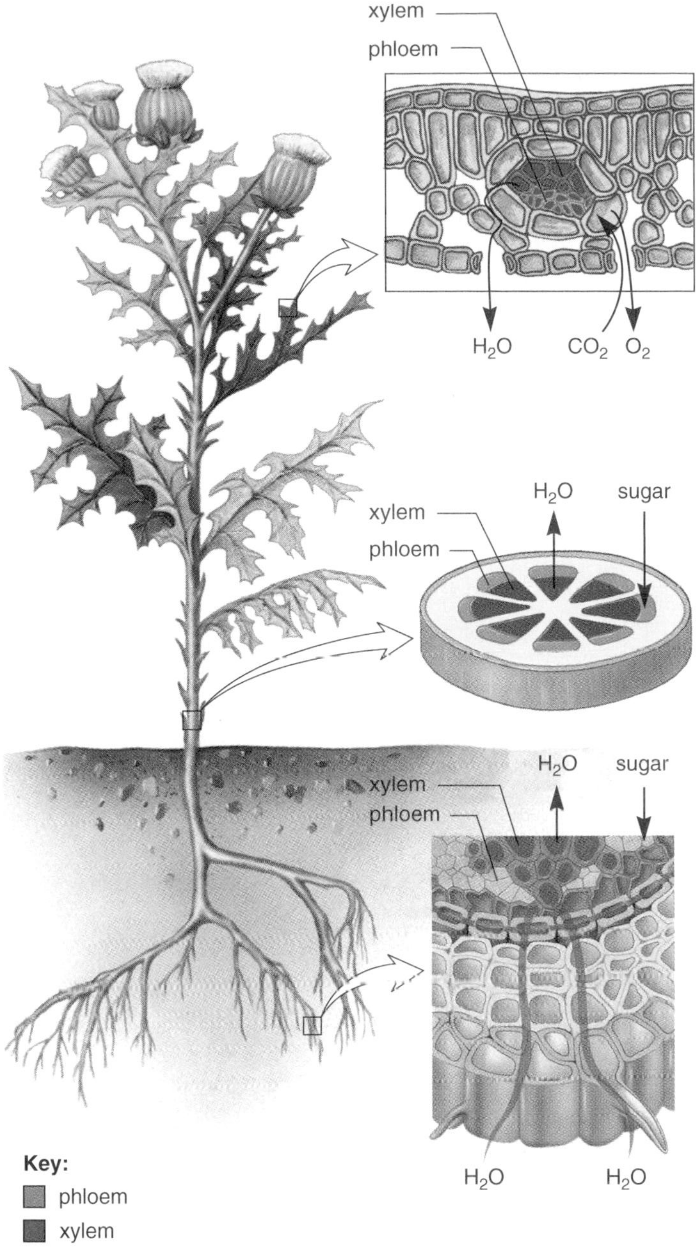

Plant transport system
Figure 26.8

xylem
phloem

H_2O CO_2 O_2

H_2O sugar
xylem
phloem

H_2O sugar
xylem
phloem

H_2O H_2O

Key:
- phloem
- xylem

**Water potential
and turgor pressure**
Figure 26A

a,b: © Dwight Kuhn

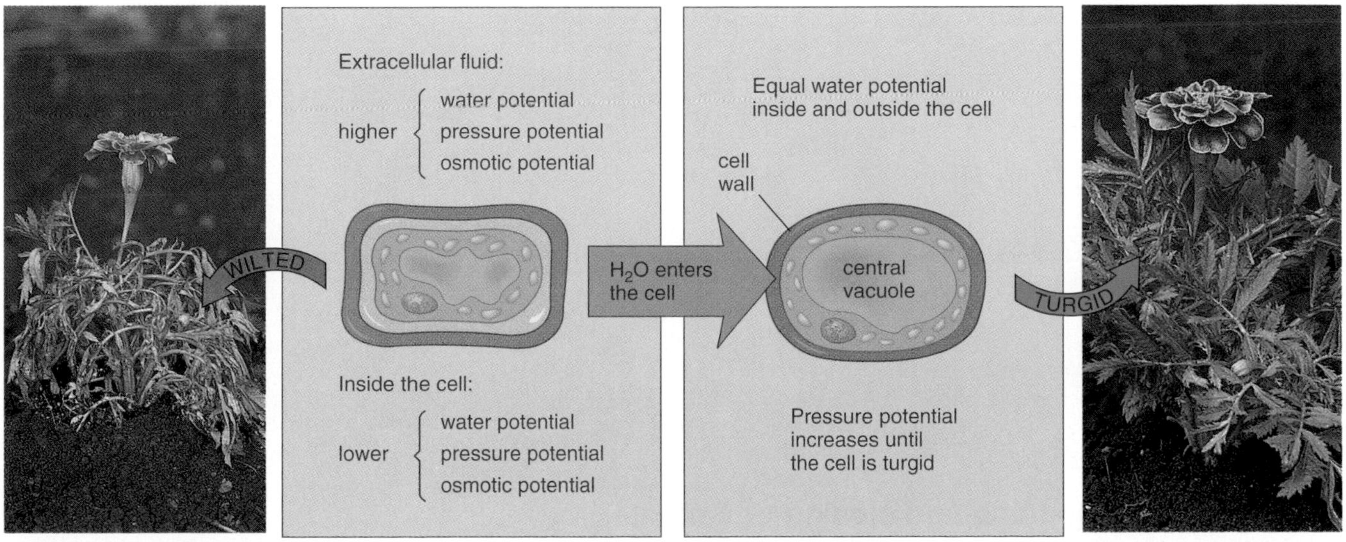

Extracellular fluid:

higher {
  water potential
  pressure potential
  osmotic potential
}

Inside the cell:

lower {
  water potential
  pressure potential
  osmotic potential
}

WILTED

H_2O enters
the cell

Equal water potential
inside and outside the cell

cell
wall

central
vacuole

Pressure potential
increases until
the cell is turgid

TURGID

a. Plant cells need water.

b. Plant cells are turgid.

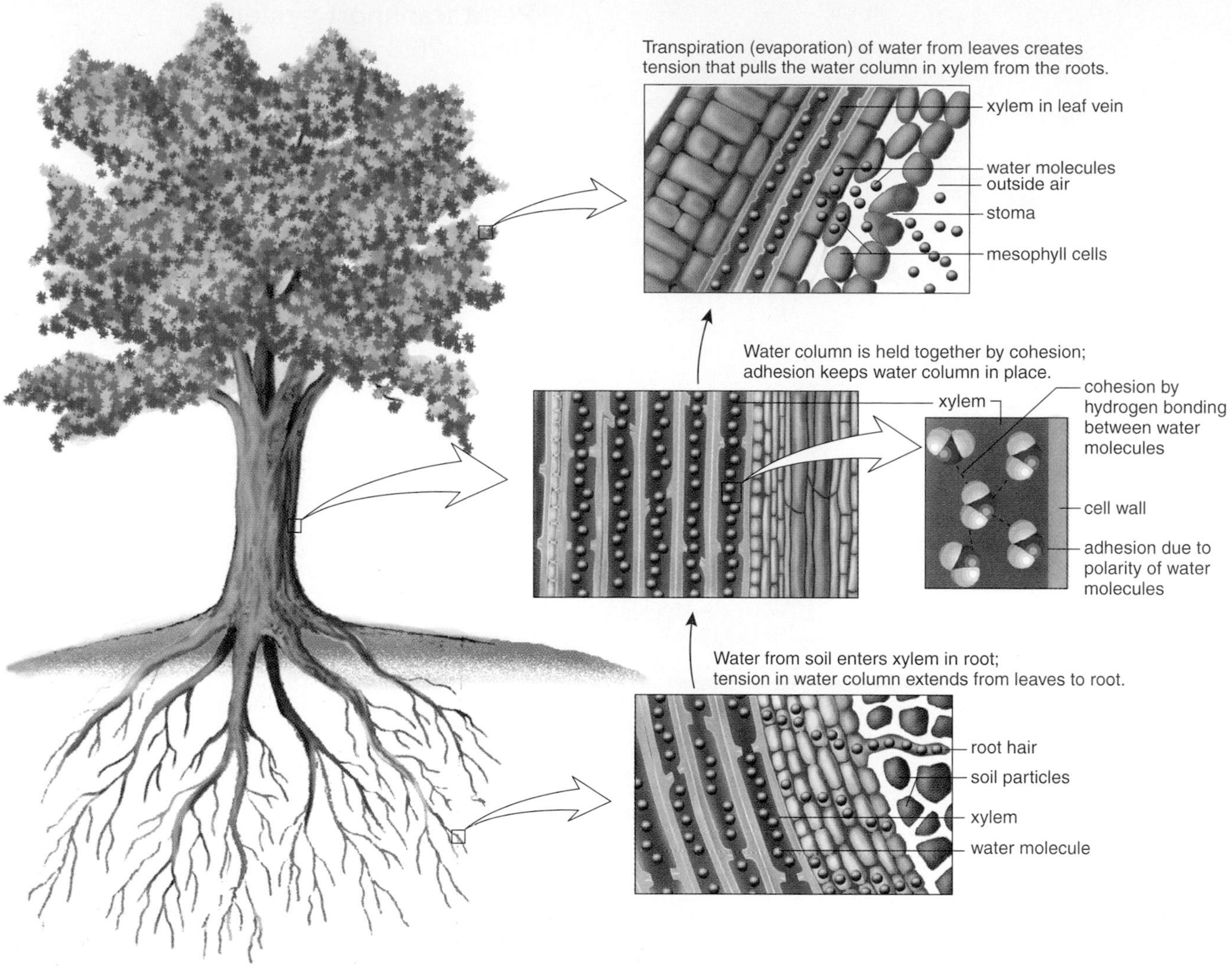

Transpiration (evaporation) of water from leaves creates tension that pulls the water column in xylem from the roots.

- xylem in leaf vein
- water molecules
- outside air
- stoma
- mesophyll cells

Water column is held together by cohesion; adhesion keeps water column in place.

- xylem
- cohesion by hydrogen bonding between water molecules
- cell wall
- adhesion due to polarity of water molecules

Water from soil enters xylem in root; tension in water column extends from leaves to root.

- root hair
- soil particles
- xylem
- water molecule

Cohesion-tension model of xylem transport
Figure 26.11

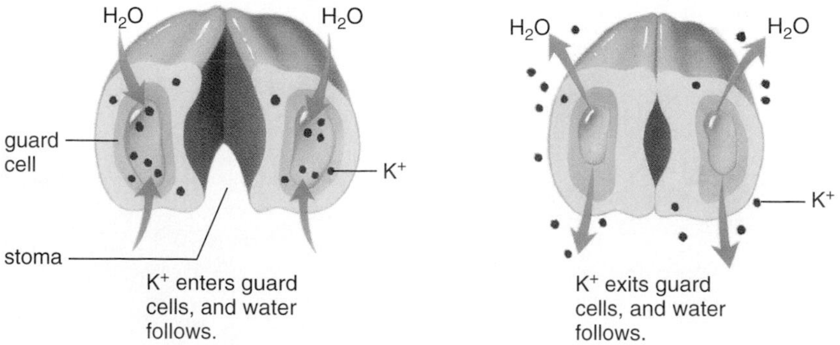

H$_2$O

H$_2$O

guard cell

stoma

K$^+$

K$^+$ enters guard cells, and water follows.

H$_2$O

H$_2$O

K$^+$

K$^+$ exits guard cells, and water follows.

Opening and closing of stomata
Figure 26.12

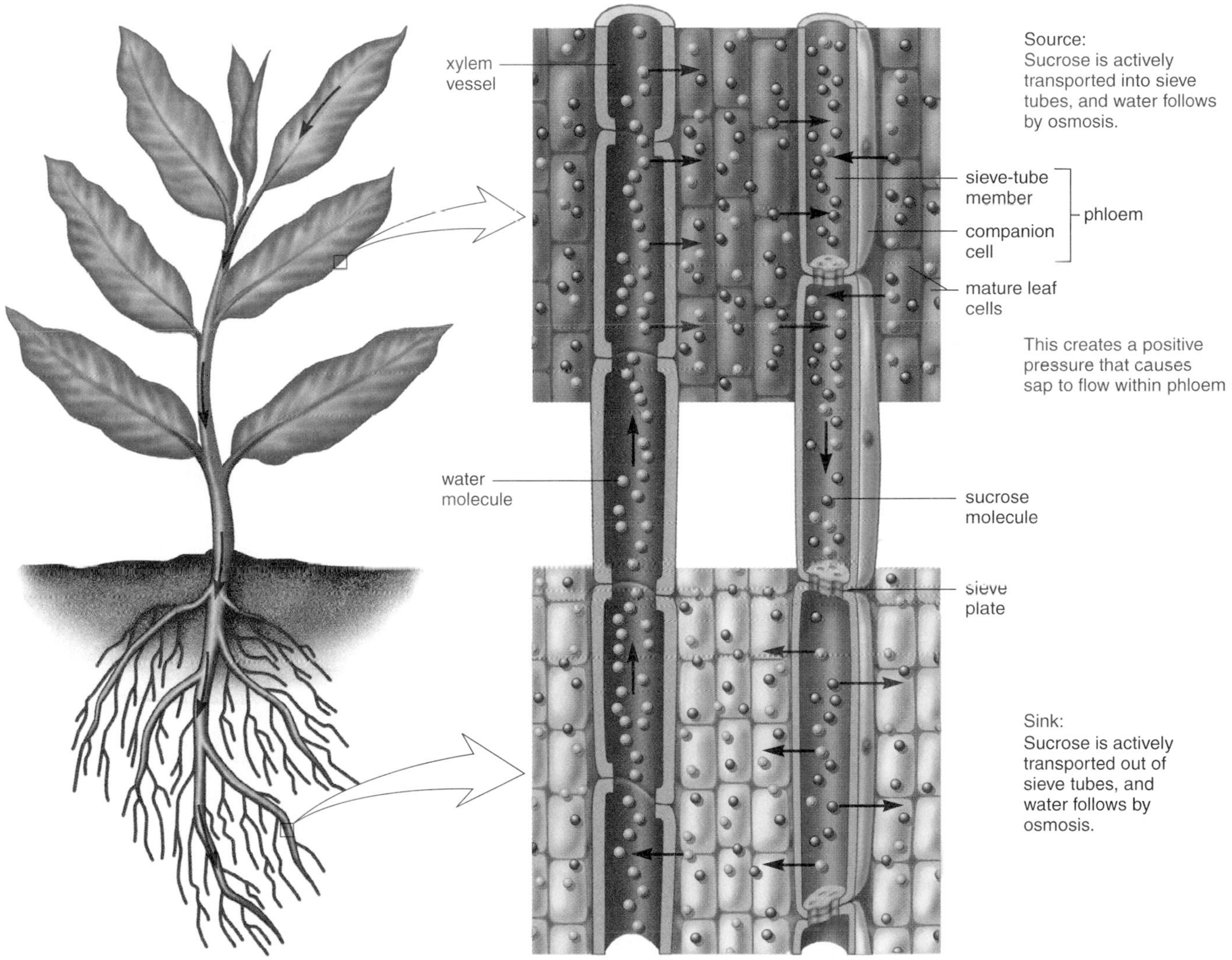

xylem vessel

Source:
Sucrose is actively transported into sieve tubes, and water follows by osmosis.

sieve-tube member

companion cell

phloem

mature leaf cells

This creates a positive pressure that causes sap to flow within phloem

water molecule

sucrose molecule

sieve plate

Sink:
Sucrose is actively transported out of sieve tubes, and water follows by osmosis.

Pressure-flow model of phloem transport
Figure 26.14

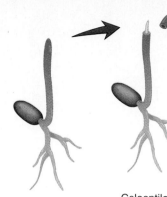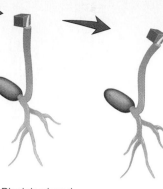

Coleoptile tip
is intact.

Coleoptile tip
is removed.

After tips are placed on agar,
agar is cut into blocks.

Block is placed
to one side of
coleoptile.

Curvature occurs
beneath block.

Demonstrating phototropism
Figure 27.7

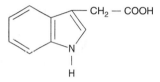

a. Indoleacetic acid (IAA)

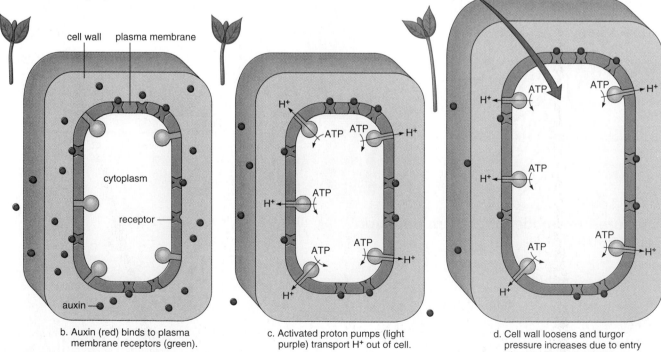

b. Auxin (red) binds to plasma
membrane receptors (green).

c. Activated proton pumps (light
purple) transport H^+ out of cell.

d. Cell wall loosens and turgor
pressure increases due to entry
of water. The cell enlarges.

Auxin structure and mode of action
Figure 27.8

Gibberellic acid structure and mode of action
Figure 27.10

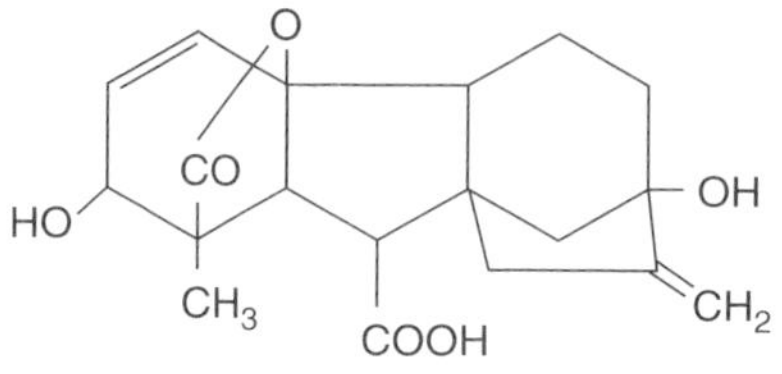

a. Gibberellic acid (GA$_3$)

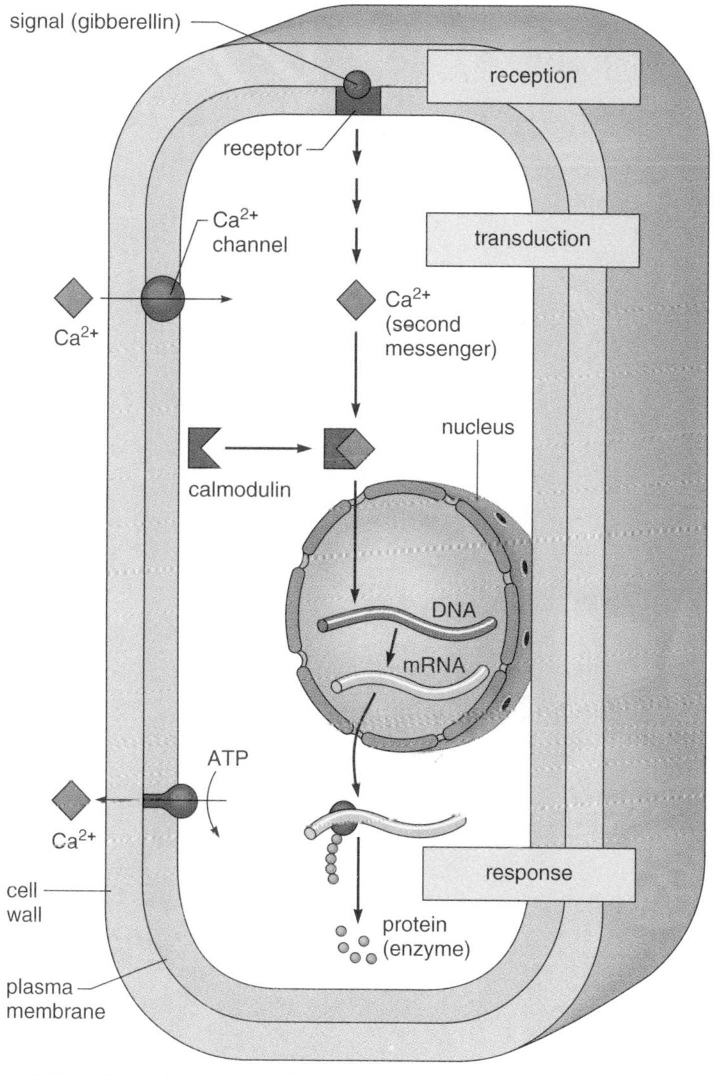

b. Gibberellic acid mode of action

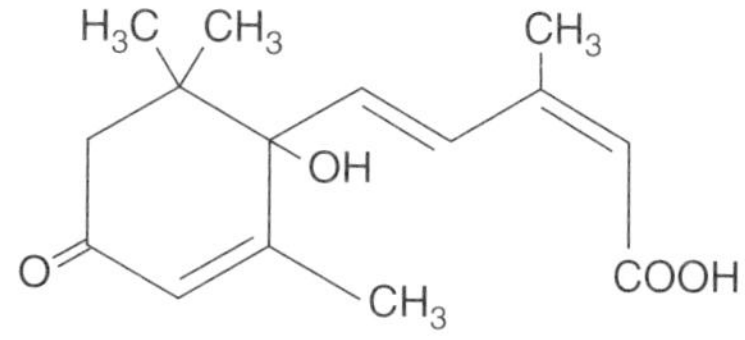

a. Abscisic acid (ABA)

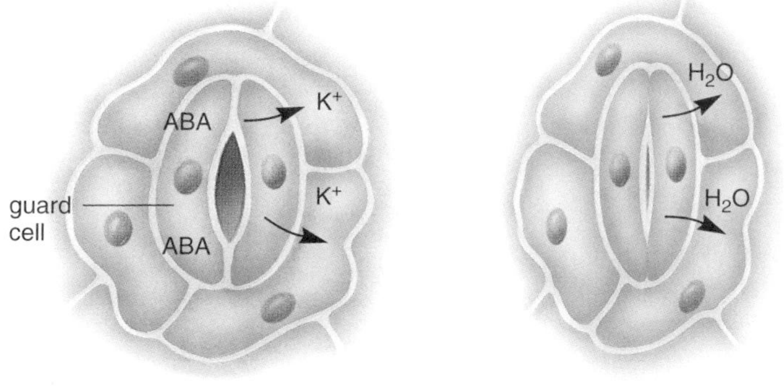

b. Open stoma

c. Closed stoma

Control of stoma opening
Figure 27.13

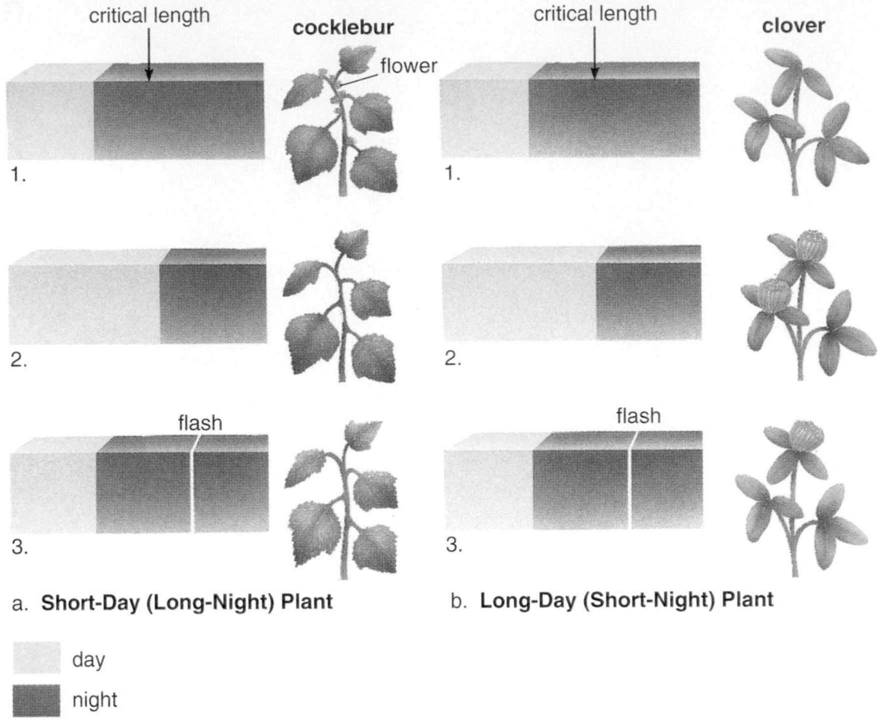

a. **Short-Day (Long-Night) Plant**   b. **Long-Day (Short-Night) Plant**

day

night

Photoperiodism and flowering
Figure 27.15

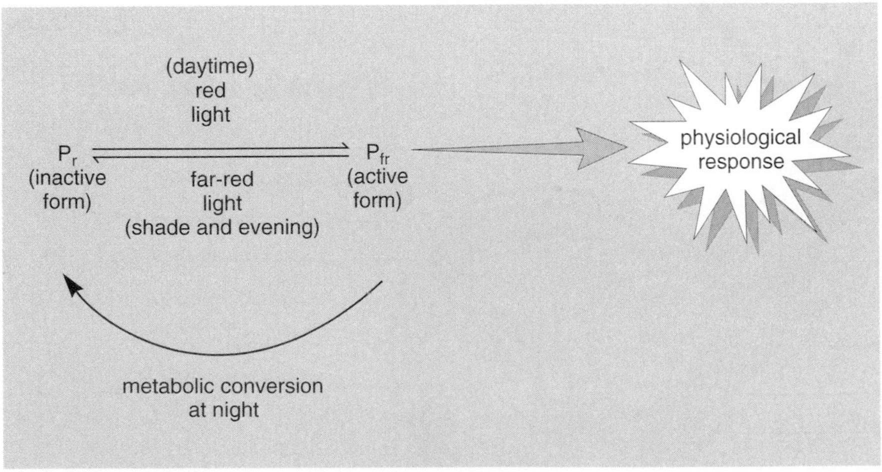

Phytochrome conversion cycle
Figure 27.16

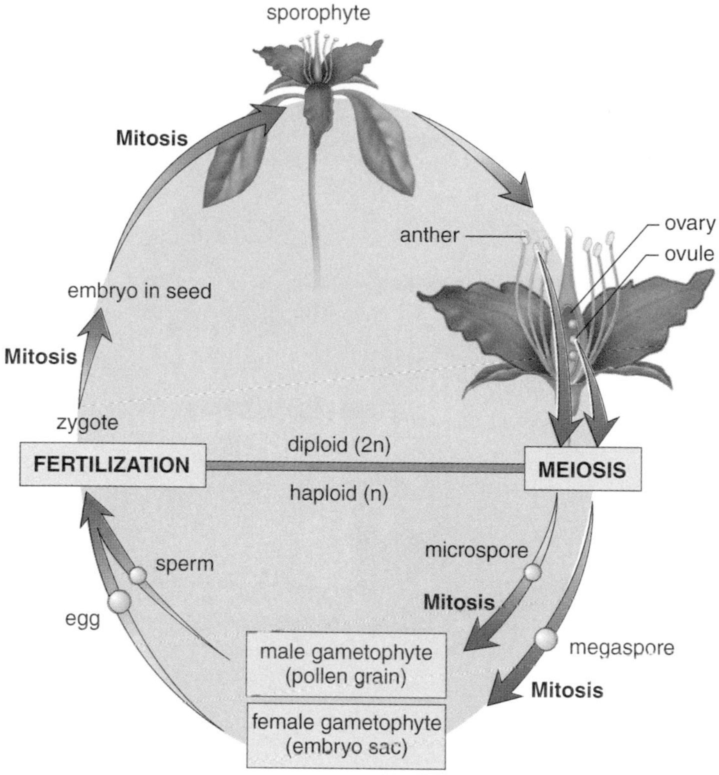

Alternation of generations in flowering plants
Figure 28.1

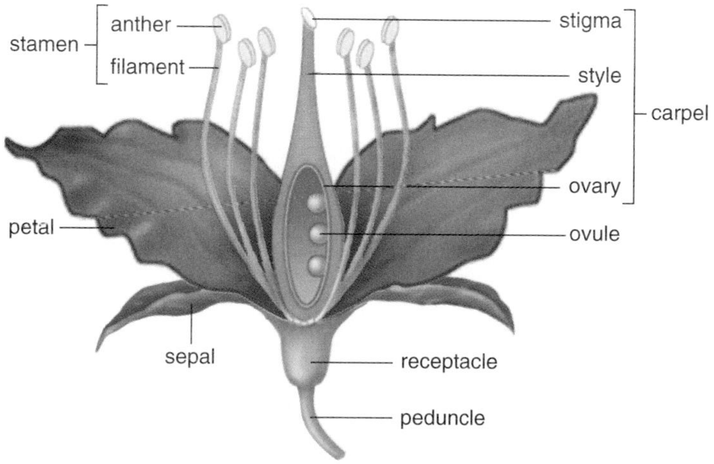

Anatomy of a flower
Figure 28.2

185

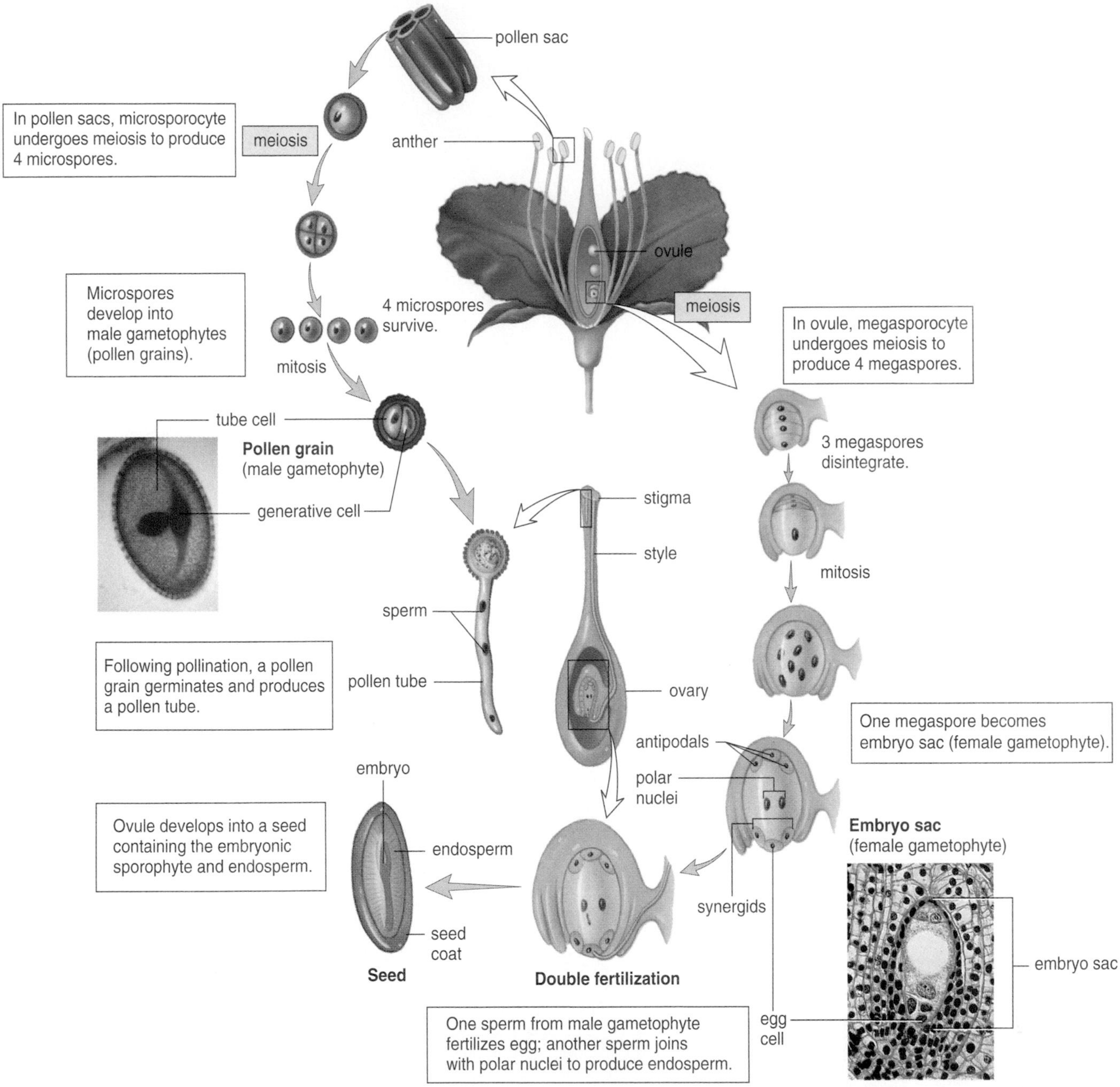

In pollen sacs, microsporocyte undergoes meiosis to produce 4 microspores.

meiosis

pollen sac

anther

ovule

meiosis

Microspores develop into male gametophytes (pollen grains).

4 microspores survive.

mitosis

In ovule, megasporocyte undergoes meiosis to produce 4 megaspores.

tube cell

Pollen grain
(male gametophyte)

generative cell

3 megaspores disintegrate.

mitosis

stigma

style

sperm

pollen tube

ovary

One megaspore becomes embryo sac (female gametophyte).

Following pollination, a pollen grain germinates and produces a pollen tube.

antipodals

polar nuclei

Embryo sac
(female gametophyte)

embryo

endosperm

Ovule develops into a seed containing the embryonic sporophyte and endosperm.

seed coat

Seed

Double fertilization

synergids

egg cell

embryo sac

One sperm from male gametophyte fertilizes egg; another sperm joins with polar nuclei to produce endosperm.

Life cycle of a flowering plant
Figure 28.5

left: Courtesy Graham Kent; right: © Ed Reschke

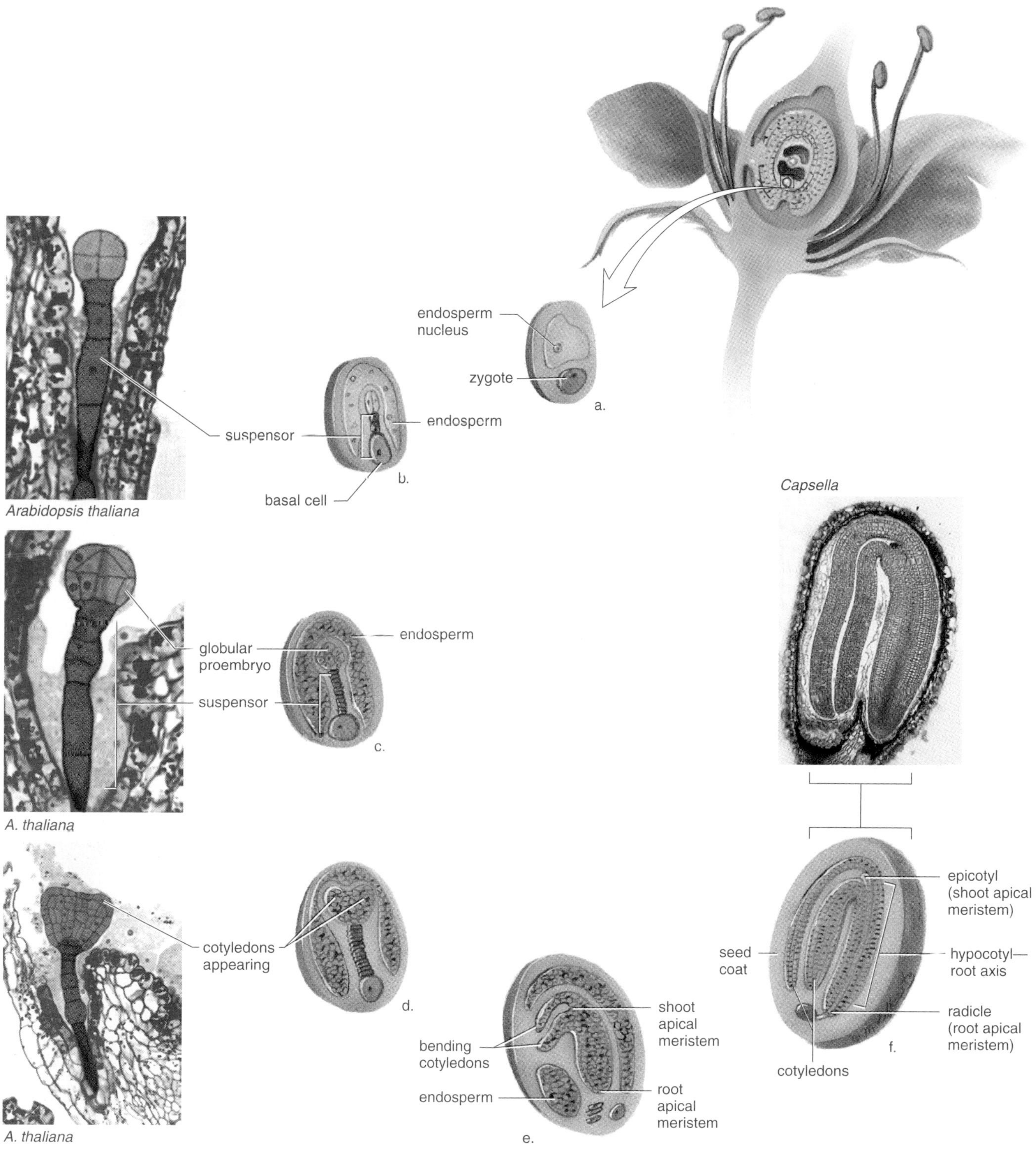

endosperm nucleus

zygote

a.

suspensor

endosperm

basal cell

b.

Arabidopsis thaliana

globular proembryo

endosperm

suspensor

c.

A. thaliana

cotyledons appearing

d.

Capsella

bending cotyledons

shoot apical meristem

endosperm

root apical meristem

e.

A. thaliana

epicotyl (shoot apical meristem)

seed coat

hypocotyl— root axis

radicle (root apical meristem)

cotyledons

f.

Development of a eudicot embryo
Figure 28.7

top, bottom, middle: Courtesy Dr. Chun-Ming Liu; right: © Jack Bostrack/Visuals Unlimited

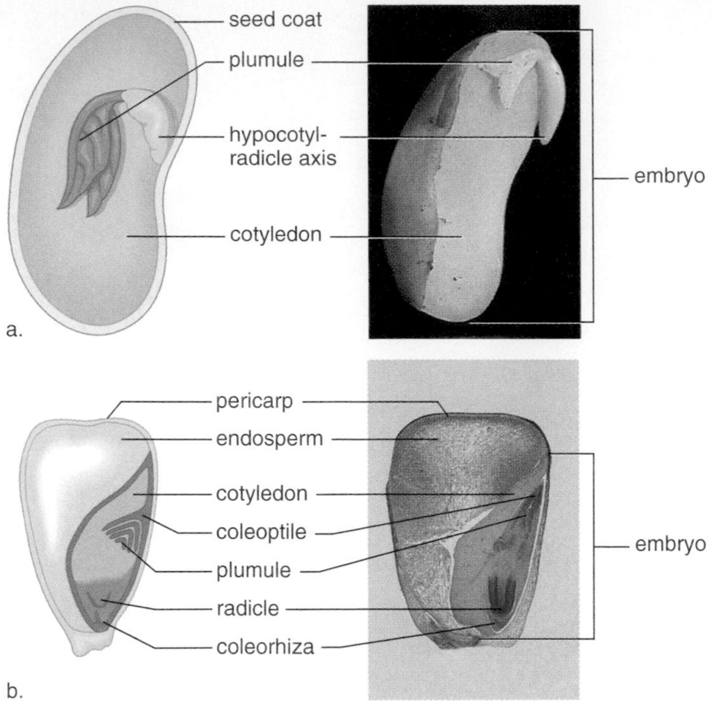

seed coat
plumule
hypocotyl-radicle axis
cotyledon
embryo

a.

pericarp
endosperm
cotyledon
coleoptile
plumule
radicle
coleorhiza
embryo

b.

Monocot versus eudicot
Figure 28.8

a: © Dwight Kuhn; b: Courtesy Ray F. Evert/University of Wisconsin Madison

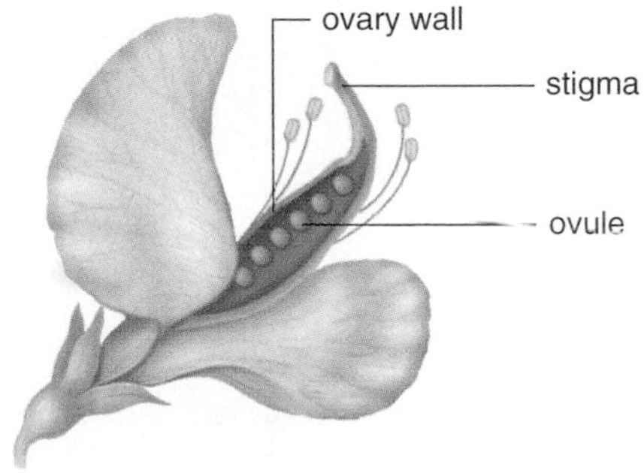

ovary wall
stigma
ovule

Ovules become
the seeds.

Pea flower and development of a pea pod
Figure 28.9

Common garden bean seed structure and germination
Figure 28.11
© Ed Reschke

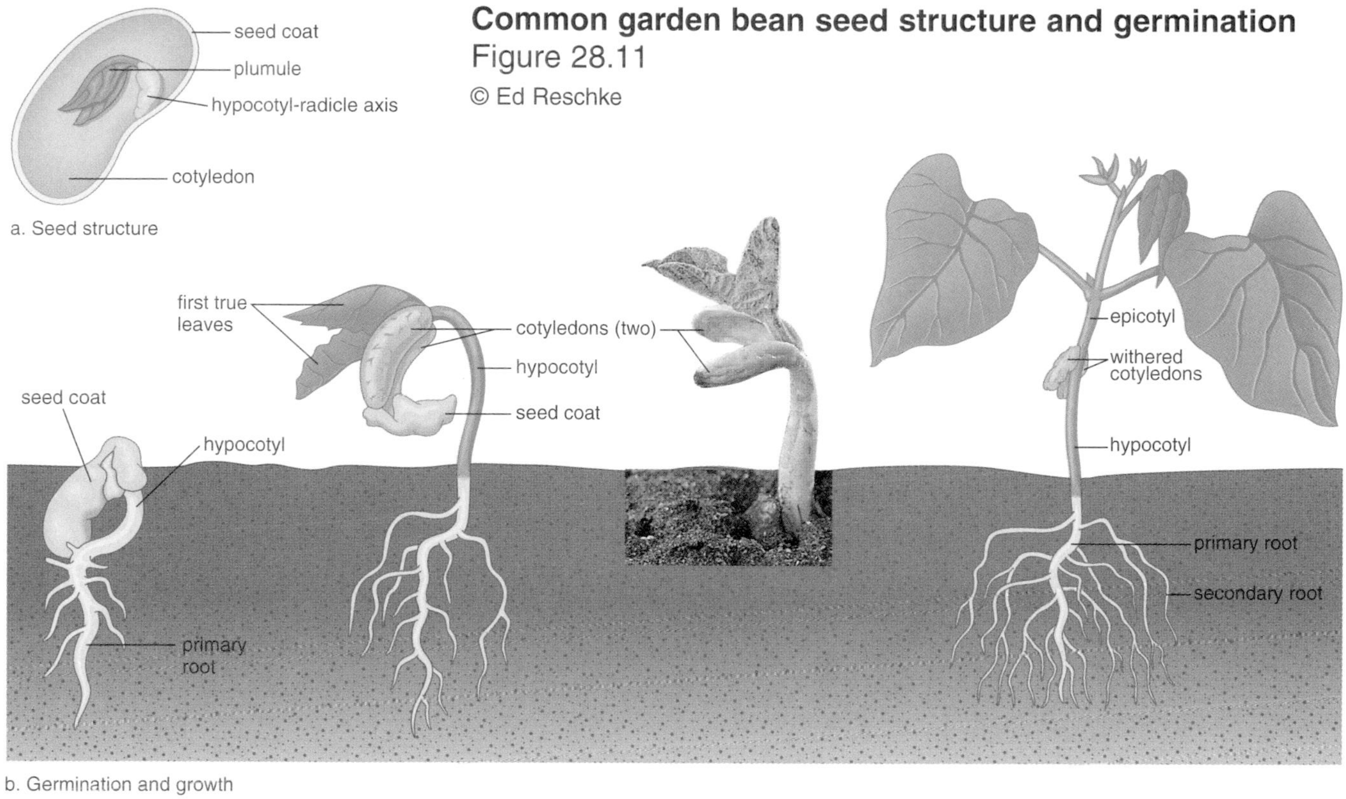

seed coat
plumule
hypocotyl-radicle axis
cotyledon

a. Seed structure

first true leaves
seed coat
hypocotyl
primary root

cotyledons (two)
hypocotyl
seed coat

epicotyl
withered cotyledons
hypocotyl
primary root
secondary root

b. Germination and growth

Corn kernel structure and germination
Figure 28.12
a (left): © James Mauseth; b (right): © Barry L. Runk/Grant Heilman, Inc.

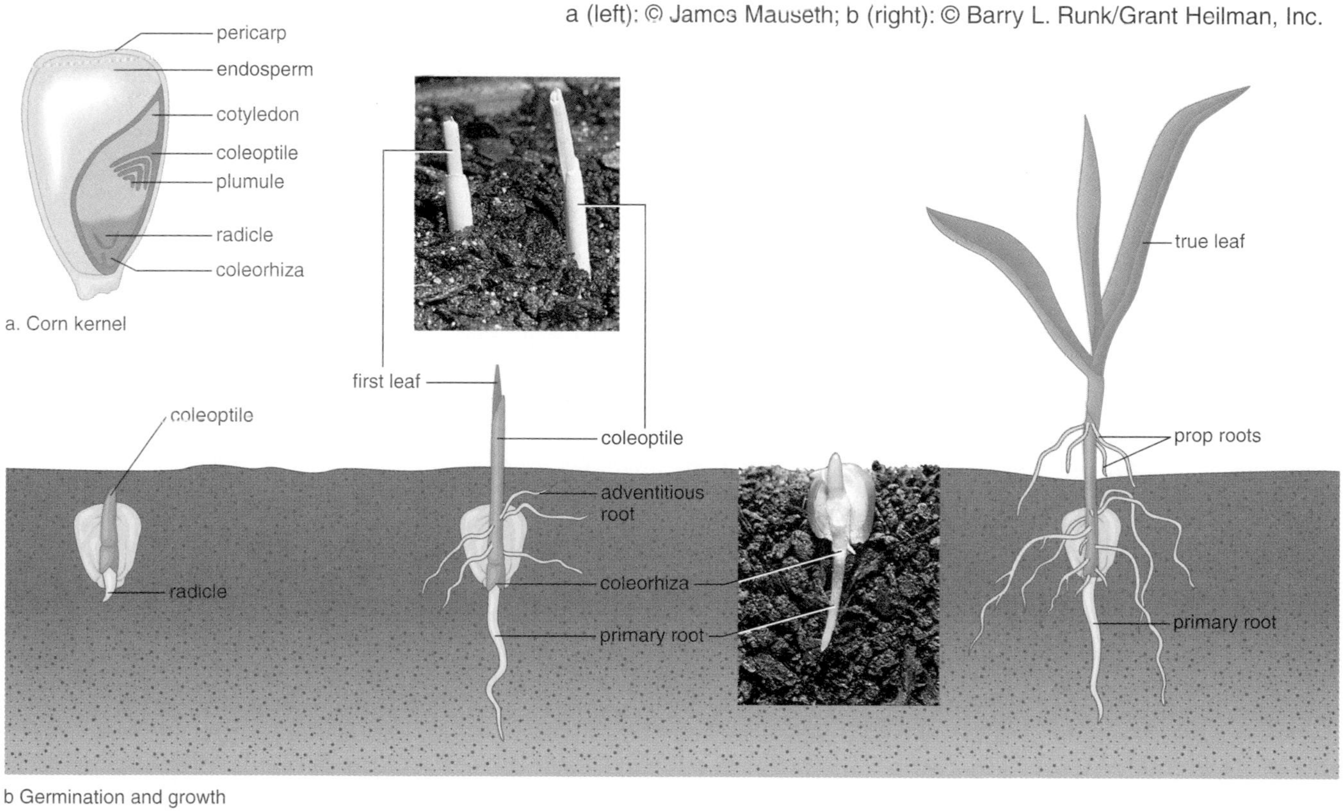

pericarp
endosperm
cotyledon
coleoptile
plumule
radicle
coleorhiza

a. Corn kernel

coleoptile
radicle

first leaf
coleoptile
adventitious root
coleorhiza
primary root

true leaf
prop roots
primary root

b Germination and growth

Overall appearance of *Arabidopsis thaliana*
Figure 28B

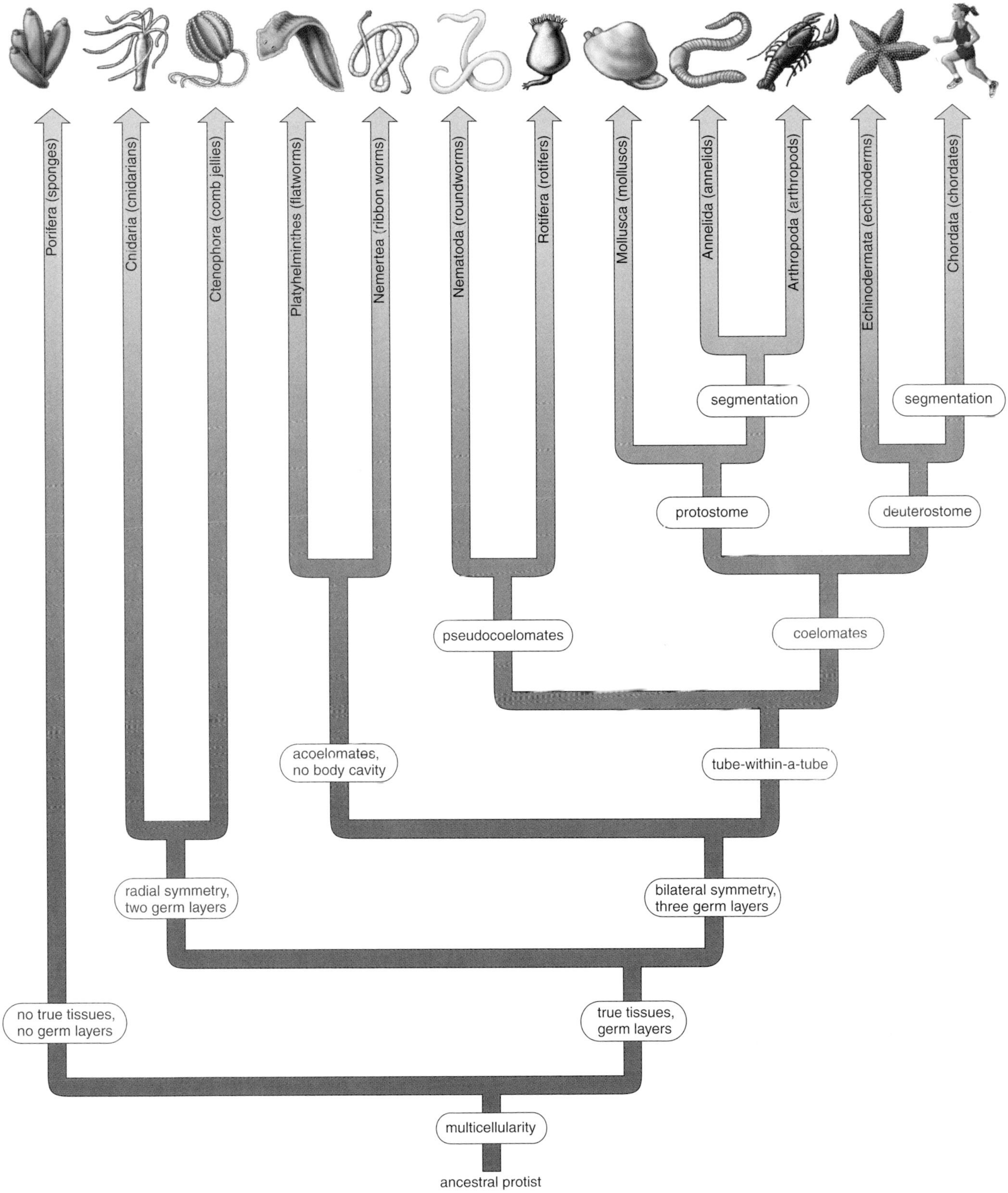
Porifera (sponges)
Cnidaria (cnidarians)
Ctenophora (comb jellies)
Platyhelminthes (flatworms)
Nemertea (ribbon worms)
Nematoda (roundworms)
Rotifera (rotifers)
Mollusca (molluscs)
Annelida (annelids)
Arthropoda (arthropods)
Echinodermata (echinoderms)
Chordata (chordates)
segmentation
segmentation
protostome
deuterostome
pseudocoelomates
coelomates
acoelomates, no body cavity
tube-within-a-tube
radial symmetry, two germ layers
bilateral symmetry, three germ layers
no true tissues, no germ layers
true tissues, germ layers
multicellularity
ancestral protist

Evolution of animals
Figure 29.2

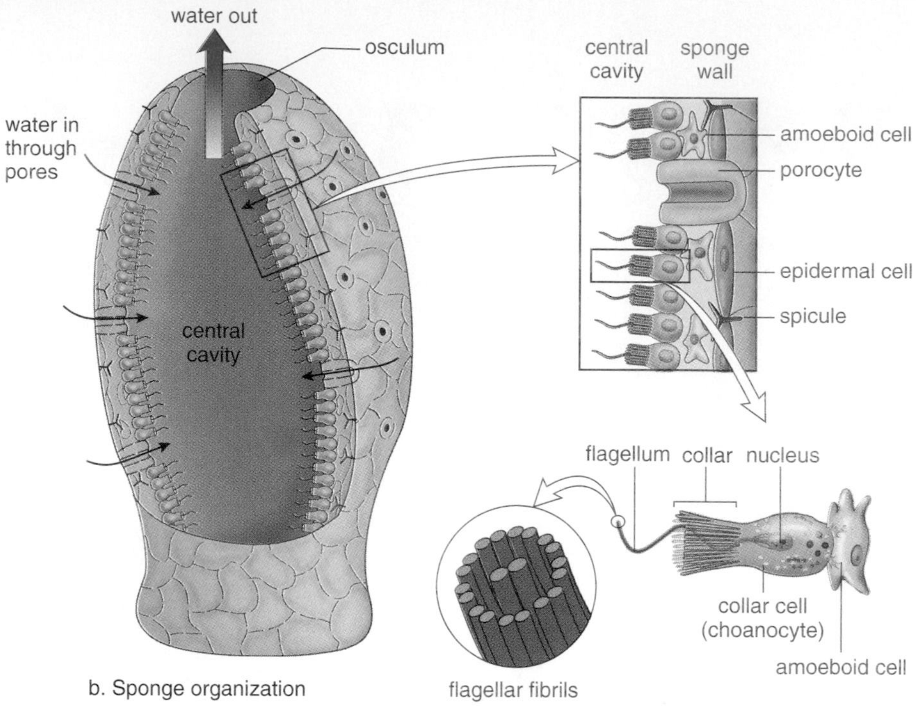

Simple sponge anatomy
Figure 29.3

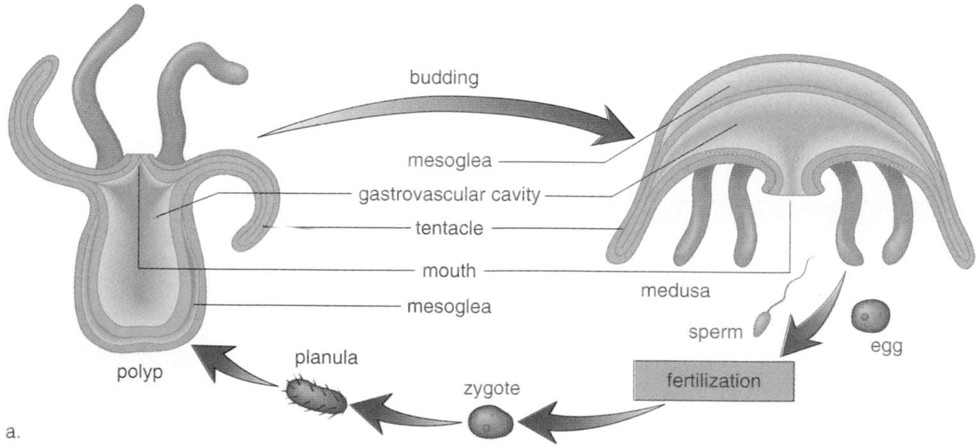

Cnidarian diversity
Figure 29.5

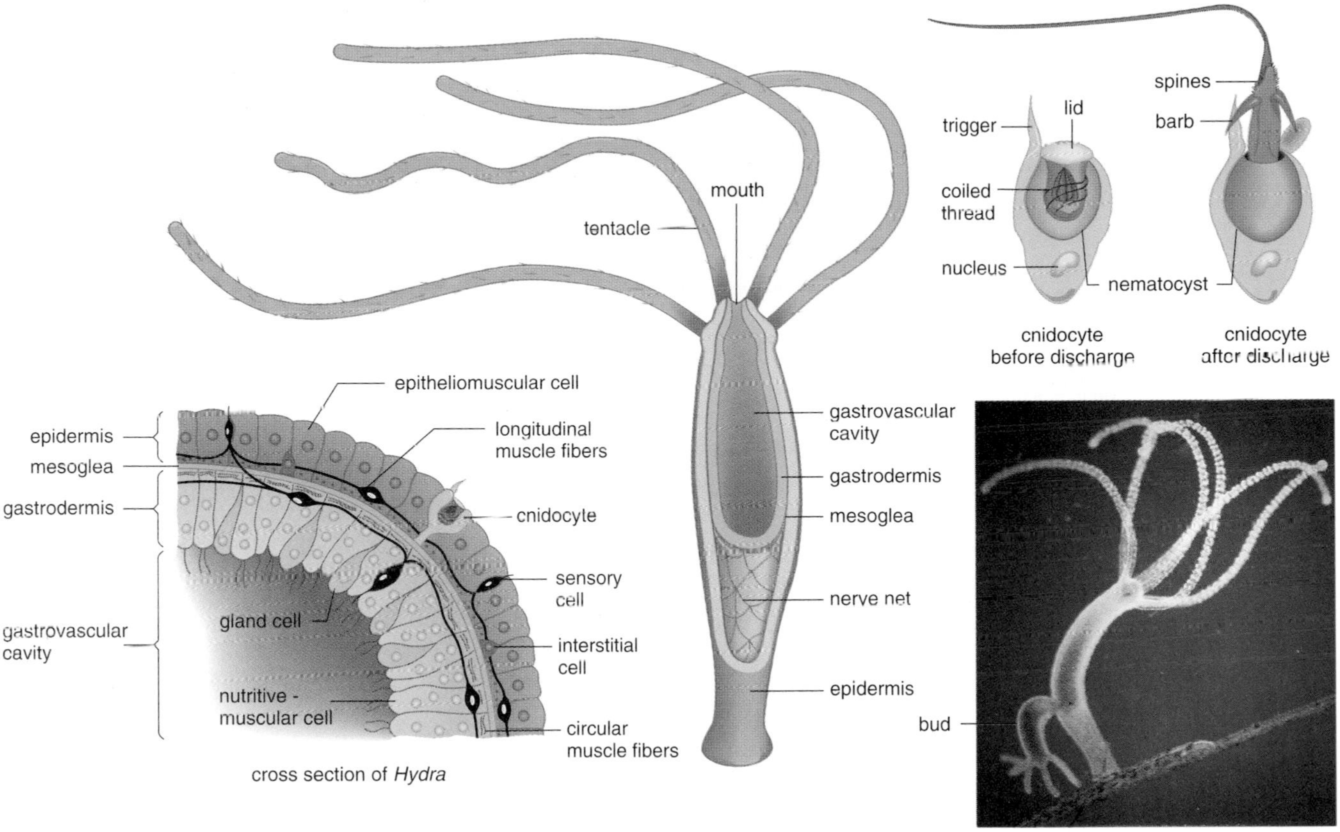

cnidocyte
before discharge

cnidocyte
after discharge

cross section of *Hydra*

Anatomy of a *Hydra*
Figure 29.6
© CABISCO/Visuals Unlimited

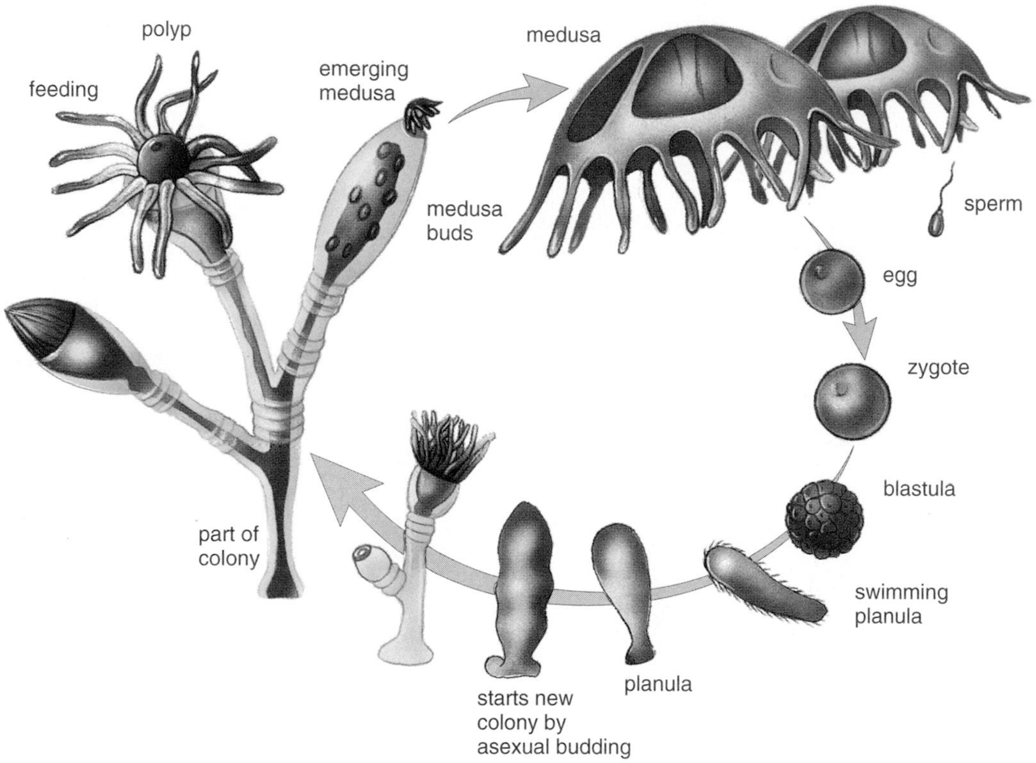

feeding

polyp

emerging medusa

medusa

medusa buds

sperm

egg

zygote

blastula

swimming planula

part of colony

starts new colony by asexual budding

planula

***Obelia* structure and life cycle**
Figure 29.7

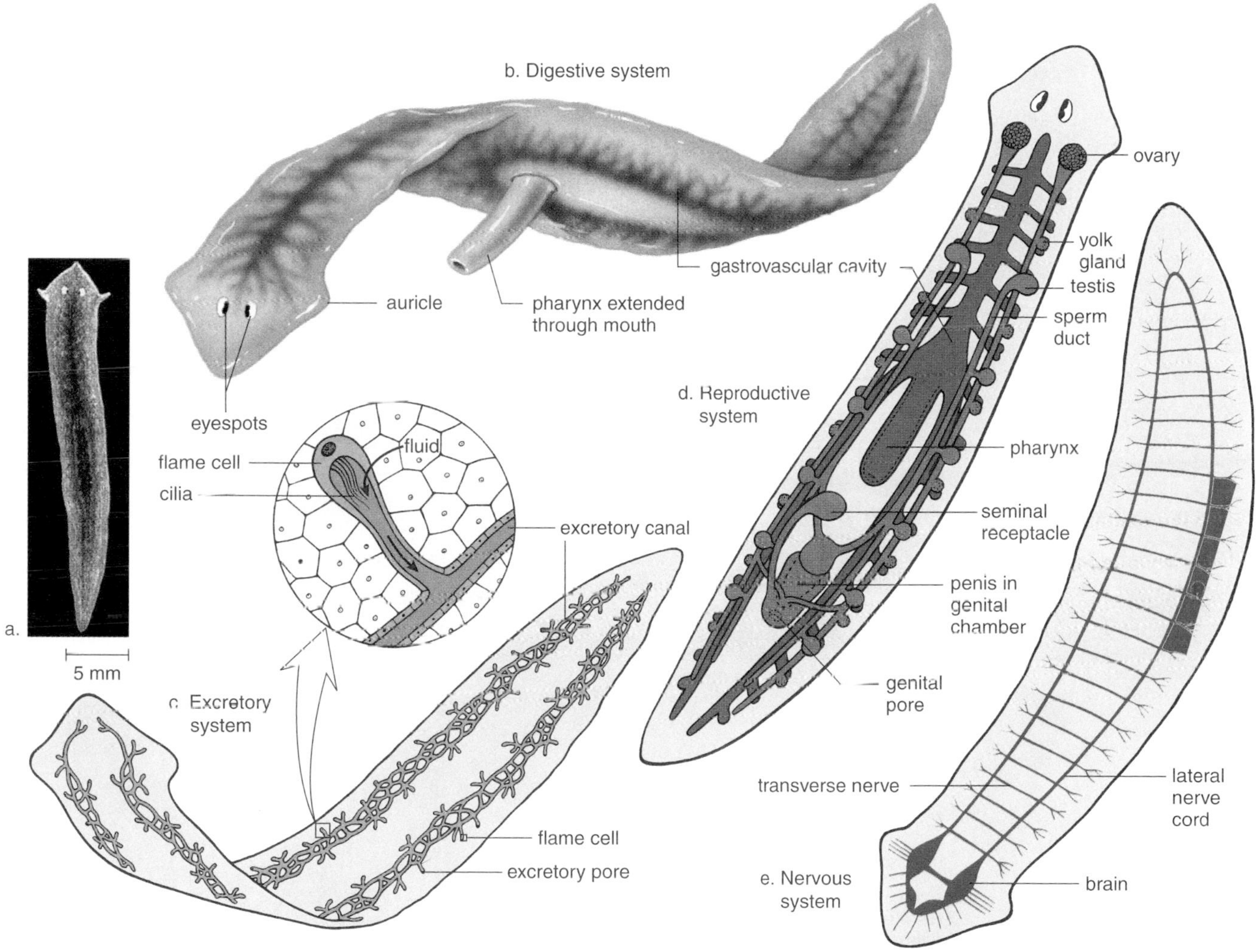

b. Digestive system

gastrovascular cavity

auricle

pharynx extended
through mouth

ovary

yolk
gland

testis

sperm
duct

d. Reproductive
system

eyespots

flame cell

cilia

fluid

excretory canal

pharynx

seminal
receptacle

penis in
genital
chamber

c. Excretory
system

genital
pore

flame cell

excretory pore

transverse nerve

lateral
nerve
cord

e. Nervous
system

brain

a.

5 mm

Planarian anatomy

Figure 29.9

© Tom E. Adams/Peter Arnold, Inc.

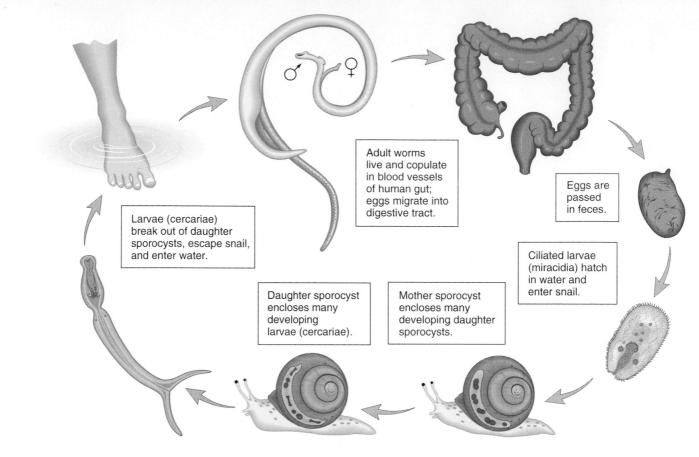

Adult worms live and copulate in blood vessels of human gut; eggs migrate into digestive tract.

Eggs are passed in feces.

Ciliated larvae (miracidia) hatch in water and enter snail.

Larvae (cercariae) break out of daughter sporocysts, escape snail, and enter water.

Daughter sporocyst encloses many developing larvae (cercariae).

Mother sporocyst encloses many developing daughter sporocysts.

Transmission of schistosomiasis
Figure 29.10

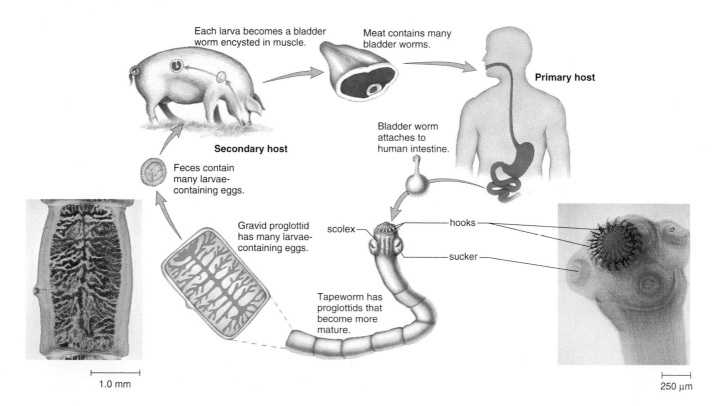

Each larva becomes a bladder worm encysted in muscle.

Meat contains many bladder worms.

Primary host

Bladder worm attaches to human intestine.

Secondary host

Feces contain many larvae-containing eggs.

Gravid proglottid has many larvae-containing eggs.

scolex

hooks

sucker

Tapeworm has proglottids that become more mature.

1.0 mm

250 μm

Life cycle of a tapeworm, *Taenia*
Figure 29.11

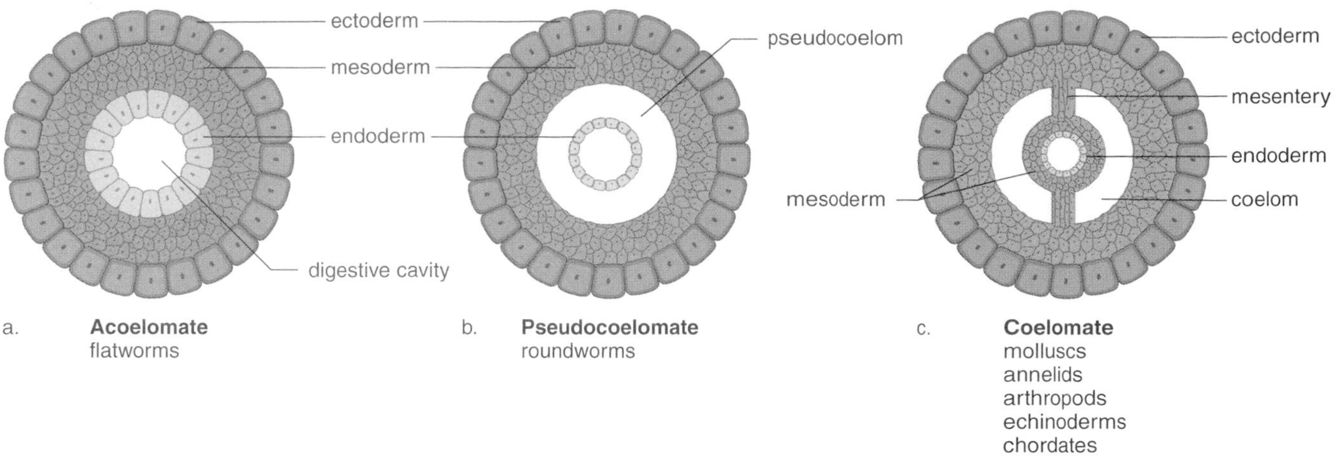

a. **Acoelomate**
flatworms

b. **Pseudocoelomate**
roundworms

c. **Coelomate**
molluscs
annelids
arthropods
echinoderms
chordates

Acoelomate, pseudocoelomate, and coelomate comparison
Figure 29A

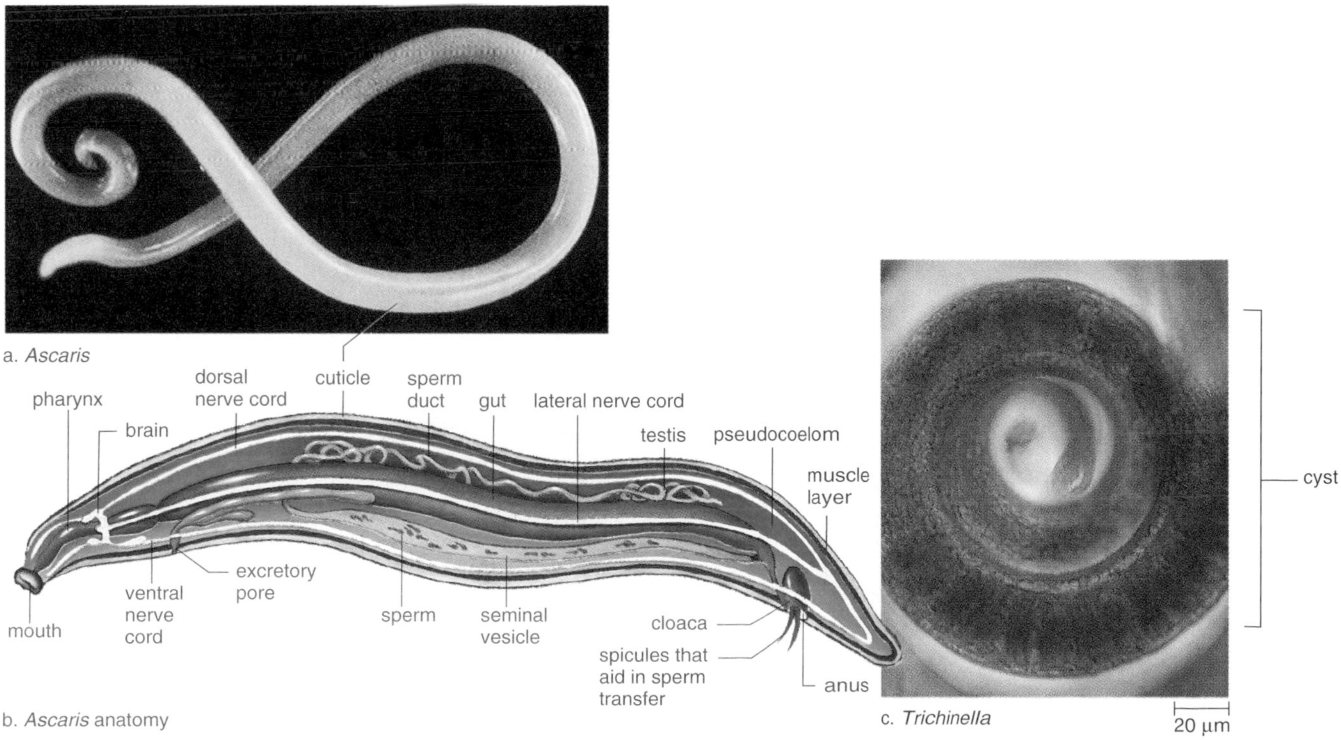

a. *Ascaris*

b. *Ascaris* anatomy

c. *Trichinella*

20 µm

Roundworm anatomy
Figure 29.12

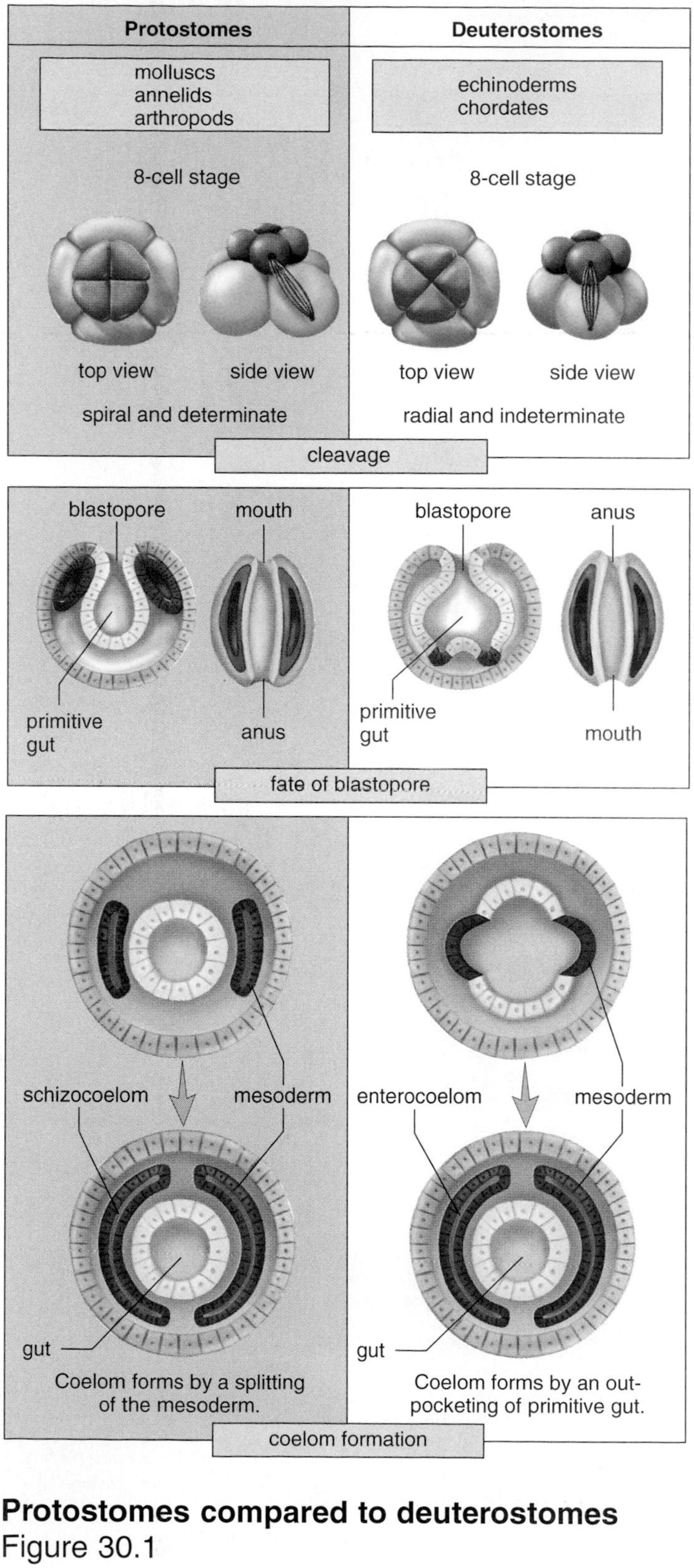

Protostomes compared to deuterostomes
Figure 30.1

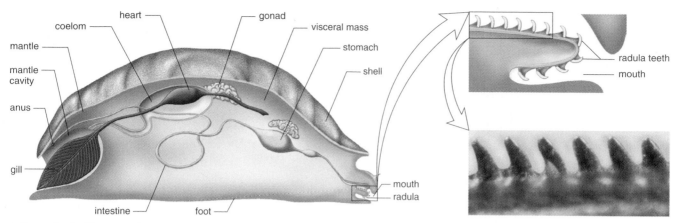

a. Generalized molluscan anatomy

b. Radula

Body plan of molluscs
Figure 30.2

b: © Kjell Sandved/Butterfly Alphabet

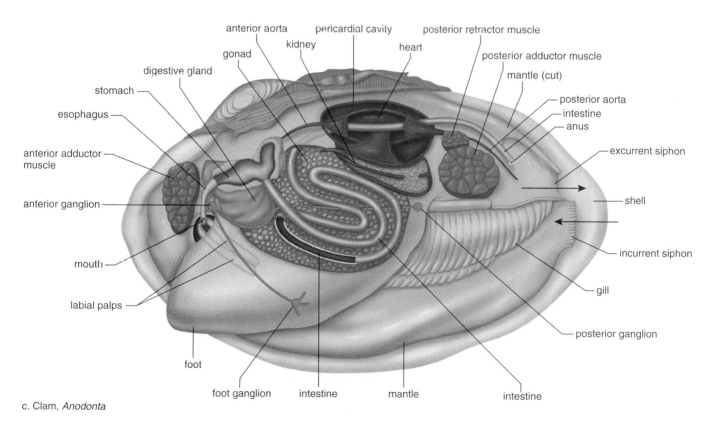

c. Clam, *Anodonta*

Bivalve diversity
Figure 30.3

Cephalopod diversity
Figure 30.4

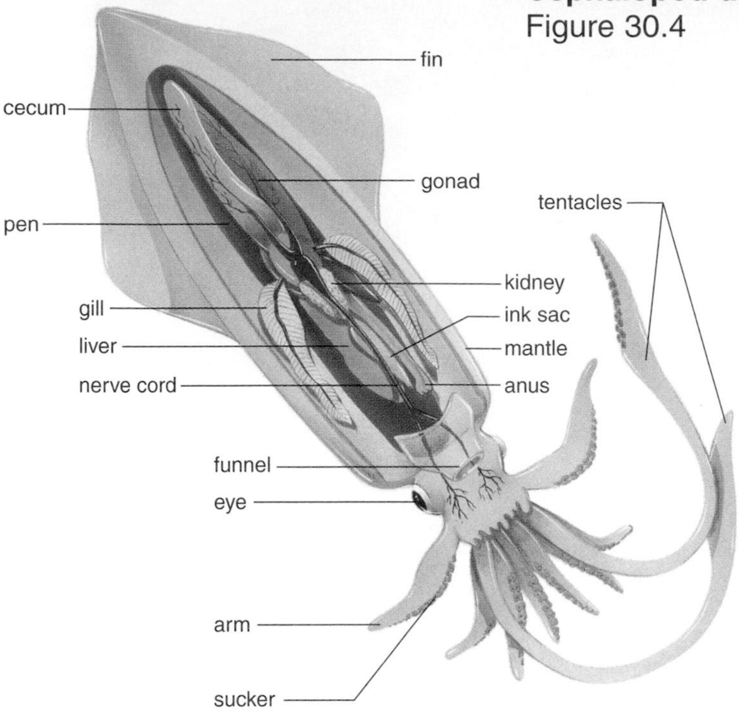

fin

cecum

gonad

pen

tentacles

kidney

gill

ink sac

liver

mantle

nerve cord

anus

funnel

eye

arm

sucker

Gastropod diversity
Figure 30.5

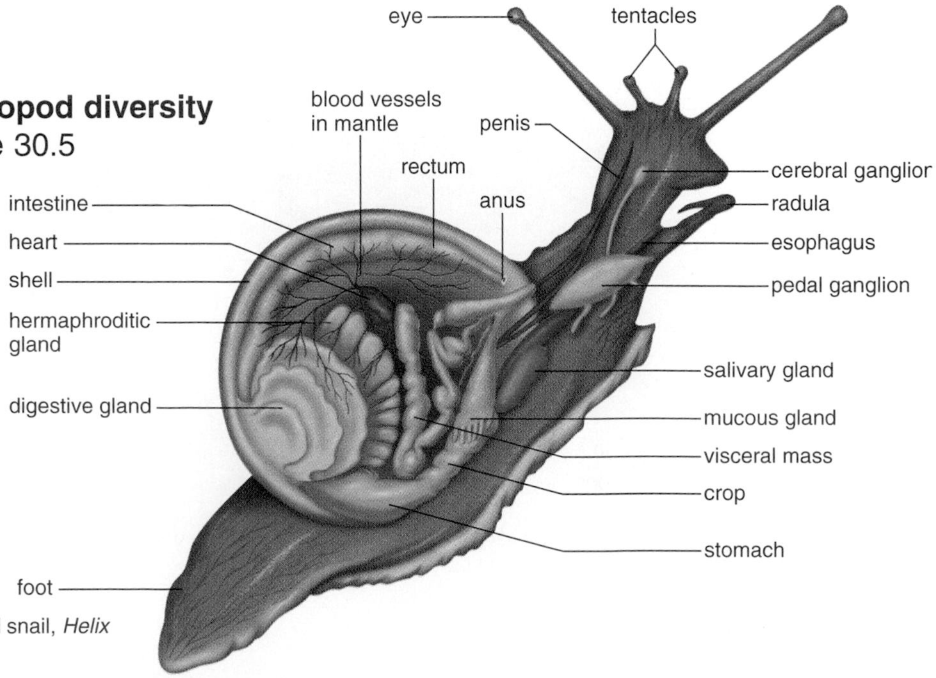

eye

tentacles

blood vessels
in mantle

penis

rectum

anus

cerebral ganglior

intestine

radula

heart

esophagus

shell

pedal ganglion

hermaphroditic
gland

digestive gland

salivary gland

mucous gland

visceral mass

crop

stomach

foot

c. Land snail, *Helix*

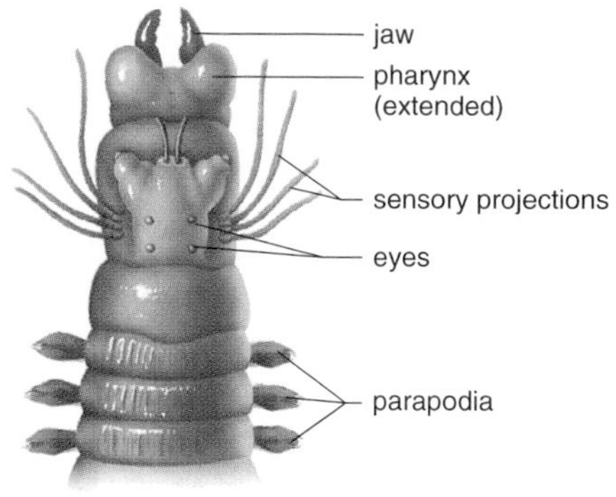

Polychaete diversity
Figure 30.6

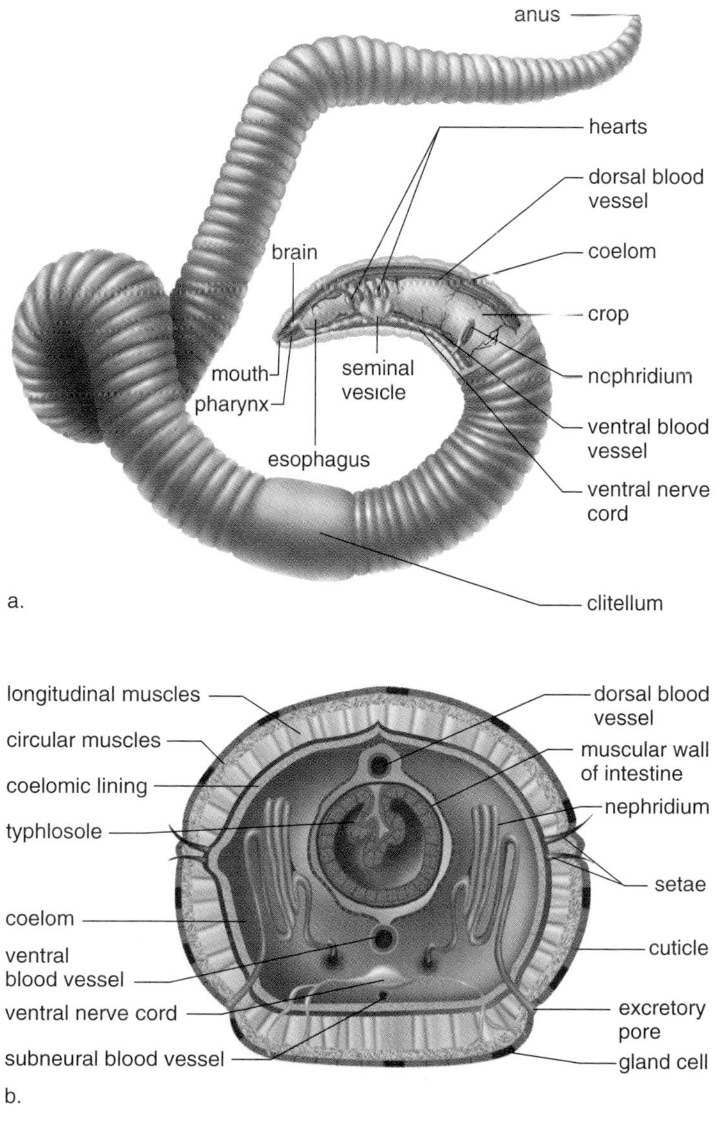

a.

b.

Earthworm, *Lumbricus*
Figure 30.7

Arthropod skeleton and eye
Figure 30.9

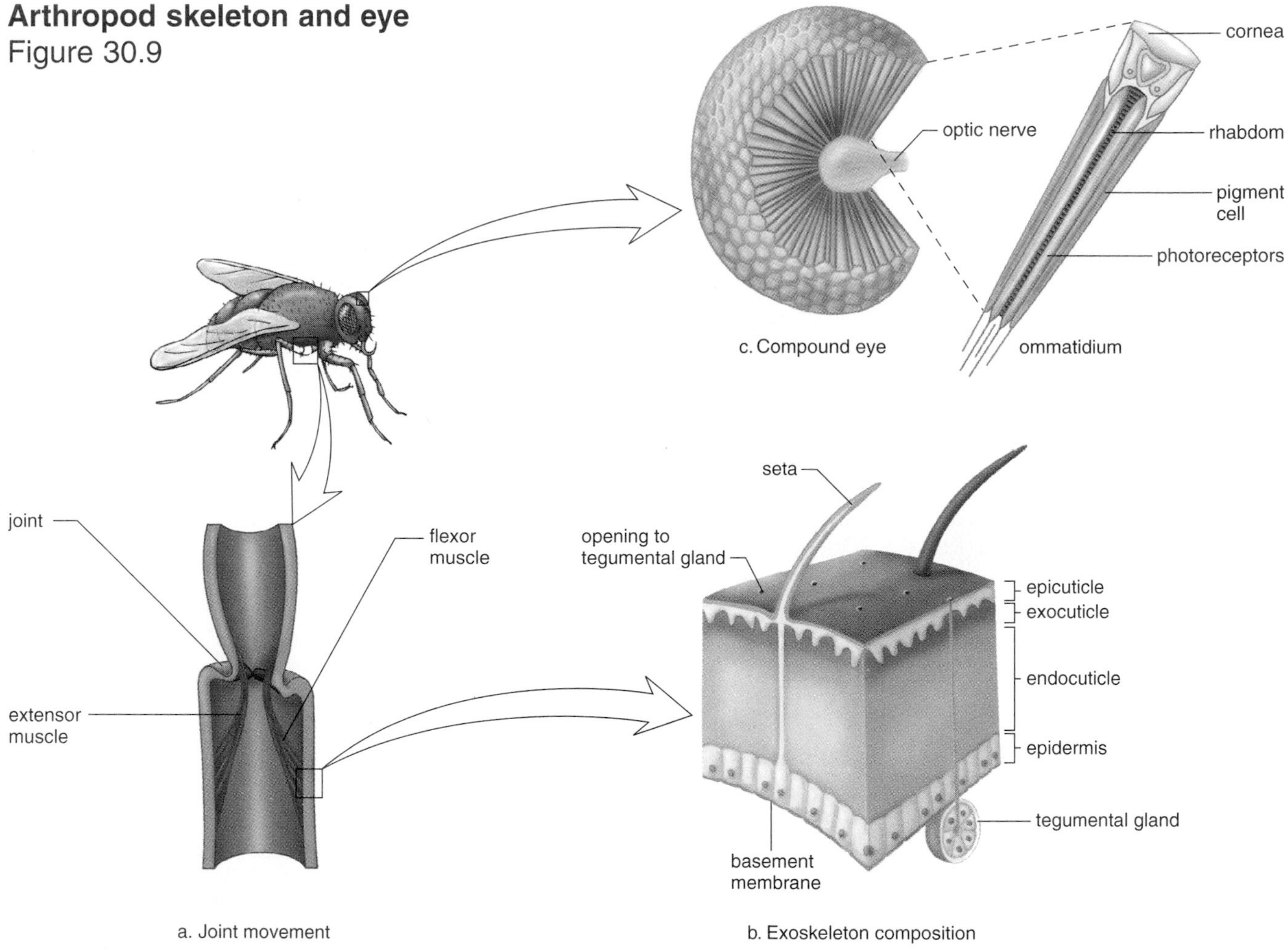

c. Compound eye

cornea

optic nerve

rhabdom

pigment cell

photoreceptors

ommatidium

joint

flexor muscle

extensor muscle

a. Joint movement

seta

opening to tegumental gland

epicuticle

exocuticle

endocuticle

epidermis

tegumental gland

basement membrane

b. Exoskeleton composition

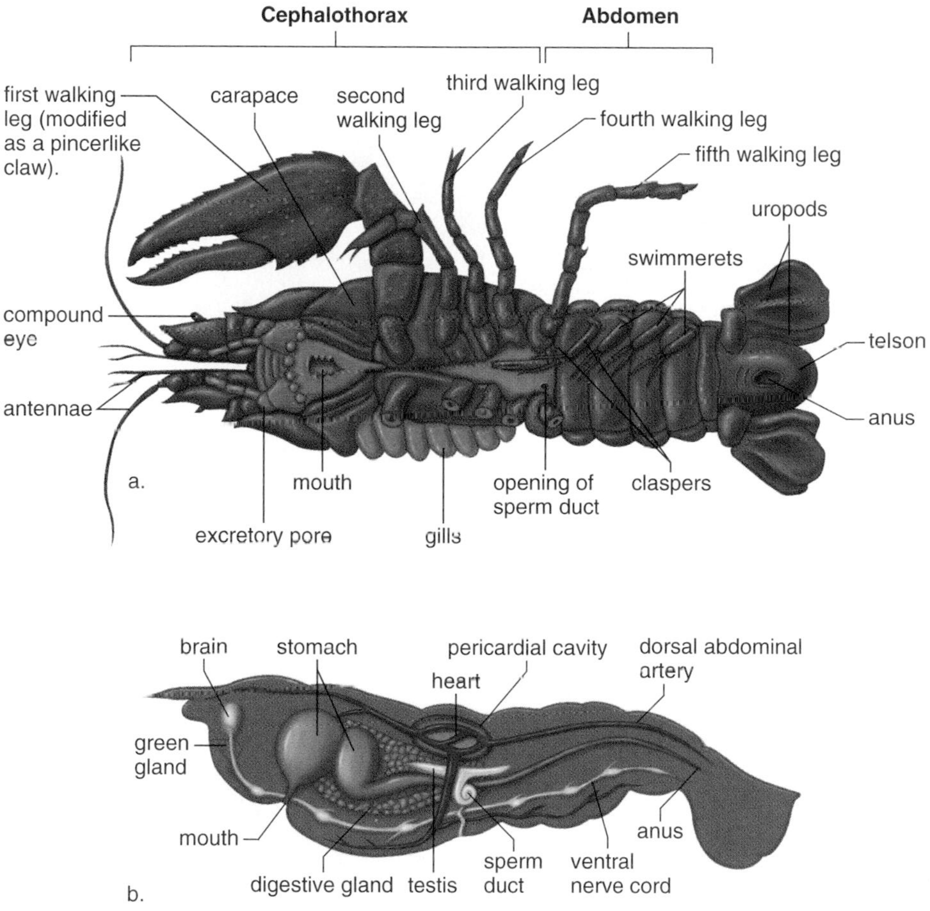

Cephalothorax　　**Abdomen**

first walking leg (modified as a pincerlike claw).

carapace

second walking leg

third walking leg

fourth walking leg

fifth walking leg

uropods

compound eye

swimmerets

telson

antennae

anus

a.

mouth

excretory pore

gills

opening of sperm duct

claspers

brain

stomach

pericardial cavity

dorsal abdominal artery

heart

green gland

mouth

anus

b.

digestive gland

testis

sperm duct

ventral nerve cord

Male crayfish, *Cambarus*
Figure 30.10

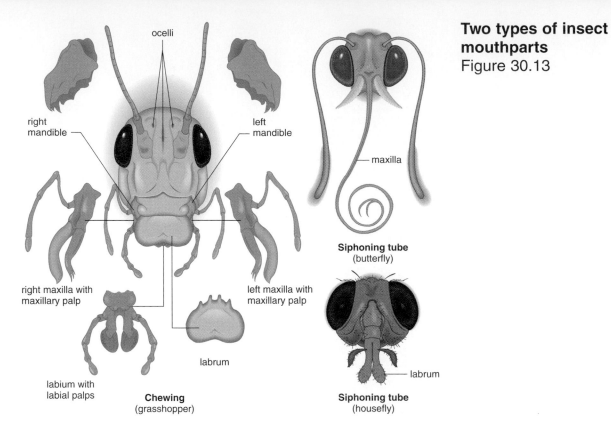

ocelli

right
mandible

left
mandible

right maxilla with
maxillary palp

left maxilla with
maxillary palp

labium with
labial palps

labrum

Chewing
(grasshopper)

Two types of insect mouthparts
Figure 30.13

maxilla

Siphoning tube
(butterfly)

labrum

Siphoning tube
(housefly)

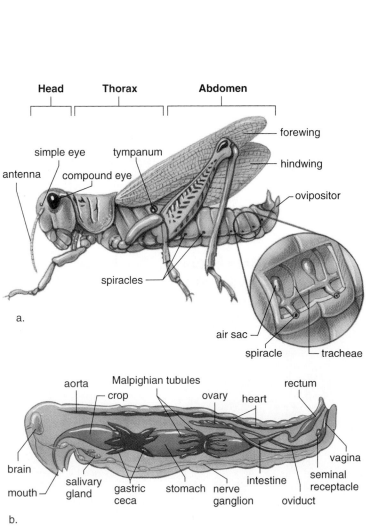

Head   Thorax   Abdomen

forewing

hindwing

ovipositor

simple eye

tympanum

antenna

compound eye

spiracles

air sac

spiracle

tracheae

a.

aorta

Malpighian tubules

crop

ovary

heart

rectum

brain

mouth

salivary
gland

gastric
ceca

stomach

nerve
ganglion

intestine

oviduct

vagina

seminal
receptacle

b.

Female grasshopper, *Romalea*
Figure 30.14

204

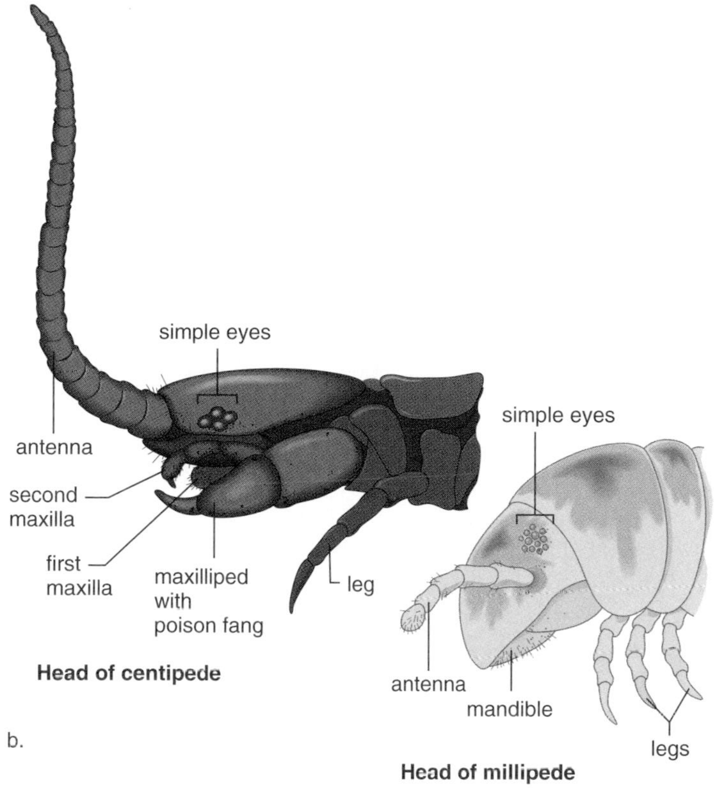

simple eyes

antenna

second
maxilla

first
maxilla

maxilliped
with
poison fang

leg

Head of centipede

simple eyes

antenna

mandible

legs

Head of millipede

b.

Centipede and millipede
Figure 30.15

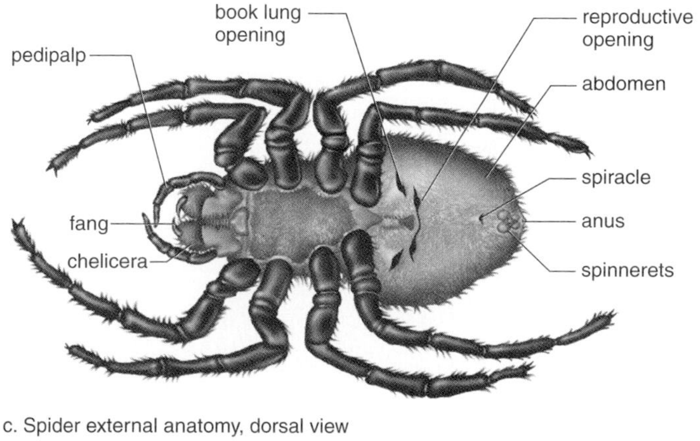

pedipalp

book lung
opening

reproductive
opening

abdomen

spiracle

fang

anus

chelicera

spinnerets

c. Spider external anatomy, dorsal view

Chelicerate diversity
Figure 30.16

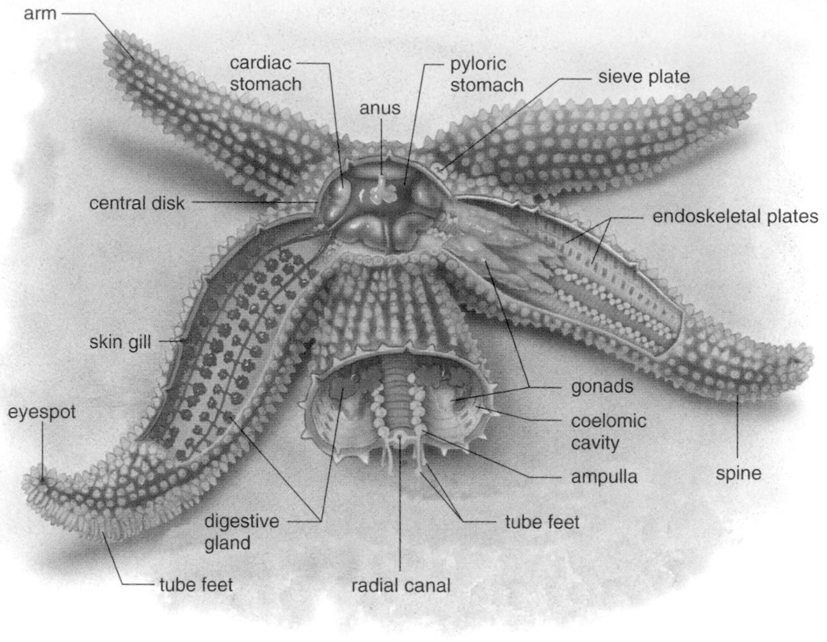

Echinoderms
Figure 31.1

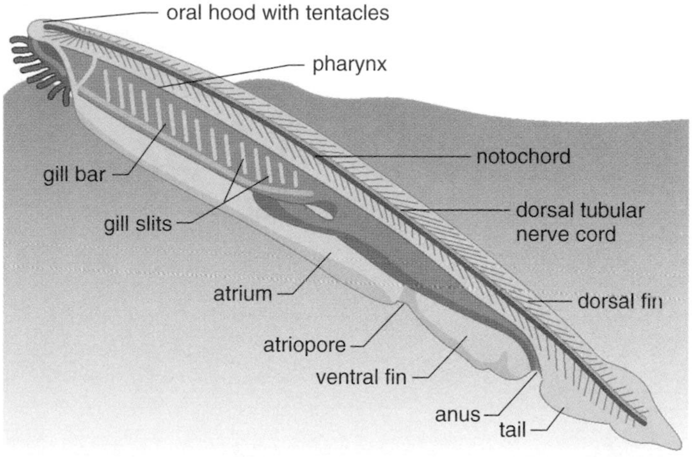

Lancelet, *Branchiostoma*
Figure 31.2

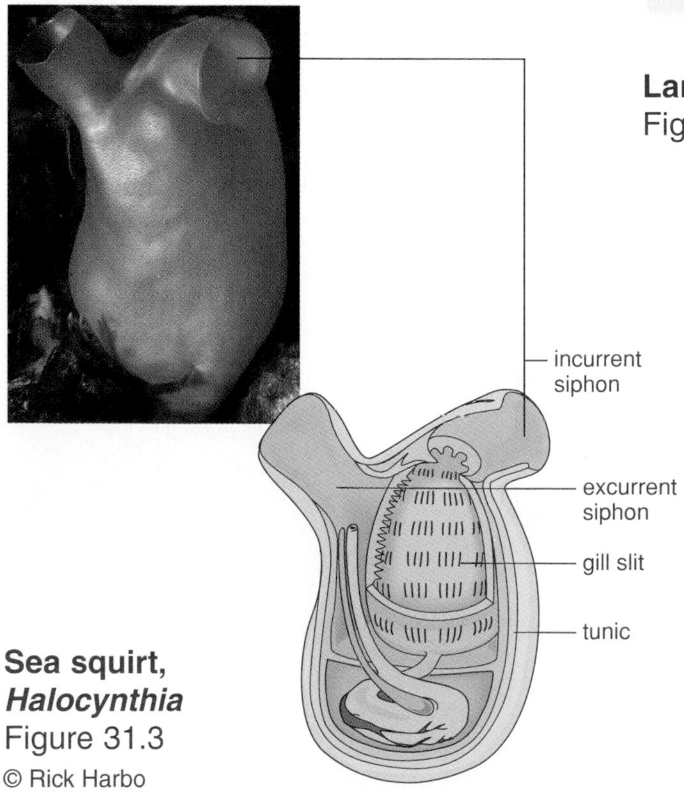

Sea squirt,
Halocynthia
Figure 31.3
© Rick Harbo

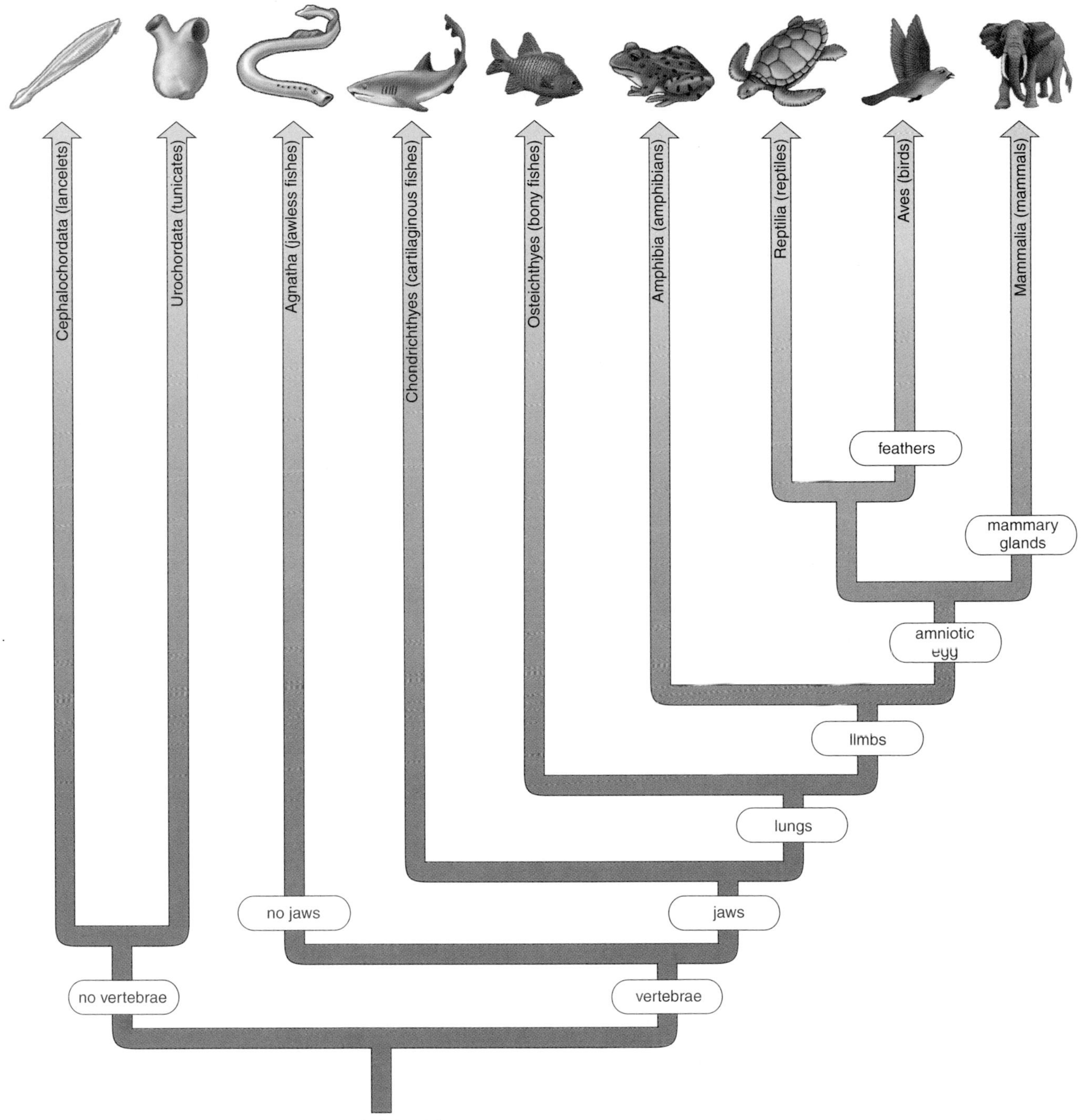
Cephalochordata (lancelets)
Urochordata (tunicates)
Agnatha (jawless fishes)
Chondrichthyes (cartilaginous fishes)
Osteichthyes (bony fishes)
Amphibia (amphibians)
Reptilia (reptiles)
Aves (birds)
Mammalia (mammals)
feathers
mammary glands
amniotic egg
limbs
lungs
no jaws
jaws
no vertebrae
vertebrae
ancestral chordate

Phylogenetic tree of the chordates
Figure 31.4

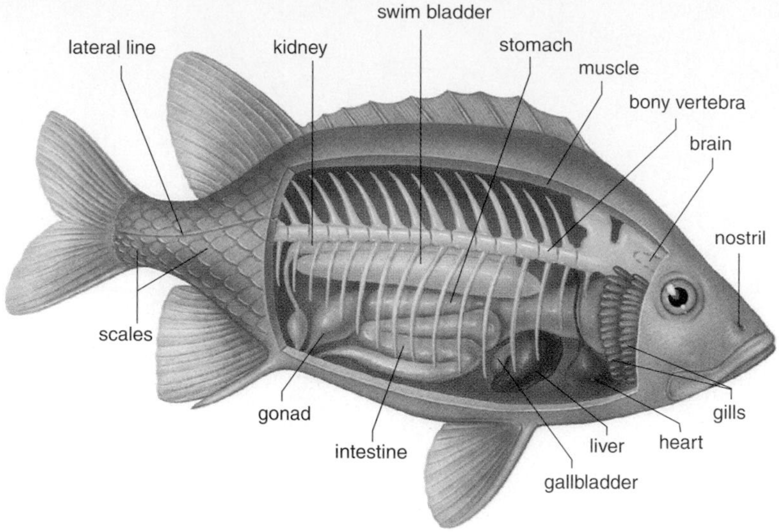

Ray-finned fishes
Figure 31.7

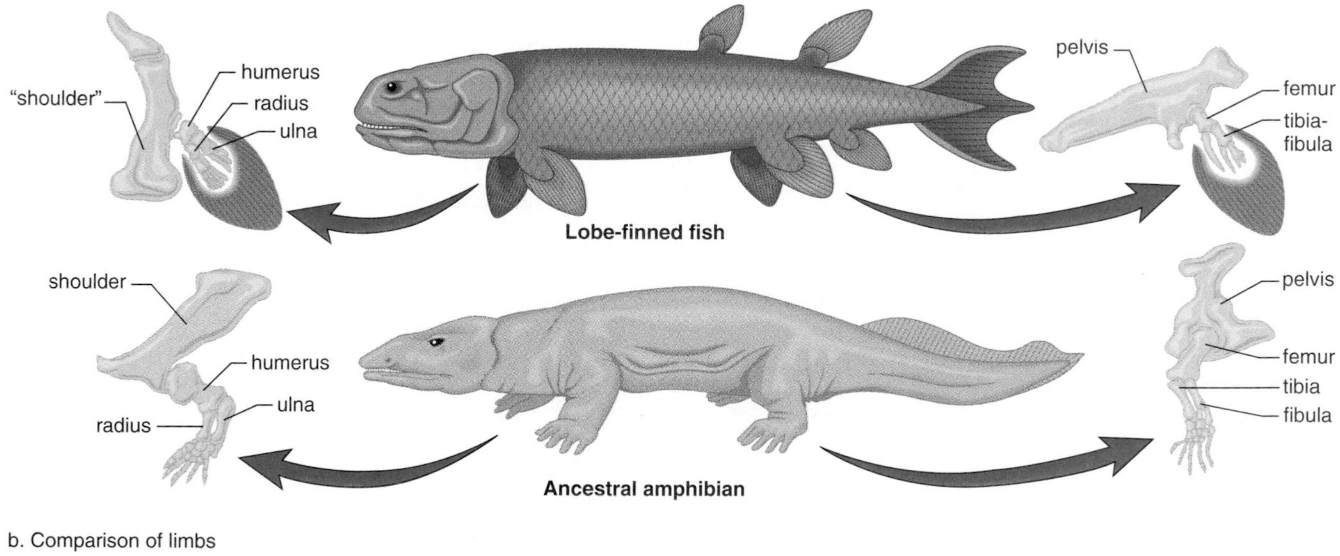

b. Comparison of limbs

Lobe-finned fish versus amphibian
Figure 31.8

208

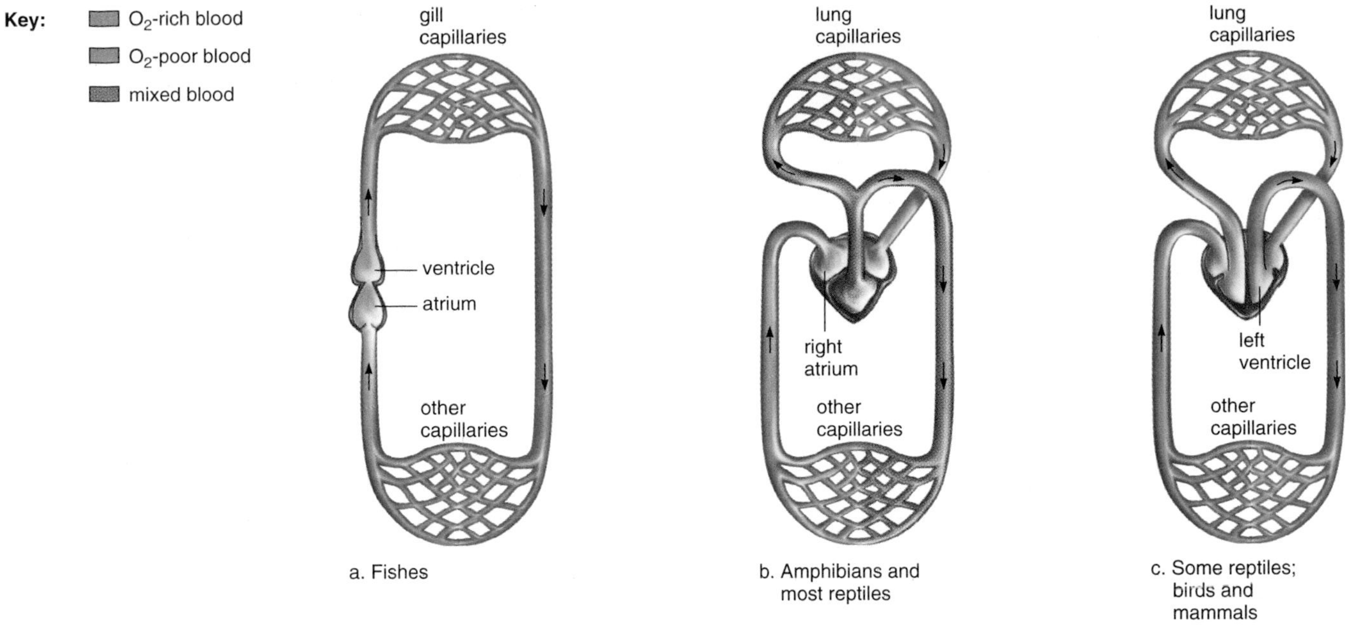

Key:
- ▨ O₂-rich blood
- ▨ O₂-poor blood
- ▨ mixed blood

gill capillaries

ventricle

atrium

other capillaries

a. Fishes

lung capillaries

right atrium

other capillaries

b. Amphibians and most reptiles

lung capillaries

left ventricle

other capillaries

c. Some reptiles; birds and mammals

Vertebrate circulatory pathways
Figure 31.11

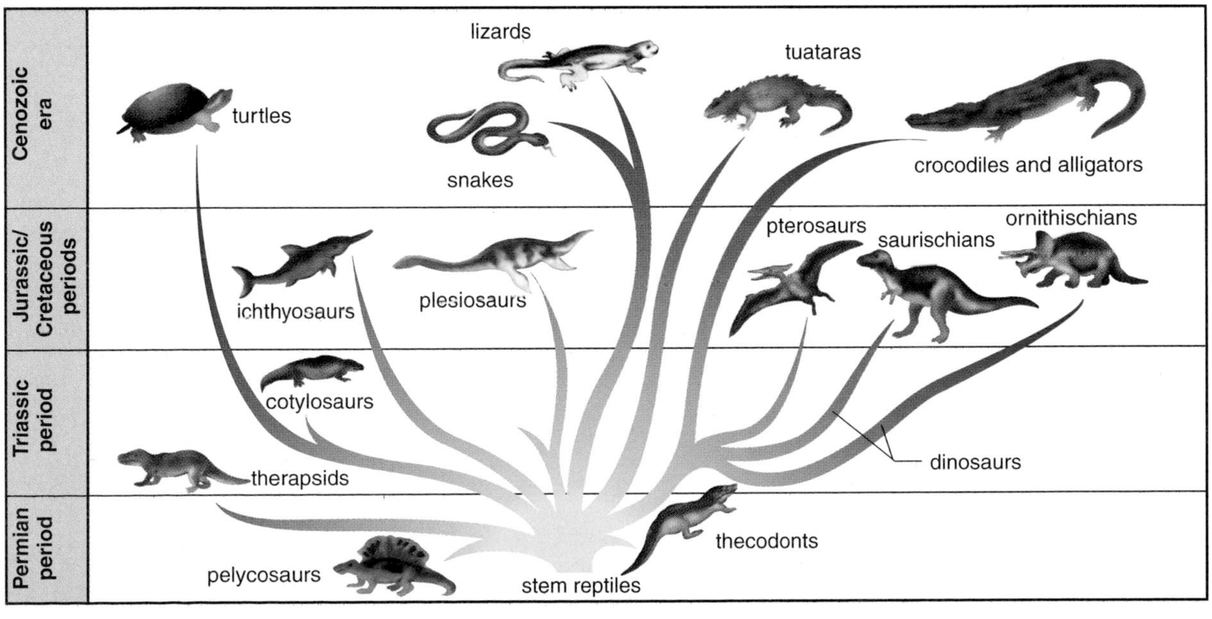

Phylogenetic tree
Figure 31.12

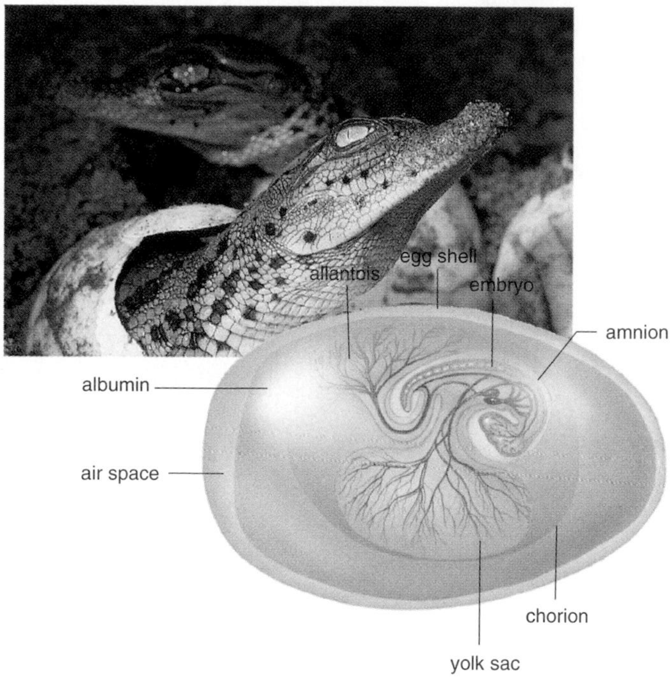

a. American crocodile, *Crocodylus acutus*

Reptilian diversity
Figure 31.13

a: © Bruce Davidson/Animals Animals/Earth Scenes

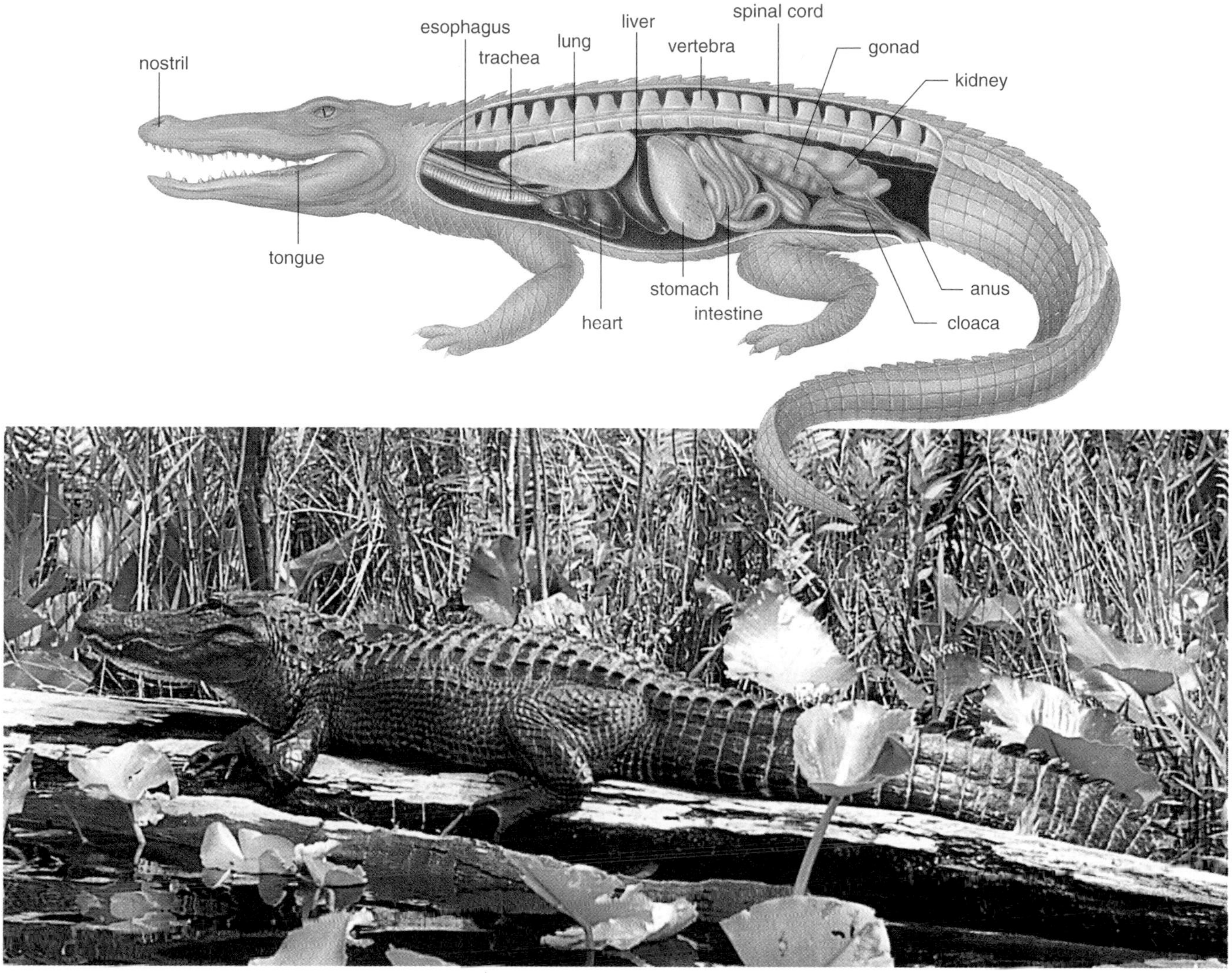

Reptilian anatomy
Figure 31.14

b: © R.F. Ashley/Visuals Unlimited

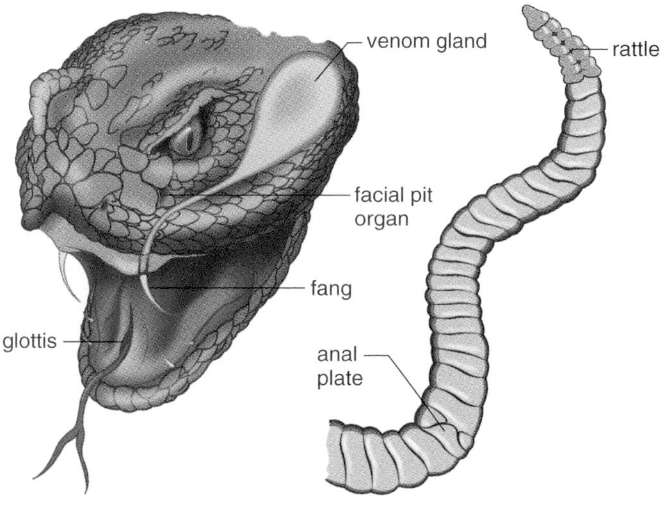

Viper characteristics
Figure 31B

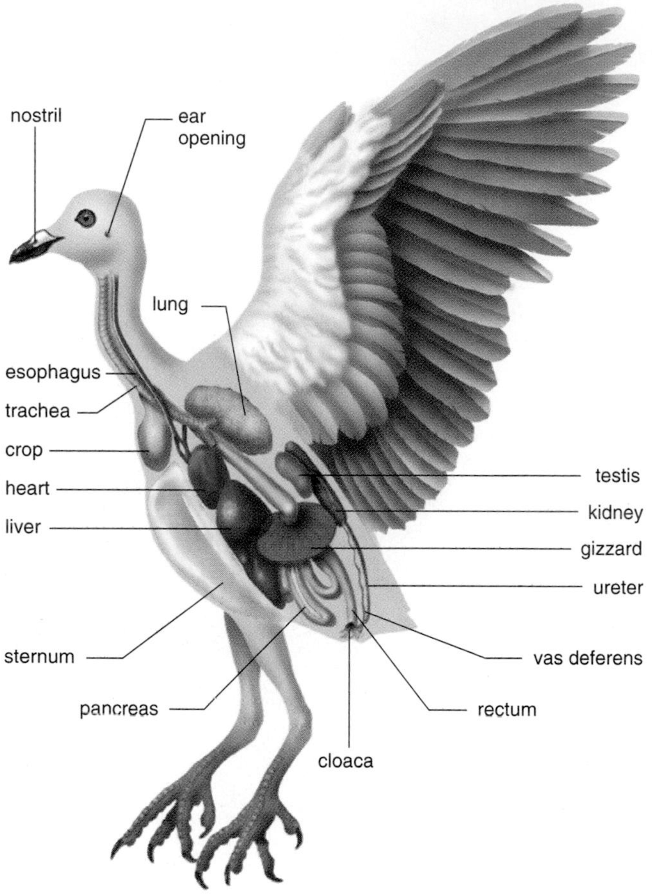

nostril

ear
opening

lung

esophagus

trachea

crop

heart

liver

sternum

pancreas

cloaca

testis

kidney

gizzard

ureter

vas deferens

rectum

a. Pigeon, *Columba*

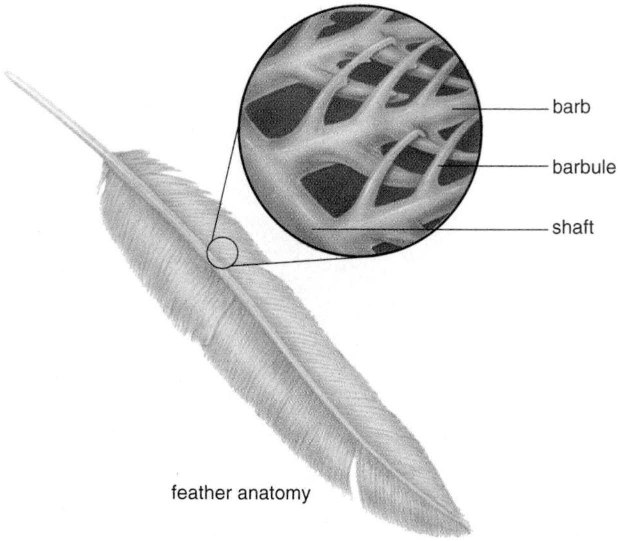

barb

barbule

shaft

feather anatomy

Bird anatomy and flight
Figure 31.15

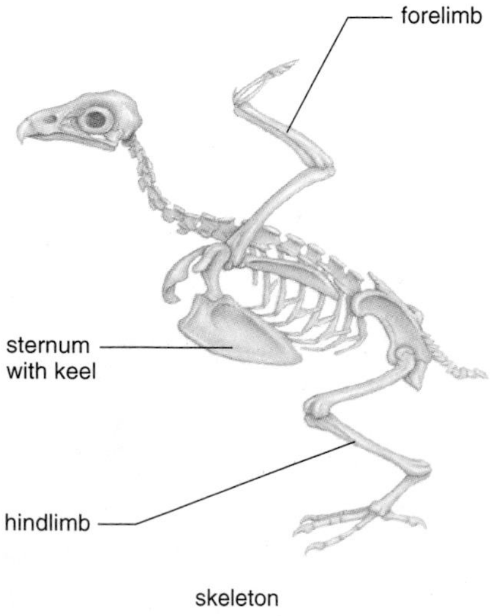

forelimb

sternum
with keel

hindlimb

skeleton

b. Bald eagle, *Haliaetus*

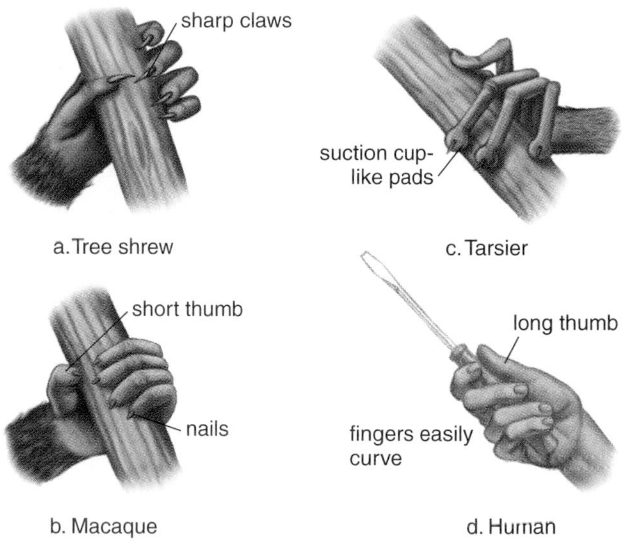

a. Tree shrew — sharp claws

c. Tarsier — suction cup-like pads

b. Macaque — short thumb, nails

d. Human — long thumb, fingers easily curve

Evolution of primate hand
Figure 32.2

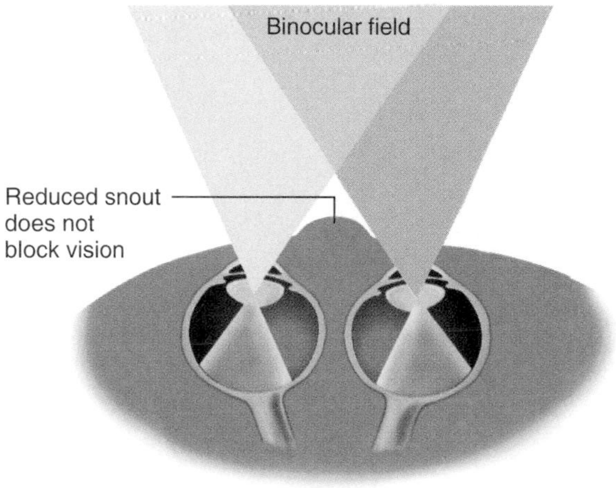

Binocular field

Reduced snout does not block vision

Binocular vision
Figure 32.3

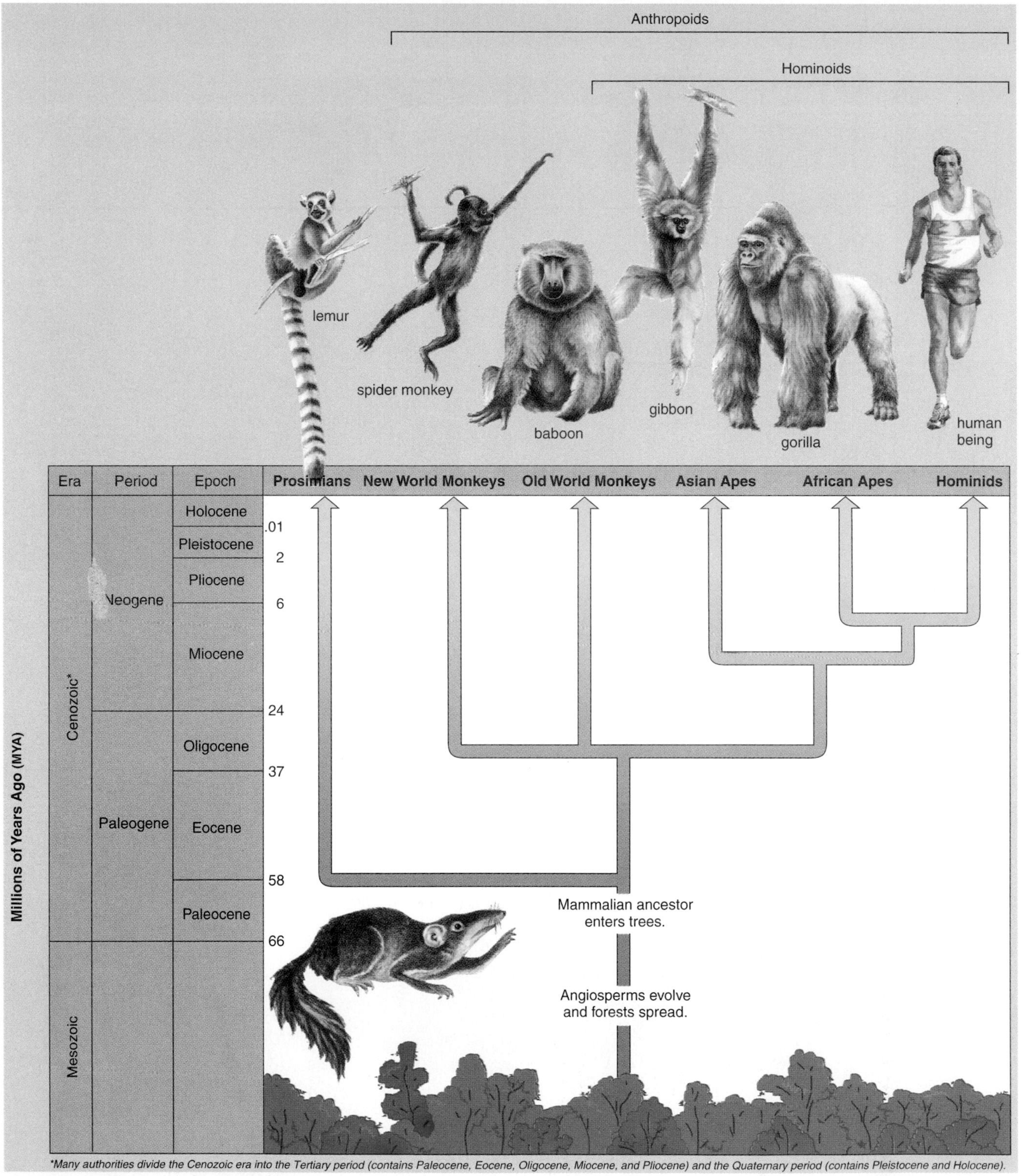
Anthropoids
Hominoids
lemur
spider monkey
baboon
gibbon
gorilla
human being
Era
Period
Epoch
Prosimians
New World Monkeys
Old World Monkeys
Asian Apes
African Apes
Hominids
Holocene
Pleistocene
Pliocene
Miocene
Oligocene
Eocene
Paleocene
Neogene
Paleogene
Cenozoic*
Mesozoic
.01
2
6
24
37
58
66
Millions of Years Ago (MYA)
Mammalian ancestor enters trees.
Angiosperms evolve and forests spread.
*Many authorities divide the Cenozoic era into the Tertiary period (contains Paleocene, Eocene, Oligocene, Miocene, and Pliocene) and the Quaternary period (contains Pleistocene and Holocene).

Evolution of primates
Figure 32.4

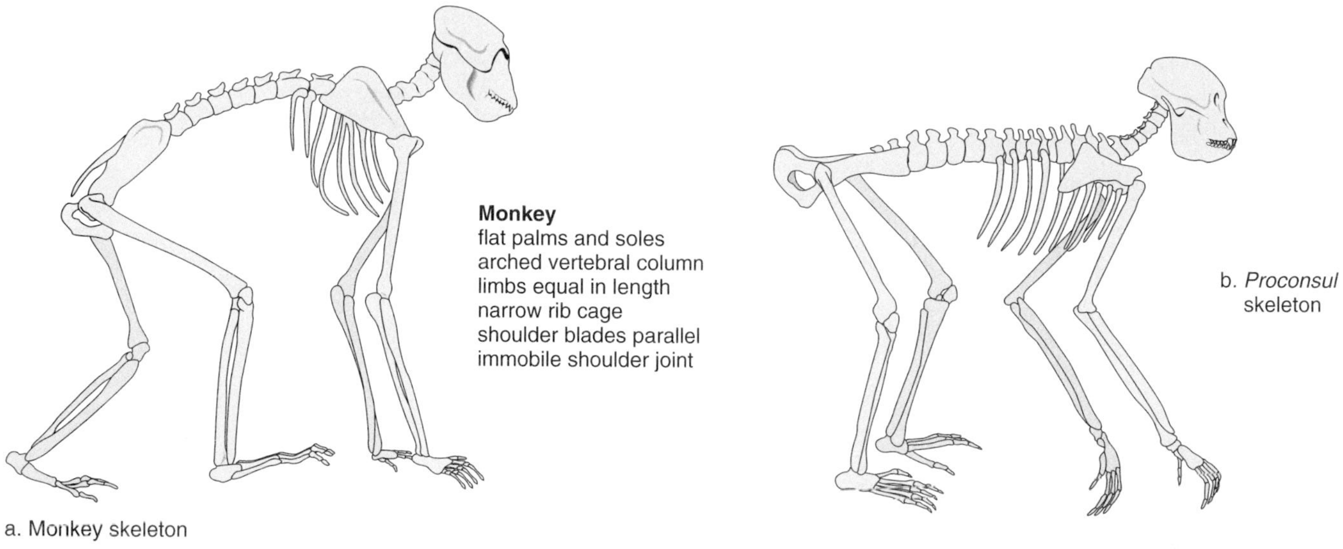

Monkey
flat palms and soles
arched vertebral column
limbs equal in length
narrow rib cage
shoulder blades parallel
immobile shoulder joint

b. *Proconsul*
skeleton

a. Monkey skeleton

Monkey skeleton compared to *Proconsul* skeleton
Figure 32.5

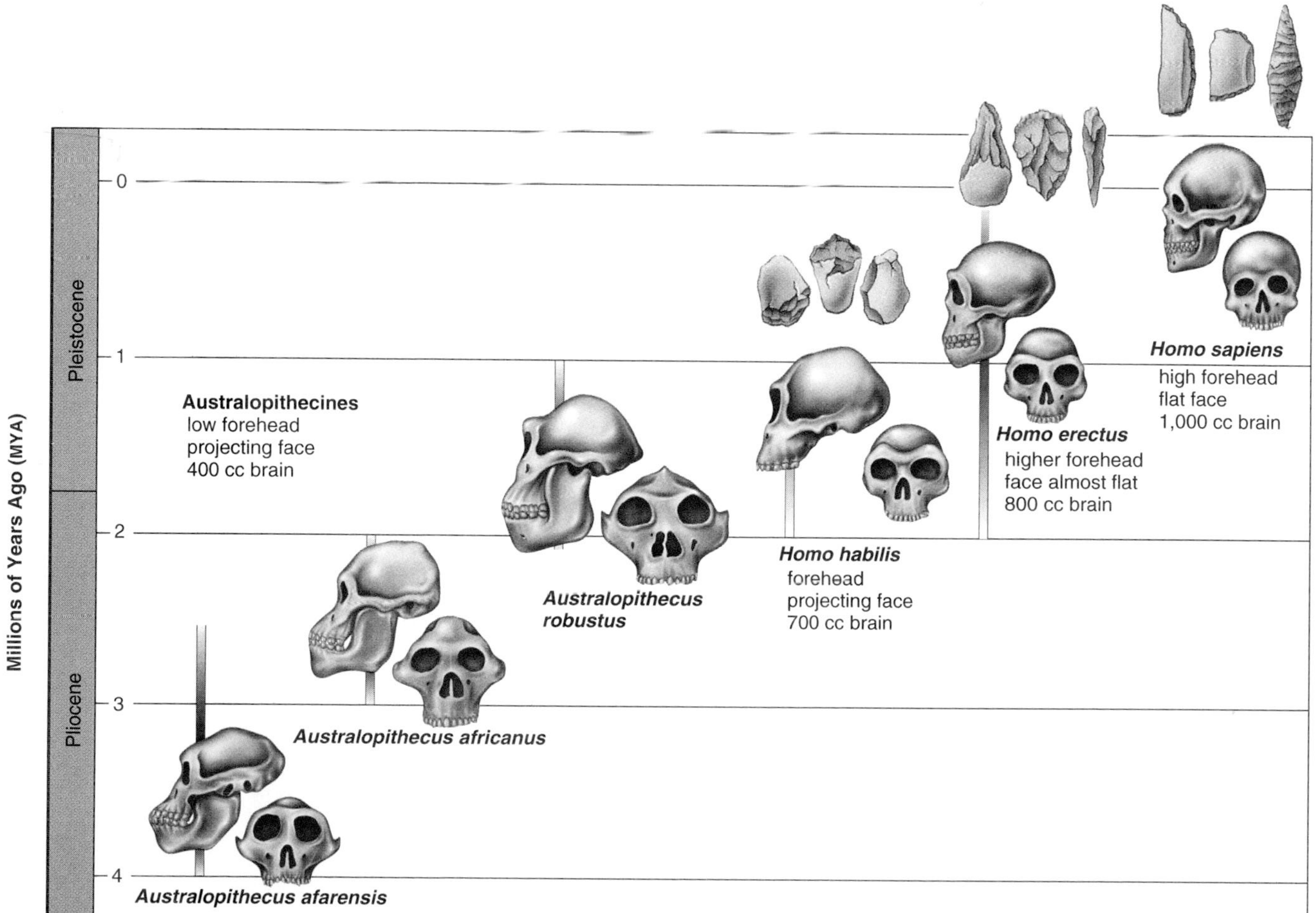

Millions of Years Ago (MYA)

Pleistocene

Pliocene

Australopithecines
low forehead
projecting face
400 cc brain

Australopithecus robustus

Australopithecus africanus

Australopithecus afarensis

Homo habilis
forehead
projecting face
700 cc brain

Homo erectus
higher forehead
face almost flat
800 cc brain

Homo sapiens
high forehead
flat face
1,000 cc brain

Human evolution
Figure 32.8

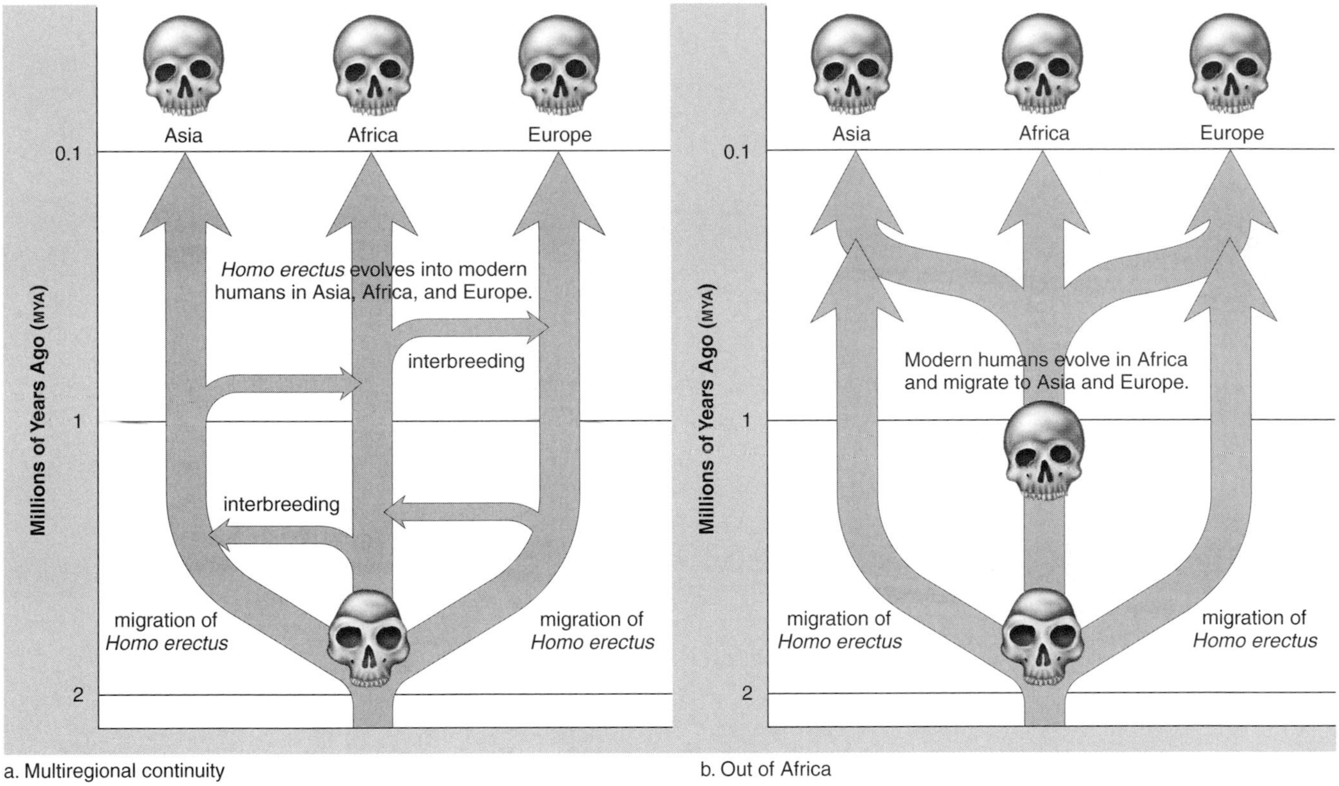

Within the figure:

0.1

Asia   Africa   Europe

Homo erectus evolves into modern humans in Asia, Africa, and Europe.

interbreeding

1

interbreeding

migration of *Homo erectus*   migration of *Homo erectus*

2

Millions of Years Ago (MYA)

a. Multiregional continuity

0.1

Asia   Africa   Europe

Modern humans evolve in Africa and migrate to Asia and Europe.

1

migration of *Homo erectus*   migration of *Homo erectus*

2

Millions of Years Ago (MYA)

b. Out of Africa

Evolution of modern humans
Figure 32.10

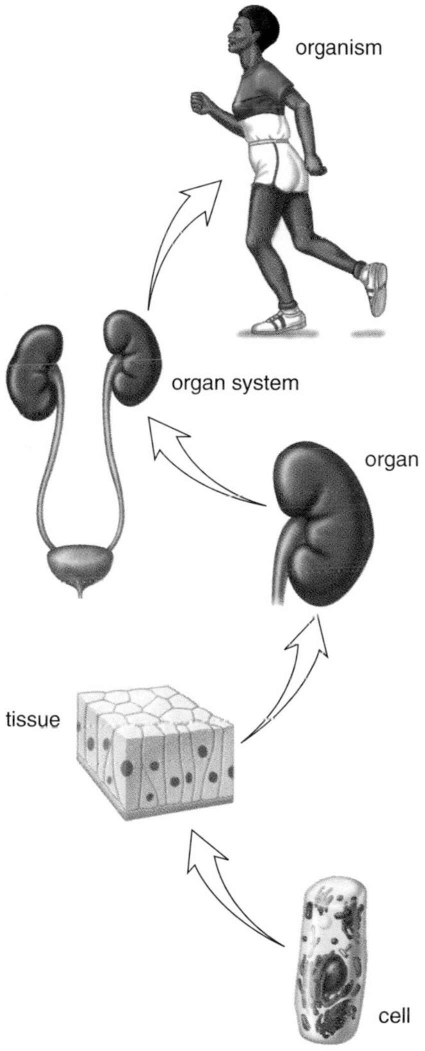

organism

organ system

organ

tissue

cell

Levels of organization
Figure 33.1

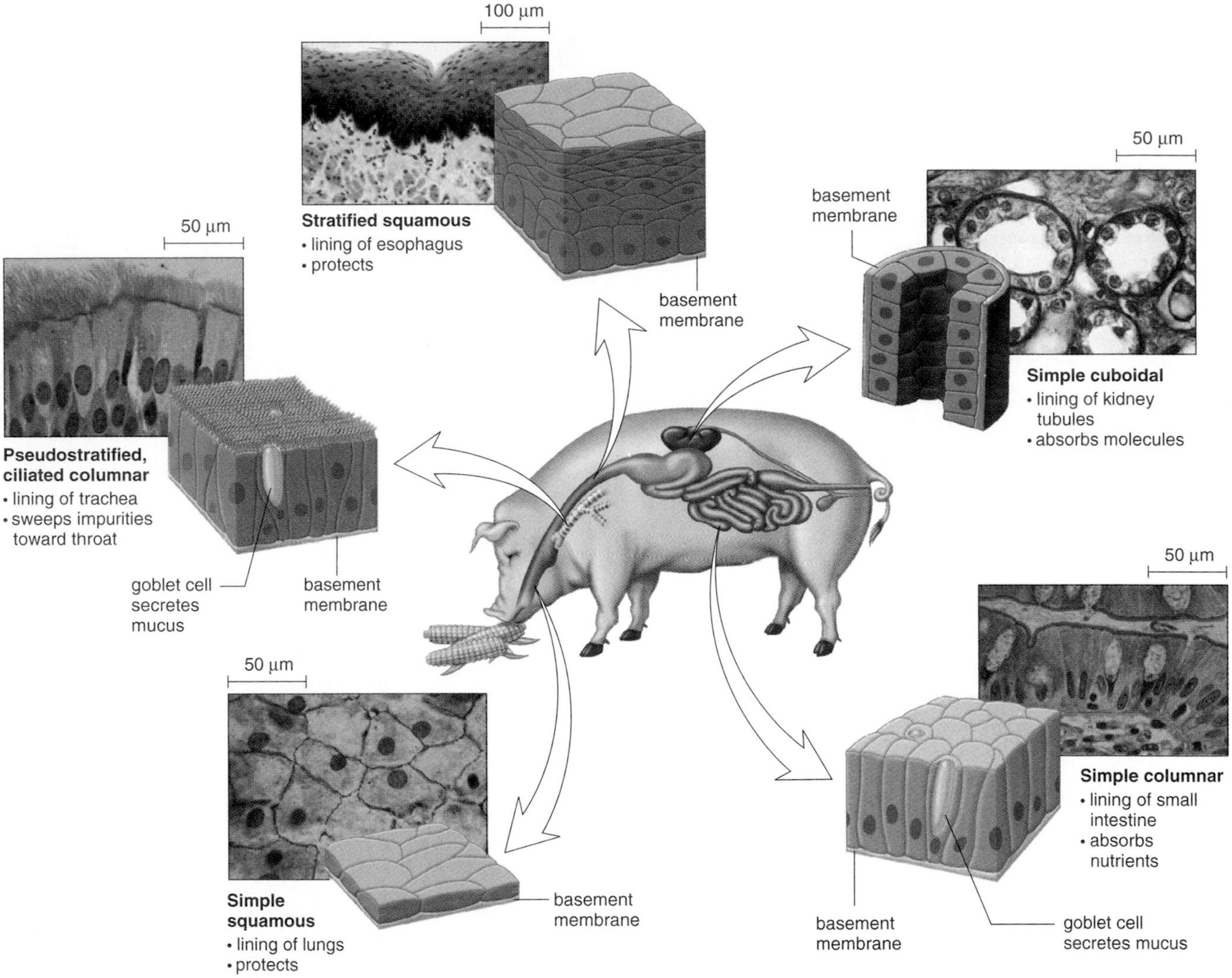

100 μm

Stratified squamous
• lining of esophagus
• protects

basement
membrane

50 μm

basement
membrane

Simple cuboidal
• lining of kidney
  tubules
• absorbs molecules

50 μm

**Pseudostratified,
ciliated columnar**
• lining of trachea
• sweeps impurities
  toward throat

goblet cell
secretes
mucus

basement
membrane

50 μm

50 μm

Simple columnar
• lining of small
  intestine
• absorbs
  nutrients

**Simple
squamous**
• lining of lungs
• protects

basement
membrane

basement
membrane

goblet cell
secretes mucus

Types of epithelial tissues in vertebrates
Figure 33.2

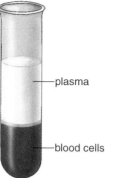

plasma

blood cells

a. Blood sample

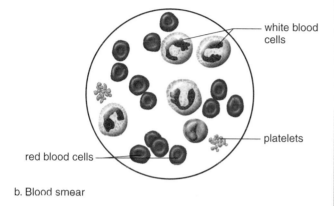

white blood cells

platelets

red blood cells

b. Blood smear

Blood, a liquid tissue
Figure 33.4

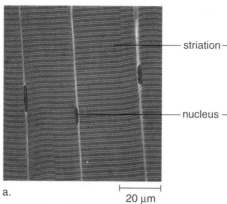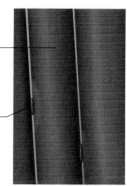

striation

nucleus

a.

20 μm

Skeletal muscle
• has striated cells with multiple nuclei.
• occurs in muscles attached to skeleton.
• functions in voluntary movement of body.
• is voluntary.

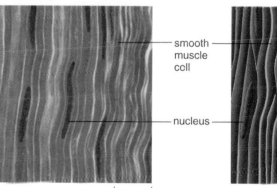

smooth muscle cell

nucleus

b.

12 μm

Smooth muscle
• has spindle-shaped cells, each with a single nucleus.
• cells have no striations.
• functions in movement of substances in lumens of body.
• is involuntary.

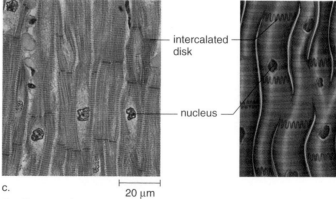

intercalated disk

nucleus

c.

20 μm

Cardiac muscle
• has branching striated cells, each with a single nucleus.
• occurs in the wall of the heart.
• functions in the pumping of blood.
• is involuntary.

Muscular tissue
Figure 33.5

Neurons and neuroglia
Figure 33.6

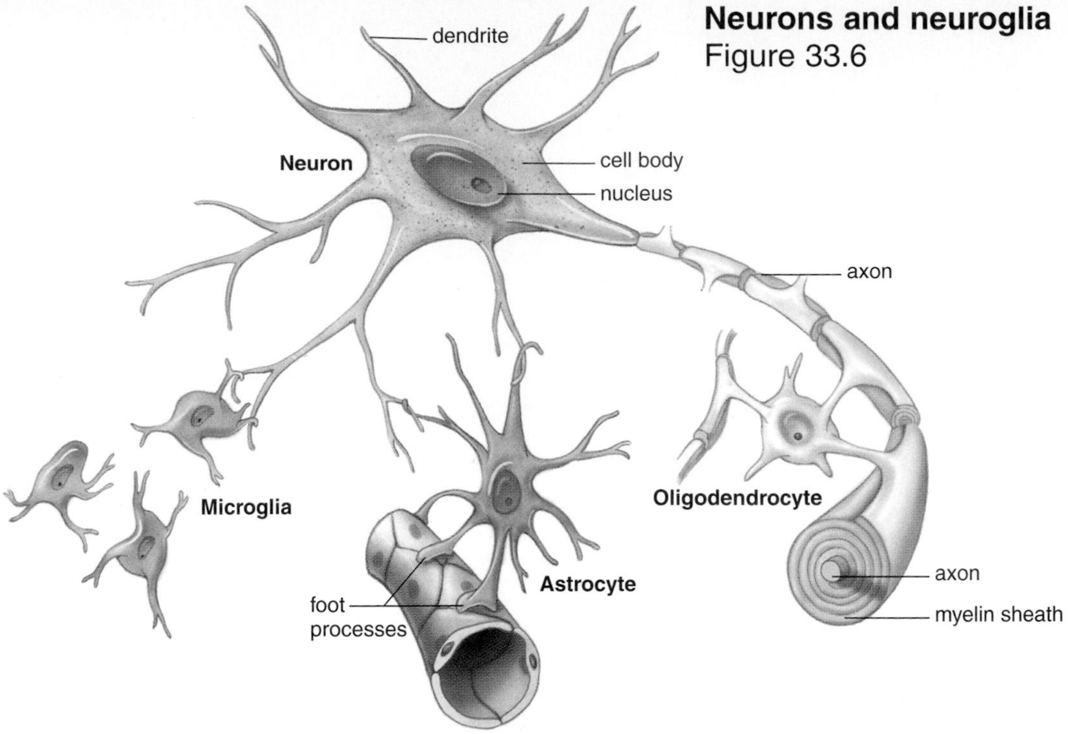

dendrite

Neuron

cell body

nucleus

axon

Microglia

Oligodendrocyte

foot processes

Astrocyte

axon

myelin sheath

Human skin anatomy
Figure 33.7

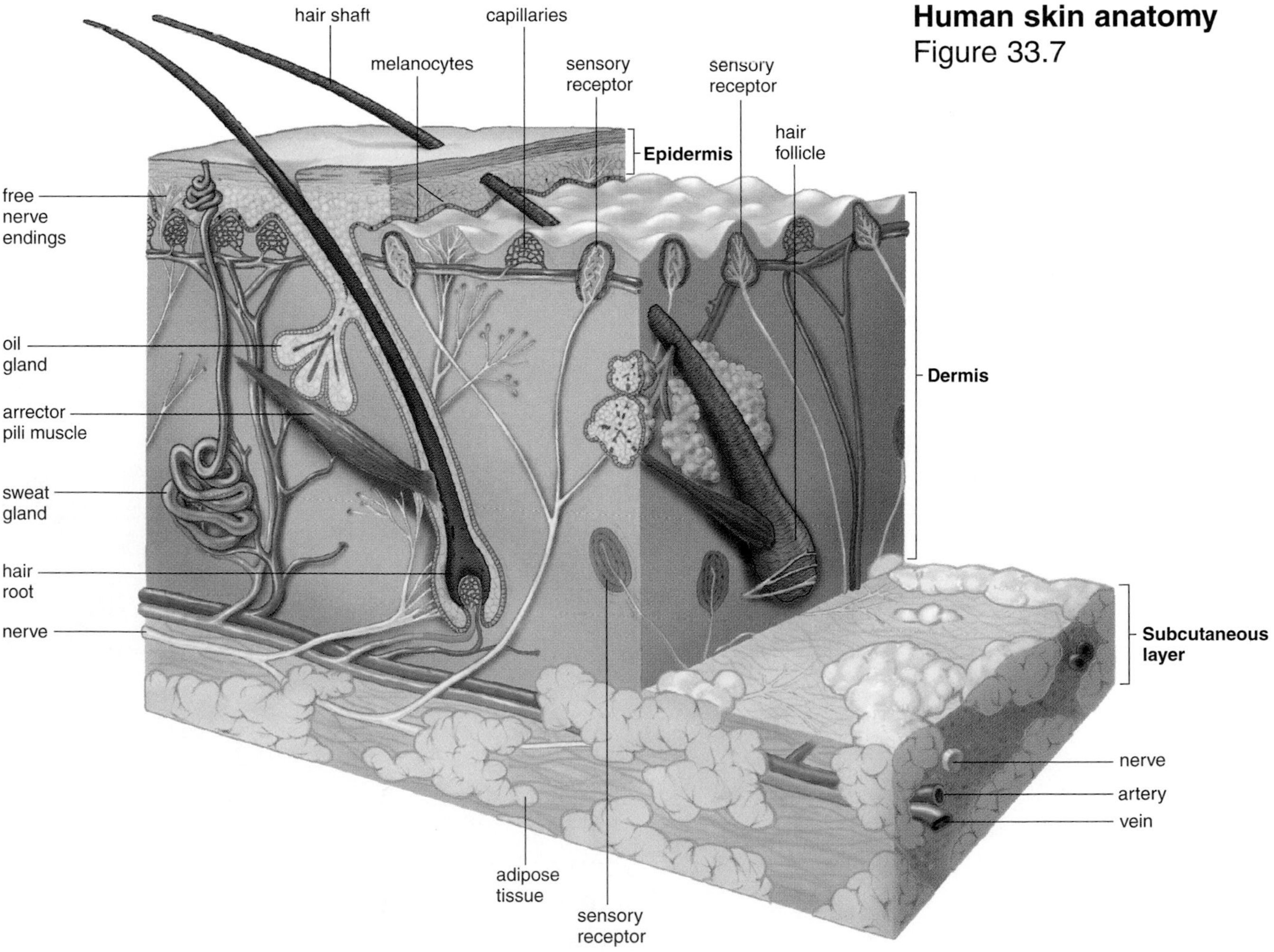

hair shaft

capillaries

melanocytes

sensory receptor

sensory receptor

hair follicle

Epidermis

free nerve endings

oil gland

arrector pili muscle

sweat gland

hair root

nerve

Dermis

Subcutaneous layer

nerve

artery

vein

adipose tissue

sensory receptor

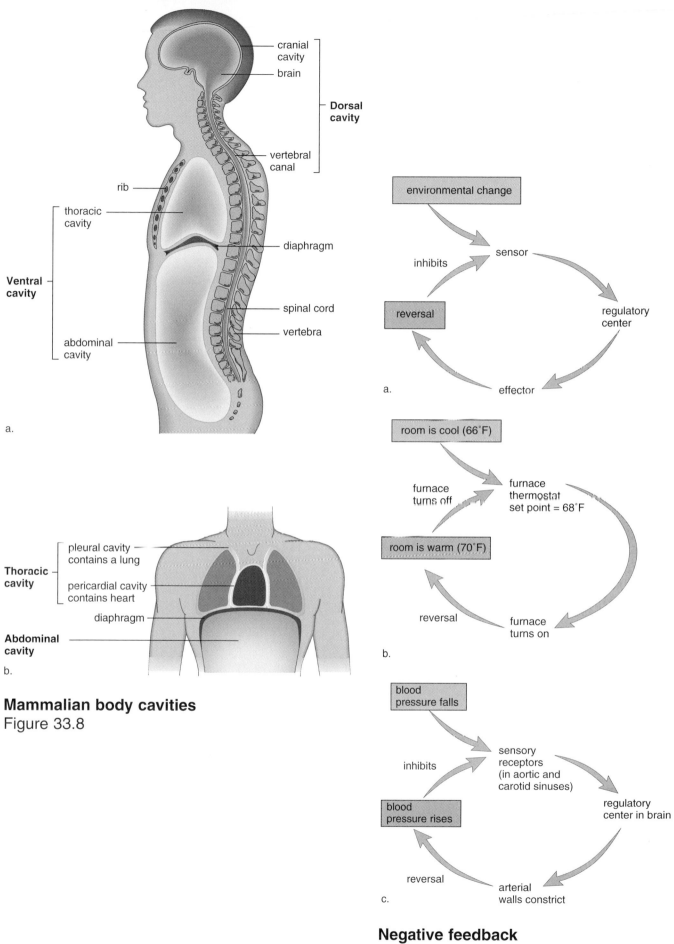

Mammalian body cavities
Figure 33.8

Negative feedback
Figure 33.9

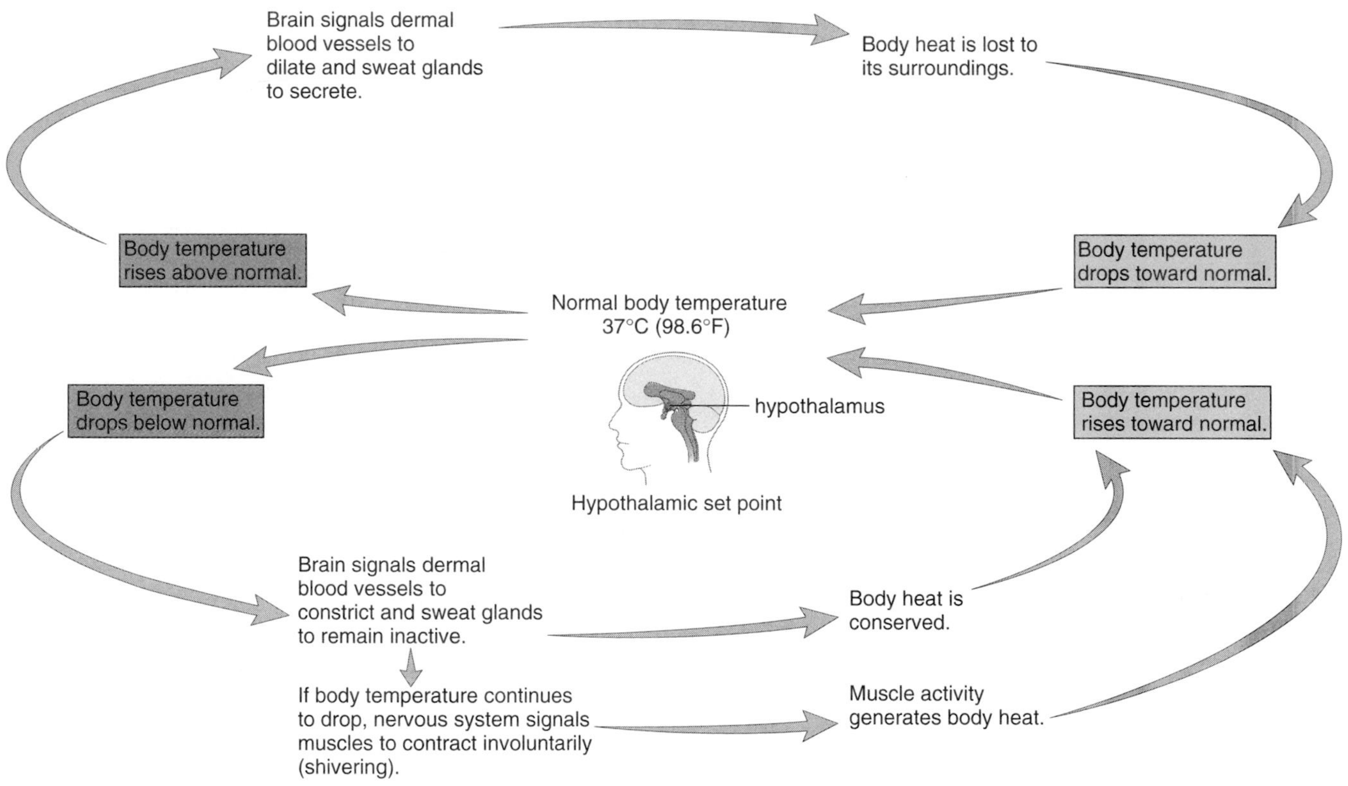

Homeostasis and body temperature regulation
Figure 33.10

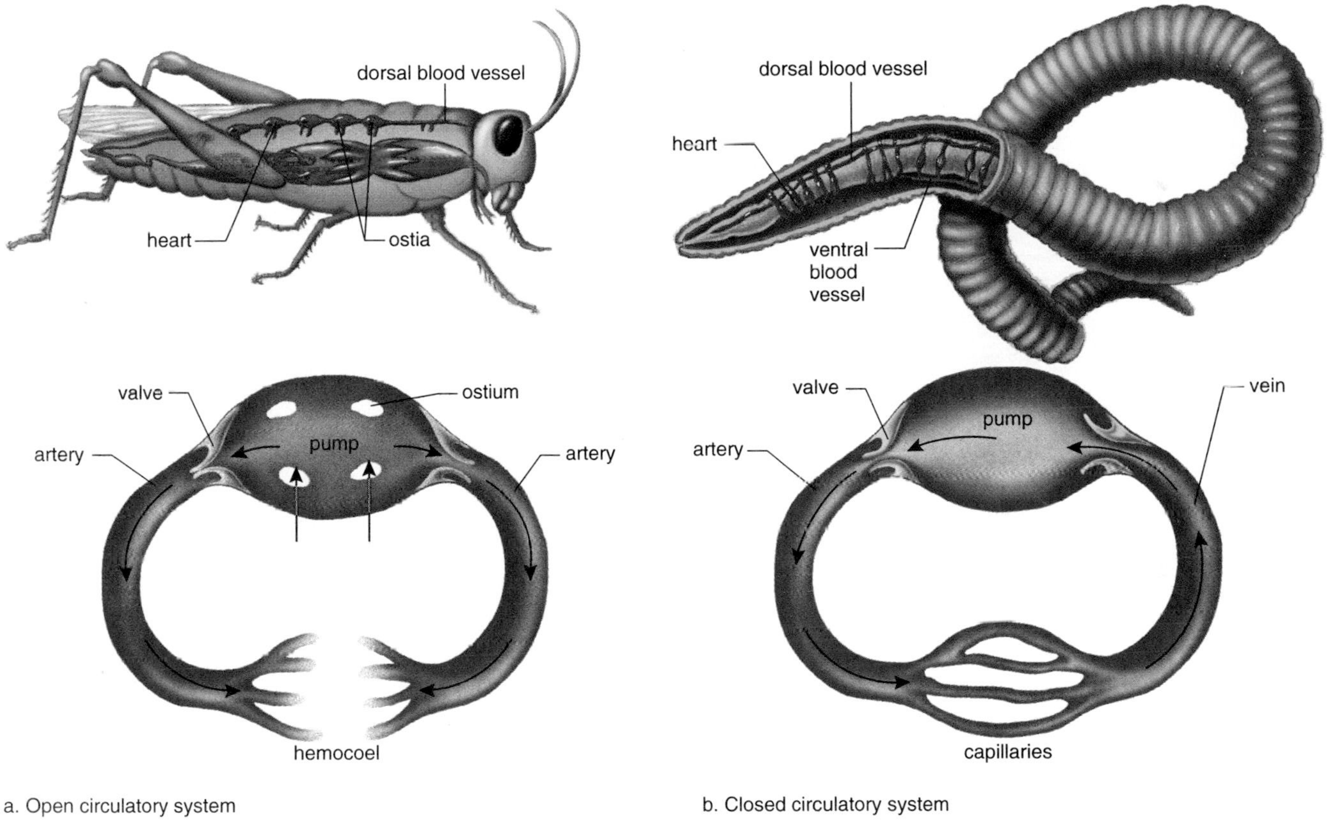

dorsal blood vessel

heart

ostia

valve

ostium

artery

pump

artery

hemocoel

a. Open circulatory system

dorsal blood vessel

heart

ventral blood vessel

valve

pump

vein

artery

capillaries

b. Closed circulatory system

Open versus closed circulatory systems
Figure 34.2

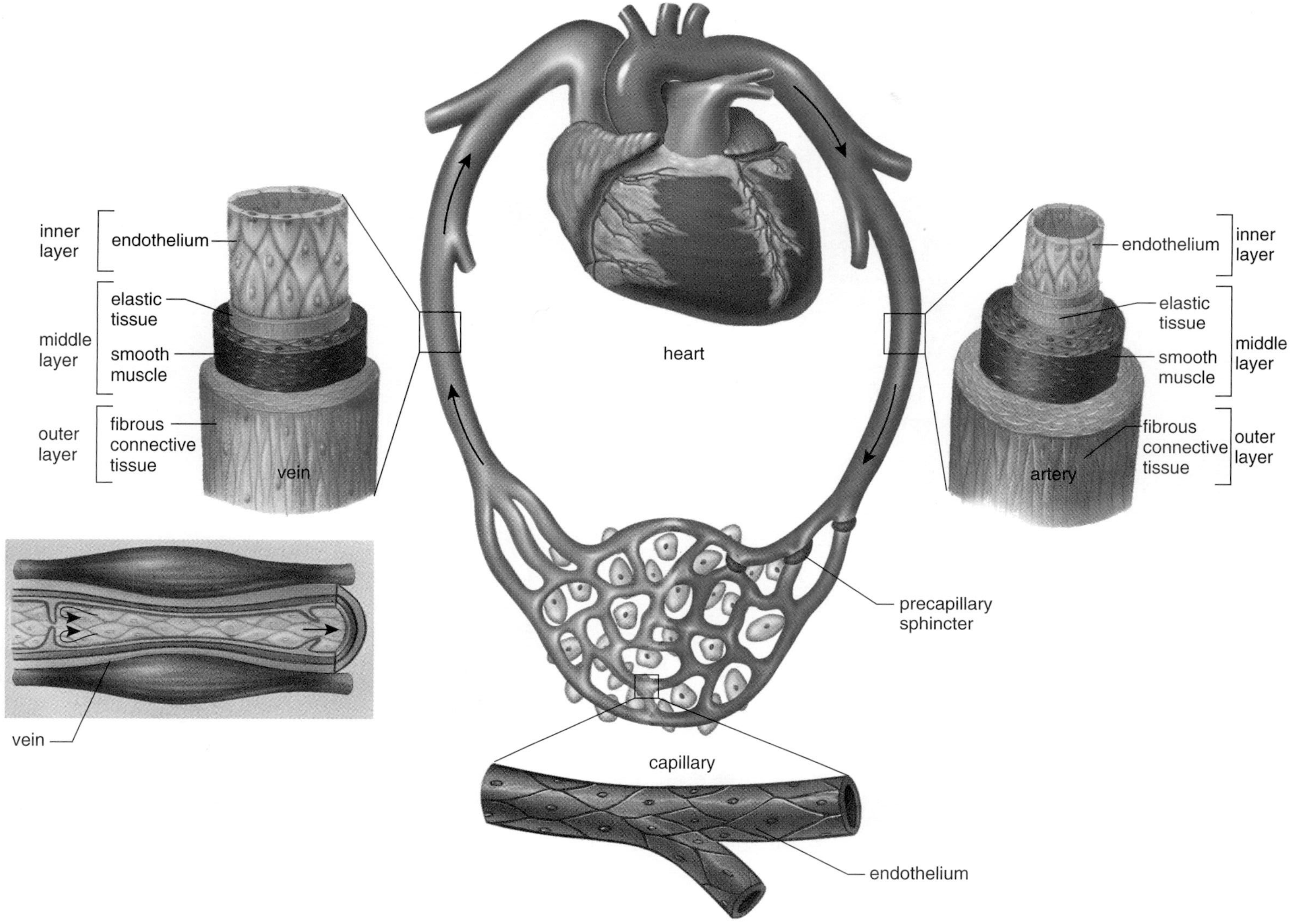
inner layer
endothelium
elastic tissue
middle layer
smooth muscle
outer layer
fibrous connective tissue
vein
vein
heart
endothelium
inner layer
elastic tissue
middle layer
smooth muscle
fibrous connective tissue
outer layer
artery
precapillary sphincter
capillary
endothelium

Transport in birds and mammals
Figure 34.3

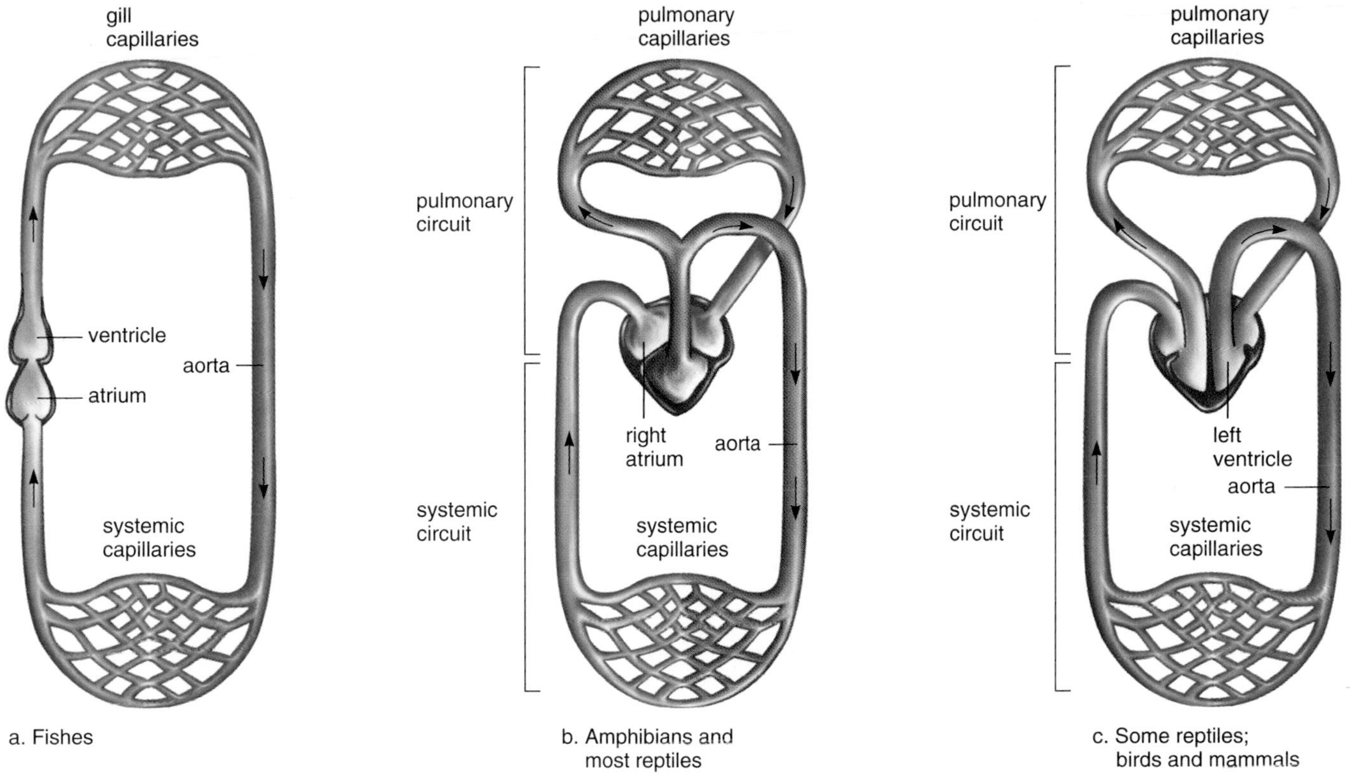

a. Fishes

b. Amphibians and most reptiles

c. Some reptiles; birds and mammals

Key:
- O$_2$-rich blood
- O$_2$-poor blood
- mixed blood

Comparison of circulatory circuits in vertebrates
Figure 34.4

External heart anatomy
Figure 34.5

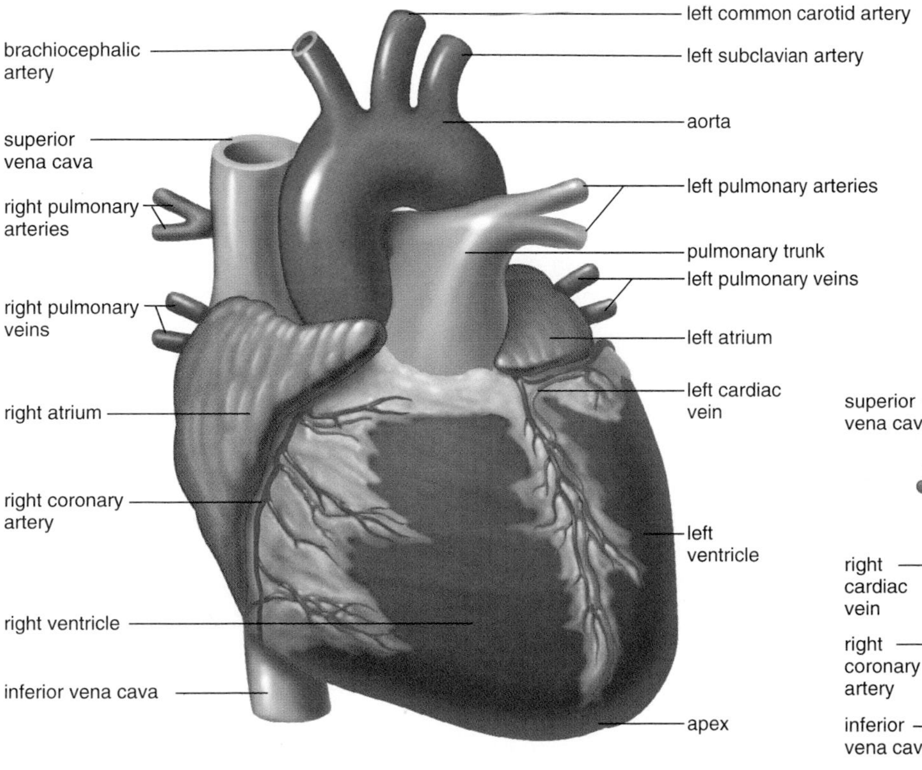

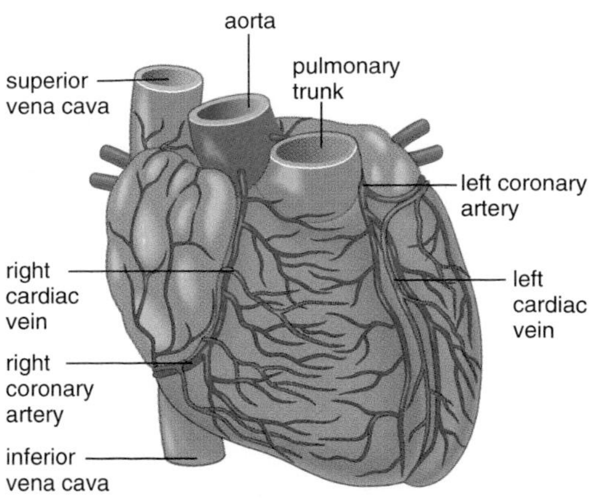

Internal view of the heart
Figure 34.6

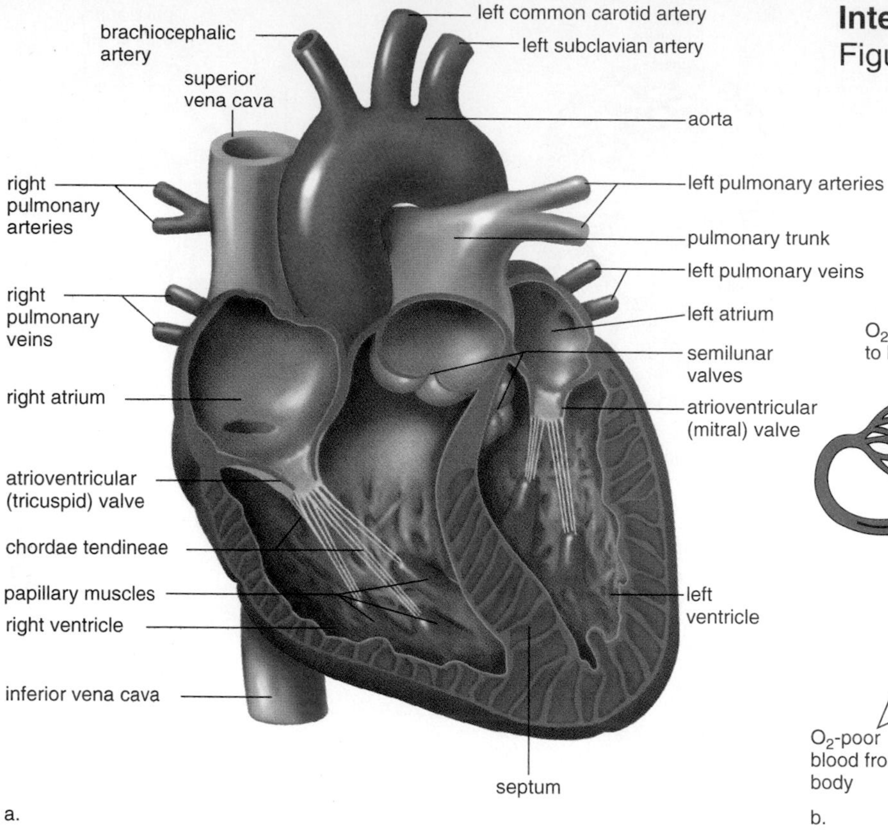

a.

- brachiocephalic artery
- superior vena cava
- left common carotid artery
- left subclavian artery
- aorta
- right pulmonary arteries
- right pulmonary veins
- right atrium
- atrioventricular (tricuspid) valve
- chordae tendineae
- papillary muscles
- right ventricle
- inferior vena cava
- septum
- left pulmonary arteries
- pulmonary trunk
- left pulmonary veins
- left atrium
- semilunar valves
- atrioventricular (mitral) valve
- left ventricle

b.

- O₂-rich blood to body
- O₂-poor blood to lungs
- O₂-rich blood from lungs
- O₂-poor blood from body

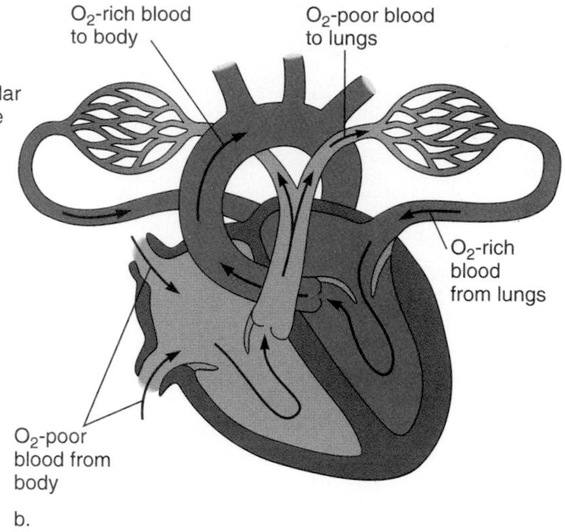

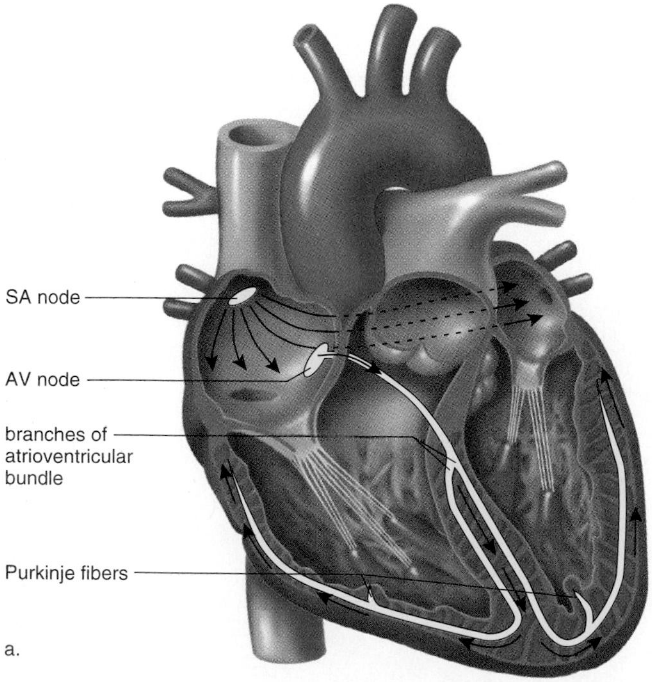

- SA node
- AV node
- branches of atrioventricular bundle
- Purkinje fibers

a.

Conduction system of the heart
Figure 34.7

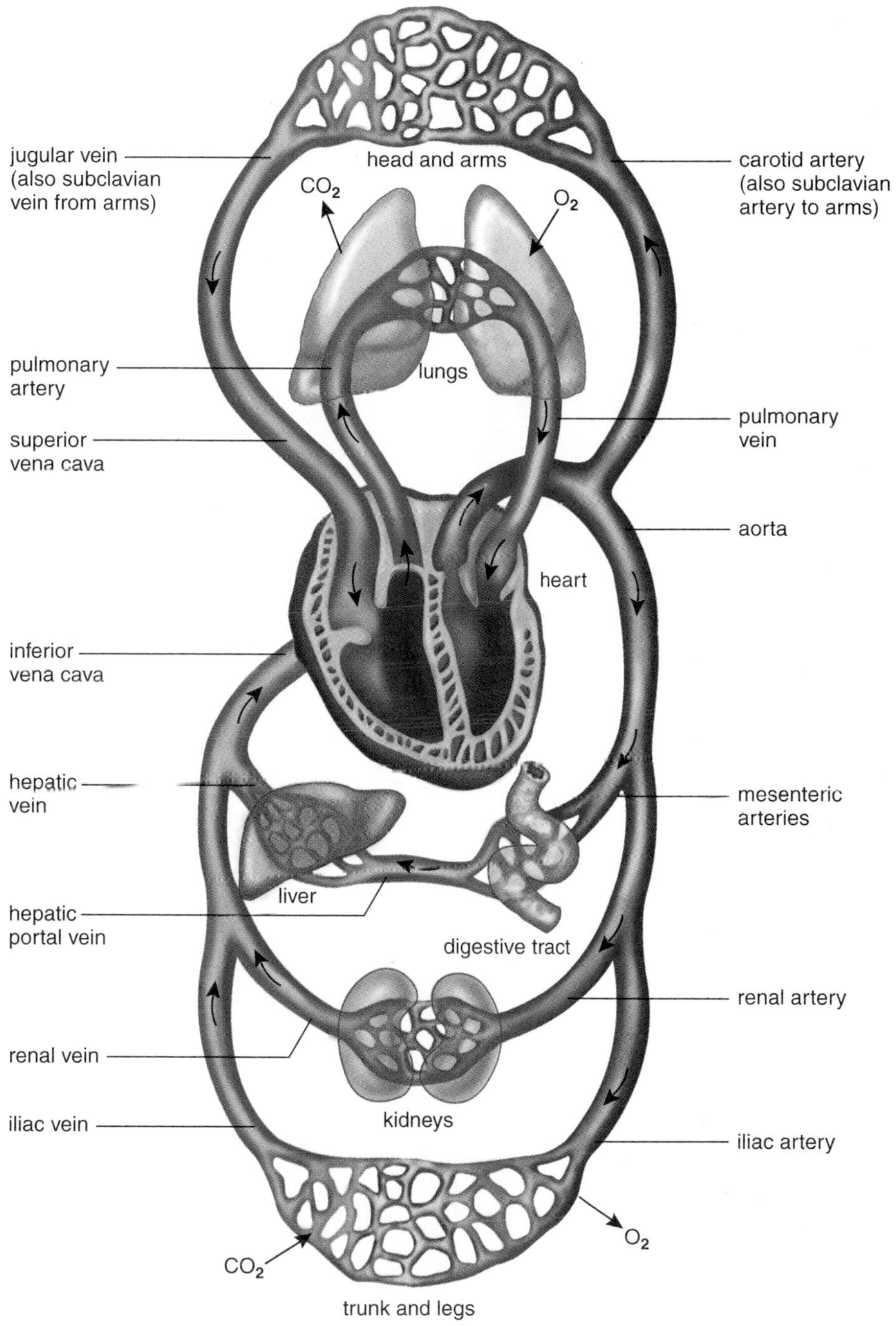

jugular vein
(also subclavian
vein from arms)

head and arms

CO_2

O_2

carotid artery
(also subclavian
artery to arms)

pulmonary
artery

lungs

superior
vena cava

pulmonary
vein

aorta

heart

inferior
vena cava

hepatic
vein

mesenteric
arteries

liver

hepatic
portal vein

digestive tract

renal artery

renal vein

kidneys

iliac vein

iliac artery

CO_2

O_2

trunk and legs

Path of blood
Figure 34.8

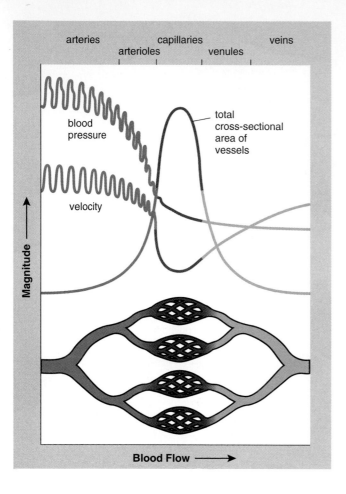

arteries · arterioles · capillaries · venules · veins

blood pressure

velocity

total cross-sectional area of vessels

Magnitude

Blood Flow

Velocity and blood pressure related to vascular cross-sectional area
Figure 34.9

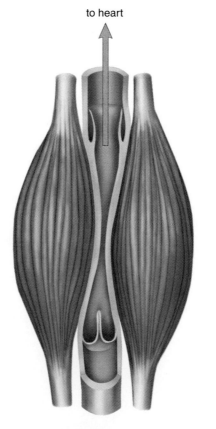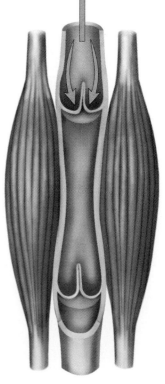

to heart

to heart

Cross section of a valve in a vein
Figure 34.10

a. Contracted skeletal muscle pushes blood past open valve.

b. Closed valve prevents backward flow of blood.

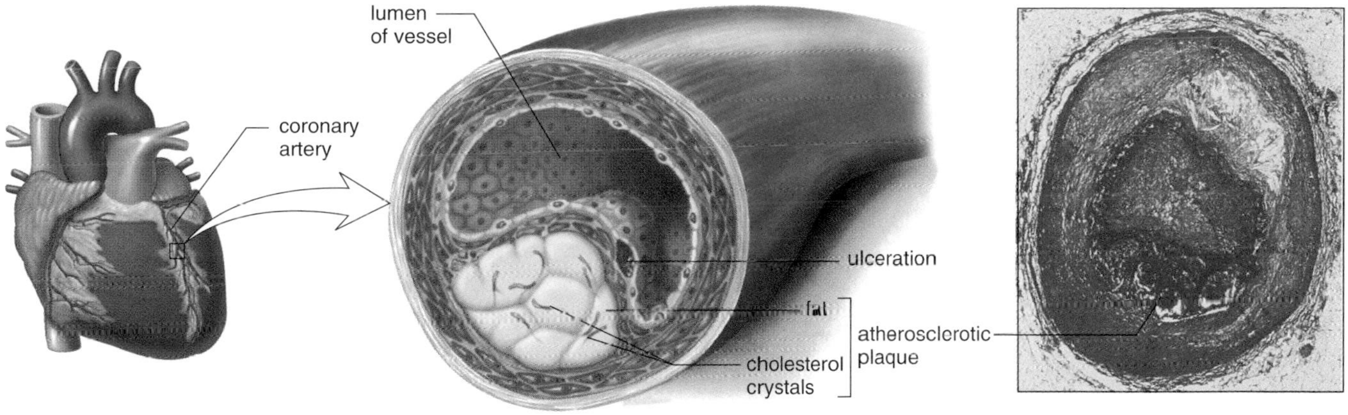

lumen
of vessel

coronary
artery

ulceration

fat

cholesterol
crystals

atherosclerotic
plaque

Coronary arteries and plaque
Figure 34B

© Biophoto Associates/Photo Researchers, Inc.

FORMED ELEMENTS	Function and Description	Source
Red Blood Cells (erythrocytes) 4 million–6 million per mm³ blood	Transport O_2 and help transport CO_2 7–8 μm in diameter Bright-red to dark-purple biconcave disks without nuclei	Red bone marrow
White Blood Cells (leukocytes) 5,000–11,000 per mm³ blood	Fight infection	Red bone marrow
Granular leukocytes		
• Neutrophils 40–70%	10–14 μm in diameter Spherical cells with multilobed nuclei; fine, pink granules in cytoplasm; phagocytize pathogens	
• Eosinophils 1–4%	10–14 μm in diameter Spherical cells with bilobed nuclei; coarse, deep-red, uniformly sized granules in cytoplasm; phagocytize antigen-antibody complexes and allergens	
• Basophils 0–1%	10–12 μm in diameter Spherical cells with lobed nuclei; large, irregularly shaped, deep-blue granules in cytoplasm; release histamine, which promotes blood flow to injured tissues	
Agranular leukocytes		
• Lymphocytes 20–45%	5–17 μm in diameter (average 9–10 μm) Spherical cells with large round nuclei; responsible for specific immunity	
• Monocytes 4–8%	10–24 μm in diameter Large spherical cells with kidney-shaped, round, or lobed nuclei; become macrophages that phagocytize pathogens and cellular debris	
Platelets (thrombocytes) 150,000–300,000 per mm³ blood	Aid clotting 2–4 μm in diameter Disk-shaped cell fragments with no nuclei; purple granules in cytoplasm	Red bone marrow

PLASMA	Function	Source
Water (90–92% of plasma)	Maintains blood volume; transports molecules	Absorbed from intestine
Plasma proteins (7–8% of plasma)	Maintain blood osmotic pressure and pH	Liver
Albumin	Maintains blood volume and pressure	
Globulins	Transport; fight infection	
Fibrinogen	Clotting	
Salts (less than 1% of plasma)	Maintain blood osmotic pressure and pH; aid metabolism	Absorbed from intestine
Gases		
Oxygen	Cellular respiration	Lungs
Carbon dioxide	End product of metabolism	Tissues
Nutrients	Food for cells	Absorbed from intestine
Lipids Glucose Amino acids		
Nitrogenous wastes	Excretion by kidneys	Liver
Urea Uric acid		
Other		
Hormones, vitamins, etc.	Aid metabolism	Varied

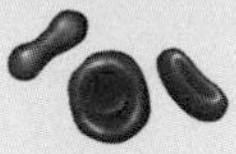

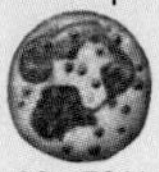

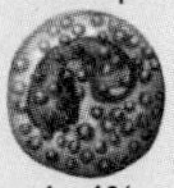

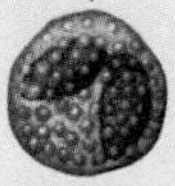

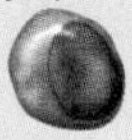

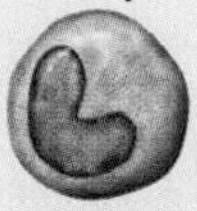

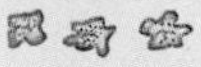

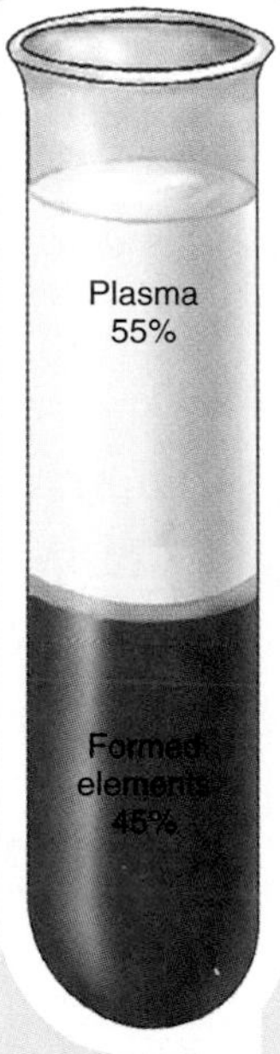

Plasma 55%

Formed elements 45%

• Appearance with Wright's stain.

Composition of blood
Figure 34.11

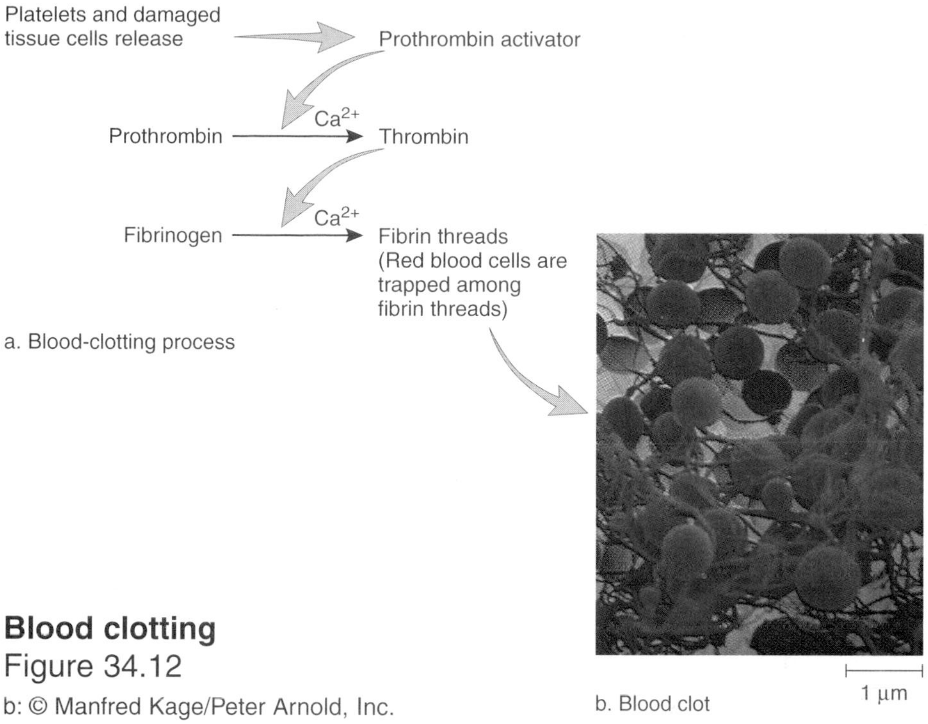

Platelets and damaged tissue cells release ⟶ Prothrombin activator

Prothrombin ⟶ Ca^{2+} ⟶ Thrombin

Fibrinogen ⟶ Ca^{2+} ⟶ Fibrin threads (Red blood cells are trapped among fibrin threads)

a. Blood-clotting process

Blood clotting
Figure 34.12

b: © Manfred Kage/Peter Arnold, Inc.

b. Blood clot

1 μm

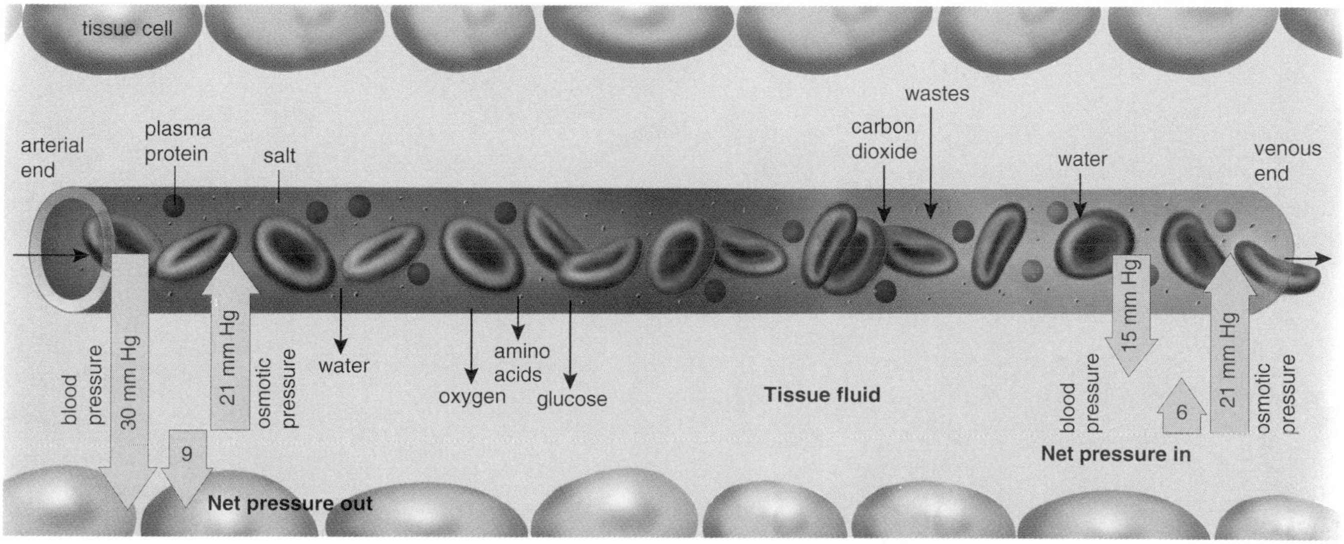

tissue cell

arterial end

plasma protein

salt

water

oxygen

amino acids

glucose

carbon dioxide

wastes

water

venous end

blood pressure 30 mm Hg

21 mm Hg osmotic pressure

9

Net pressure out

Tissue fluid

blood pressure 15 mm Hg

6

21 mm Hg osmotic pressure

Net pressure in

Capillary exchange
Figure 34.13

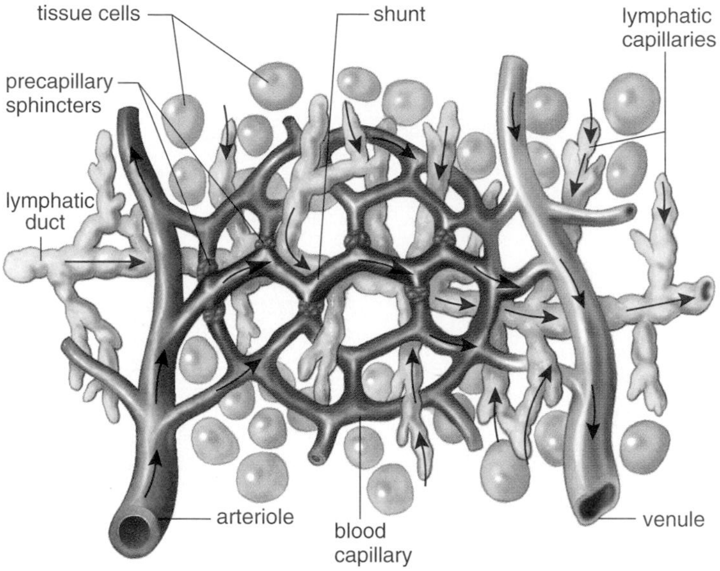

tissue cells

shunt

lymphatic capillaries

precapillary sphincters

lymphatic duct

arteriole

blood capillary

venule

Capillary bed
Figure 34.14

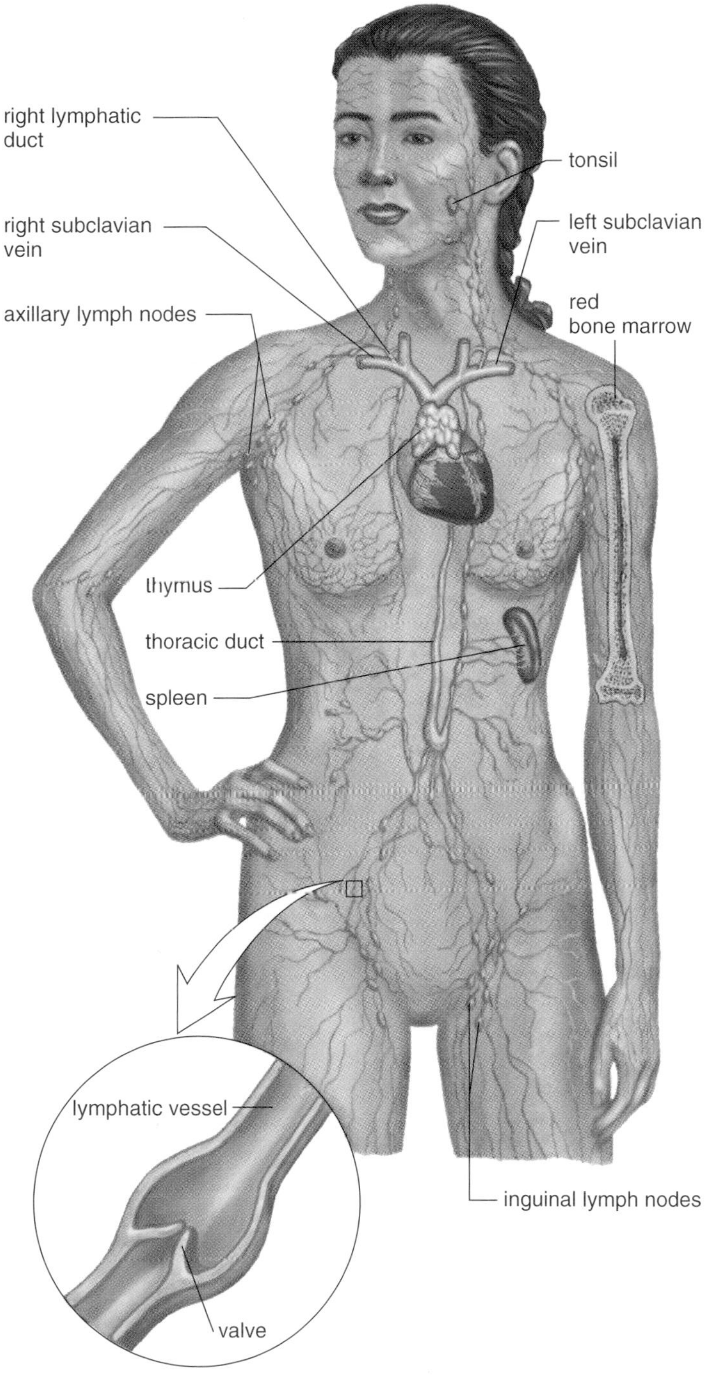

right lymphatic duct

tonsil

right subclavian vein

left subclavian vein

axillary lymph nodes

red bone marrow

thymus

thoracic duct

spleen

lymphatic vessel

valve

inguinal lymph nodes

Lymphatic system
Figure 35.1

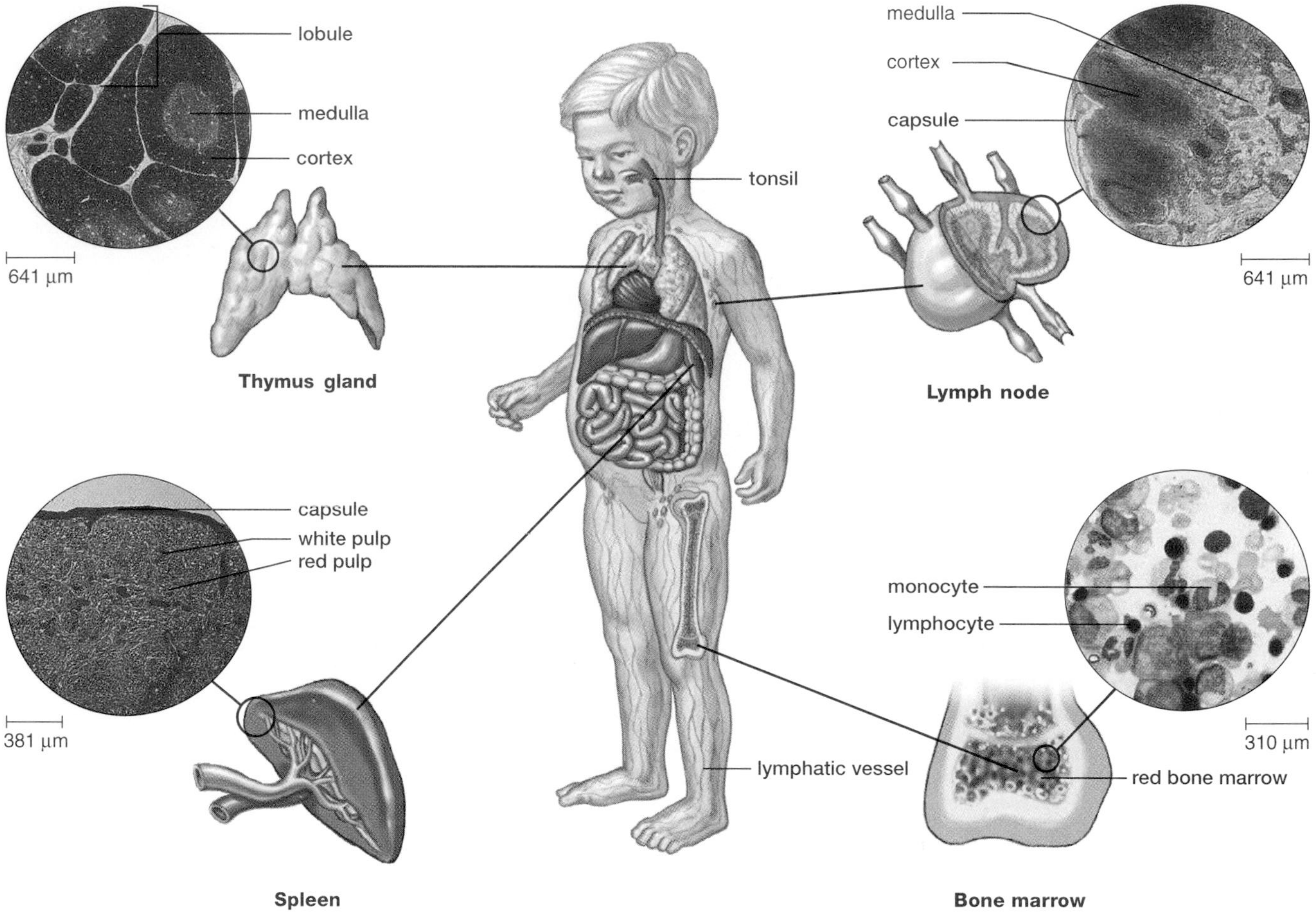

Thymus gland

- lobule
- medulla
- cortex

641 μm

Lymph node

- medulla
- cortex
- capsule

641 μm

- tonsil

Spleen

- capsule
- white pulp
- red pulp

381 μm

Bone marrow

- monocyte
- lymphocyte

310 μm

- lymphatic vessel
- red bone marrow

The lymphoid organs
Figure 35.2

b (thymus): © Ed Reschke/Peter Arnold, Inc.; c (spleen): © Ed Reschke/Peter Arnold, Inc.; d (bone marrow): © R. Calentine/ Visuals Unlimited; e (lymph node): © Fred E. Hossler/Visuals Unlimited

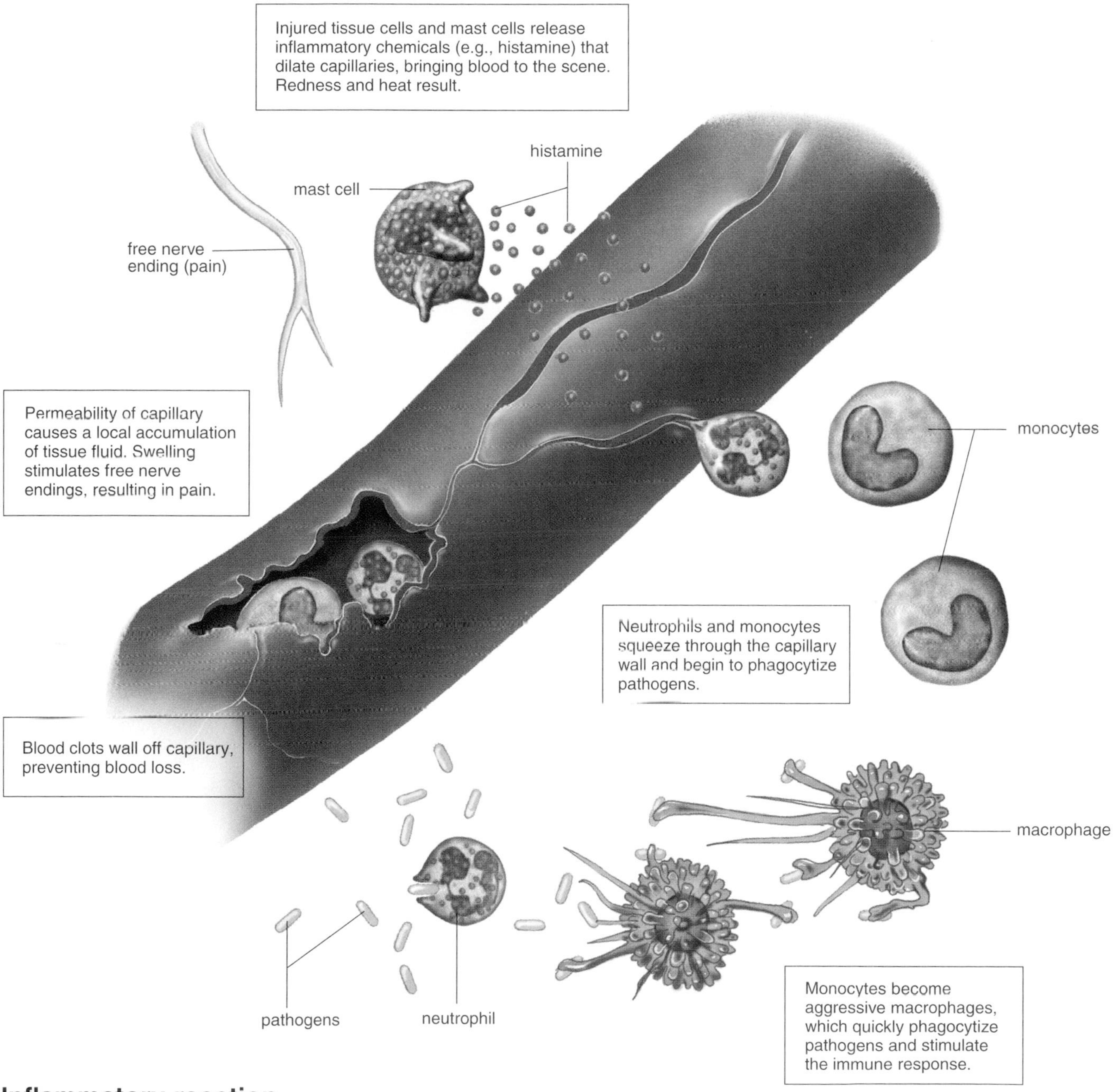

Injured tissue cells and mast cells release inflammatory chemicals (e.g., histamine) that dilate capillaries, bringing blood to the scene. Redness and heat result.

histamine

mast cell

free nerve ending (pain)

Permeability of capillary causes a local accumulation of tissue fluid. Swelling stimulates free nerve endings, resulting in pain.

monocytes

Neutrophils and monocytes squeeze through the capillary wall and begin to phagocytize pathogens.

Blood clots wall off capillary, preventing blood loss.

macrophage

pathogens

neutrophil

Monocytes become aggressive macrophages, which quickly phagocytize pathogens and stimulate the immune response.

Inflammatory reaction
Figure 35.3

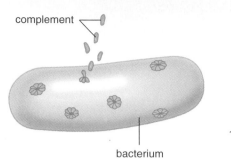
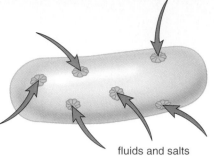

complement

bacterium

fluids and salts

Complement proteins form holes in the bacterial cell wall and membrane.

Holes allow fluids and salts to enter the bacterium.

Bacterium expands until it bursts.

Action of the complement system against a bacterium
Figure 35.4

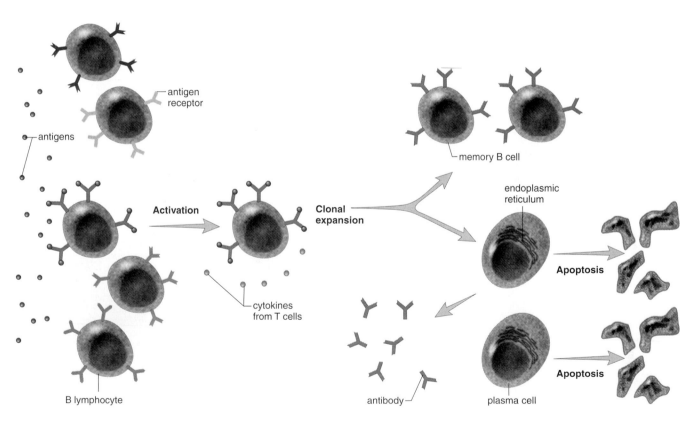

antigen receptor

antigens

Activation

Clonal expansion

memory B cell

endoplasmic reticulum

Apoptosis

cytokines from T cells

B lymphocyte

antibody

plasma cell

Apoptosis

Clonal selection theory as it applies to B cells
Figure 35.5

Structure of the most common antibody (IgG)
Figure 35.6

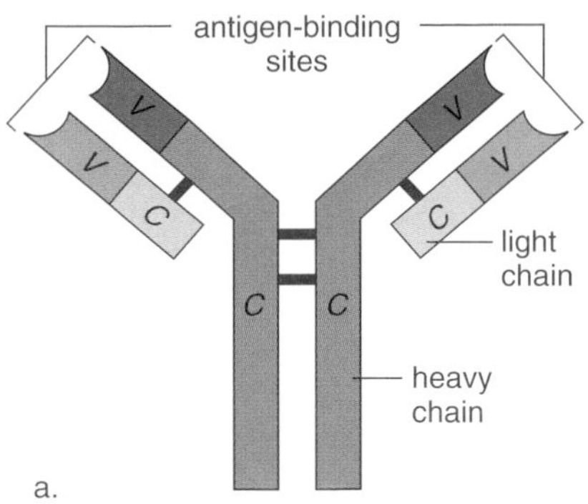

antigen-binding sites

V

V

C

C

C

light chain

heavy chain

a.

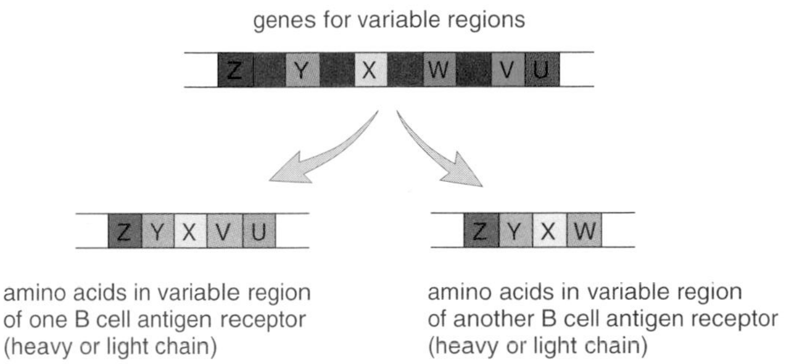

genes for variable regions

Z Y X W V U

Z Y X V U

amino acids in variable region
of one B cell antigen receptor
(heavy or light chain)

Z Y X W

amino acids in variable region
of another B cell antigen receptor
(heavy or light chain)

Antibody diversity
Figure 35A

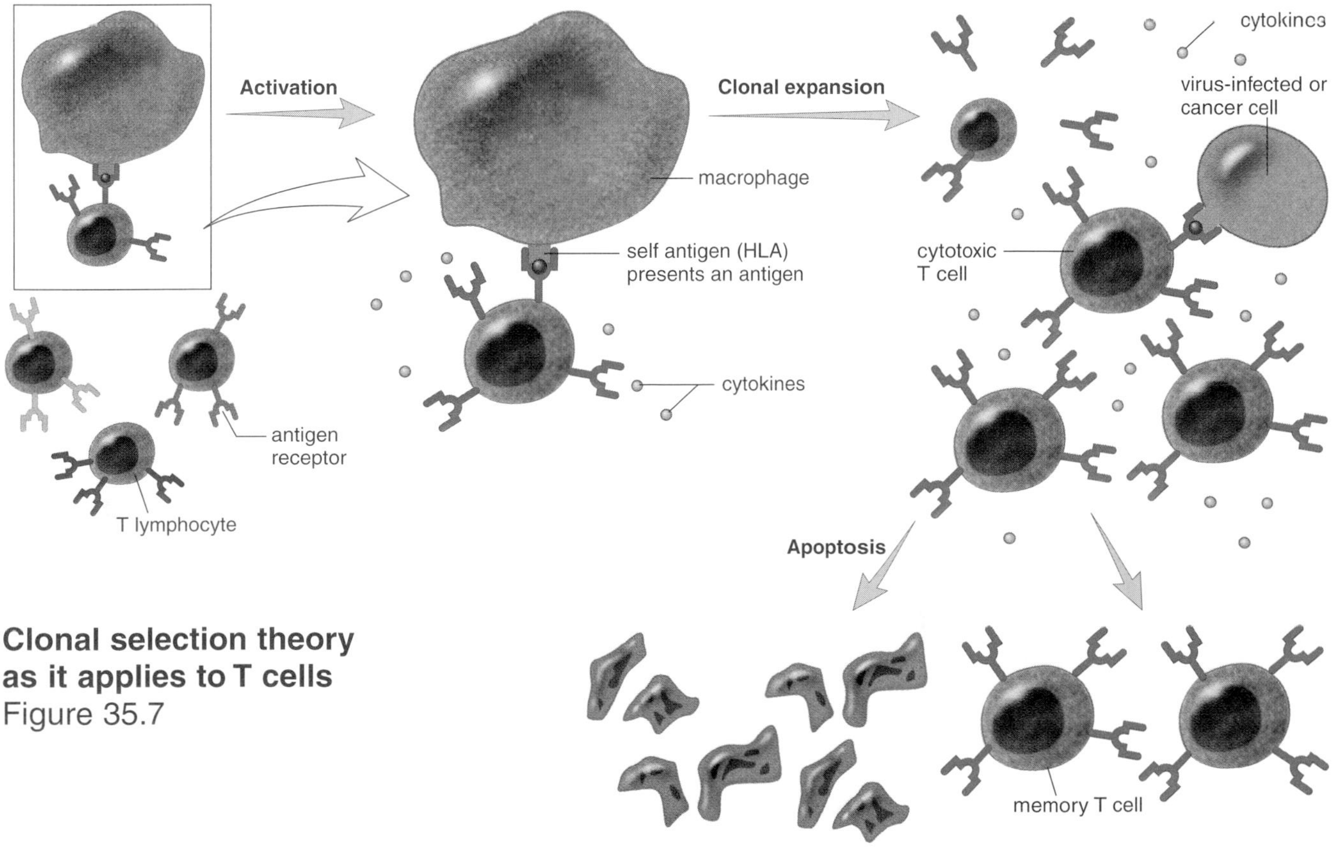

Activation

Clonal expansion

cytokines

virus-infected or cancer cell

macrophage

self antigen (HLA) presents an antigen

cytotoxic T cell

antigen receptor

cytokines

T lymphocyte

Apoptosis

Clonal selection theory as it applies to T cells
Figure 35.7

memory T cell

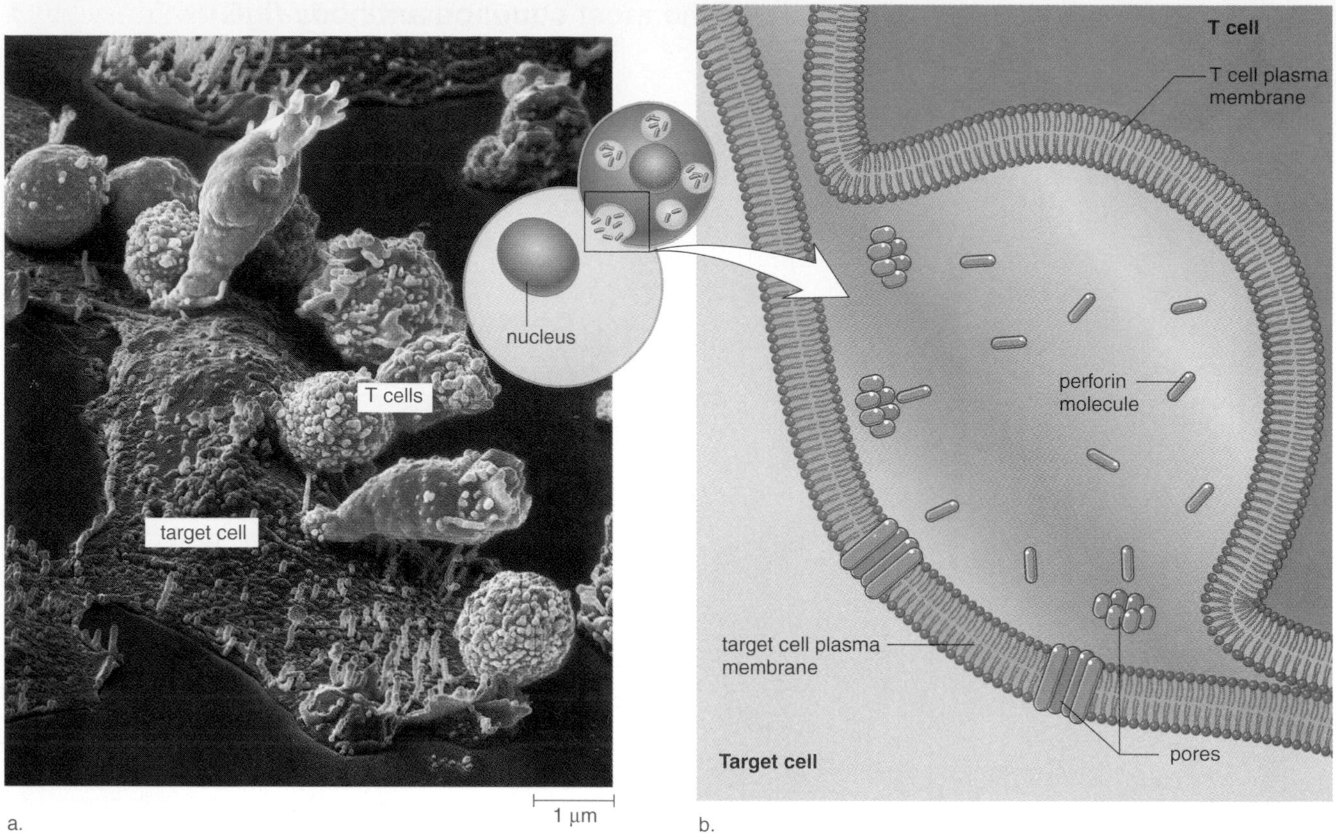

a.

b.

Cell-mediated immunity
Figure 35.8

© Bohringer Ingelheim International, photo by Lennart Nilsson

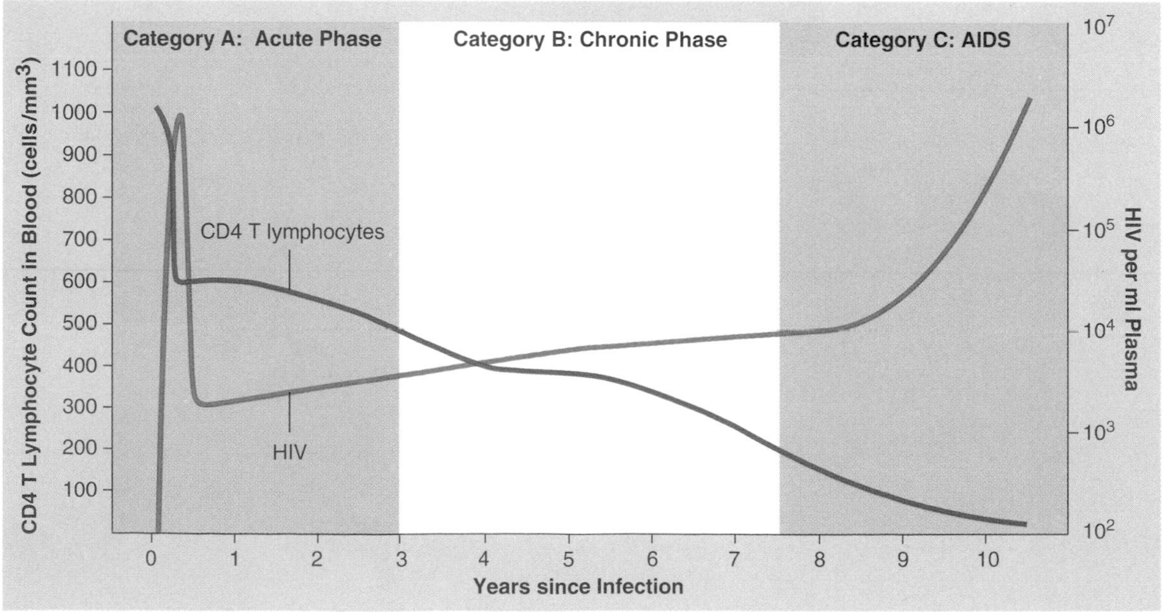

Stages of an HIV-1 infection
Figure 35B

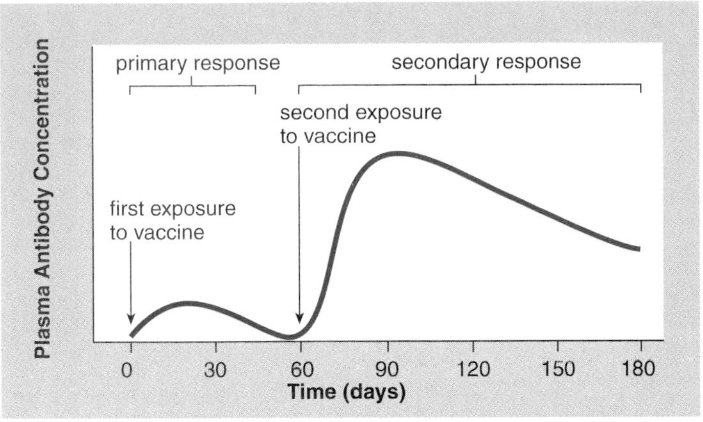

Active immunity due to immunizations
Figure 35.9

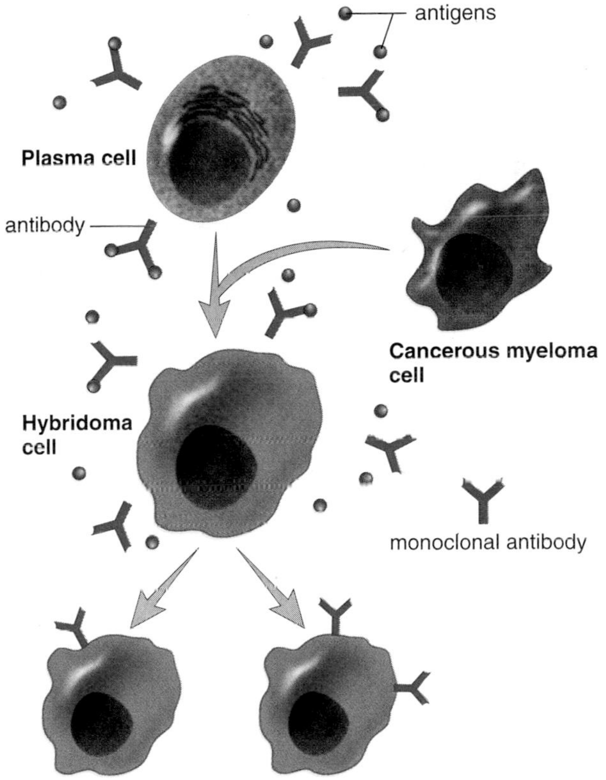

Production of monoclonal antibodies
Figure 35.11

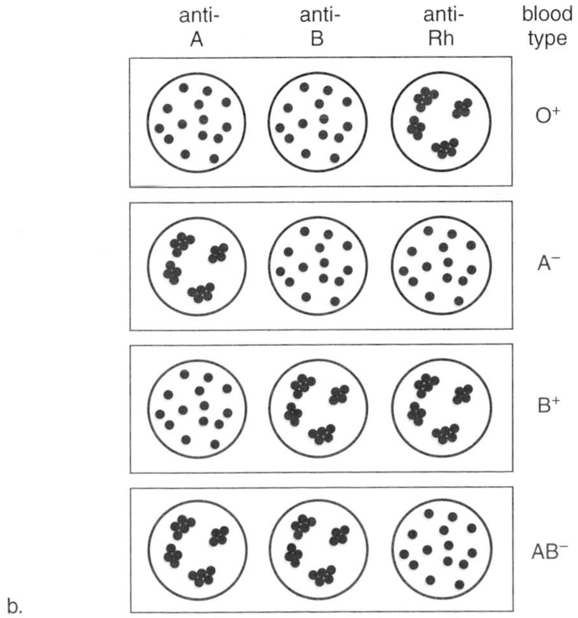

b.

Blood typing
Figure 35.12

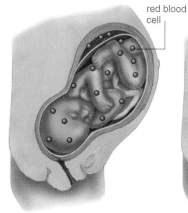

red blood cell

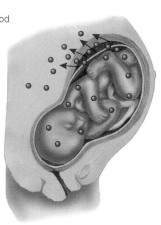

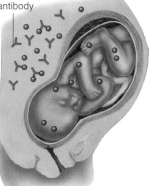

anti-Rh antibody

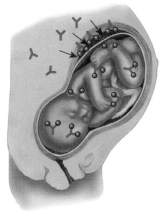

Child is Rh positive; mother is Rh negative.

Red blood cells leak across placenta.

Mother makes anti-Rh antibodies.

Antibodies attack Rh-positive red blood cells in child.

Hemolytic disease of the newborn
Figure 35.13

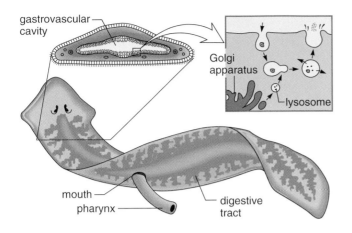

gastrovascular cavity

Golgi apparatus

lysosome

mouth

pharynx

digestive tract

Incomplete digestive tract of a planarian
Figure 36.1

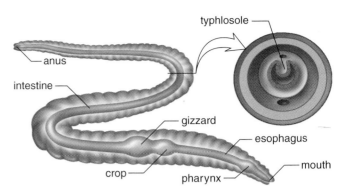

typhlosole

anus

intestine

gizzard

esophagus

crop

pharynx

mouth

Complete digestive tract of an earthworm
Figure 36.2

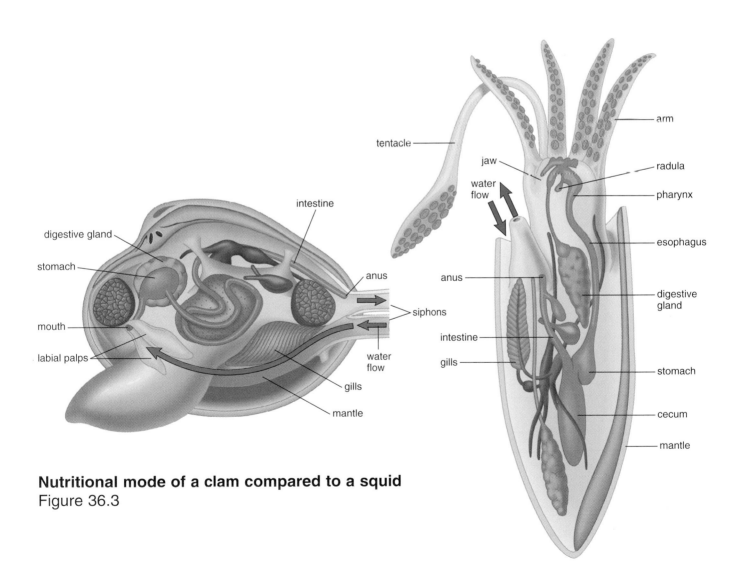

digestive gland

stomach

mouth

labial palps

intestine

anus

siphons

water flow

gills

mantle

tentacle

jaw

water flow

anus

intestine

gills

arm

radula

pharynx

esophagus

digestive gland

stomach

cecum

mantle

Nutritional mode of a clam compared to a squid
Figure 36.3

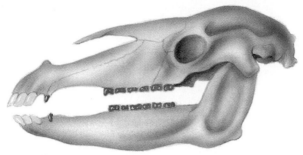

a. Horse

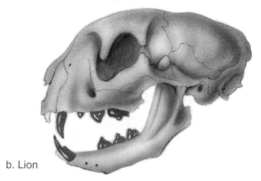

b. Lion

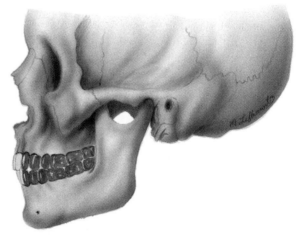

c. Human

Key:
☐ = incisors
■ = canines
■ = premolars
■ = molars

Dentition among mammals
Figure 36.4

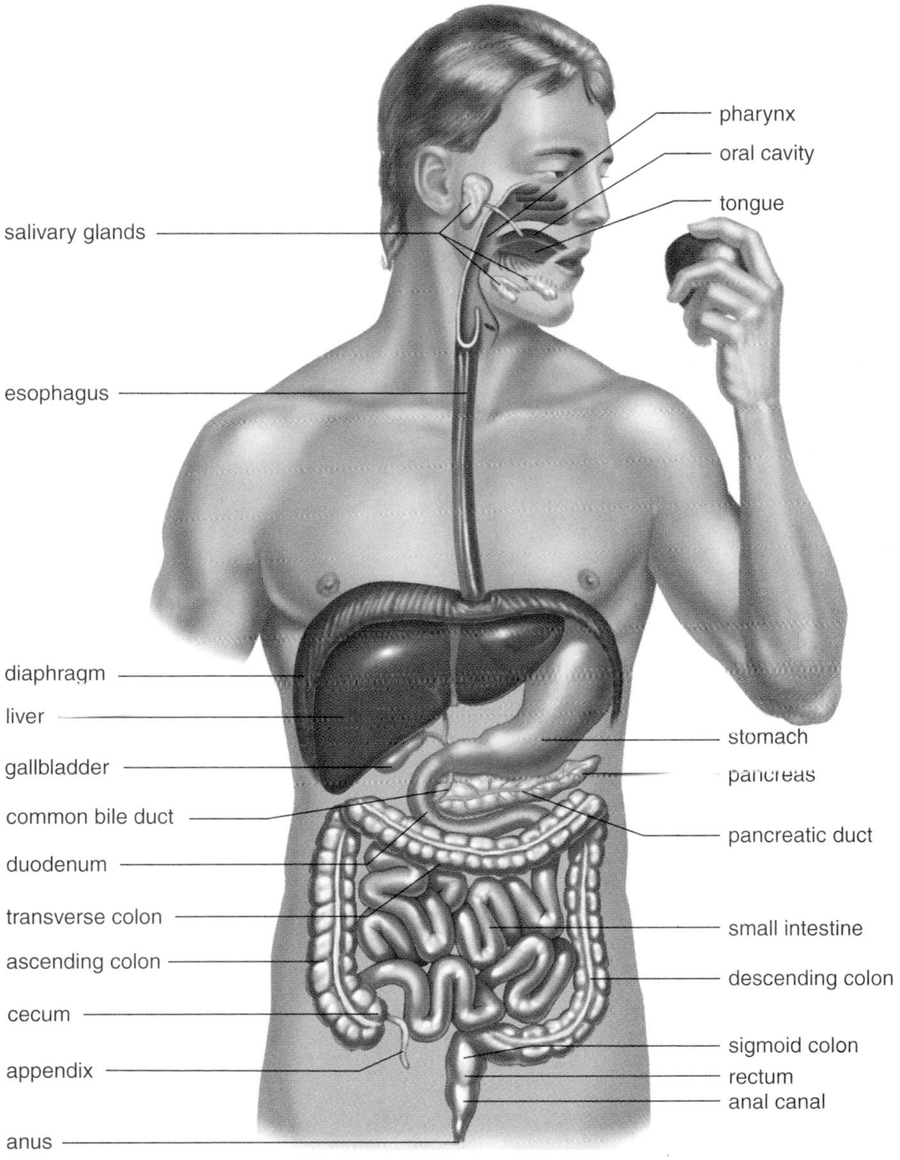
pharynx
oral cavity
tongue
salivary glands
esophagus
diaphragm
liver
gallbladder
common bile duct
duodenum
transverse colon
ascending colon
cecum
appendix
anus
stomach
pancreas
pancreatic duct
small intestine
descending colon
sigmoid colon
rectum
anal canal

The human digestive tract
Figure 36.5

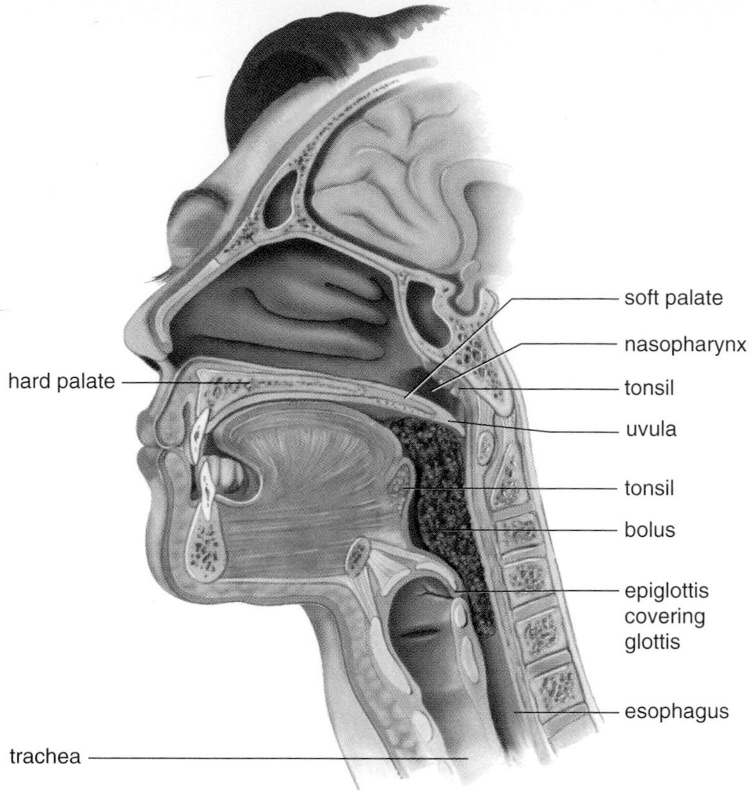

hard palate

soft palate

nasopharynx

tonsil

uvula

tonsil

bolus

epiglottis covering glottis

esophagus

trachea

Swallowing
Figure 36.6

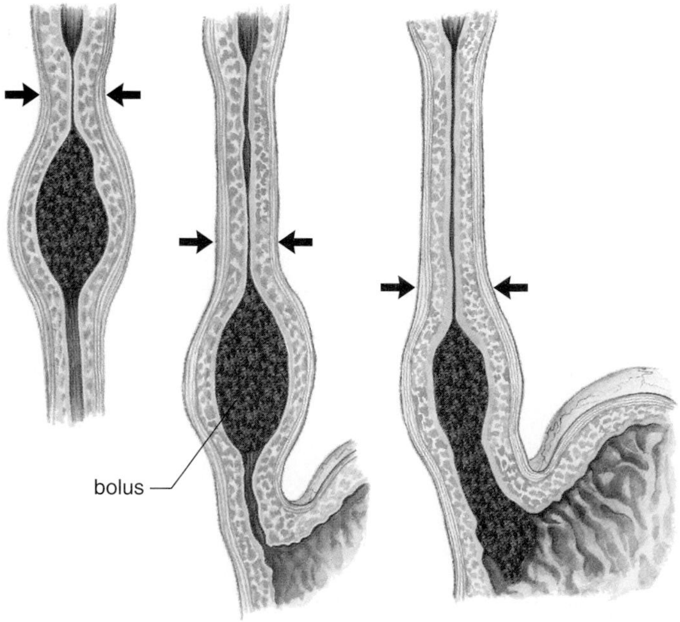

bolus

Peristalsis in the digestive tract
Figure 36.7

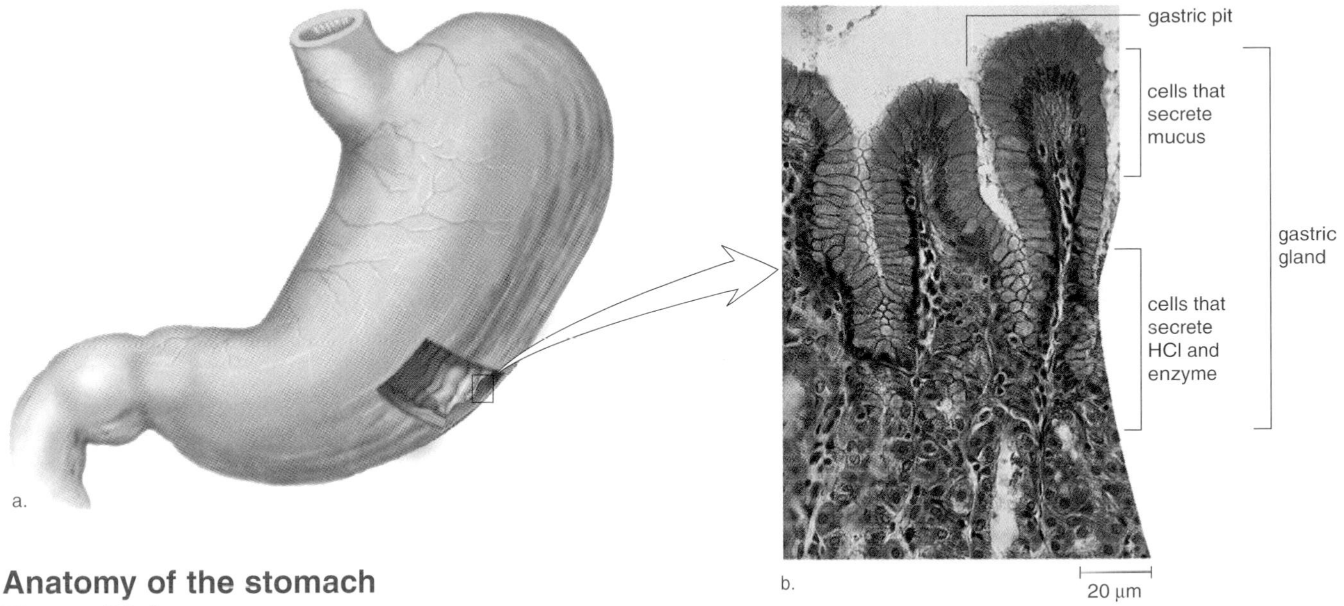

Anatomy of the stomach
Figure 36.8

b: © Ed Reschke/Peter Arnold, Inc.

Labels in figure: gastric pit; cells that secrete mucus; cells that secrete HCl and enzyme; gastric gland; 20 μm

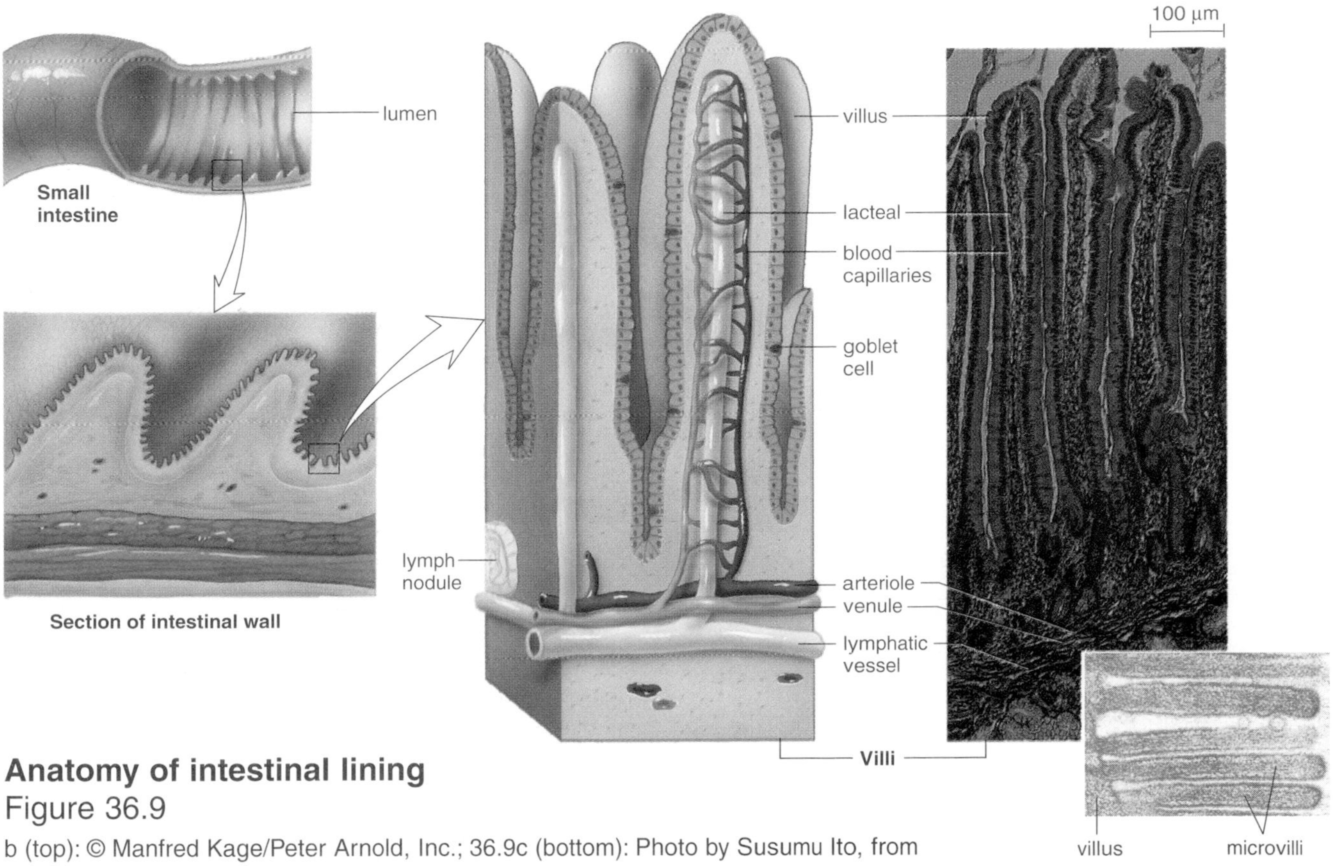

Anatomy of intestinal lining
Figure 36.9

b (top): © Manfred Kage/Peter Arnold, Inc.; 36.9c (bottom): Photo by Susumu Ito, from Charles Flickinger, *Medical Cellular Biology,* W.B. Saunders, 1979

Labels in figure: lumen; Small intestine; Section of intestinal wall; villus; lacteal; blood capillaries; goblet cell; lymph nodule; arteriole; venule; lymphatic vessel; Villi; 100 μm; villus; microvilli

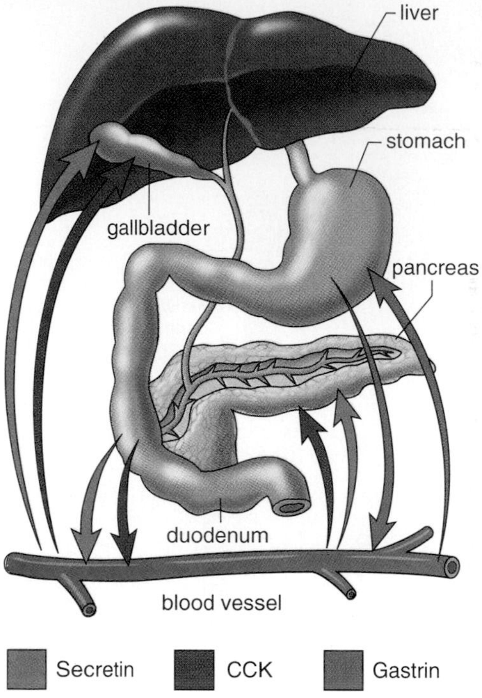

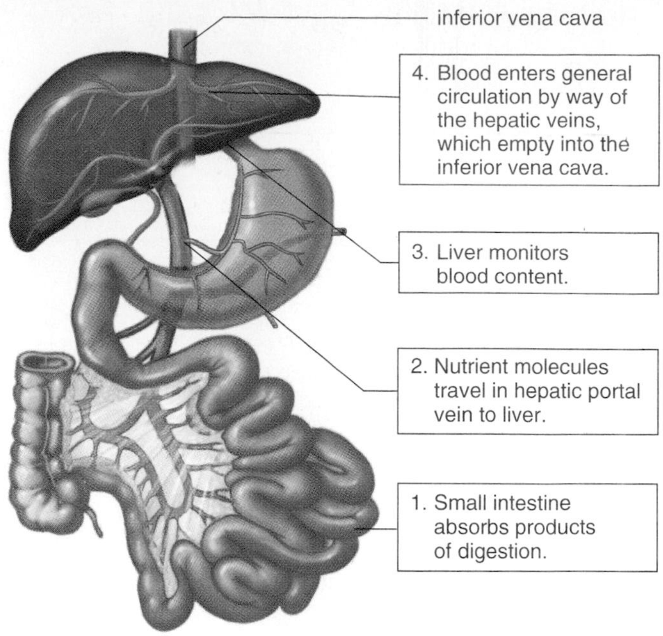

| | Secretin | | CCK | | Gastrin |

Hormonal control of digestive gland secretions
Figure 36A

Hepatic portal system
Figure 36.10

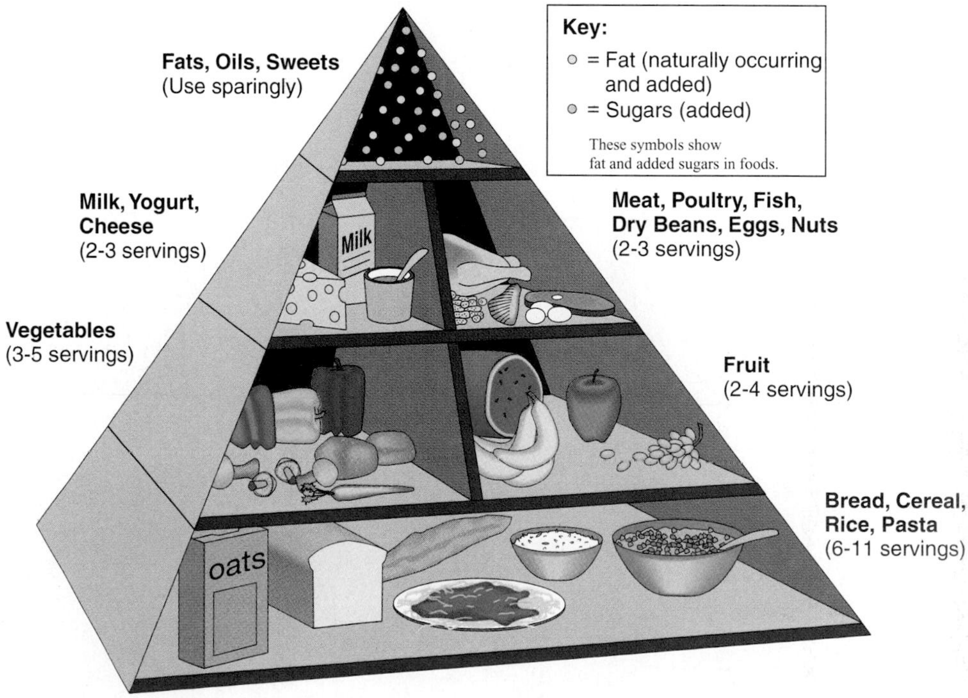

Fats, Oils, Sweets
(Use sparingly)

Key:
○ = Fat (naturally occurring and added)
◑ = Sugars (added)

These symbols show
fat and added sugars in foods.

Milk, Yogurt,
Cheese
(2-3 servings)

Meat, Poultry, Fish,
Dry Beans, Eggs, Nuts
(2-3 servings)

Vegetables
(3-5 servings)

Fruit
(2-4 servings)

Bread, Cereal,
Rice, Pasta
(6-11 servings)

Ideal American diet
Figure 36.11

Animal shapes and gas exchange
Figure 37.1

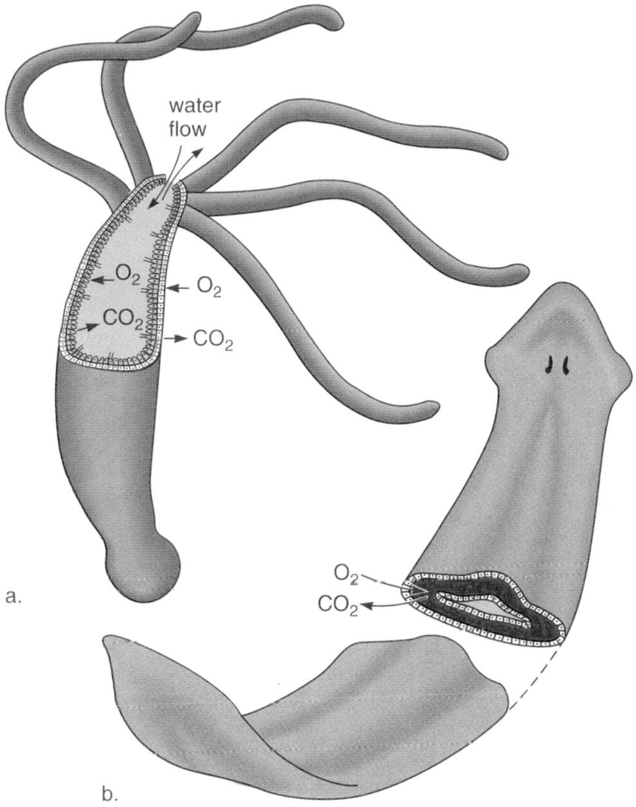

water flow

O_2

O_2

CO_2

CO_2

a.

O_2

CO_2

b.

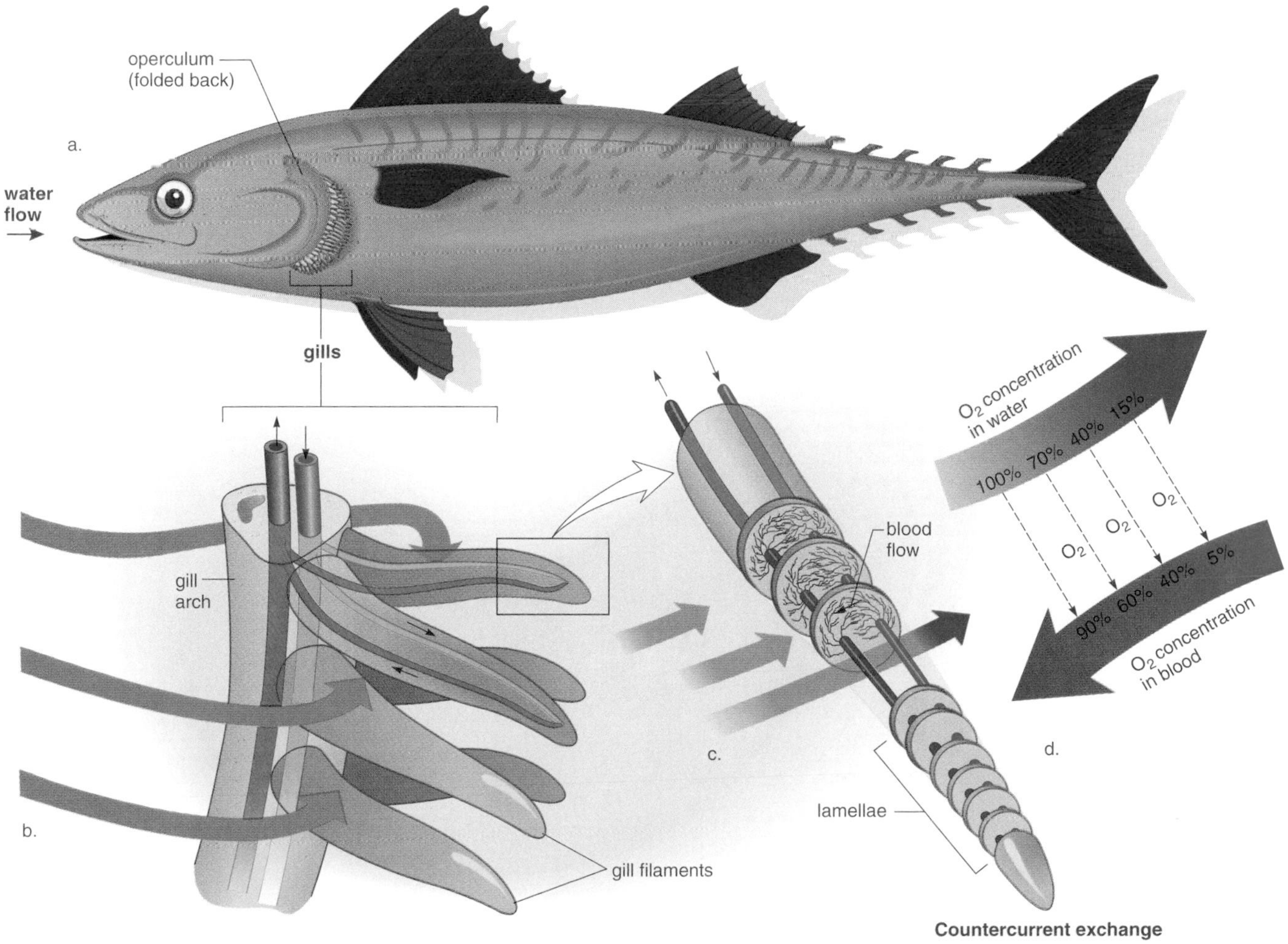

operculum
(folded back)

a.

water
flow

gills

gill
arch

b.

gill filaments

c.

blood
flow

lamellae

O_2 concentration
in water

100% 70% 40% 15%

O_2

O_2

O_2

90% 60% 40% 5%

O_2 concentration
in blood

d.

Countercurrent exchange

Anatomy of gills in bony fishes
Figure 37.2

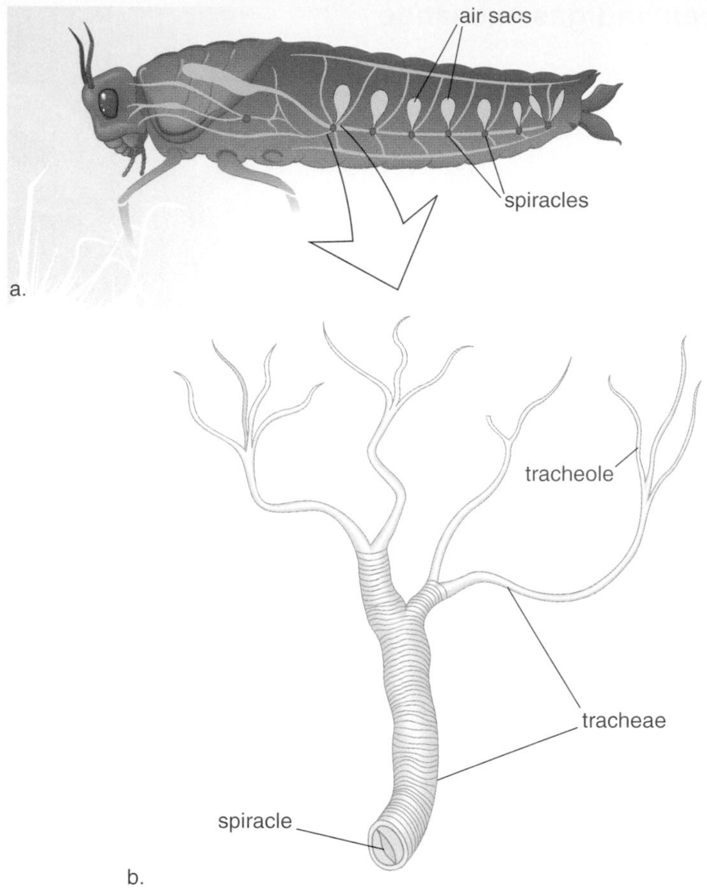

air sacs

spiracles

tracheole

tracheae

spiracle

a.

b.

Tracheae of insects
Figure 37.3

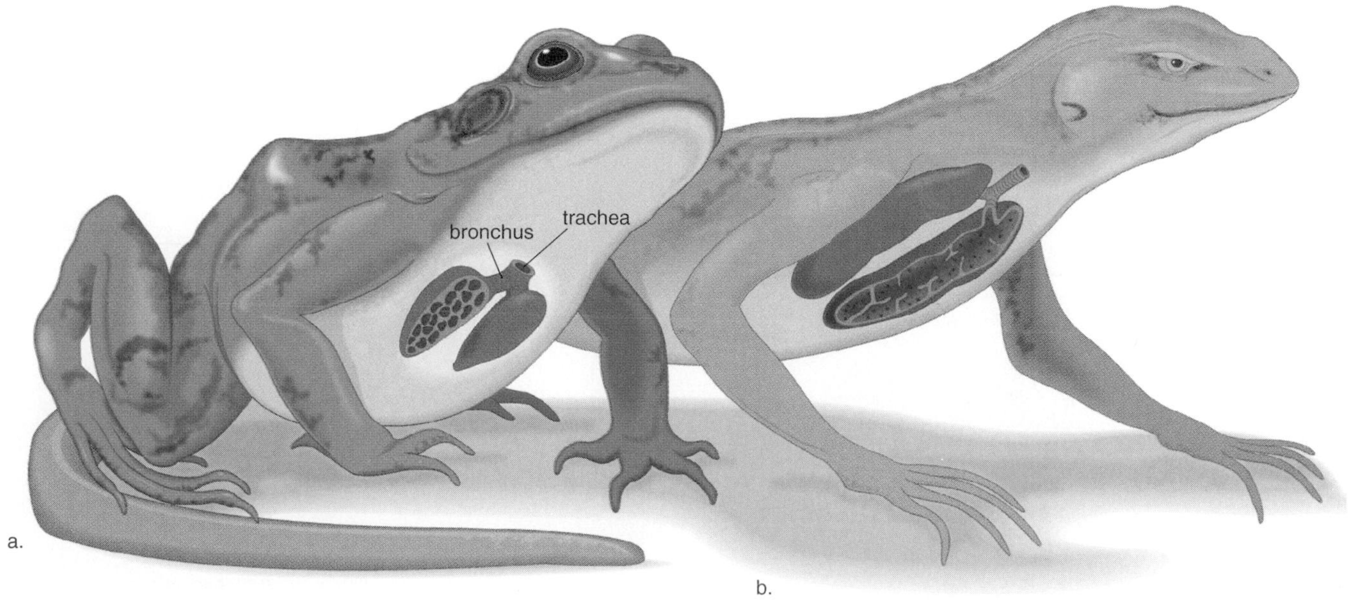

bronchus   trachea

a.

b.

Respiration in amphibians compared to reptiles
Figure 37.4

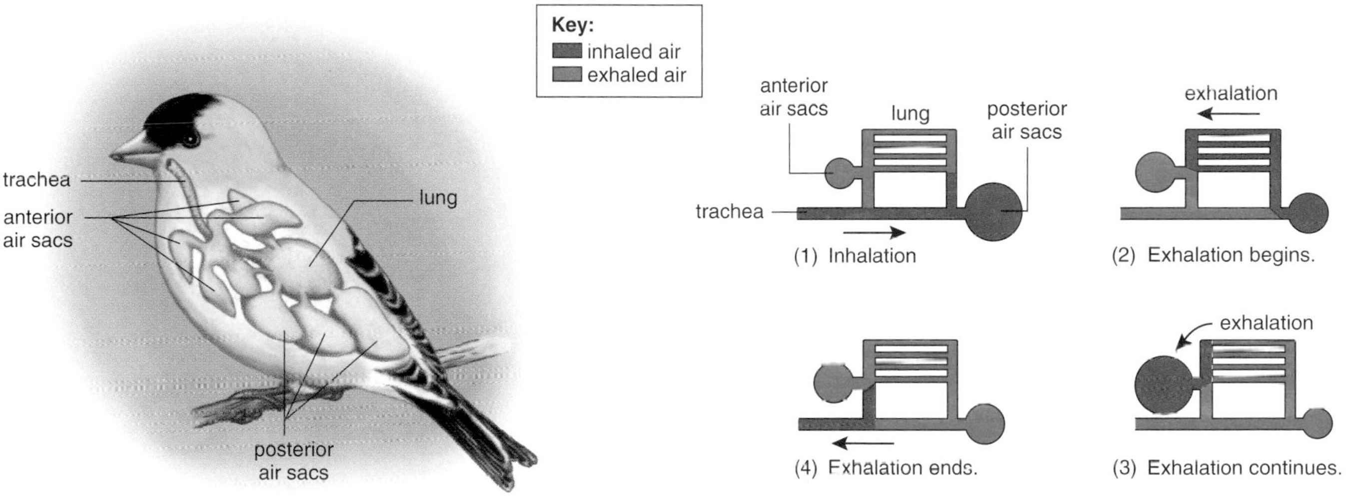

Key:
- inhaled air
- exhaled air

trachea
anterior air sacs
lung
posterior air sacs

a.

anterior air sacs
lung
posterior air sacs
trachea
(1) Inhalation

exhalation
(2) Exhalation begins.

(4) Exhalation ends.

exhalation
(3) Exhalation continues.

b.

Respiratory system in birds
Figure 37.5

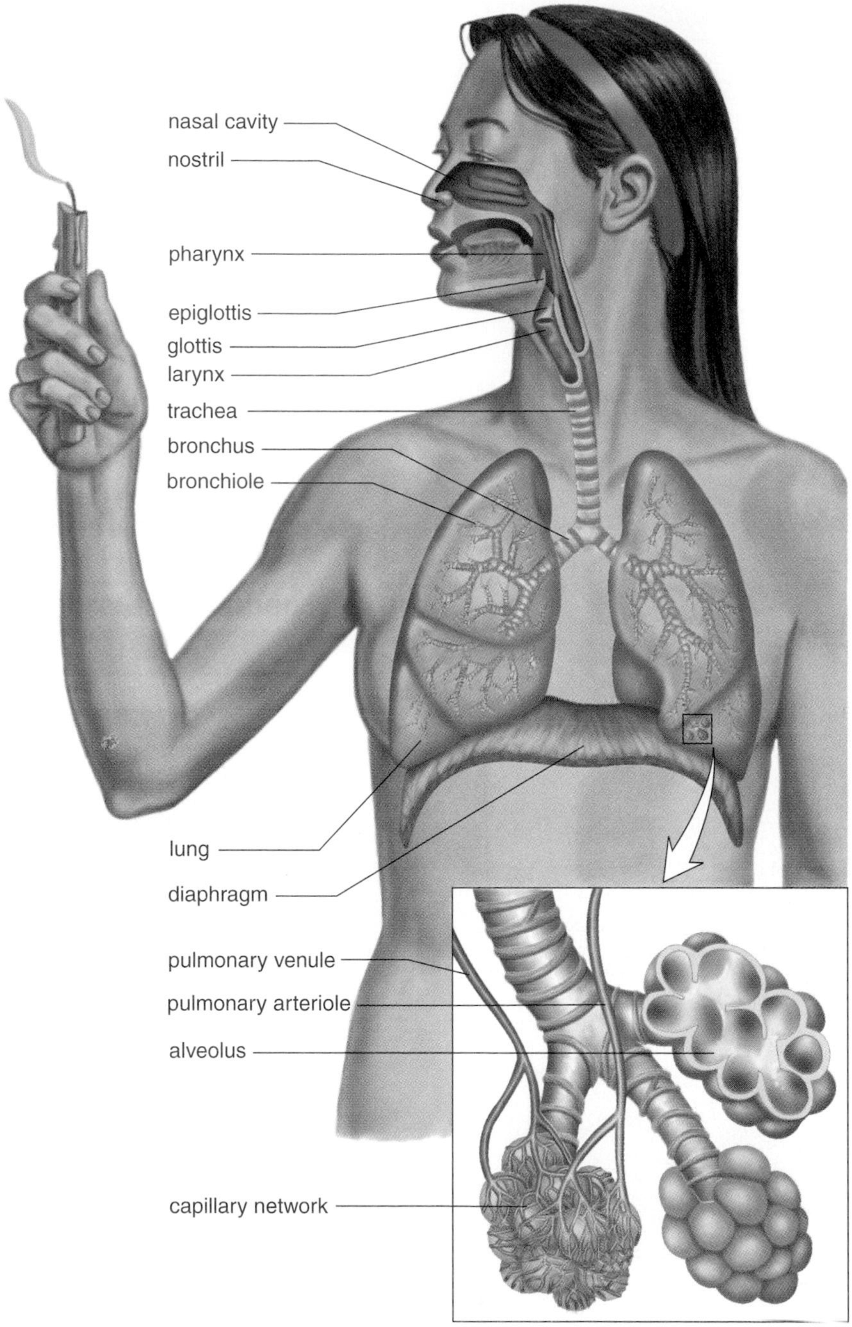

nasal cavity
nostril
pharynx
epiglottis
glottis
larynx
trachea
bronchus
bronchiole
lung
diaphragm
pulmonary venule
pulmonary arteriole
alveolus
capillary network

The human respiratory tract
Figure 37.6

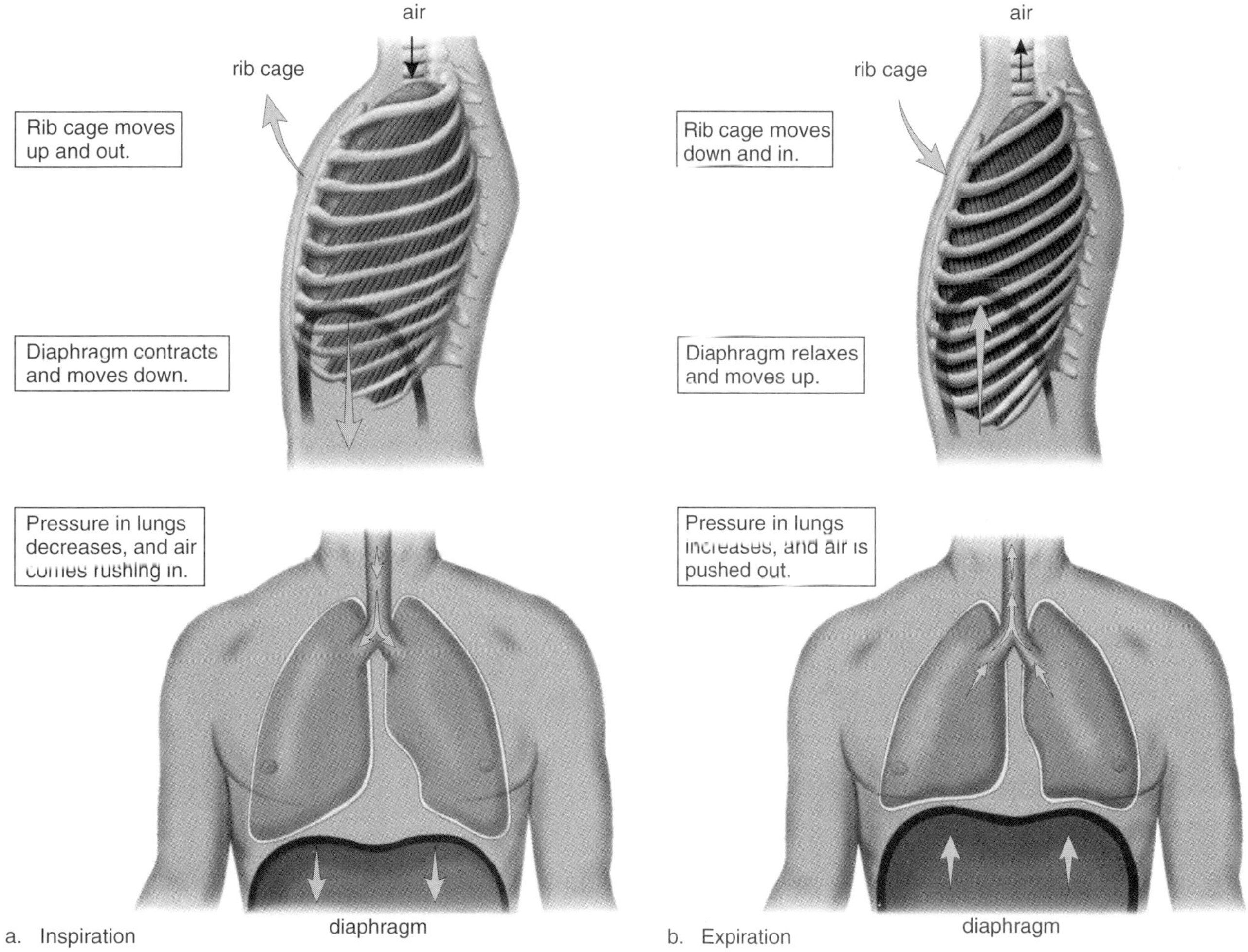

air

rib cage

Rib cage moves
up and out.

Diaphragm contracts
and moves down.

Pressure in lungs
decreases, and air
comes rushing in.

air

rib cage

Rib cage moves
down and in.

Diaphragm relaxes
and moves up.

Pressure in lungs
increases, and air is
pushed out.

a. Inspiration   diaphragm

b. Expiration   diaphragm

Inspiration versus expiration
Figure 37.7

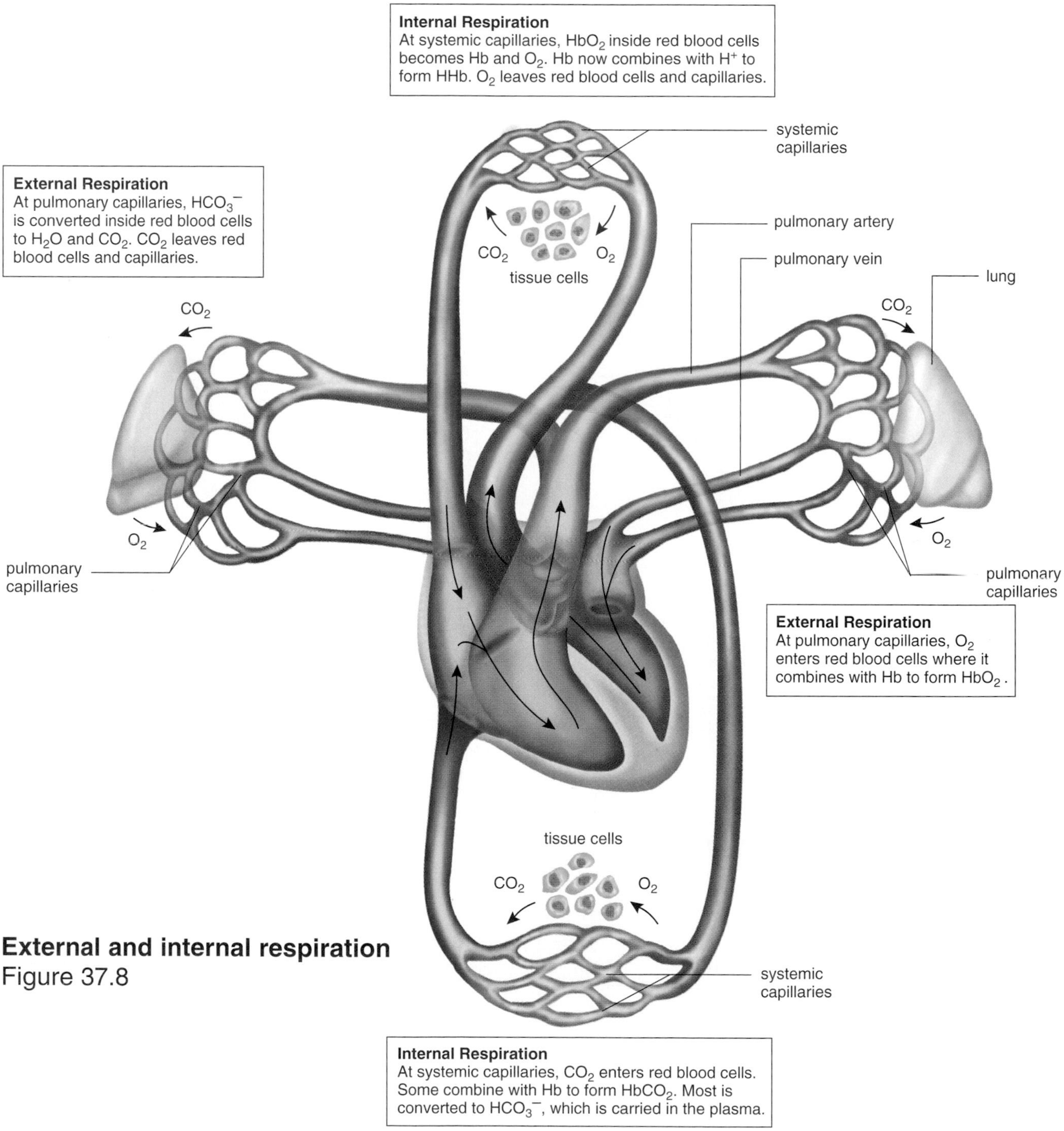

Internal Respiration
At systemic capillaries, HbO$_2$ inside red blood cells becomes Hb and O$_2$. Hb now combines with H$^+$ to form HHb. O$_2$ leaves red blood cells and capillaries.

systemic capillaries

pulmonary artery

pulmonary vein

lung

CO$_2$

CO$_2$

O$_2$

tissue cells

External Respiration
At pulmonary capillaries, HCO$_3^-$ is converted inside red blood cells to H$_2$O and CO$_2$. CO$_2$ leaves red blood cells and capillaries.

CO$_2$

O$_2$

O$_2$

pulmonary capillaries

pulmonary capillaries

External Respiration
At pulmonary capillaries, O$_2$ enters red blood cells where it combines with Hb to form HbO$_2$.

tissue cells

CO$_2$

O$_2$

systemic capillaries

Internal Respiration
At systemic capillaries, CO$_2$ enters red blood cells. Some combine with Hb to form HbCO$_2$. Most is converted to HCO$_3^-$, which is carried in the plasma.

External and internal respiration
Figure 37.8

Hemoglobin saturation in relation to temperature and acidity

Figure 37.9

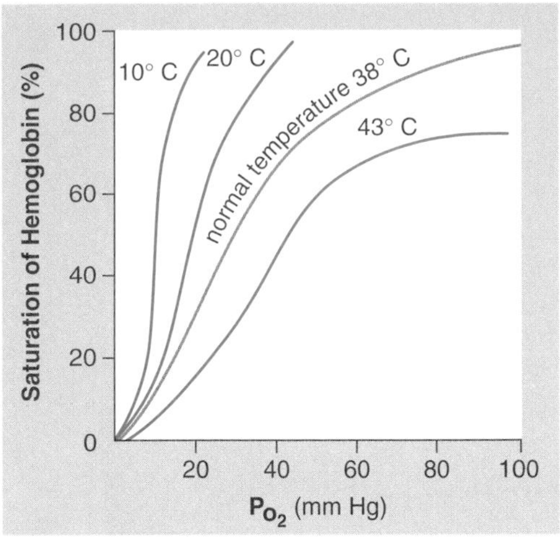

a. Saturation of Hb relative to temperature

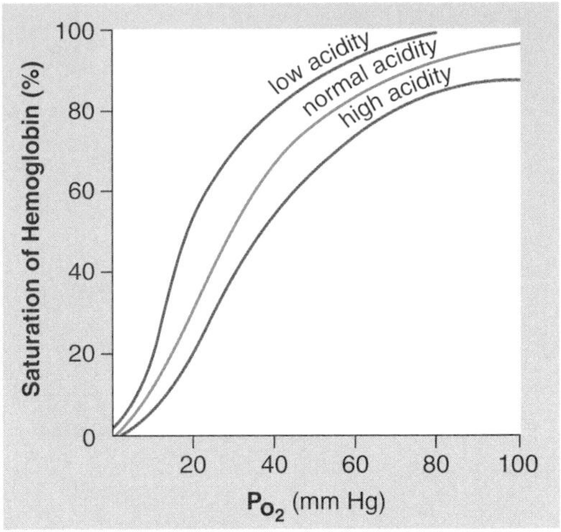

b. Saturation of Hb relative to pH

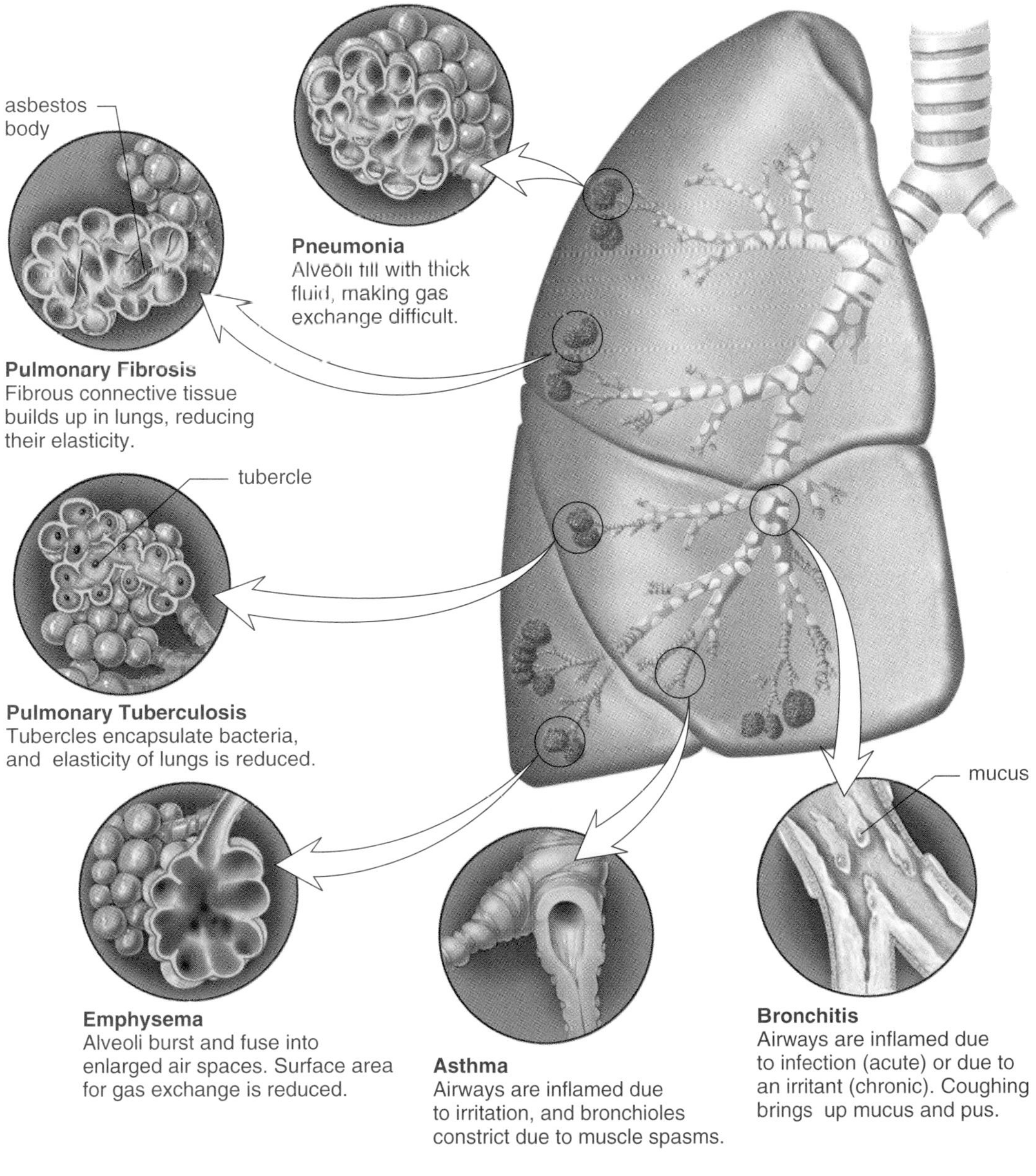

asbestos body

Pneumonia
Alveoli fill with thick fluid, making gas exchange difficult.

Pulmonary Fibrosis
Fibrous connective tissue builds up in lungs, reducing their elasticity.

tubercle

Pulmonary Tuberculosis
Tubercles encapsulate bacteria, and elasticity of lungs is reduced.

Emphysema
Alveoli burst and fuse into enlarged air spaces. Surface area for gas exchange is reduced.

Asthma
Airways are inflamed due to irritation, and bronchioles constrict due to muscle spasms.

mucus

Bronchitis
Airways are inflamed due to infection (acute) or due to an irritant (chronic). Coughing brings up mucus and pus.

Common bronchial and pulmonary diseases

Figure 37.10

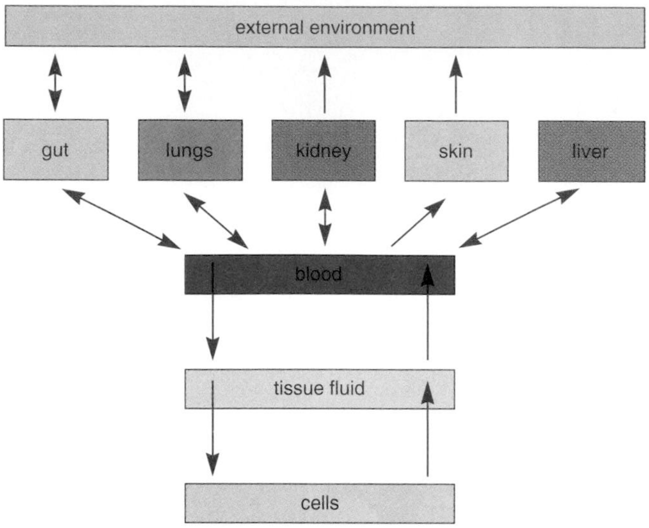

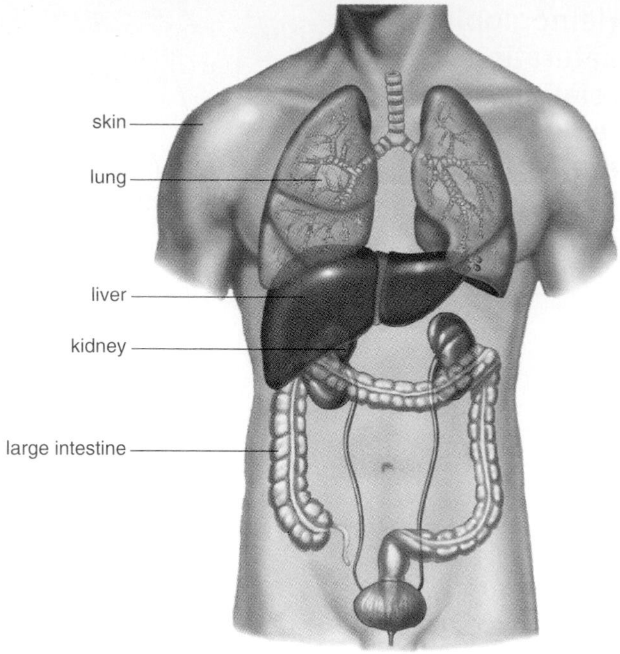

Excretory functions
Figure 38.1

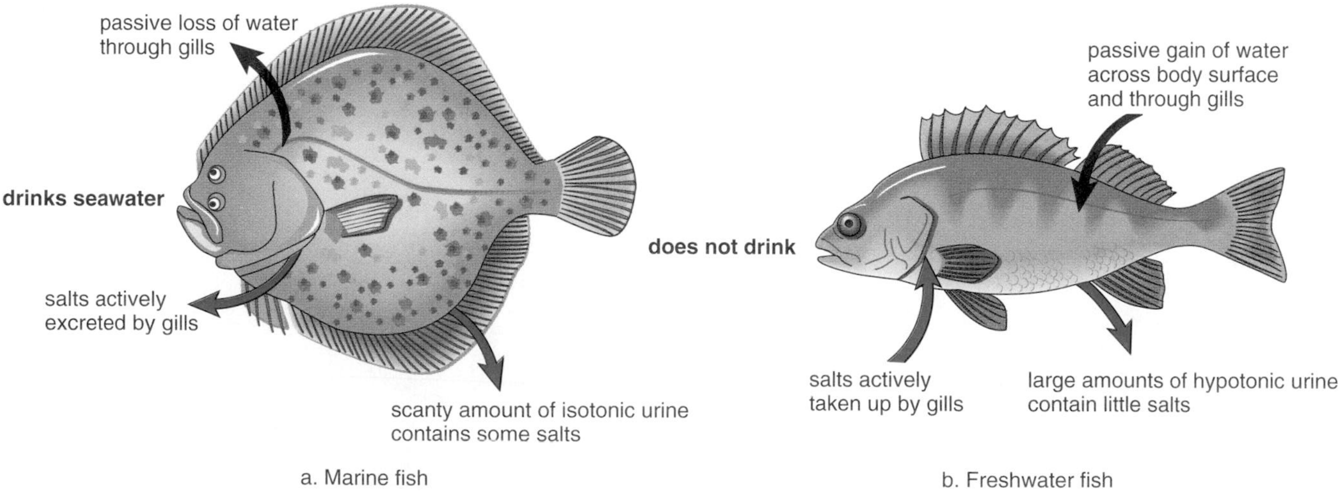

Body fluid regulation in bony fishes
Figure 38.2

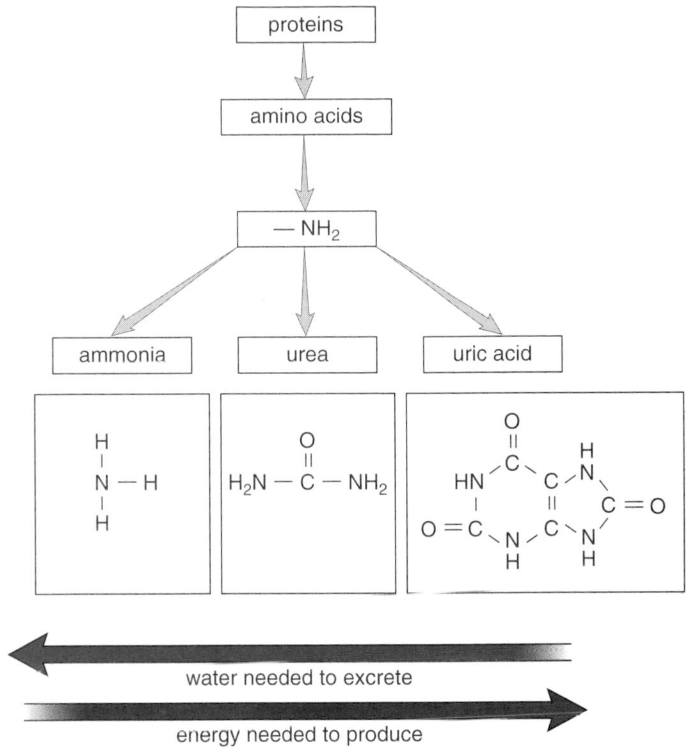

Nitrogenous wastes
Figure 38.3

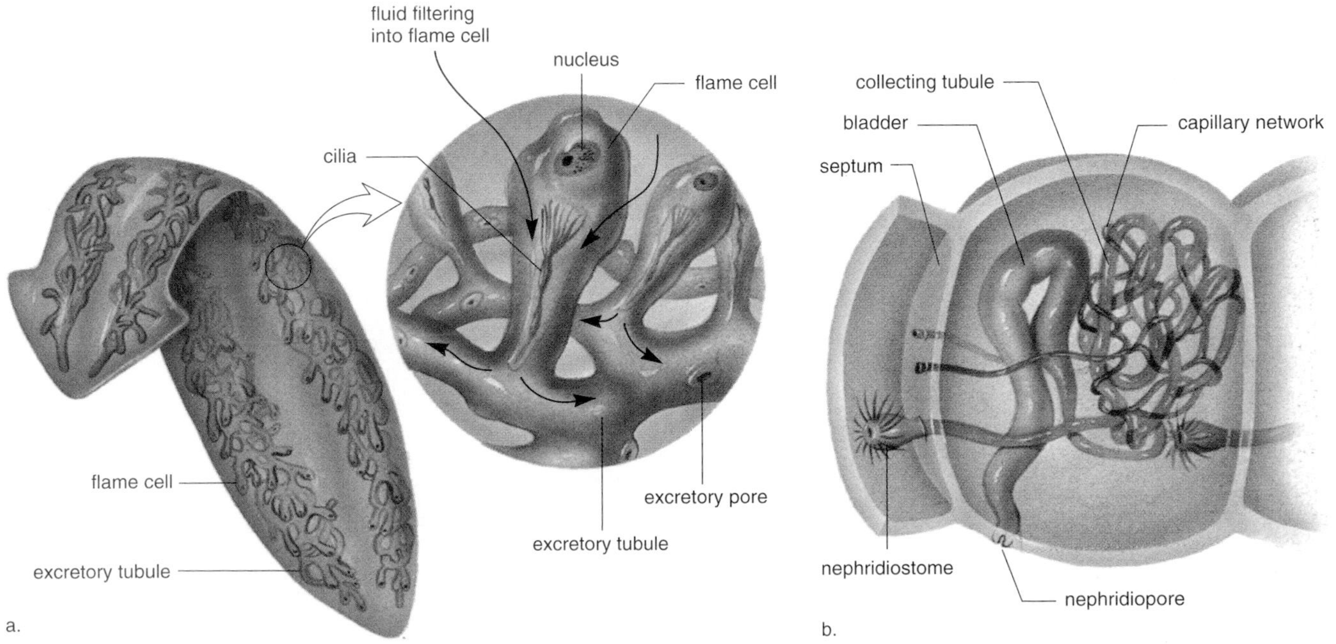

Excretory organs in animals
Figure 38.4

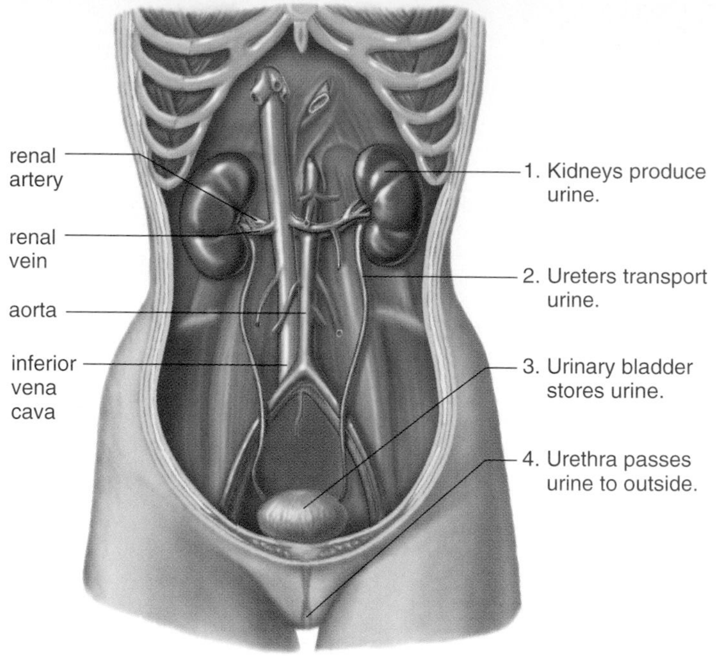

renal
artery

renal
vein

aorta

inferior
vena
cava

1. Kidneys produce
urine.

2. Ureters transport
urine.

3. Urinary bladder
stores urine.

4. Urethra passes
urine to outside.

The human urinary system
Figure 38.5

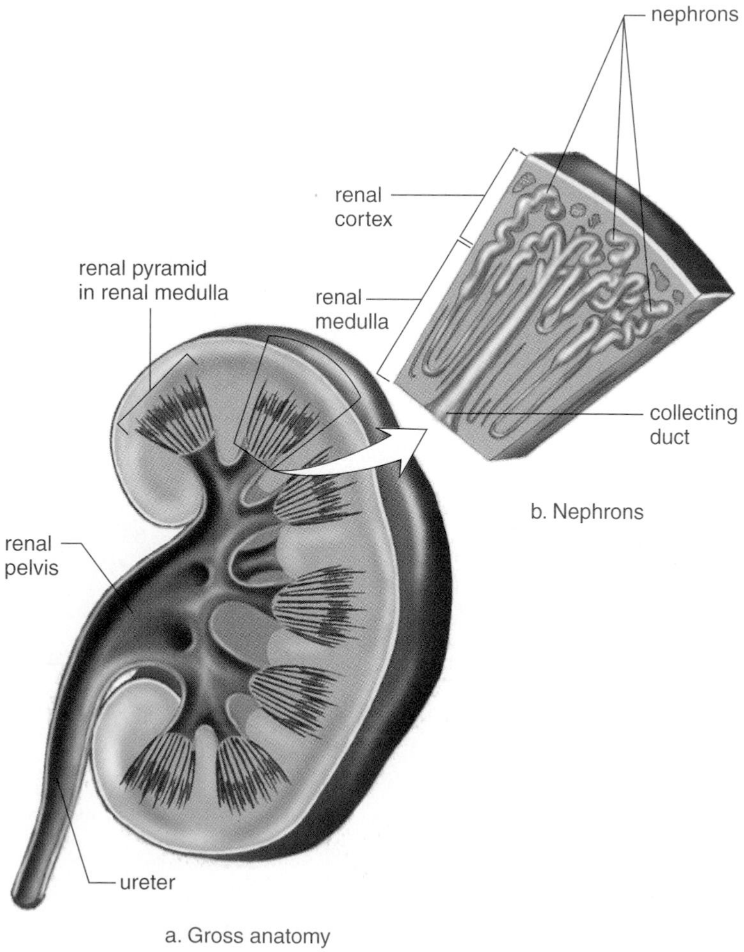

nephrons

renal
cortex

renal pyramid
in renal medulla

renal
medulla

collecting
duct

renal
pelvis

ureter

b. Nephrons

a. Gross anatomy

Macroscopic and microscopic anatomy of the kidney
Figure 38.6

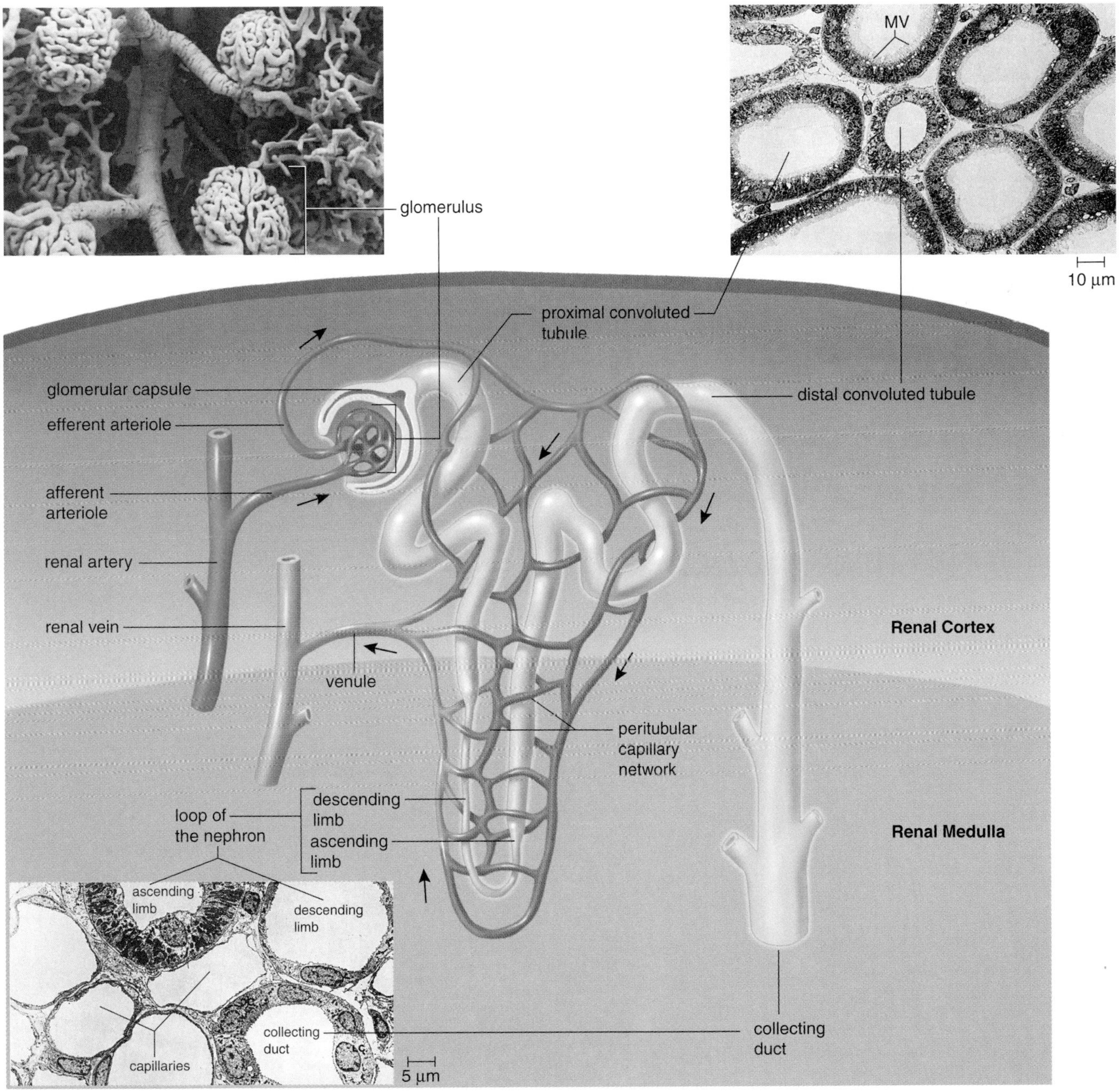

MV

10 μm

glomerulus

proximal convoluted tubule

glomerular capsule

efferent arteriole

afferent arteriole

renal artery

renal vein

venule

distal convoluted tubule

Renal Cortex

peritubular capillary network

loop of the nephron

descending limb

ascending limb

Renal Medulla

ascending limb

descending limb

collecting duct

capillaries

collecting duct

5 μm

Nephron anatomy
Figure 38.7

top left: © R.G. Kessel and R H. Kardon, *Tissues and Organs: A Text-Atlas of Scanning Electron Microscopy*, 1979; top right: © 1966 Academic Press, from A.B. Maunsbach, *J. Ultrastruct. Res.* 15:242–282; bottom right: © 1966 Academic Press, from A.B. Maunsbach, *J. Ultrastruct. Res.* 15:242–282.

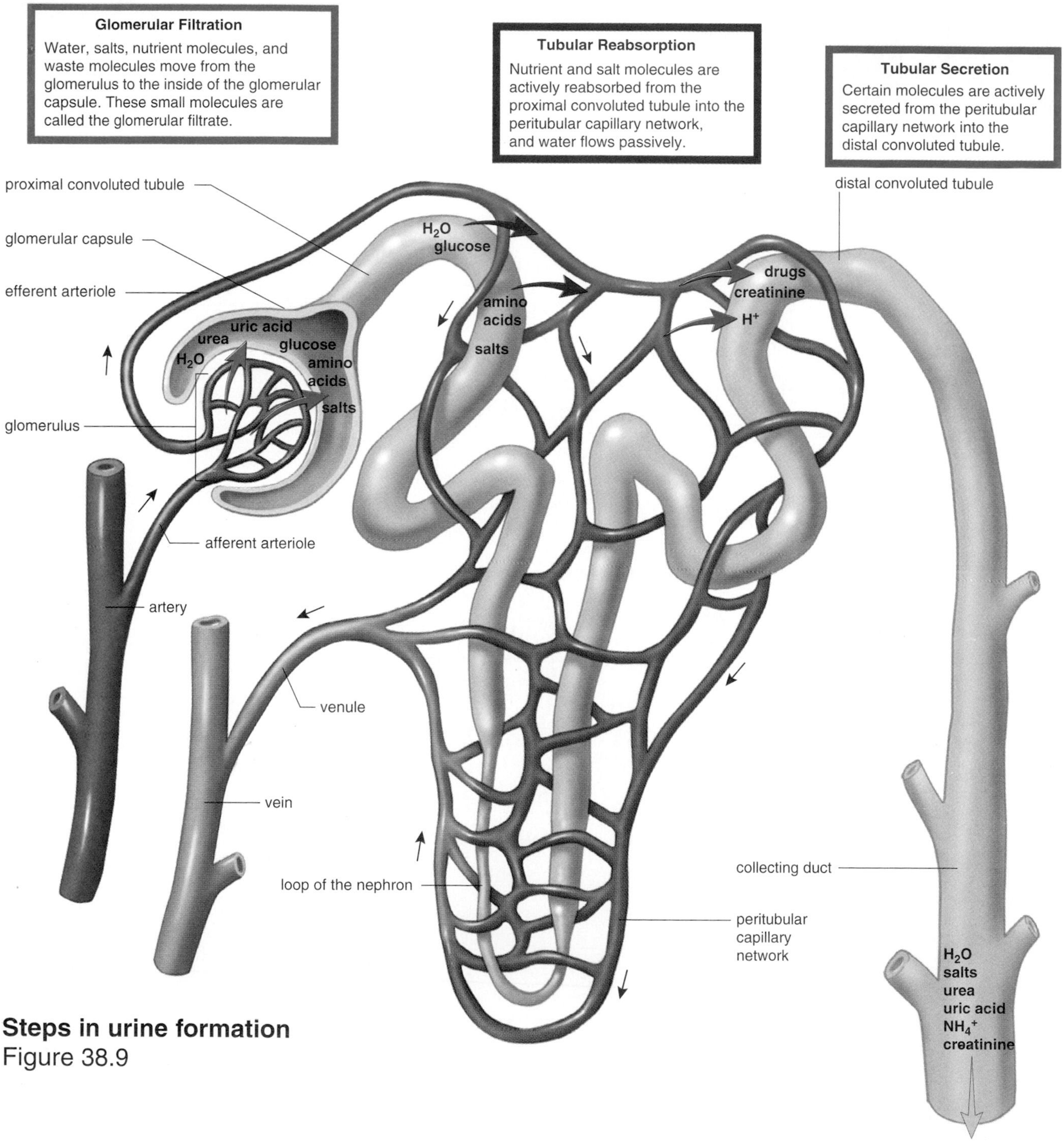

Glomerular Filtration
Water, salts, nutrient molecules, and waste molecules move from the glomerulus to the inside of the glomerular capsule. These small molecules are called the glomerular filtrate.

Tubular Reabsorption
Nutrient and salt molecules are actively reabsorbed from the proximal convoluted tubule into the peritubular capillary network, and water flows passively.

Tubular Secretion
Certain molecules are actively secreted from the peritubular capillary network into the distal convoluted tubule.

proximal convoluted tubule

glomerular capsule

efferent arteriole

glomerulus

afferent arteriole

artery

venule

vein

loop of the nephron

distal convoluted tubule

collecting duct

peritubular capillary network

H_2O
glucose

amino acids

salts

uric acid

urea

H_2O

glucose

amino acids

salts

drugs
creatinine

H^+

H_2O
salts
urea
uric acid
NH_4^+
creatinine

Steps in urine formation
Figure 38.9

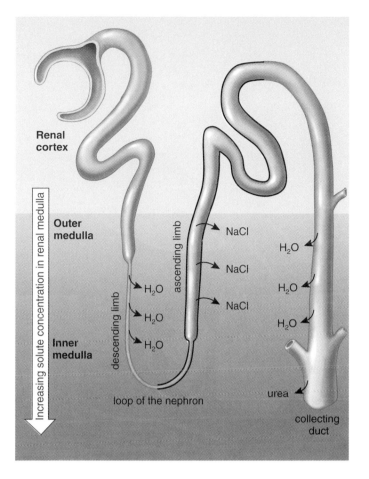

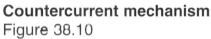

Countercurrent mechanism
Figure 38.10

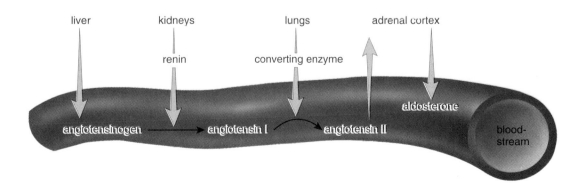

The renin-angiotensin-aldosterone system
Figure 38.11

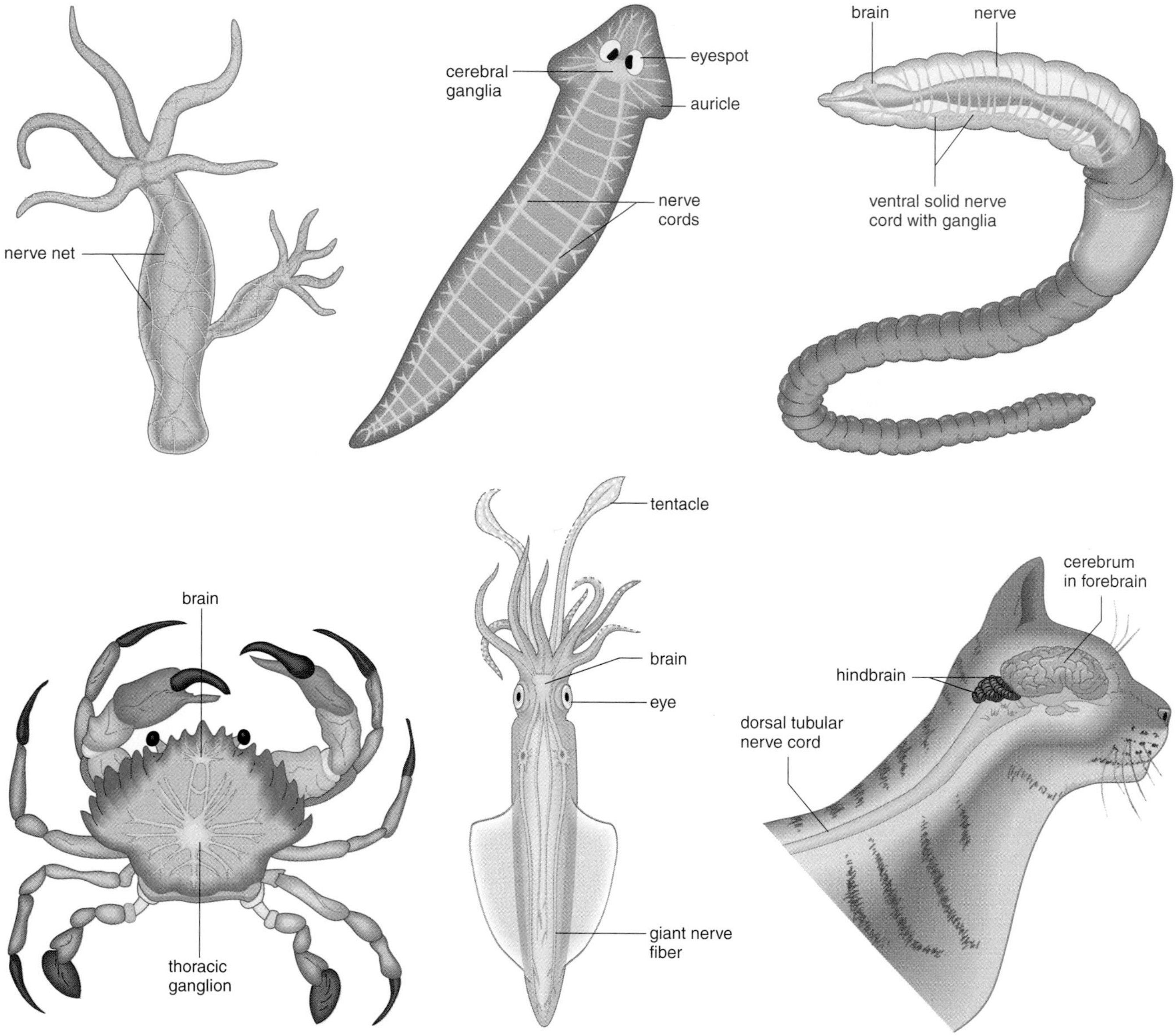
nerve net
cerebral ganglia
eyespot
auricle
nerve cords
brain
nerve
ventral solid nerve cord with ganglia
tentacle
brain
eye
giant nerve fiber
brain
thoracic ganglion
cerebrum in forebrain
hindbrain
dorsal tubular nerve cord

Evolution of the nervous system
Figure 39.1

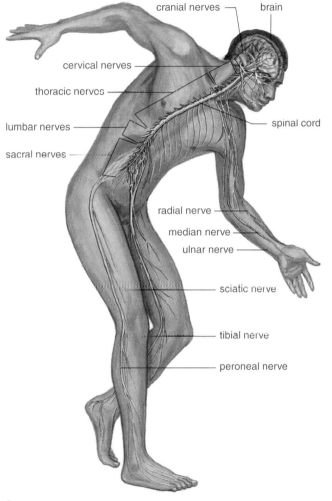

a.

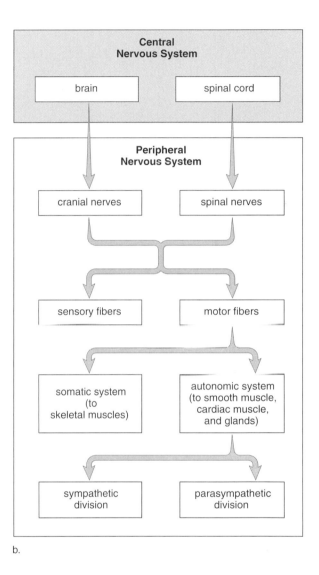

b.

Organization of the nervous system in humans
Figure 39.2

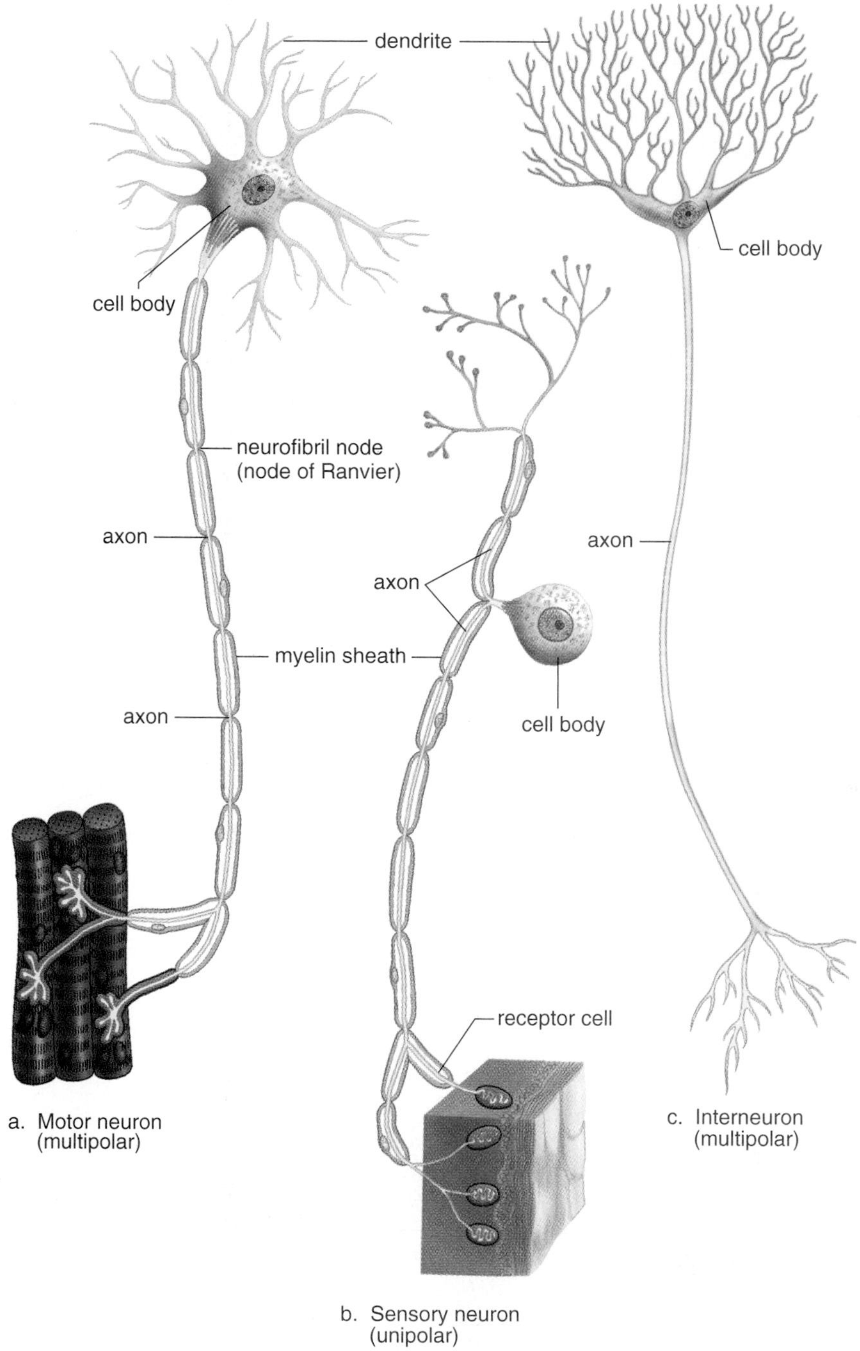

dendrite

cell body

cell body

neurofibril node
(node of Ranvier)

axon

axon

axon

myelin sheath

axon

cell body

receptor cell

a. Motor neuron
(multipolar)

b. Sensory neuron
(unipolar)

c. Interneuron
(multipolar)

Neuron anatomy
Figure 39.3

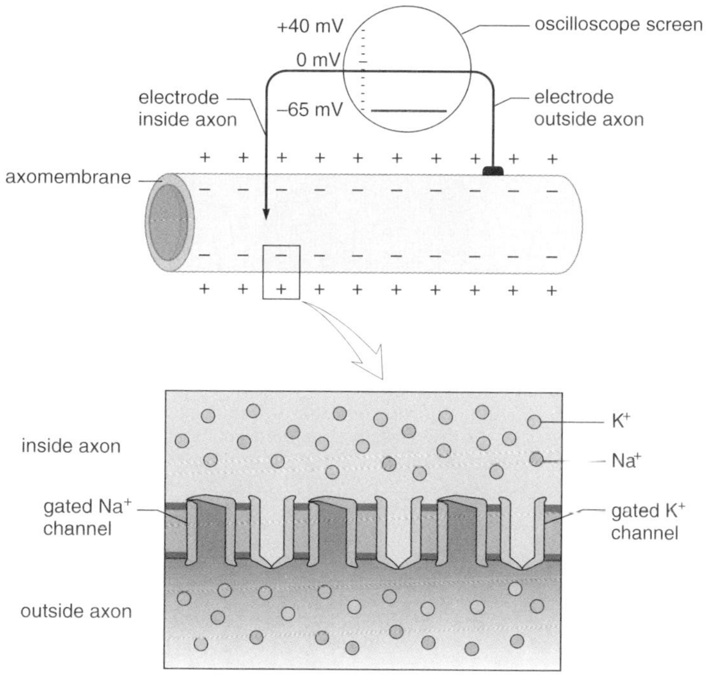

a. Resting potential

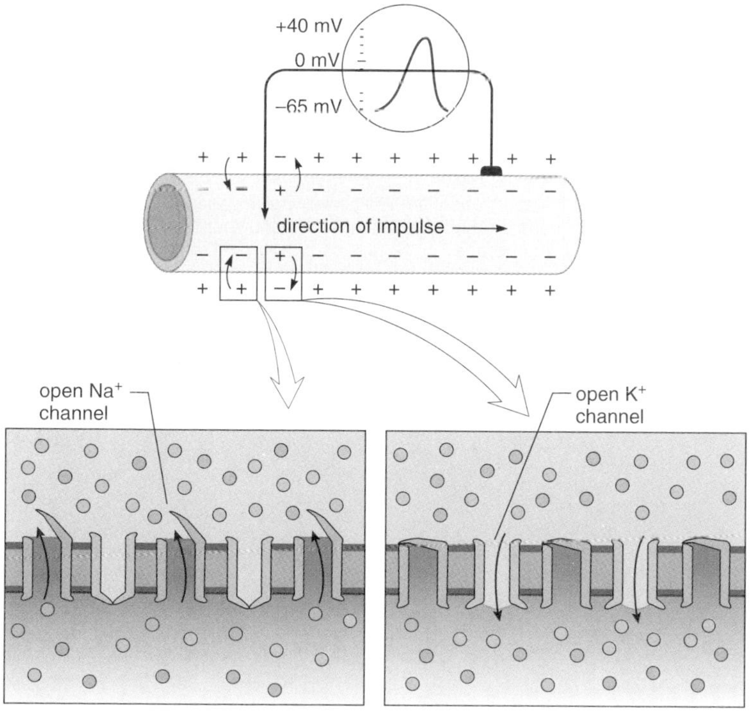

c. Enlargement of action potential

b. Action potential

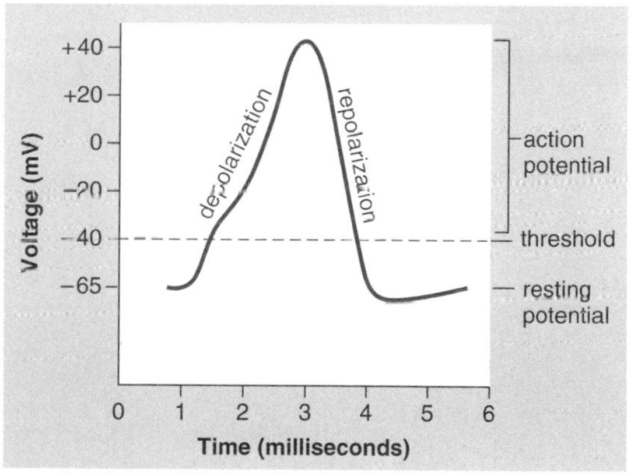

Resting and action potential of the axomembrane
Figure 39.4

Synapse structure and function
Figure 39.5

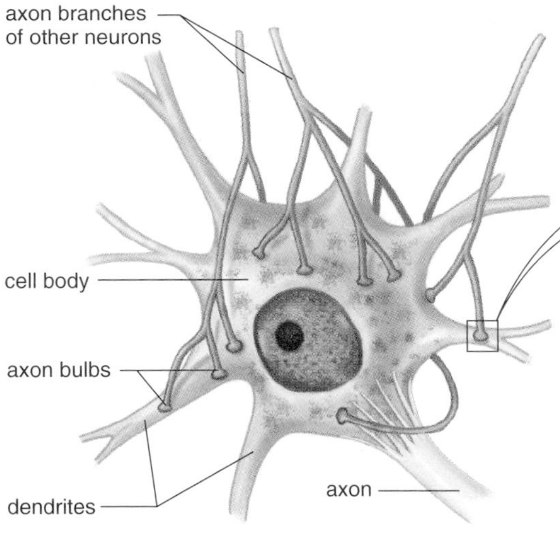

axon branches of other neurons

cell body

axon bulbs

dendrites

axon

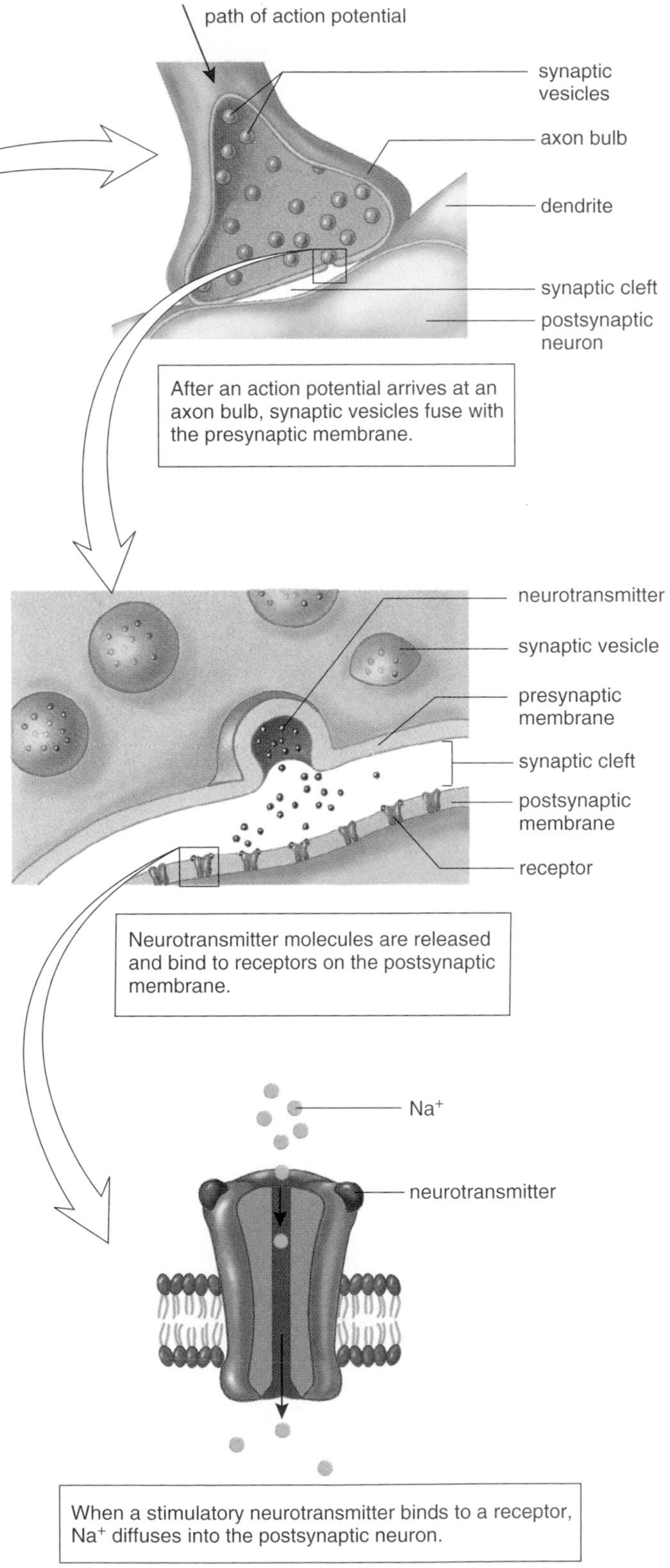

path of action potential

synaptic vesicles

axon bulb

dendrite

synaptic cleft

postsynaptic neuron

After an action potential arrives at an axon bulb, synaptic vesicles fuse with the presynaptic membrane.

neurotransmitter

synaptic vesicle

presynaptic membrane

synaptic cleft

postsynaptic membrane

receptor

Neurotransmitter molecules are released and bind to receptors on the postsynaptic membrane.

Na⁺

neurotransmitter

When a stimulatory neurotransmitter binds to a receptor, Na⁺ diffuses into the postsynaptic neuron.

Integration
Figure 39.6

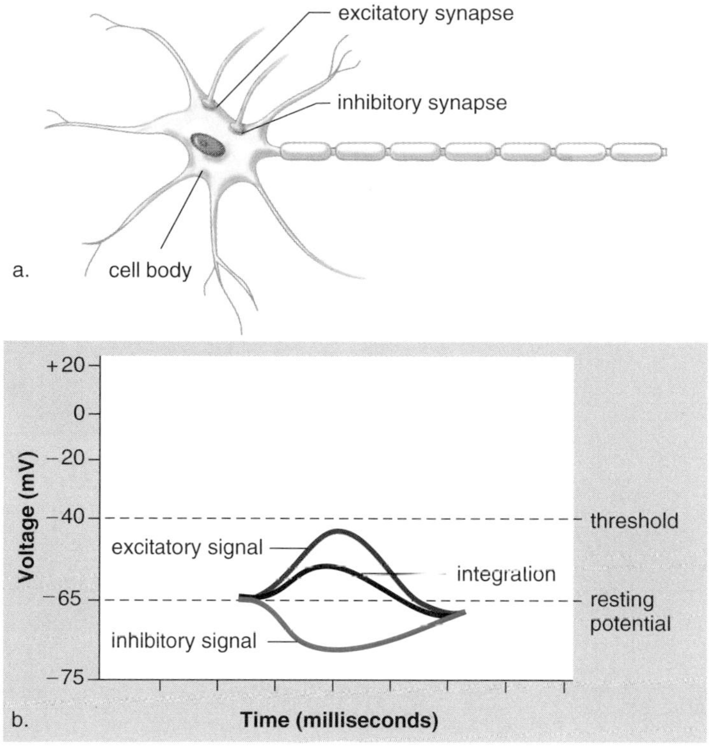

a.

- excitatory synapse
- inhibitory synapse
- cell body

b.

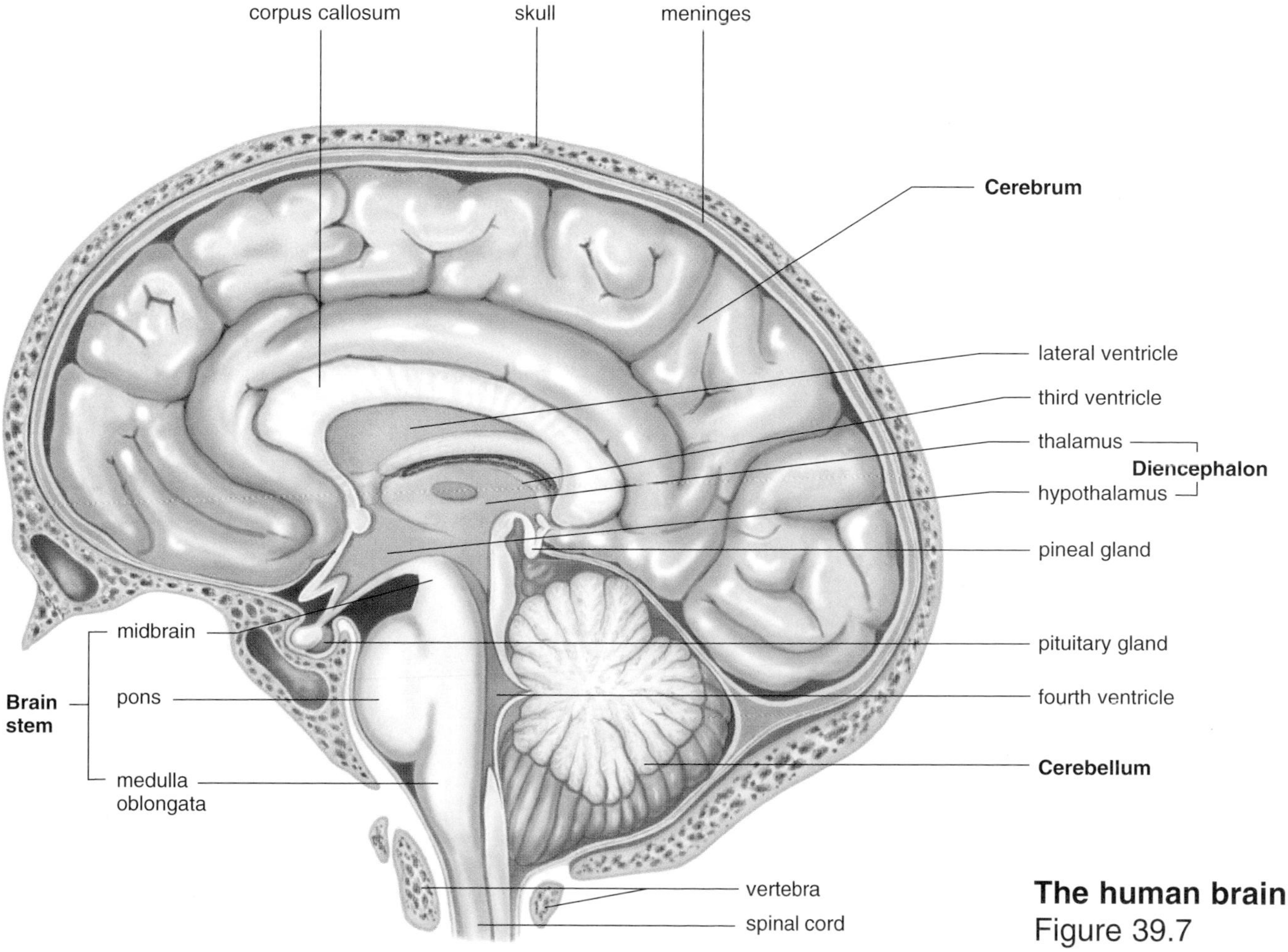

corpus callosum   skull   meninges

Cerebrum

lateral ventricle

third ventricle

thalamus ⎤
           ⎥ **Diencephalon**
hypothalamus ⎦

pineal gland

pituitary gland

fourth ventricle

Cerebellum

midbrain

pons

Brain stem

medulla oblongata

vertebra

spinal cord

The human brain
Figure 39.7

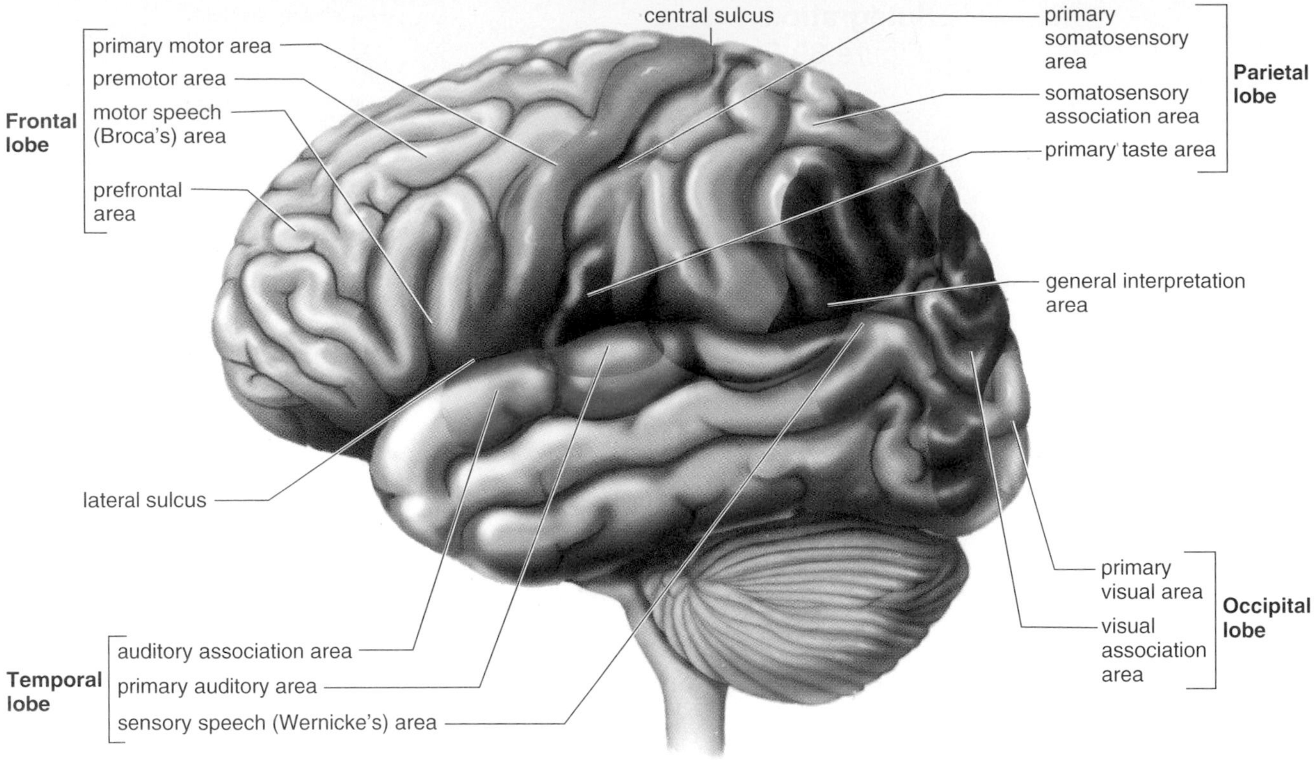

The lobes of a cerebral hemisphere
Figure 39.8

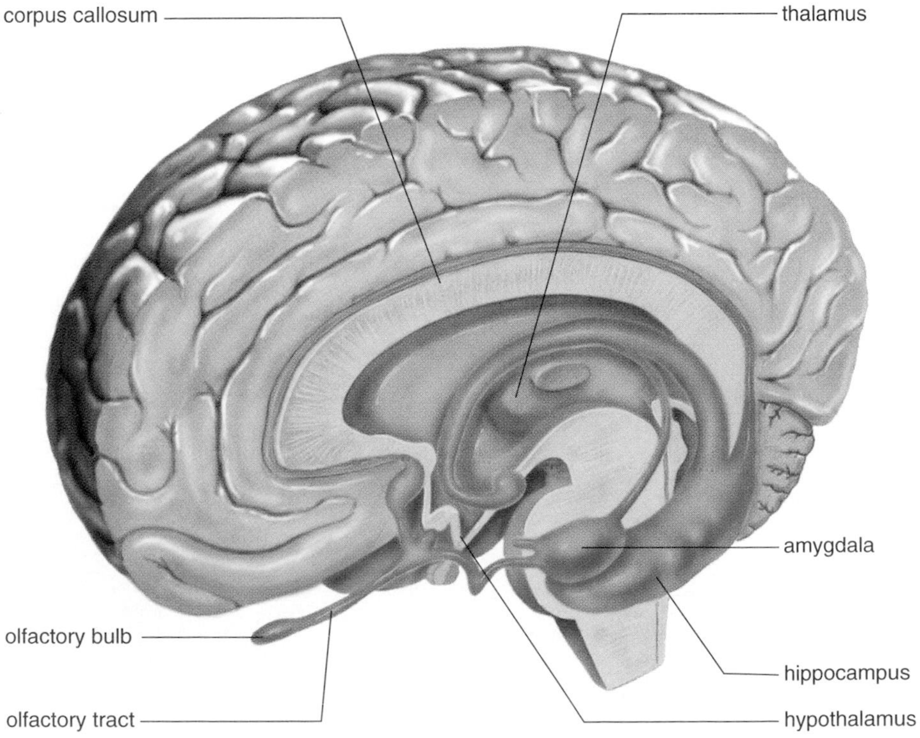

The limbic system
Figure 39.9

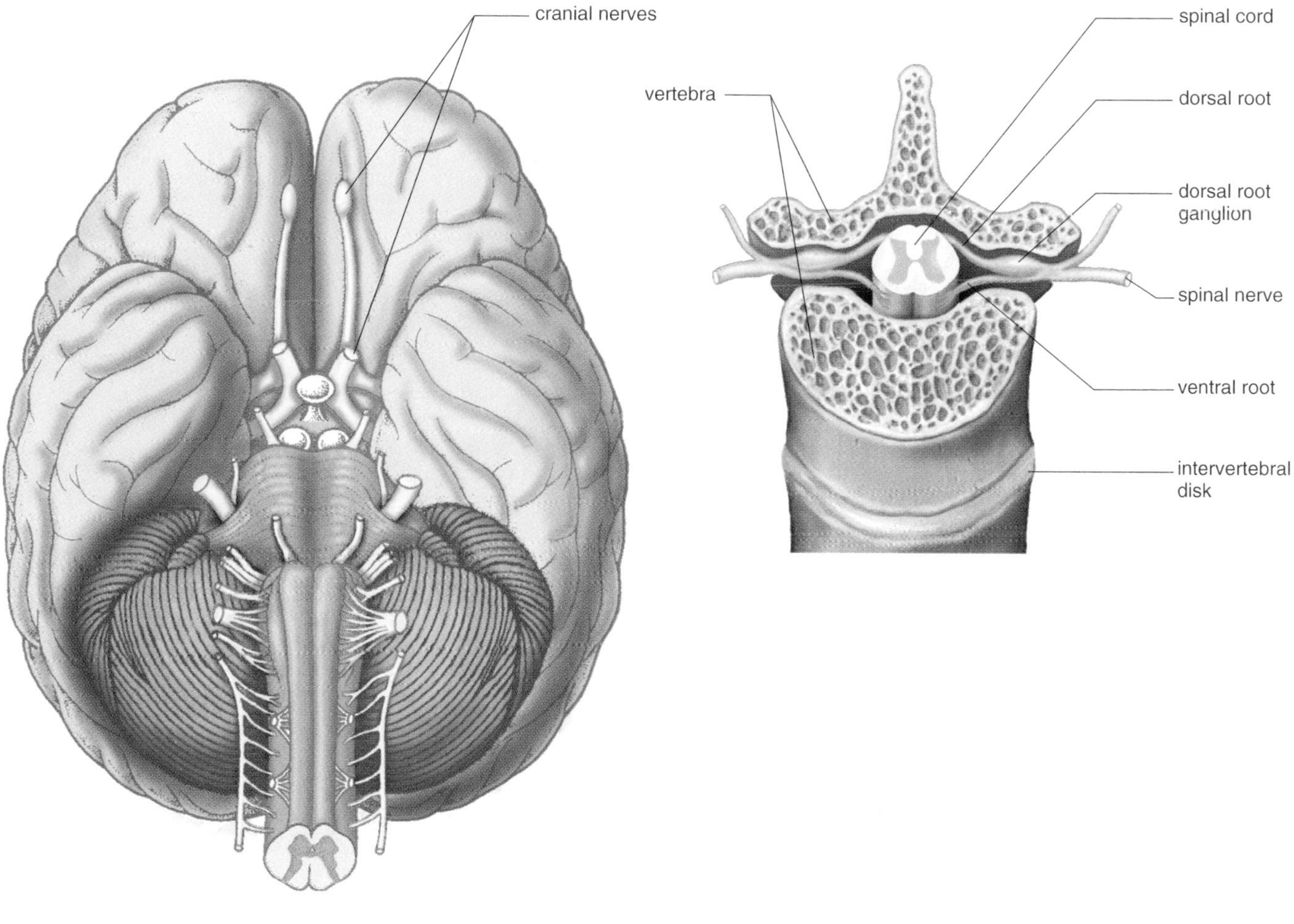

cranial nerves

vertebra

spinal cord

dorsal root

dorsal root ganglion

spinal nerve

ventral root

intervertebral disk

Cranial and spinal nerves
Figure 39.10

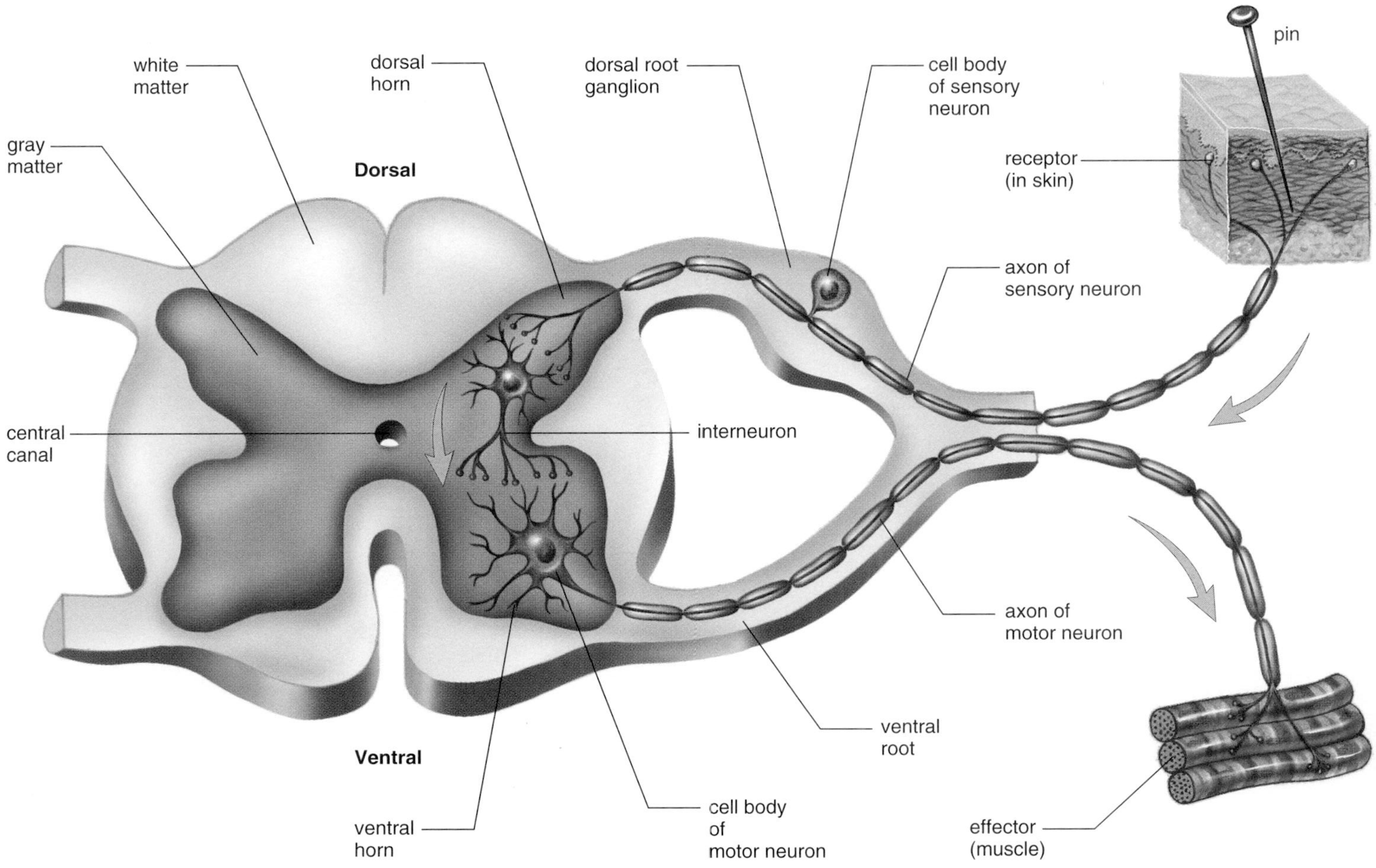

pin

receptor
(in skin)

cell body
of sensory
neuron

axon of
sensory neuron

axon of
motor neuron

effector
(muscle)

dorsal root
ganglion

interneuron

ventral
root

cell body
of
motor neuron

dorsal
horn

Dorsal

white
matter

gray
matter

central
canal

Ventral

ventral
horn

A reflex arc showing the path of a spinal reflex
Figure 39.11

Sympathetic Division

Parasympathetic Division

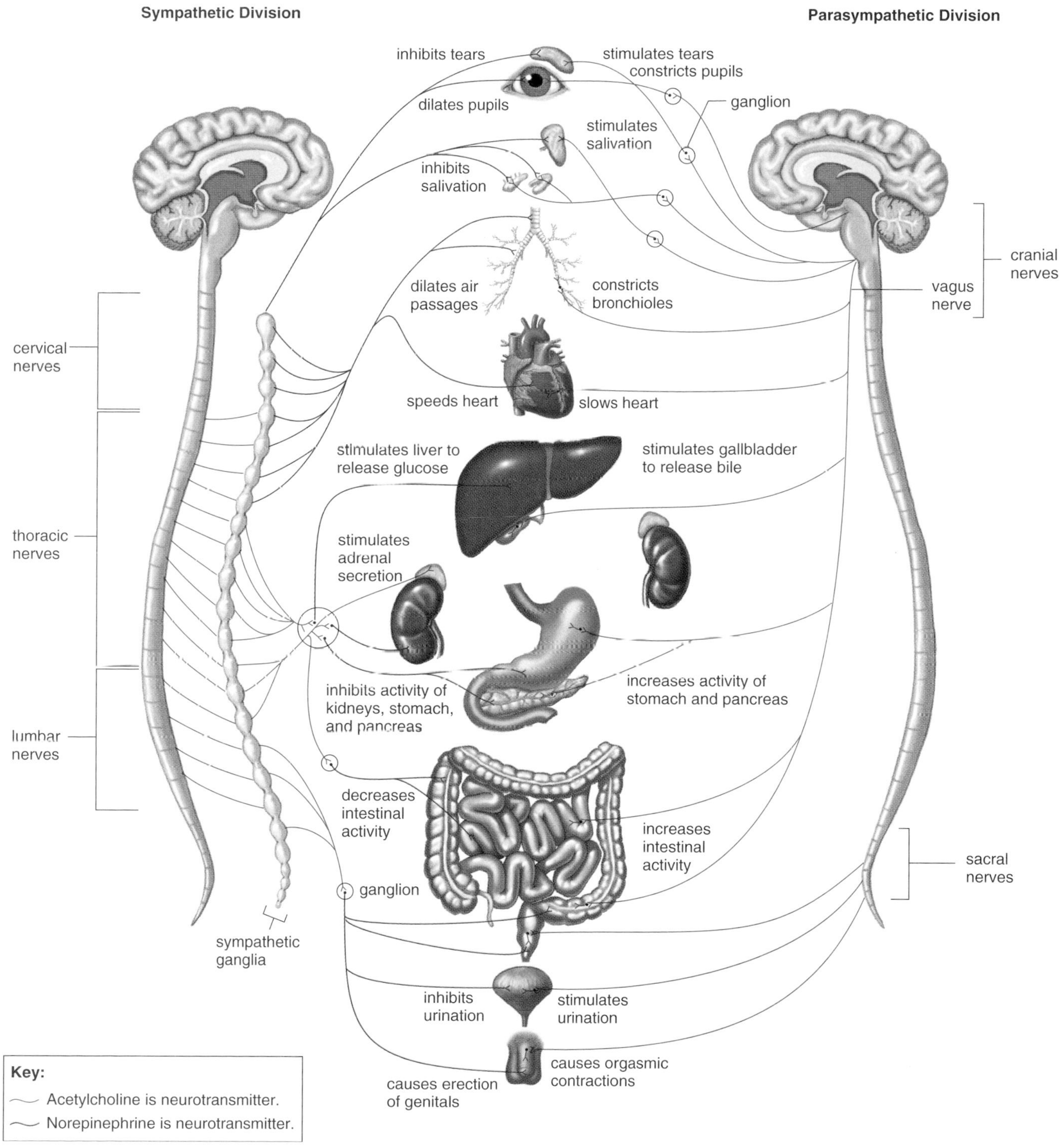

inhibits tears

stimulates tears
constricts pupils

dilates pupils

ganglion

stimulates
salivation

inhibits
salivation

dilates air
passages

constricts
bronchioles

cranial
nerves

vagus
nerve

cervical
nerves

speeds heart

slows heart

stimulates liver to
release glucose

stimulates gallbladder
to release bile

thoracic
nerves

stimulates
adrenal
secretion

inhibits activity of
kidneys, stomach,
and pancreas

increases activity of
stomach and pancreas

lumbar
nerves

decreases
intestinal
activity

increases
intestinal
activity

ganglion

sacral
nerves

sympathetic
ganglia

inhibits
urination

stimulates
urination

causes erection
of genitals

causes orgasmic
contractions

Key:
⌇ Acetylcholine is neurotransmitter.
⌇ Norepinephrine is neurotransmitter.

Autonomic system structure and function
Figure 39.12

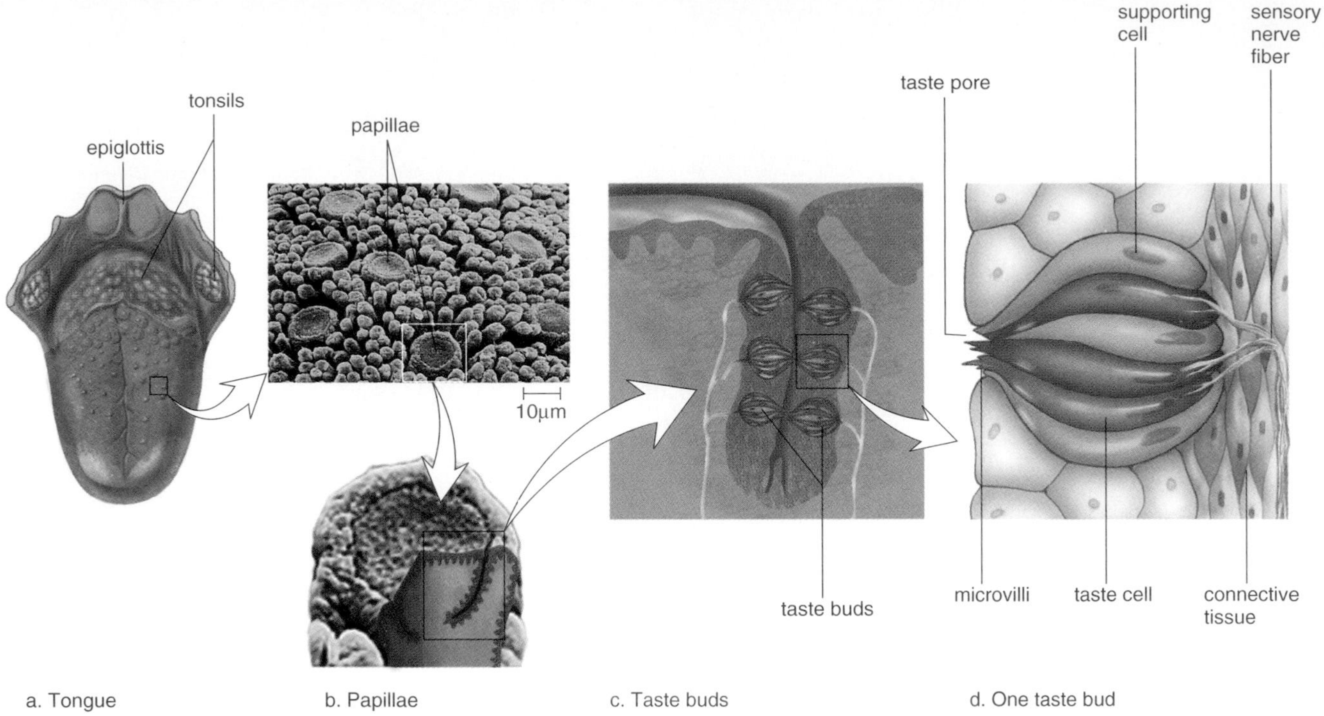

a. Tongue   b. Papillae   c. Taste buds   d. One taste bud

Taste buds in humans

Figure 40.1

b: © Omikron/SPL/Photo Researchers, Inc.

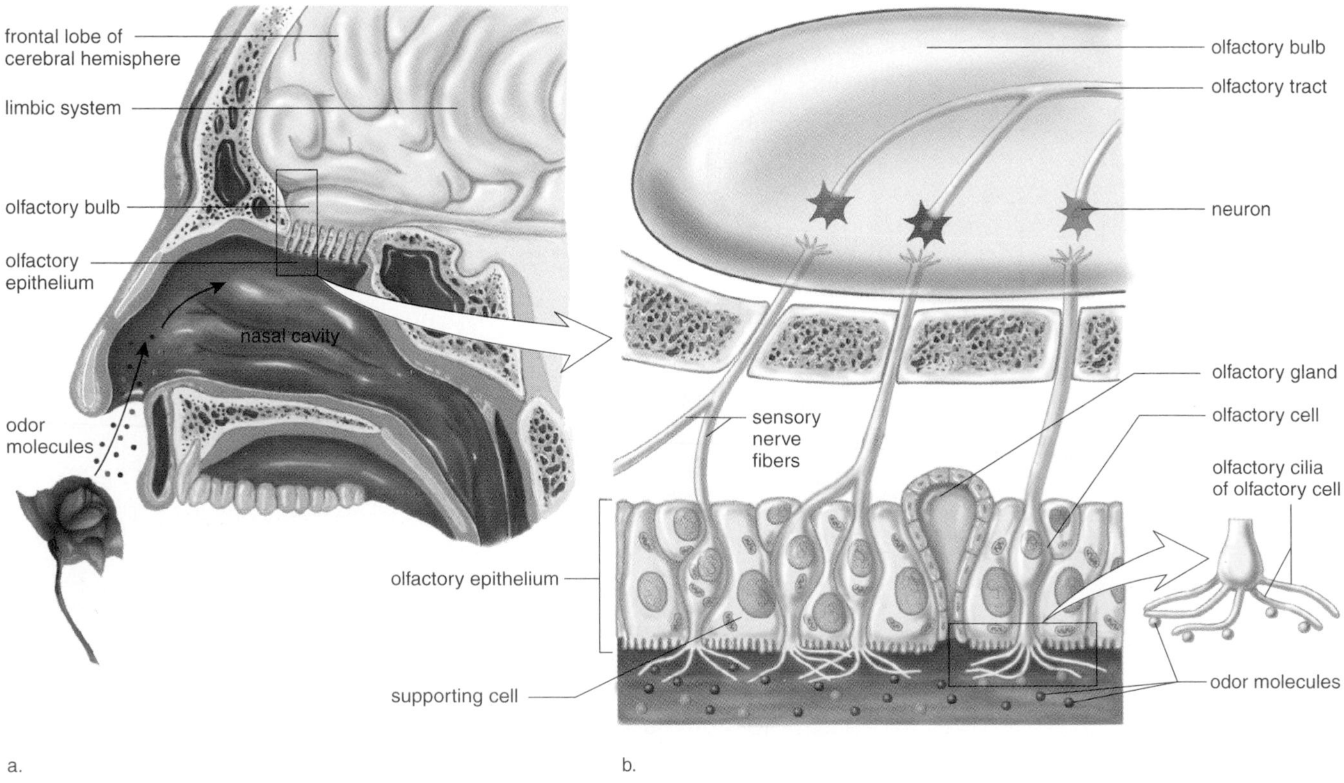

Olfactory cell location and anatomy

Figure 40.2

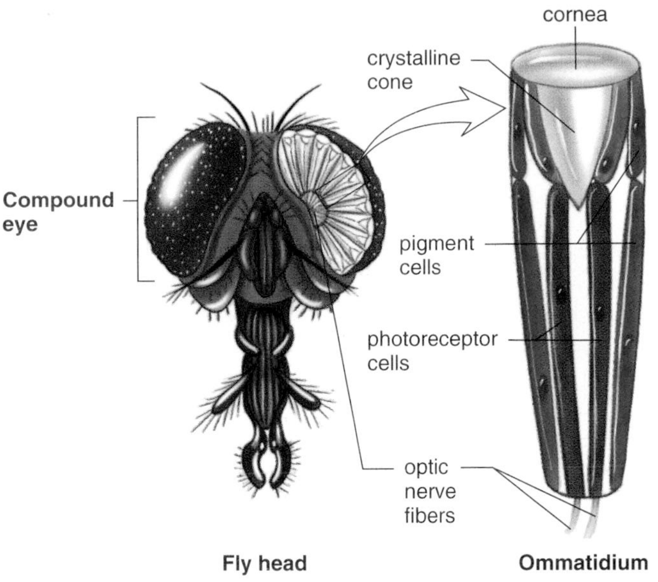

Compound eye
Figure 40.3

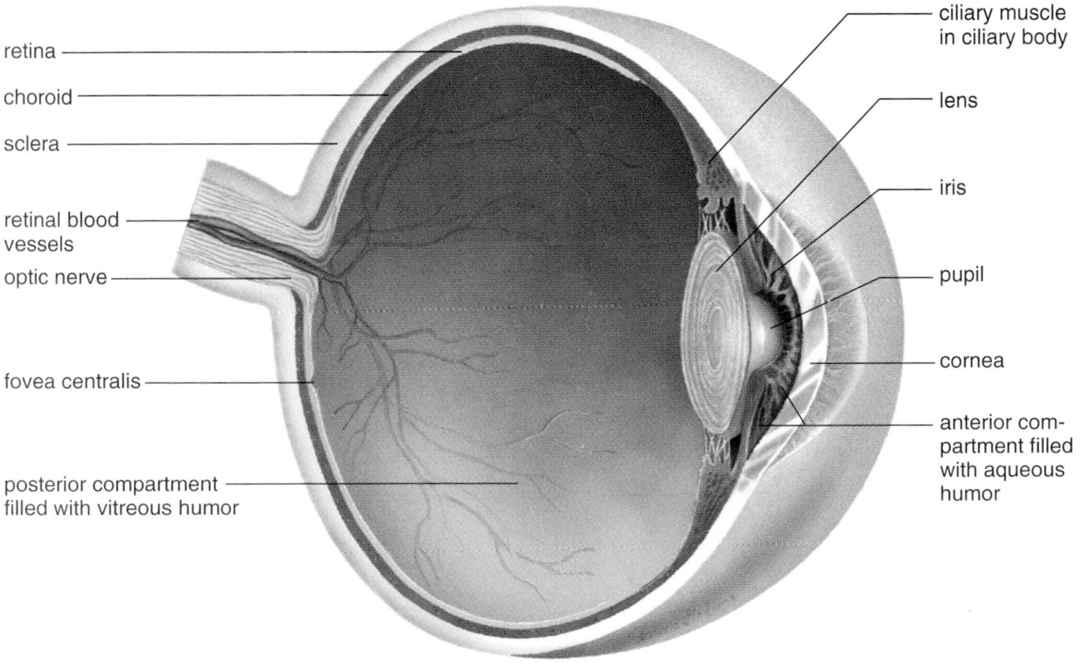

Anatomy of the human eye
Figure 40.5

Focusing of the human eye
Figure 40.6

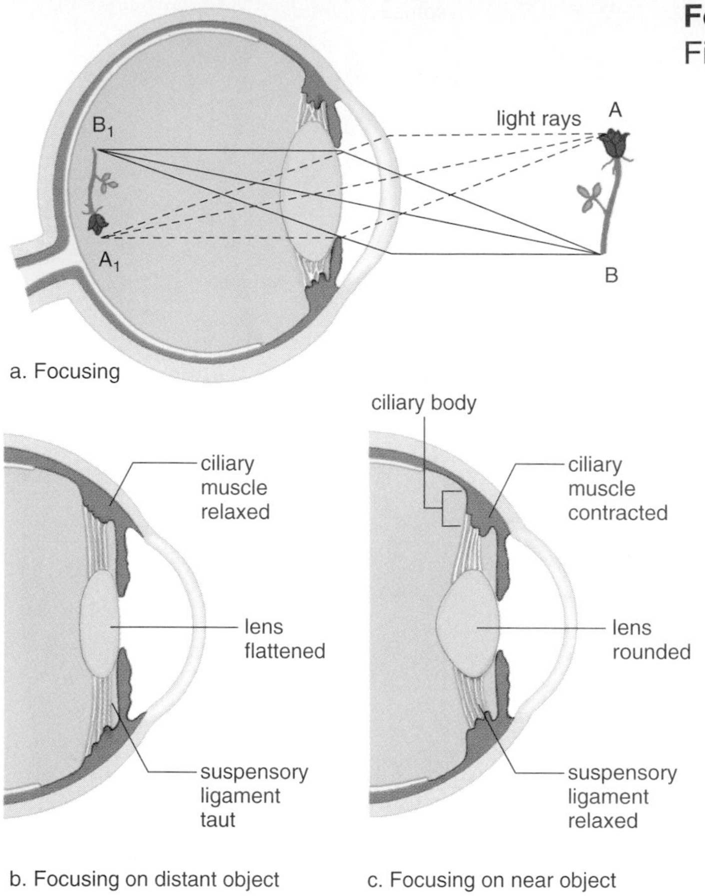

light rays

A

B₁

A₁

B

a. Focusing

ciliary body

ciliary muscle relaxed

lens flattened

suspensory ligament taut

b. Focusing on distant object

ciliary muscle contracted

lens rounded

suspensory ligament relaxed

c. Focusing on near object

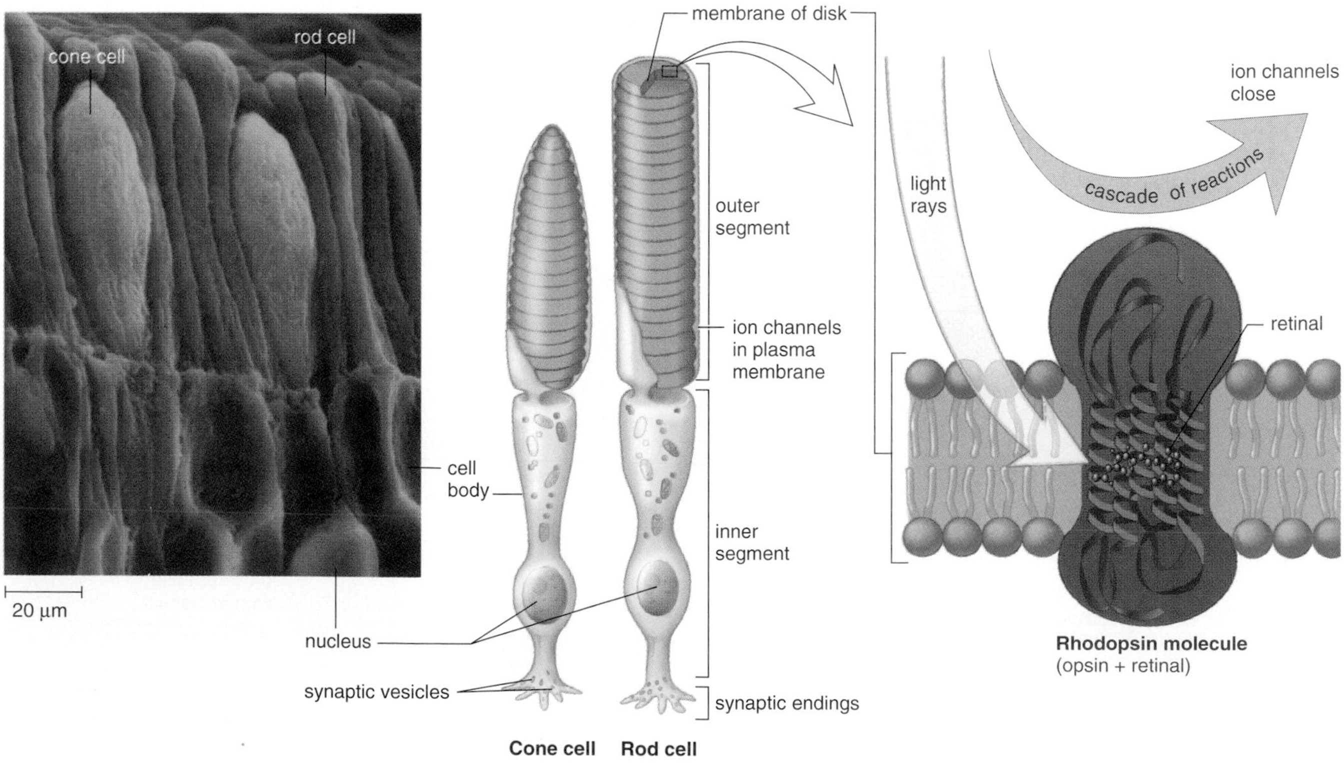

cone cell

rod cell

20 μm

membrane of disk

outer segment

ion channels in plasma membrane

inner segment

cell body

nucleus

synaptic vesicles

synaptic endings

Cone cell   Rod cell

light rays

ion channels close

cascade of reactions

retinal

Rhodopsin molecule
(opsin + retinal)

Photoreceptors in the eye
Figure 40.7
© Lennart Nilsson, from *The Incredible Machine*

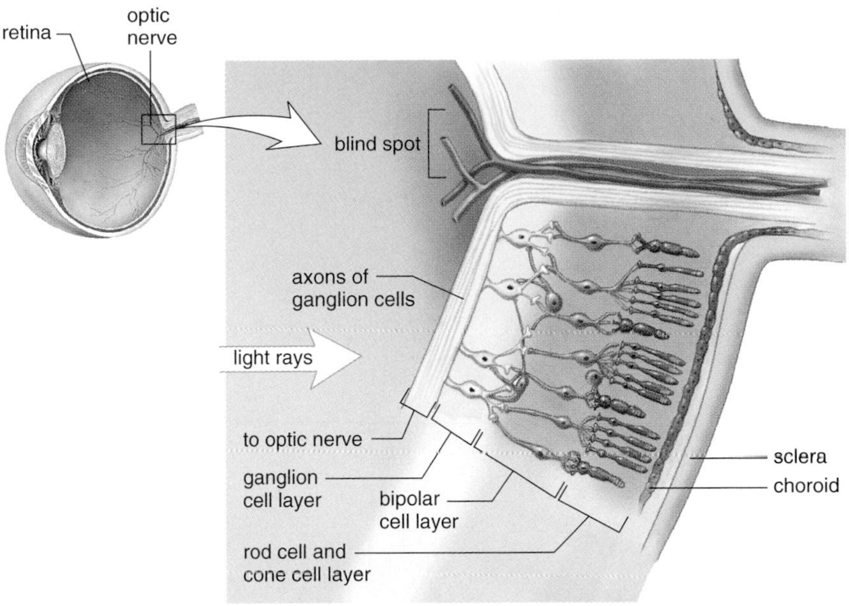

a. Drawing of retina

Structure and function of the retina
Figure 40.8

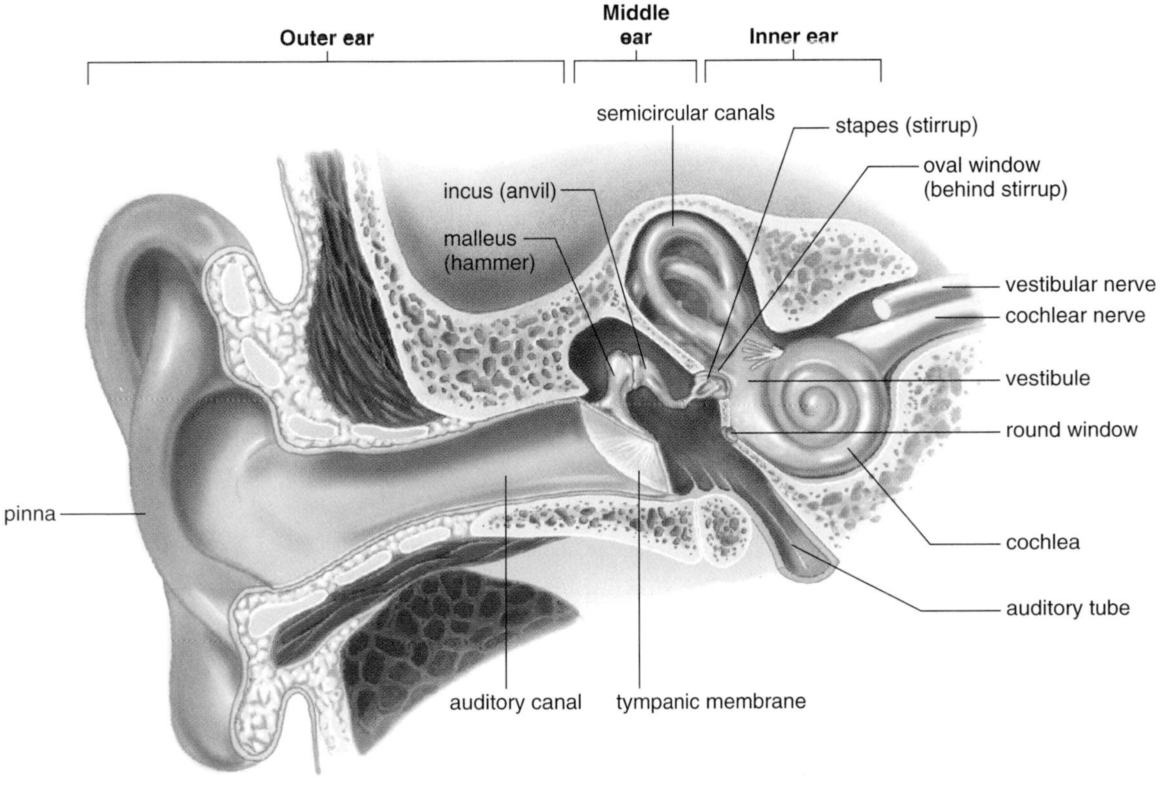

Anatomy of the human ear
Figure 40.9

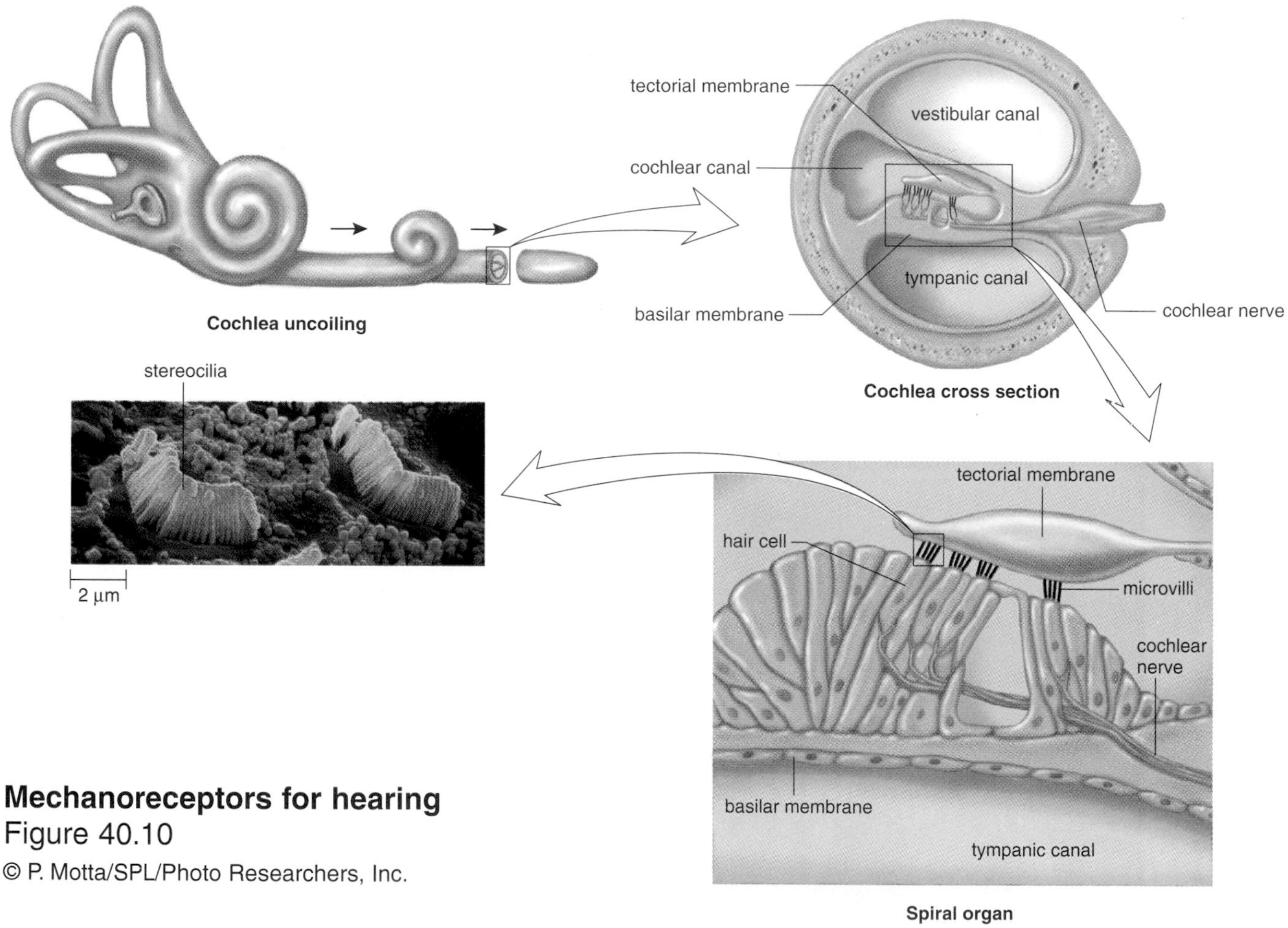

Cochlea uncoiling

tectorial membrane

vestibular canal

cochlear canal

tympanic canal

basilar membrane

cochlear nerve

Cochlea cross section

stereocilia

2 μm

tectorial membrane

hair cell

microvilli

cochlear nerve

basilar membrane

tympanic canal

Spiral organ

Mechanoreceptors for hearing
Figure 40.10
© P. Motta/SPL/Photo Researchers, Inc.

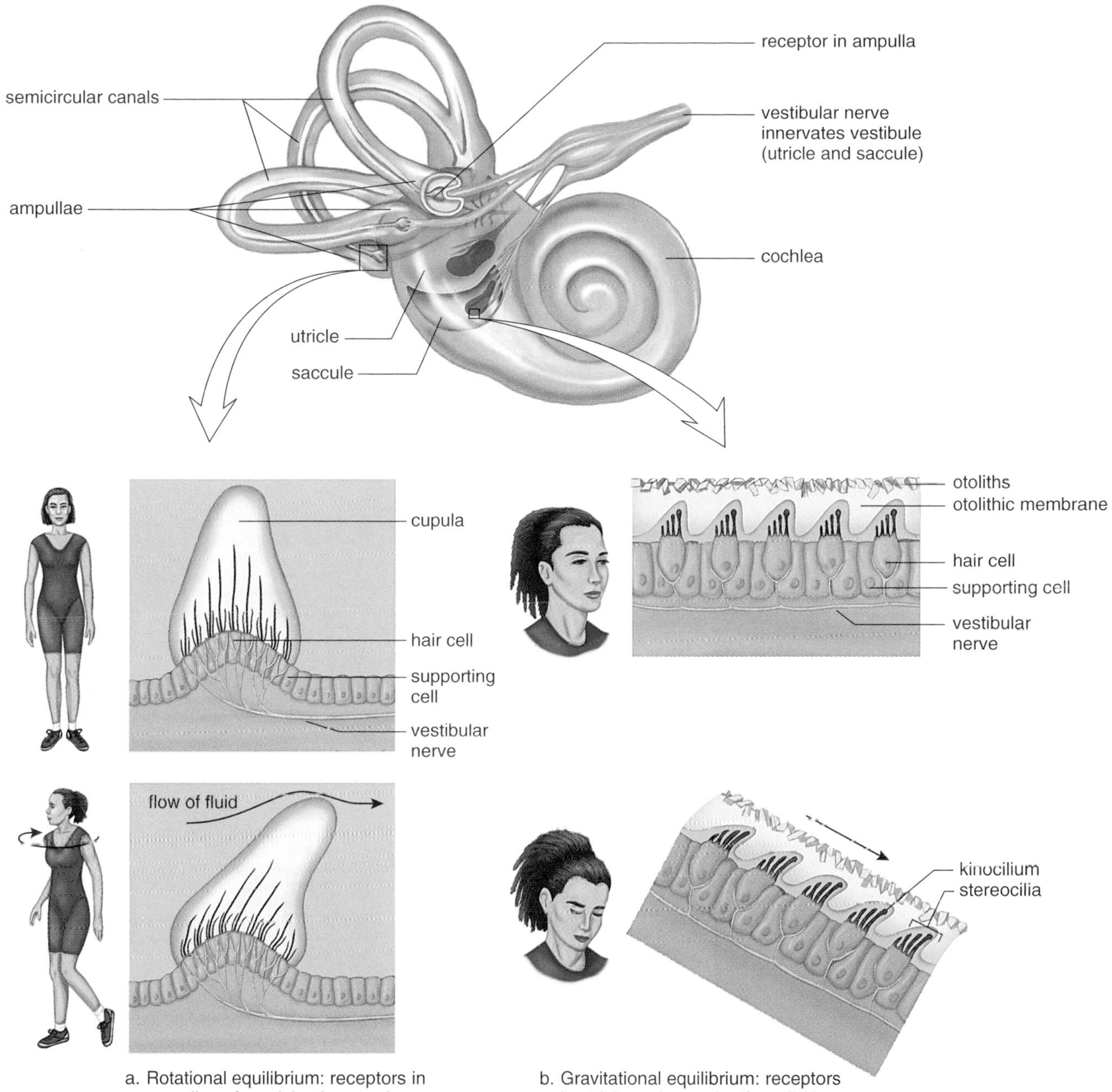

semicircular canals

ampullae

receptor in ampulla

vestibular nerve
innervates vestibule
(utricle and saccule)

cochlea

utricle

saccule

cupula

hair cell

supporting
cell

vestibular
nerve

flow of fluid

a. Rotational equilibrium: receptors in
ampullae of semicircular canal.

otoliths
otolithic membrane

hair cell
supporting cell

vestibular
nerve

kinocilium
stereocilia

b. Gravitational equilibrium: receptors
in utricle and saccule of vestibule.

Mechanoreceptors for equilibrium
Figure 40.11

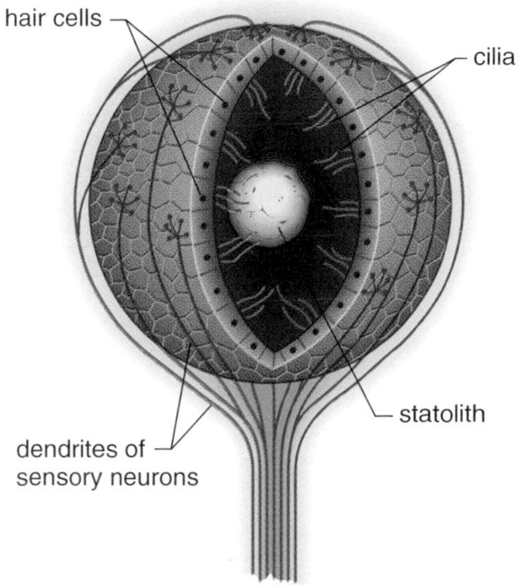

hair cells

cilia

dendrites of
sensory neurons

statolith

Statocysts
Figure 40.12

Locomotion in an earthworm
Figure 41.1

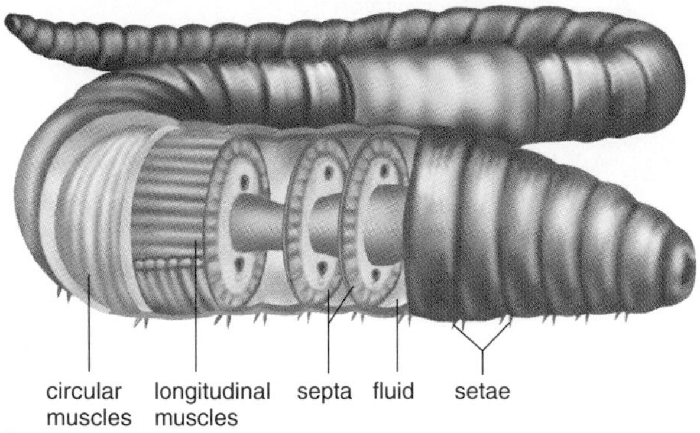

a.

circular | longitudinal | septa | fluid | setae
muscles | muscles

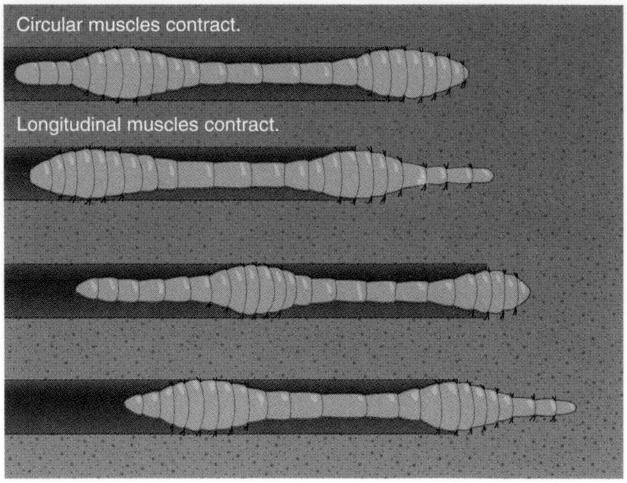

Circular muscles contract.

Longitudinal muscles contract.

b.

The vertebrate endoskeleton
Figure 41.3

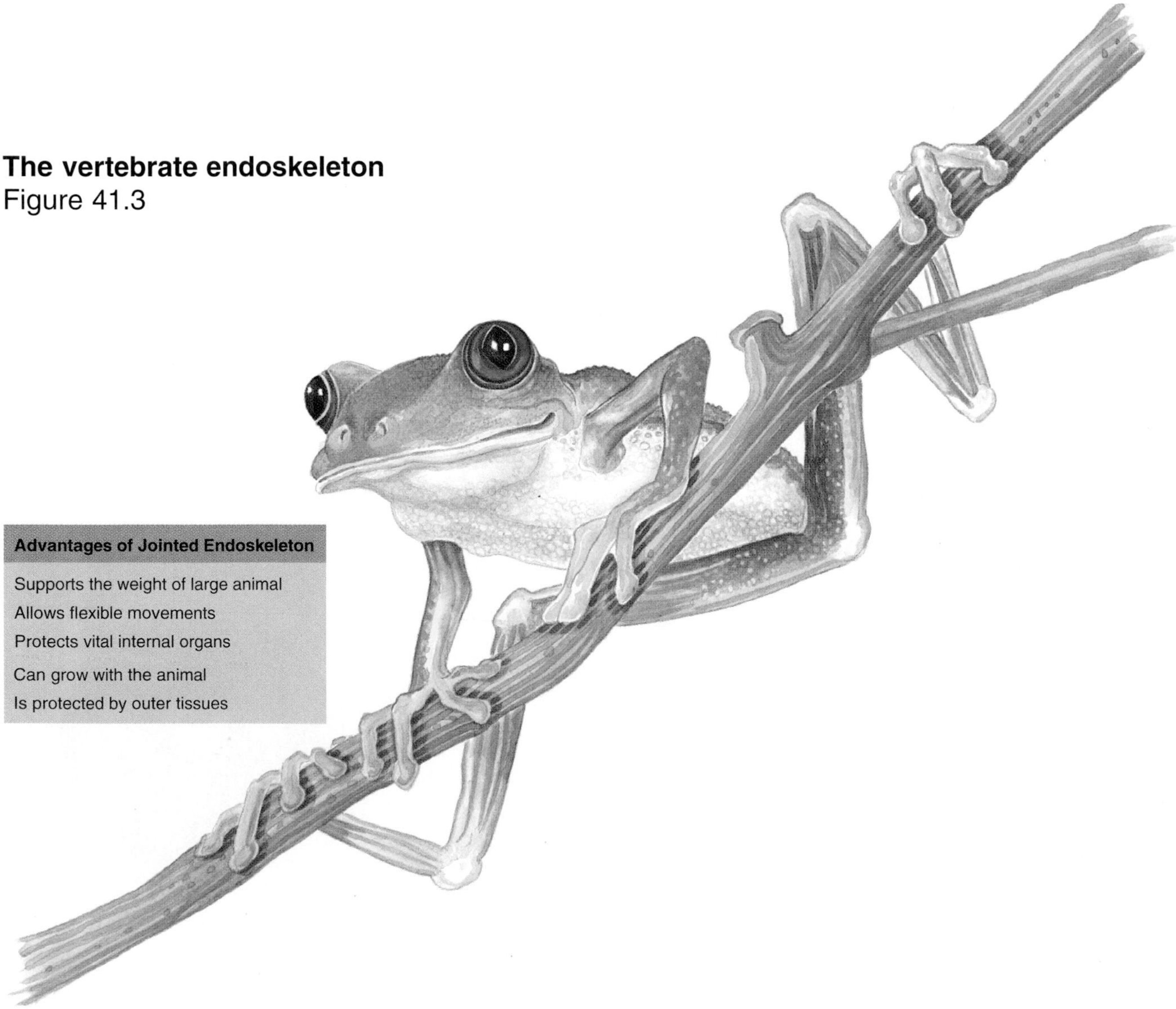

Advantages of Jointed Endoskeleton

Supports the weight of large animal

Allows flexible movements

Protects vital internal organs

Can grow with the animal

Is protected by outer tissues

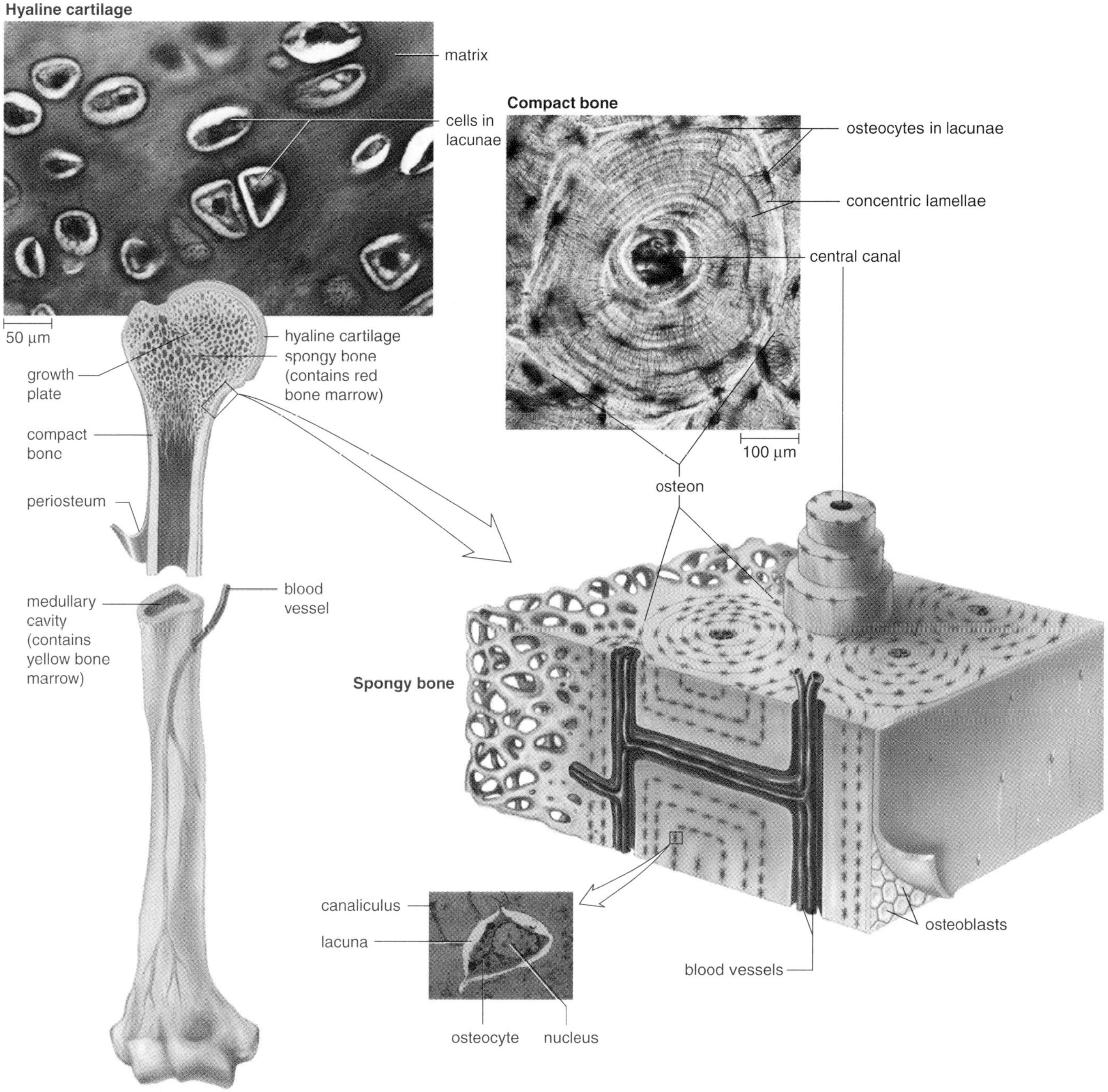

Hyaline cartilage

matrix

cells in lacunae

50 μm

growth plate

compact bone

periosteum

hyaline cartilage

spongy bone (contains red bone marrow)

medullary cavity (contains yellow bone marrow)

blood vessel

Compact bone

osteocytes in lacunae

concentric lamellae

central canal

100 μm

osteon

Spongy bone

canaliculus

lacuna

osteocyte   nucleus

blood vessels

osteoblasts

Anatomy of a long bone
Figure 41.4

a (hyaline cartilage): © Ed Reschke; b (compact bone): © Ed Reschke; c (osteocyte): © Biophoto Associates/Photo Researchers, Inc.

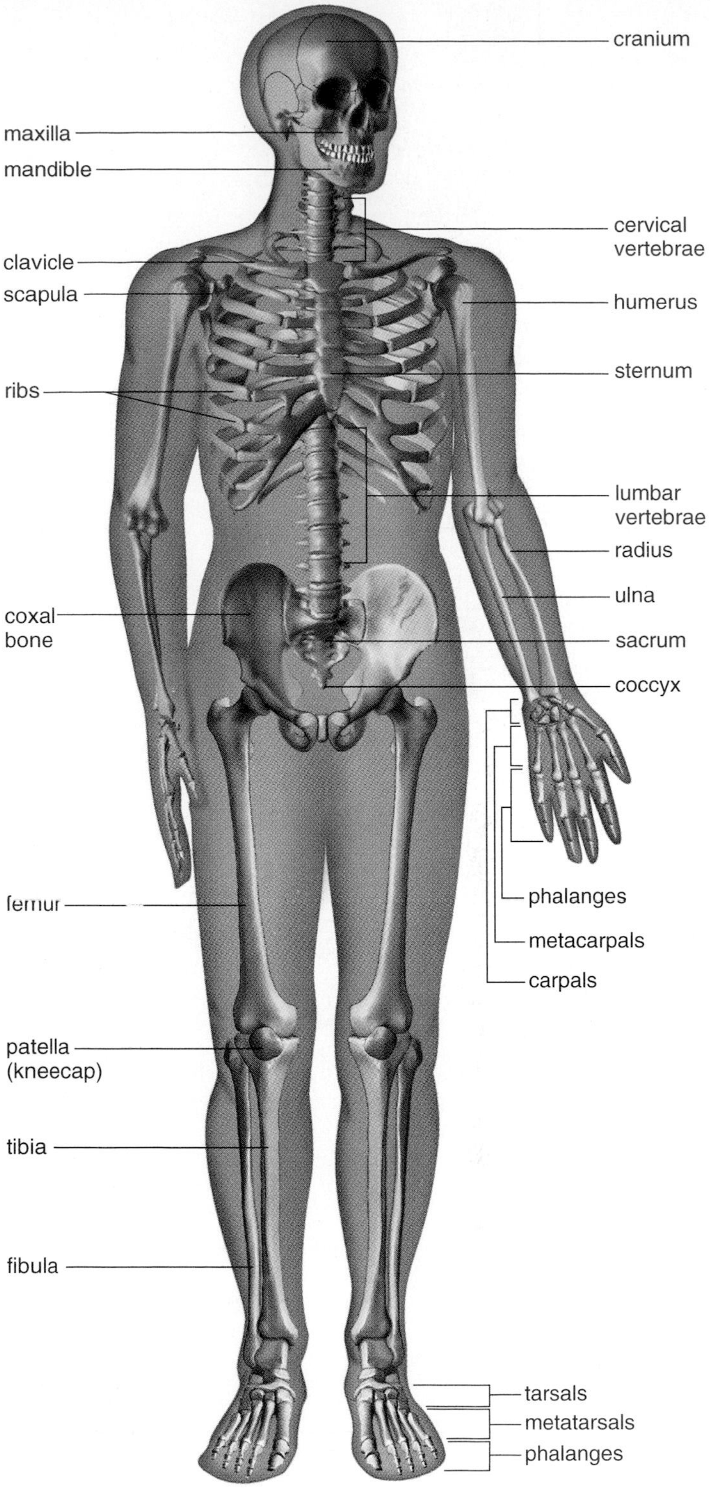

cranium

maxilla
mandible

cervical
vertebrae

clavicle
scapula

humerus

sternum

ribs

lumbar
vertebrae
radius

ulna

coxal
bone

sacrum

coccyx

phalanges

metacarpals

carpals

femur

patella
(kneecap)

tibia

fibula

tarsals
metatarsals
phalanges

The human skeleton
Figure 41.5

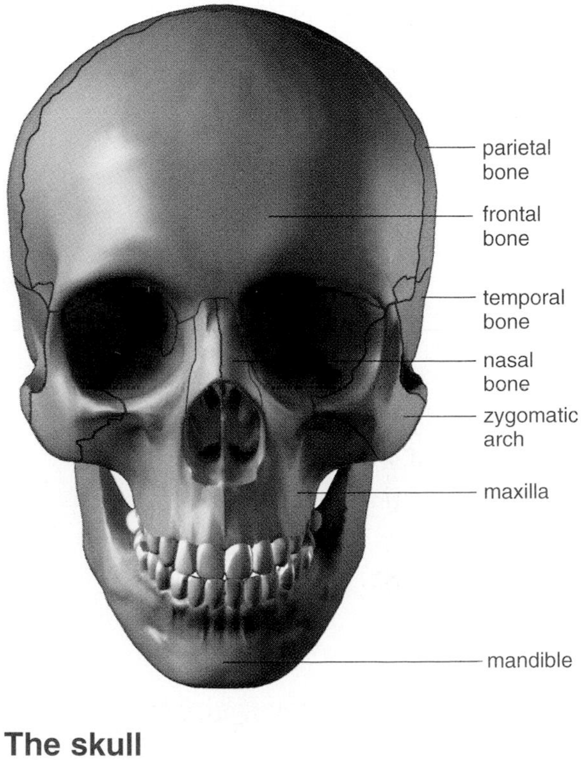

parietal
bone

frontal
bone

temporal
bone

nasal
bone

zygomatic
arch

maxilla

mandible

The skull
Figure 41.6

The rib cage
Figure 41.7

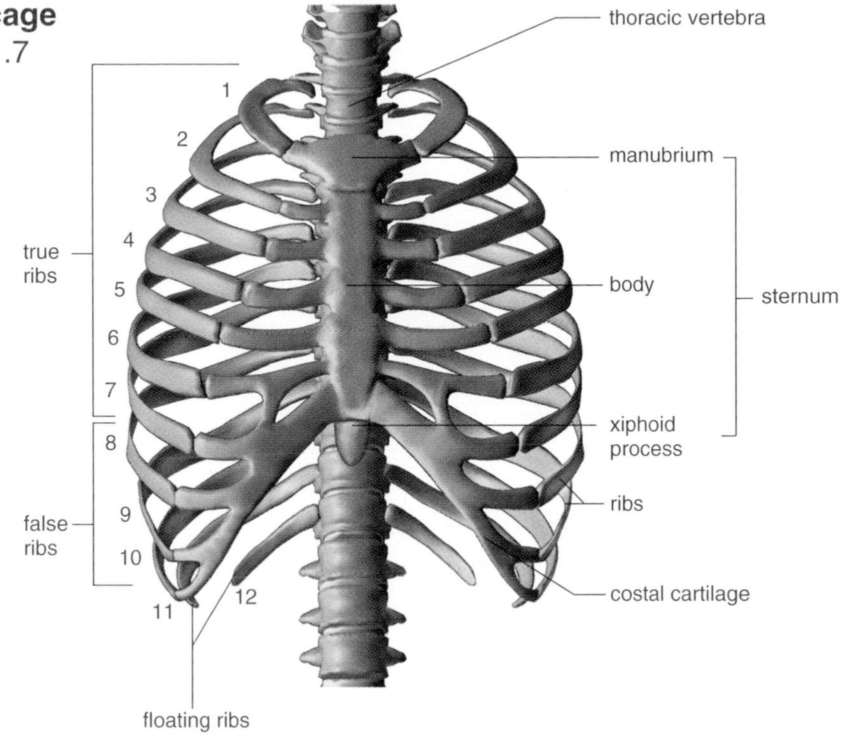

- thoracic vertebra
- manubrium
- body
- sternum
- xiphoid process
- ribs
- costal cartilage
- true ribs
- false ribs
- floating ribs

1
2
3
4
5
6
7
8
9
10
11
12

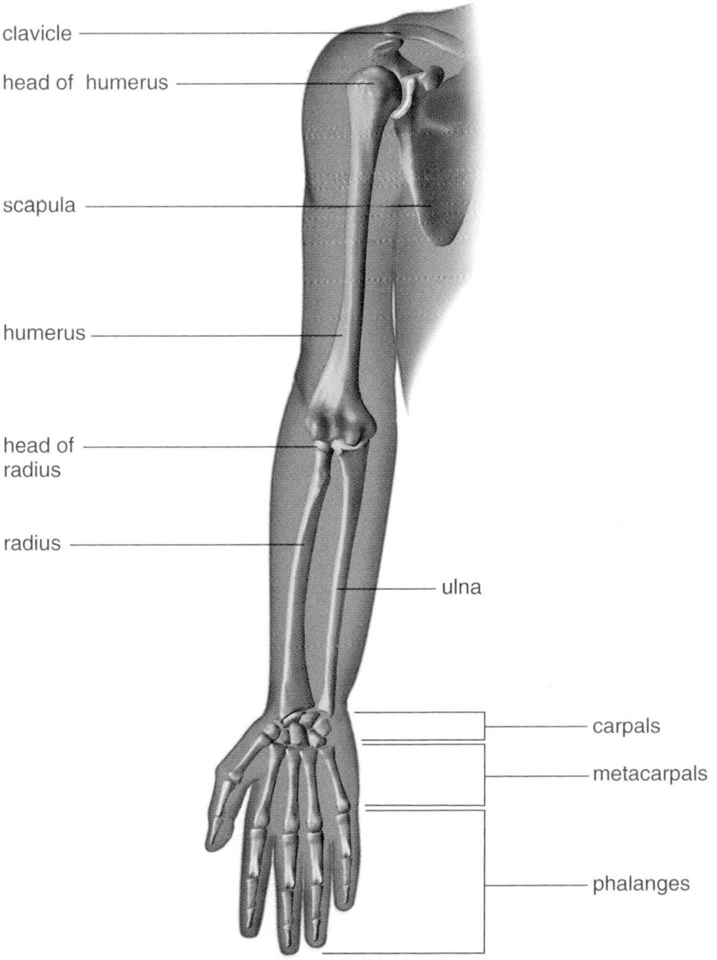

- clavicle
- head of humerus
- scapula
- humerus
- head of radius
- radius
- ulna
- carpals
- metacarpals
- phalanges

Bones of the pectoral girdle, the arm, and the hand
Figure 41.8

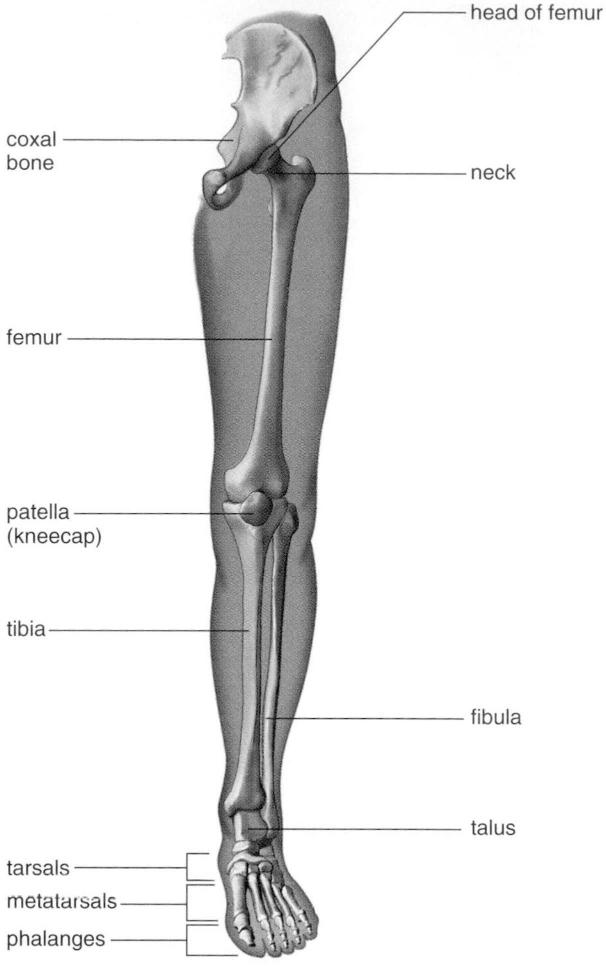

coxal
bone

head of femur

neck

femur

patella
(kneecap)

tibia

fibula

talus

tarsals

metatarsals

phalanges

**Bones of the pelvic girdle, the leg,
and the foot**
Figure 41.9

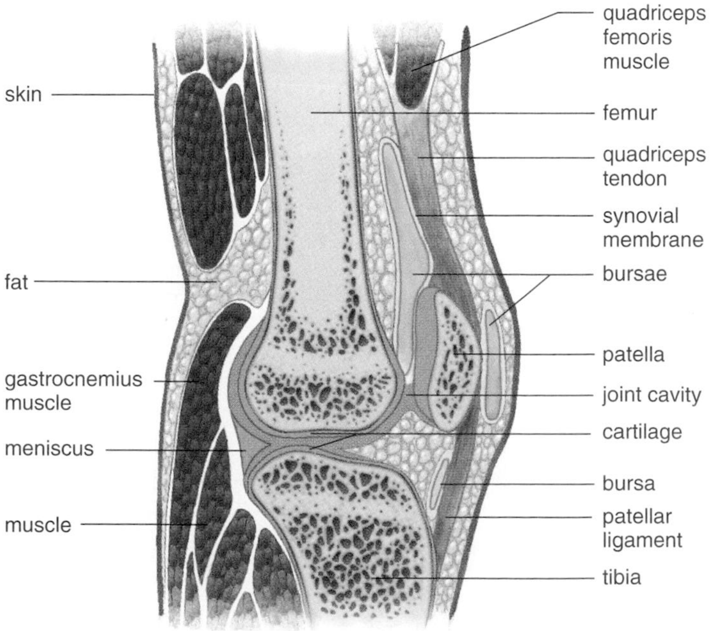

skin

fat

gastrocnemius
muscle

meniscus

muscle

quadriceps
femoris
muscle

femur

quadriceps
tendon

synovial
membrane

bursae

patella

joint cavity

cartilage

bursa

patellar
ligament

tibia

Knee joint
Figure 41.10

Human musculature
Figure 41.11

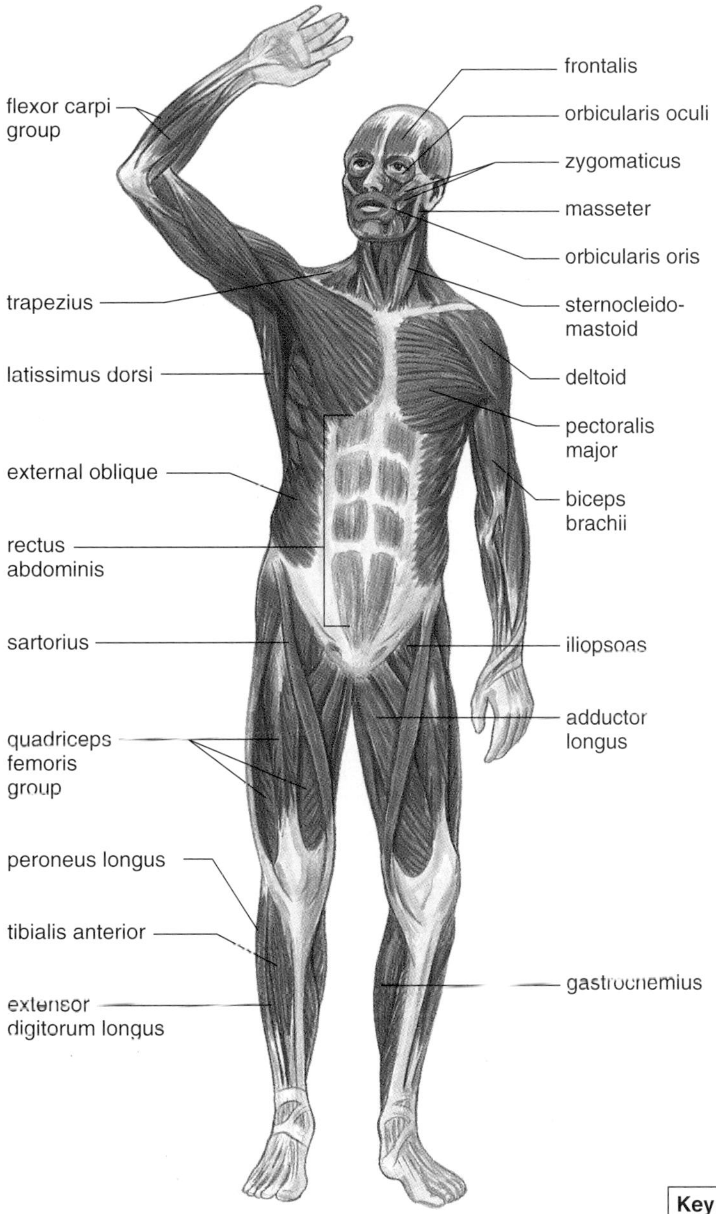

flexor carpi group

frontalis

orbicularis oculi

zygomaticus

masseter

orbicularis oris

sternocleido-mastoid

trapezius

latissimus dorsi

deltoid

external oblique

pectoralis major

biceps brachii

rectus abdominis

sartorius

iliopsoas

adductor longus

quadriceps femoris group

peroneus longus

tibialis anterior

extensor digitorum longus

gastrocnemius

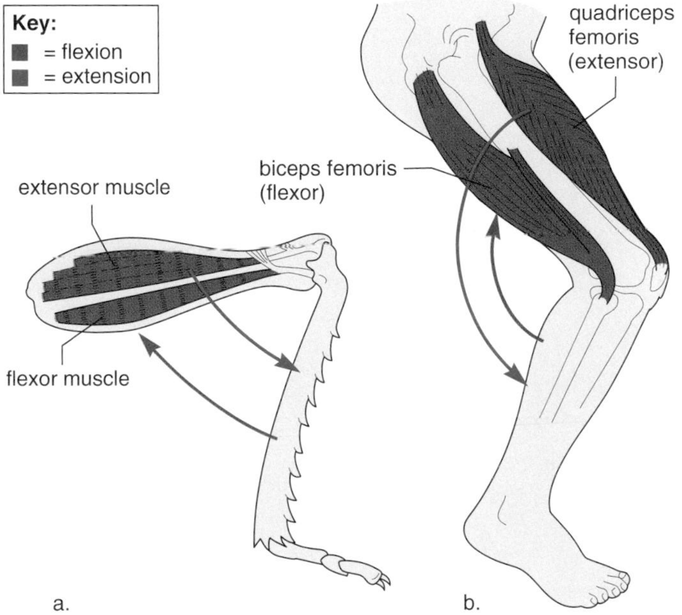

Key:
■ = flexion
■ = extension

extensor muscle

biceps femoris (flexor)

quadriceps femoris (extensor)

flexor muscle

a.

b.

Antagonistic muscle pairs
Figure 41.12

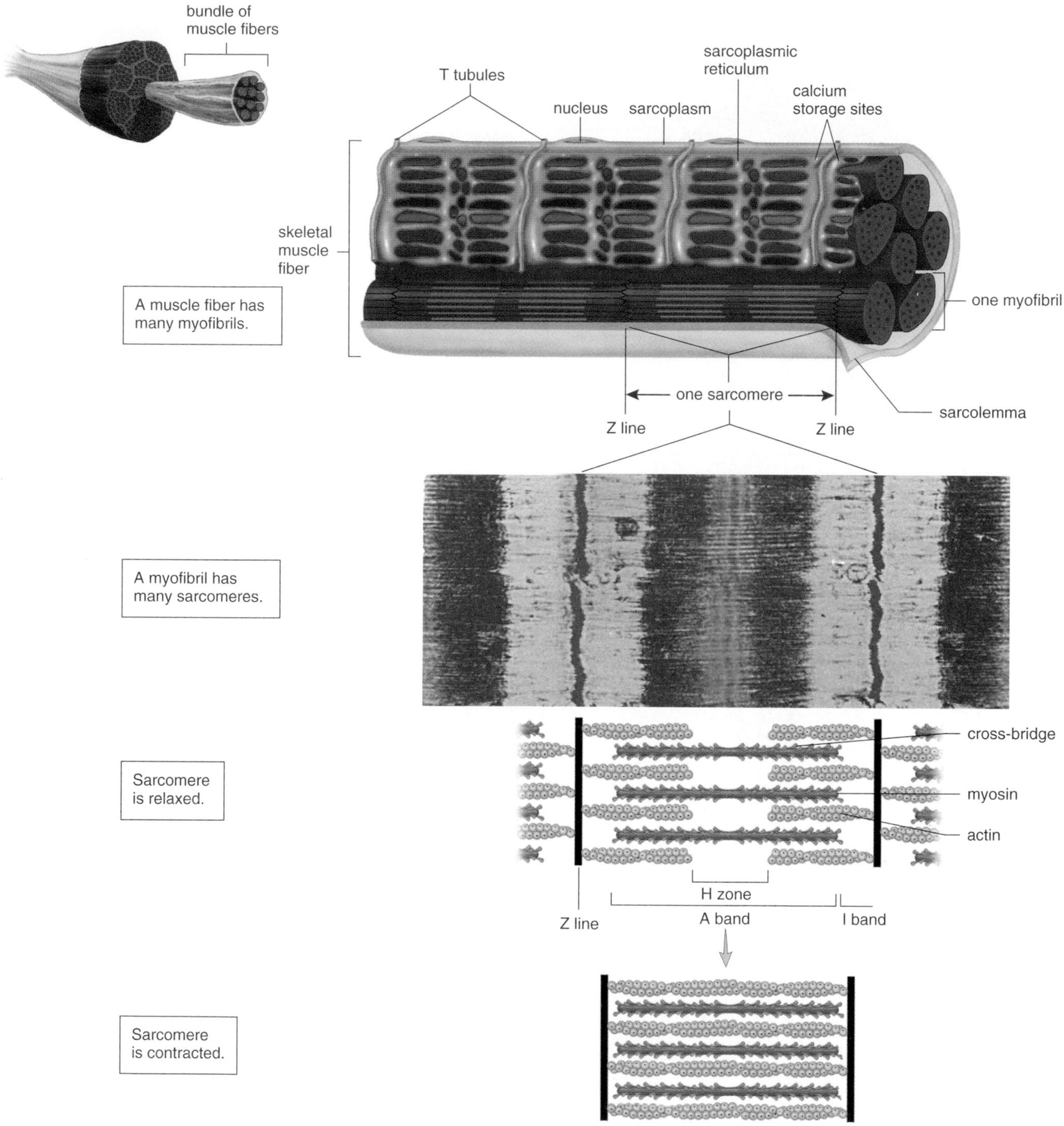

bundle of muscle fibers

T tubules

nucleus

sarcoplasm

sarcoplasmic reticulum

calcium storage sites

skeletal muscle fiber

A muscle fiber has many myofibrils.

one myofibril

one sarcomere

Z line

Z line

sarcolemma

A myofibril has many sarcomeres.

Sarcomere is relaxed.

cross-bridge

myosin

actin

Z line

A band

H zone

I band

Sarcomere is contracted.

Skeletal muscle fiber structure and function
Figure 41.13
Courtesy of Hugh E. Huxley

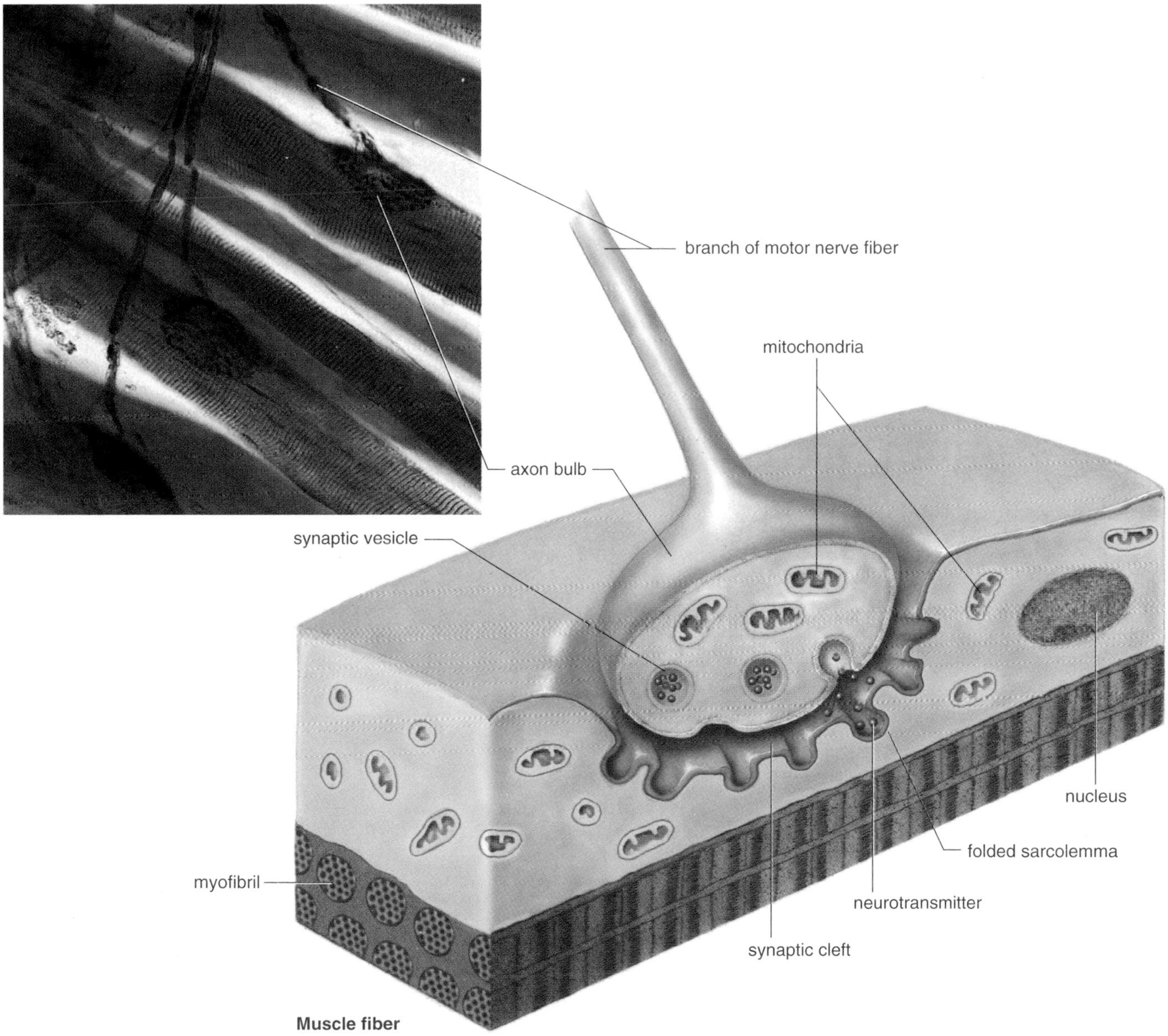

branch of motor nerve fiber

mitochondria

axon bulb

synaptic vesicle

nucleus

folded sarcolemma

myofibril

neurotransmitter

synaptic cleft

Muscle fiber

Neuromuscular junction
Figure 41.14
© Victor B. Eichler

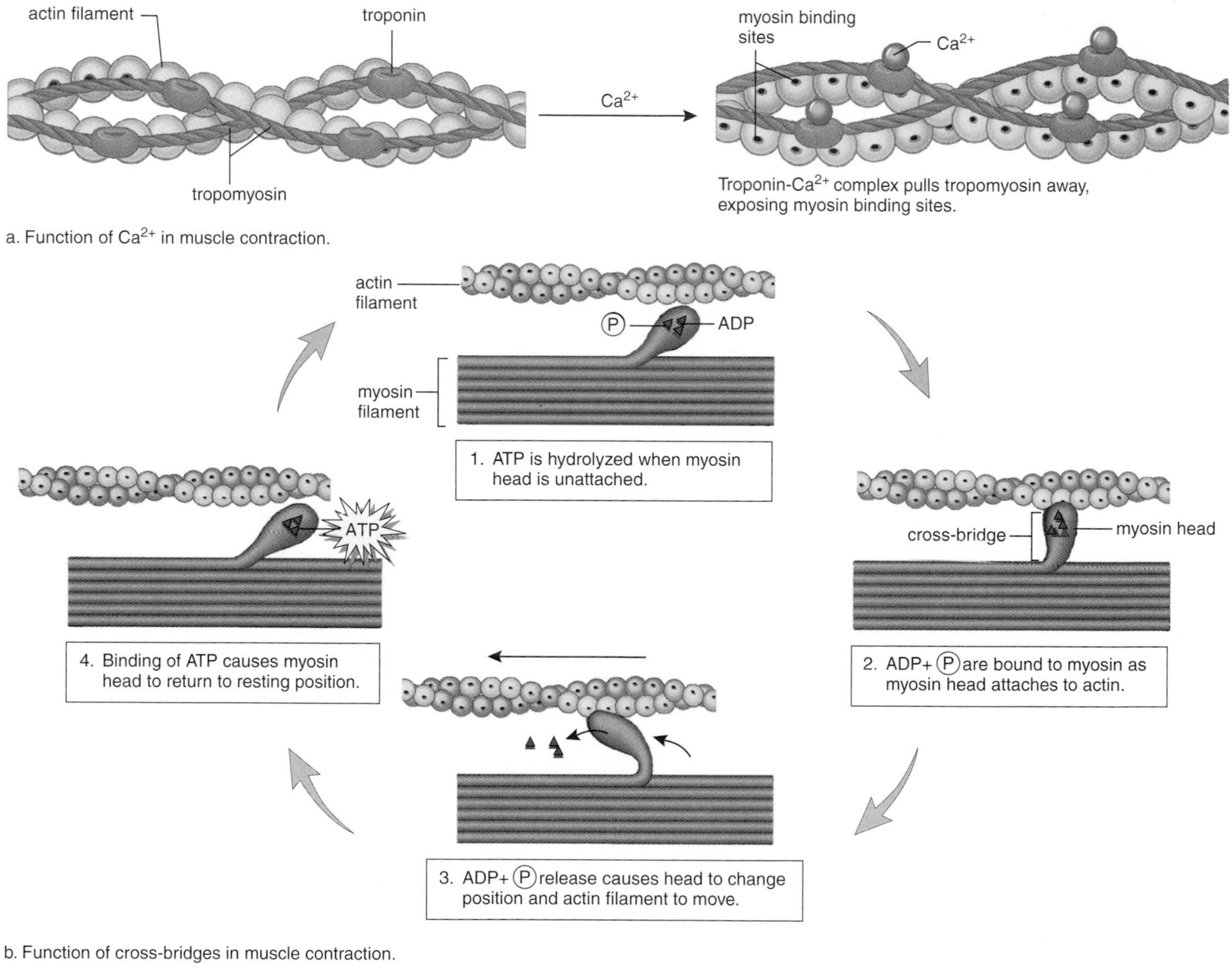

actin filament — troponin

tropomyosin

a. Function of Ca²⁺ in muscle contraction.

myosin binding sites — Ca²⁺

Troponin-Ca²⁺ complex pulls tropomyosin away, exposing myosin binding sites.

actin filament

P — ADP

myosin filament

1. ATP is hydrolyzed when myosin head is unattached.

cross-bridge — myosin head

2. ADP+ P are bound to myosin as myosin head attaches to actin.

3. ADP+ P release causes head to change position and actin filament to move.

ATP

4. Binding of ATP causes myosin head to return to resting position.

b. Function of cross-bridges in muscle contraction.

The role of calcium and myosin in muscle contraction
Figure 41.15

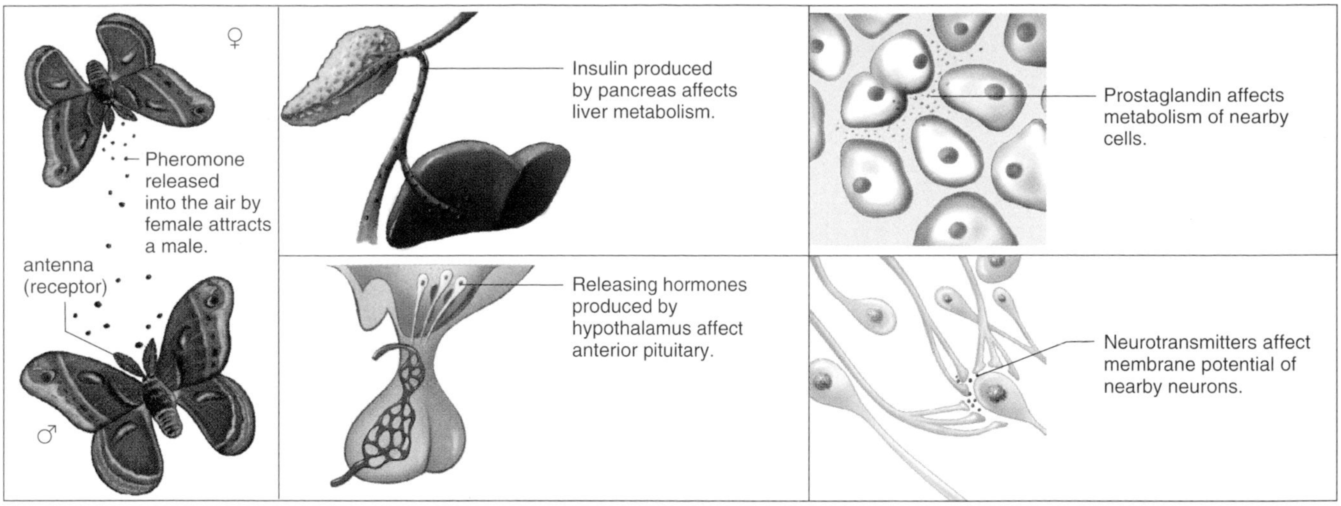

a. Signal acts at a distance between individuals.

b. Signal acts at a distance between body parts.

c. Signal acts locally between adjacent cells.

Chemical signals
Figure 42.1

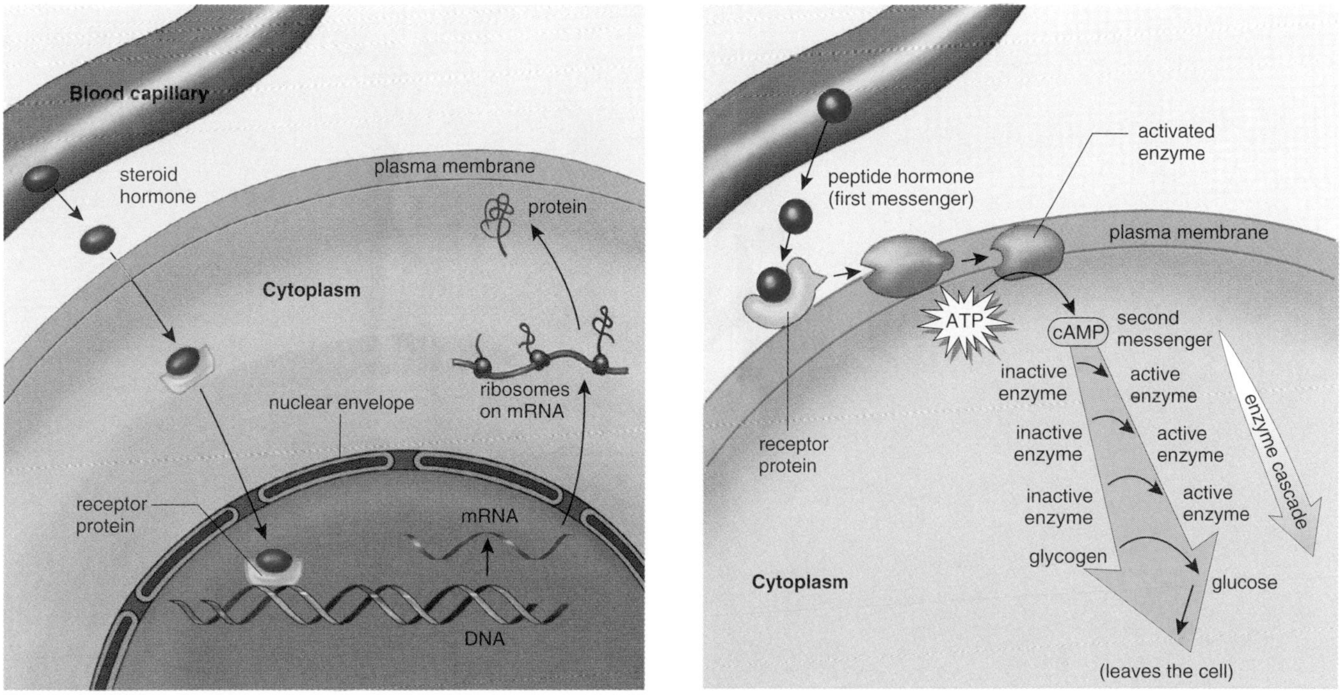

a. Action of steroid hormone

b. Action of peptide hormone

Cellular activity of hormones
Figure 42.2

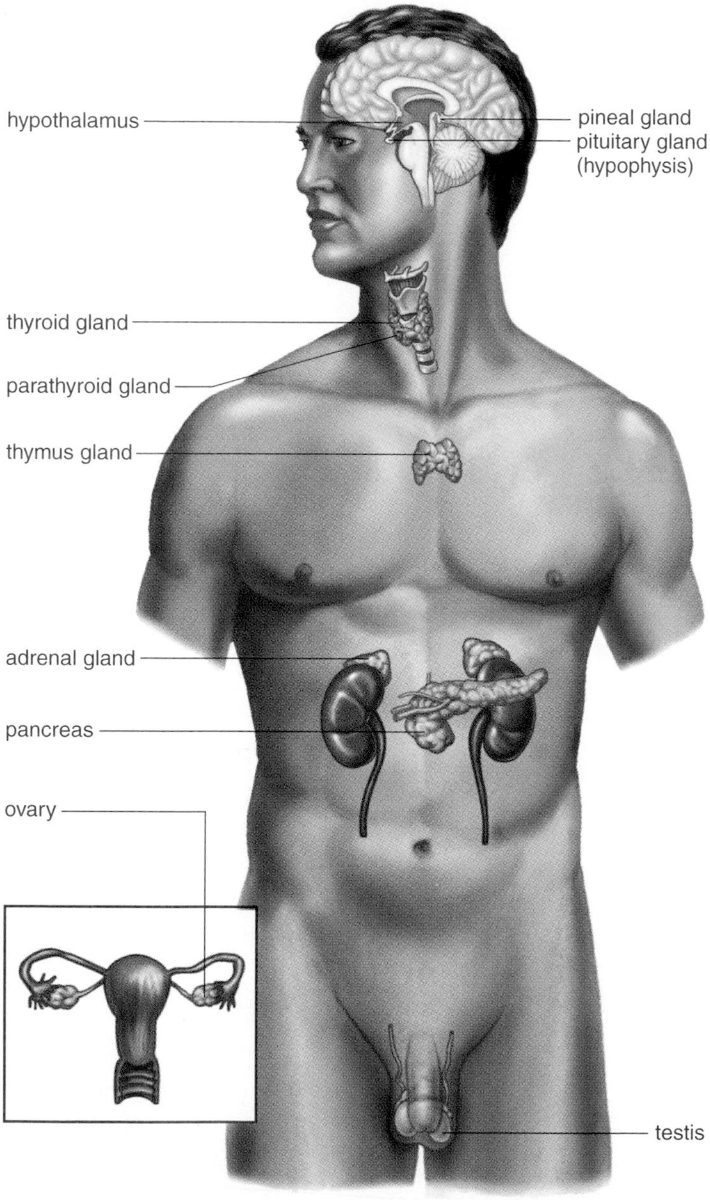

hypothalamus

pineal gland
pituitary gland
(hypophysis)

thyroid gland

parathyroid gland

thymus gland

adrenal gland

pancreas

ovary

testis

The endocrine system
Figure 42.3

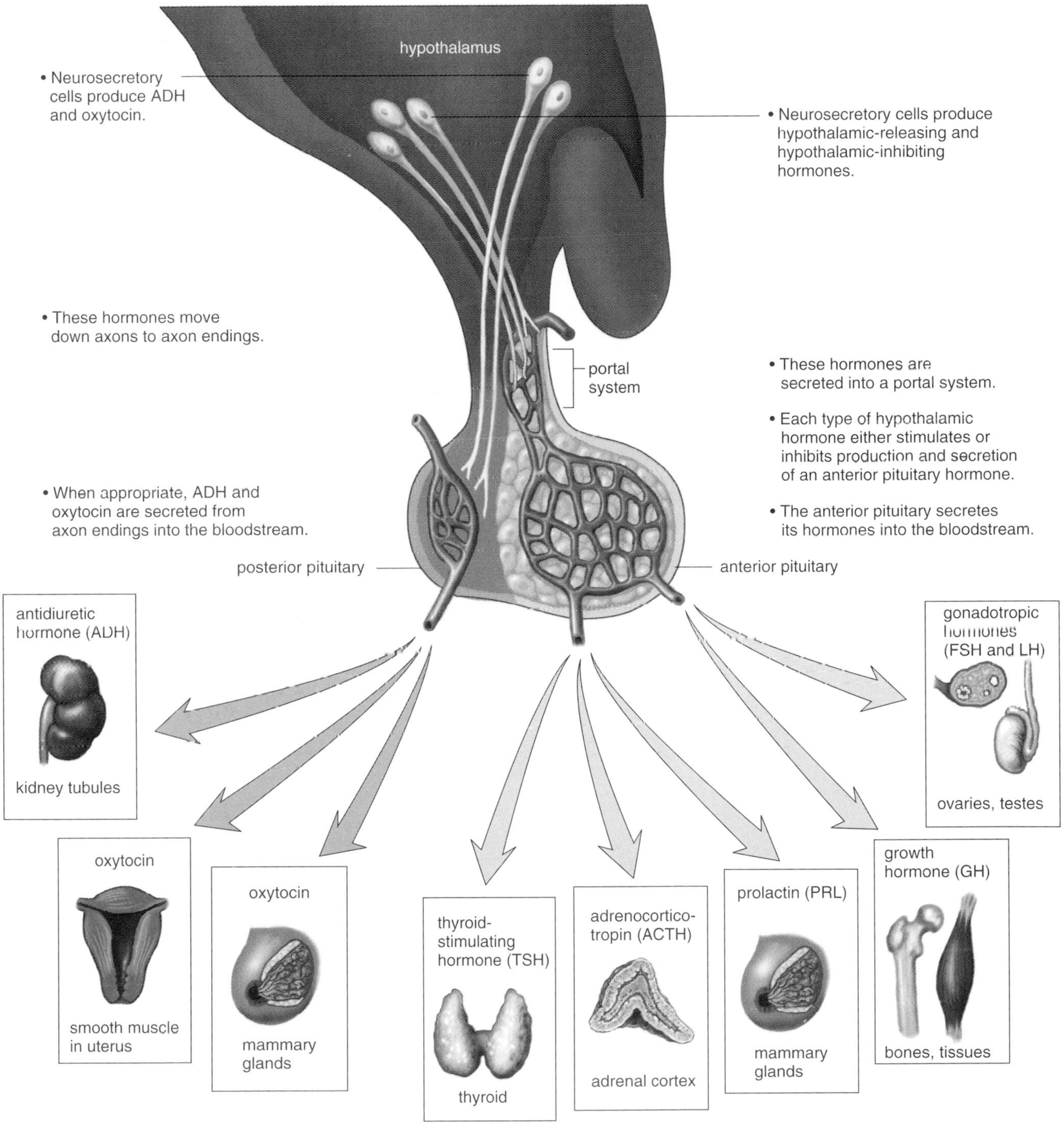

- Neurosecretory cells produce ADH and oxytocin.

hypothalamus

- Neurosecretory cells produce hypothalamic-releasing and hypothalamic-inhibiting hormones.

- These hormones move down axons to axon endings.

portal system

- These hormones are secreted into a portal system.

- Each type of hypothalamic hormone either stimulates or inhibits production and secretion of an anterior pituitary hormone.

- When appropriate, ADH and oxytocin are secreted from axon endings into the bloodstream.

- The anterior pituitary secretes its hormones into the bloodstream.

posterior pituitary

anterior pituitary

antidiuretic hormone (ADH)

kidney tubules

gonadotropic hormones (FSH and LH)

ovaries, testes

oxytocin

smooth muscle in uterus

oxytocin

mammary glands

thyroid-stimulating hormone (TSH)

thyroid

adrenocortico-tropin (ACTH)

adrenal cortex

prolactin (PRL)

mammary glands

growth hormone (GH)

bones, tissues

Hypothalamus and the pituitary
Figure 42.4

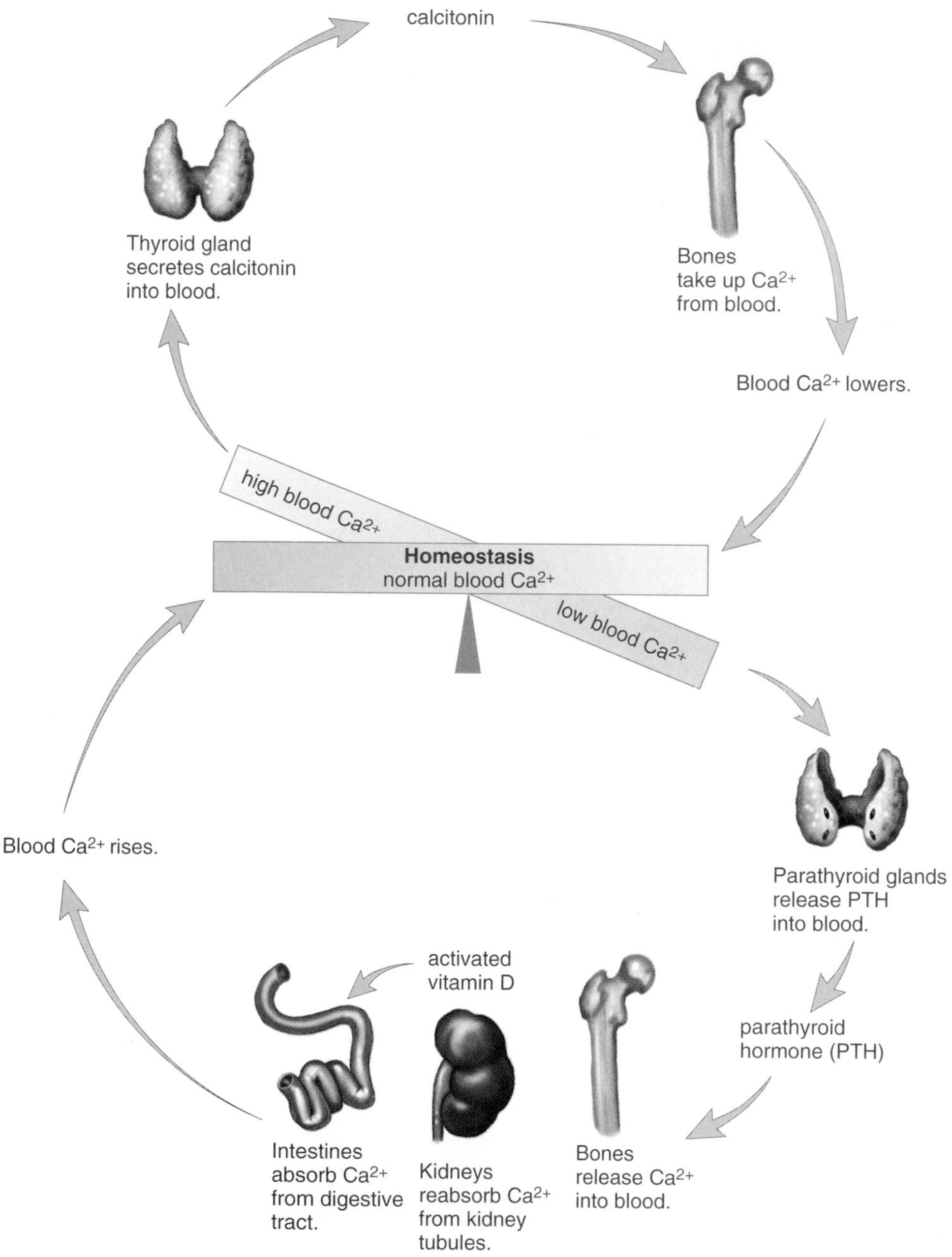

calcitonin

Thyroid gland
secretes calcitonin
into blood.

Bones
take up Ca^{2+}
from blood.

Blood Ca^{2+} lowers.

high blood Ca^{2+}

Homeostasis
normal blood Ca^{2+}

low blood Ca^{2+}

Blood Ca^{2+} rises.

Parathyroid glands
release PTH
into blood.

parathyroid
hormone (PTH)

activated
vitamin D

Intestines
absorb Ca^{2+}
from digestive
tract.

Kidneys
reabsorb Ca^{2+}
from kidney
tubules.

Bones
release Ca^{2+}
into blood.

Regulation of blood calcium level
Figure 42.9

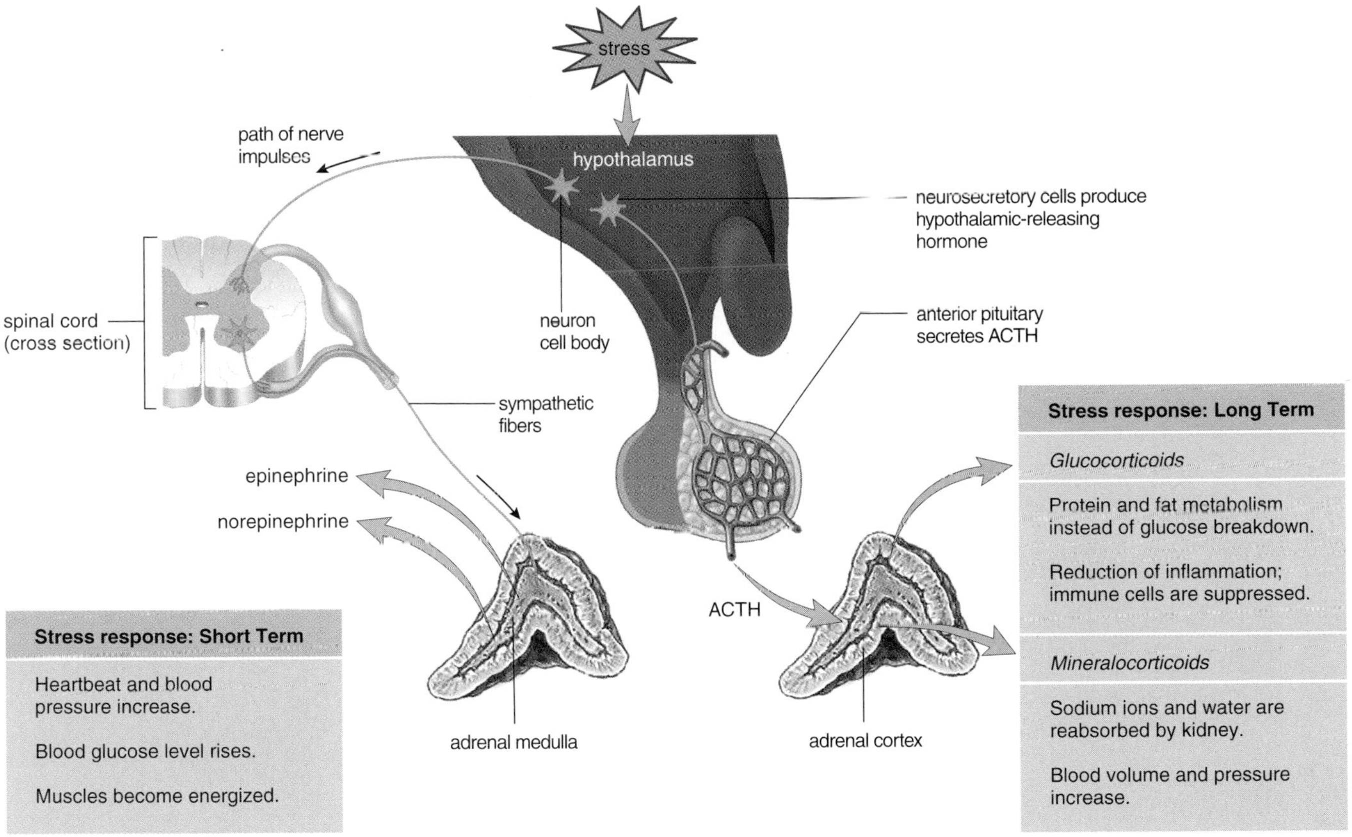

stress

path of nerve impulses

hypothalamus

neurosecretory cells produce hypothalamic-releasing hormone

spinal cord (cross section)

neuron cell body

anterior pituitary secretes ACTH

sympathetic fibers

epinephrine

norepinephrine

Stress response: Long Term

Glucocorticoids

Protein and fat metabolism instead of glucose breakdown.

Reduction of inflammation; immune cells are suppressed.

ACTH

Stress response: Short Term

Heartbeat and blood pressure increase.

Blood glucose level rises.

Muscles become energized.

adrenal medulla

adrenal cortex

Mineralocorticoids

Sodium ions and water are reabsorbed by kidney.

Blood volume and pressure increase.

Adrenal glands
Figure 42.10

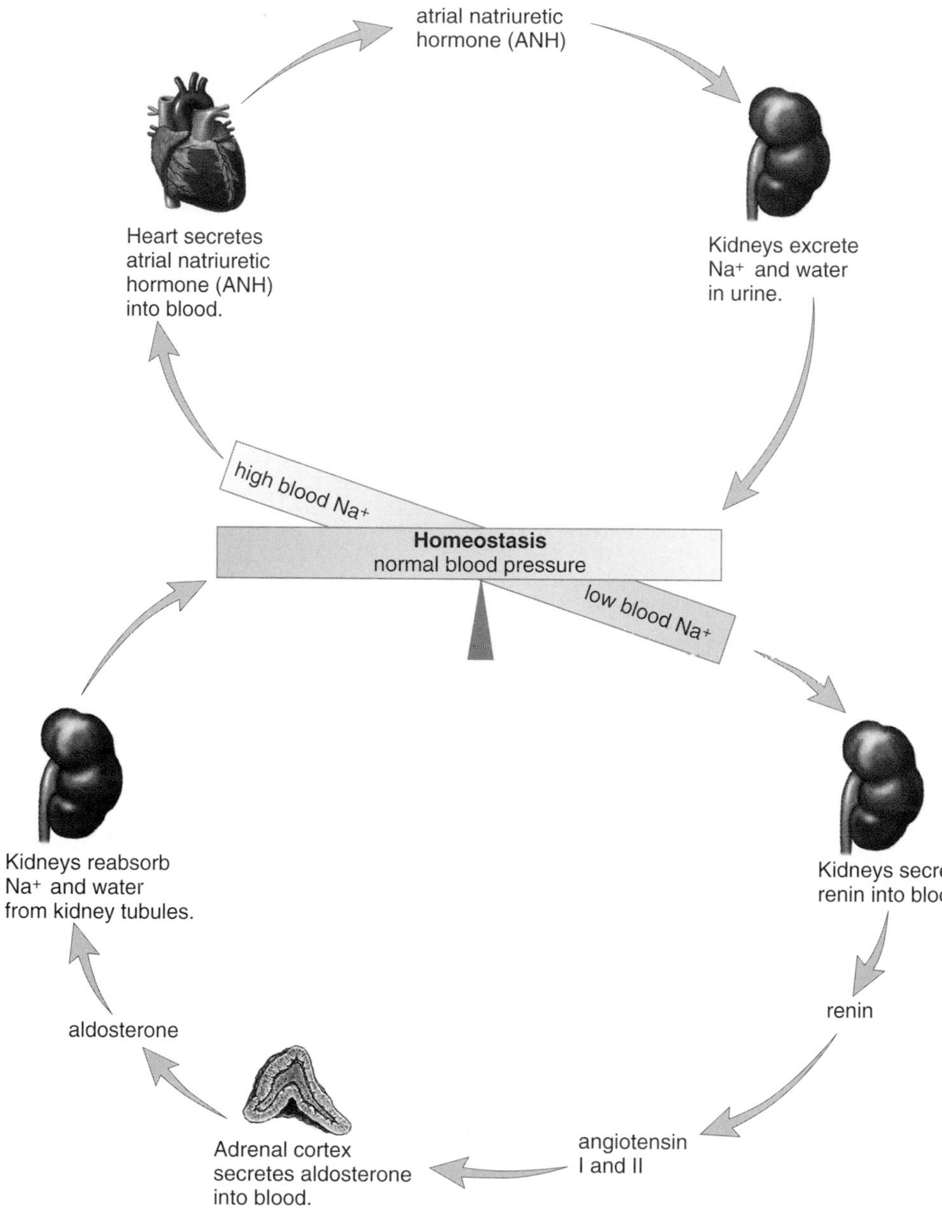

atrial natriuretic
hormone (ANH)

Heart secretes
atrial natriuretic
hormone (ANH)
into blood.

Kidneys excrete
Na+ and water
in urine.

high blood Na+

Homeostasis
normal blood pressure

low blood Na+

Kidneys reabsorb
Na+ and water
from kidney tubules.

Kidneys secrete
renin into blood.

aldosterone

renin

Adrenal cortex
secretes aldosterone
into blood.

angiotensin
I and II

Regulation of blood pressure and volume
Figure 42.11

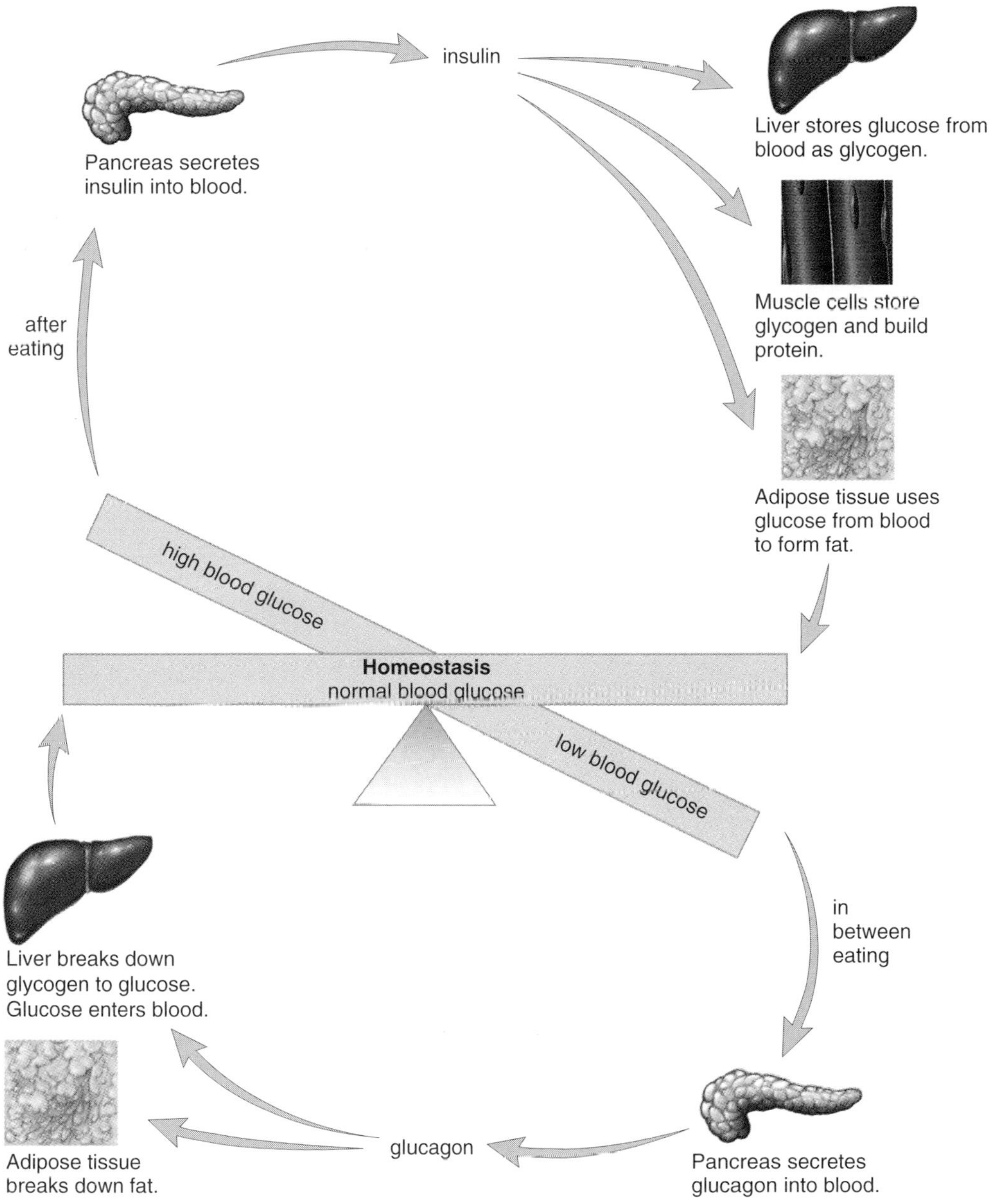

insulin

Pancreas secretes insulin into blood.

Liver stores glucose from blood as glycogen.

Muscle cells store glycogen and build protein.

Adipose tissue uses glucose from blood to form fat.

after eating

high blood glucose

Homeostasis
normal blood glucose

low blood glucose

in between eating

Liver breaks down glycogen to glucose. Glucose enters blood.

Adipose tissue breaks down fat.

glucagon

Pancreas secretes glucagon into blood.

Regulation of blood glucose level
Figure 42.14

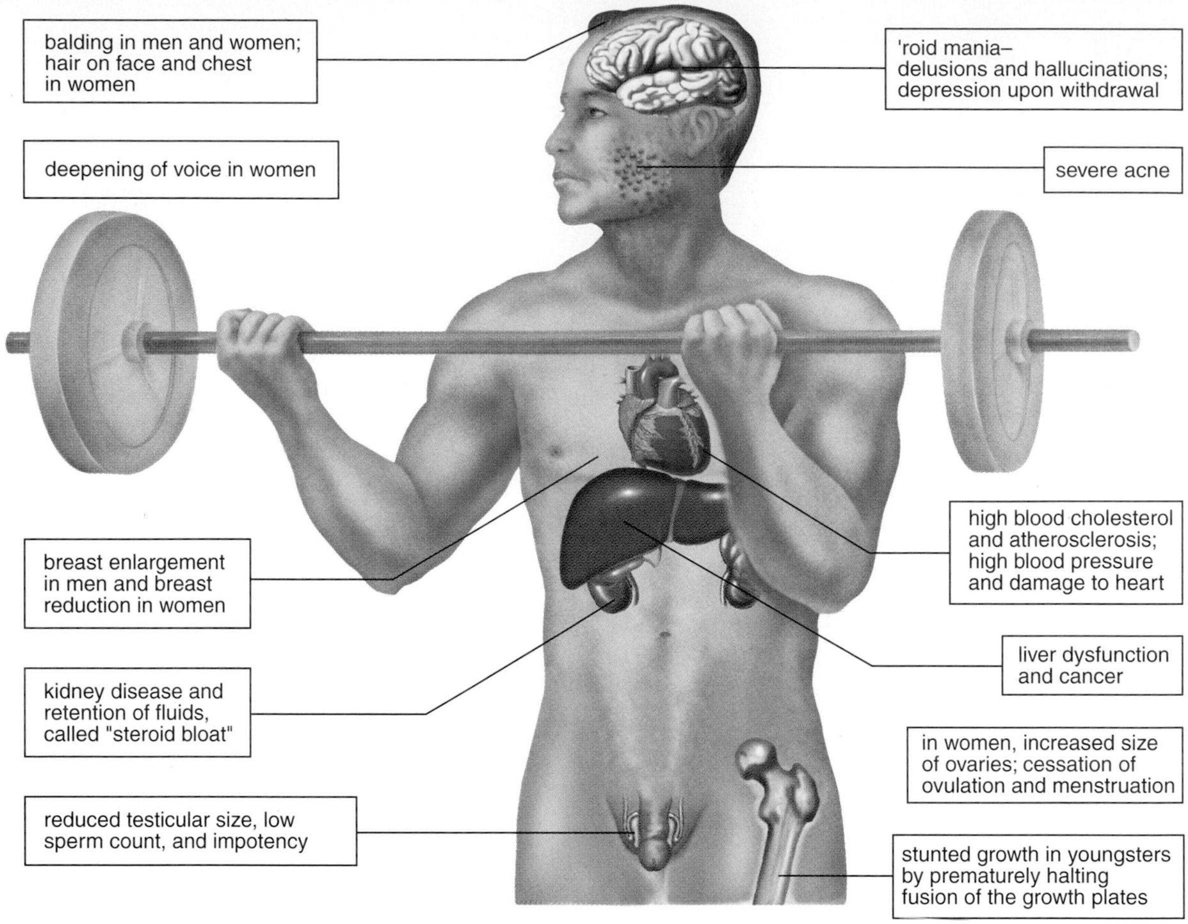

The effects of anabolic steroid use
Figure 42.15

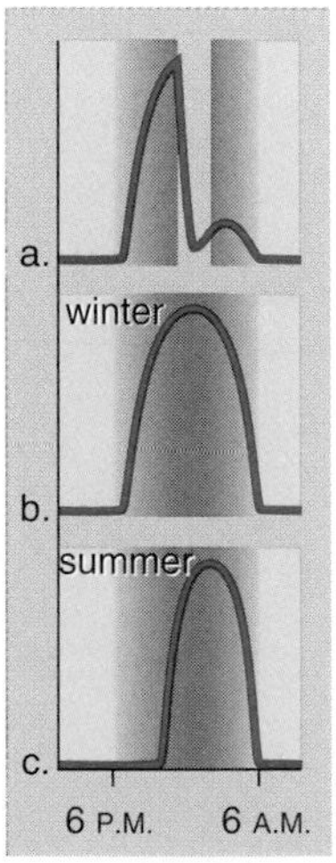

Melatonin production
Figure 42.16

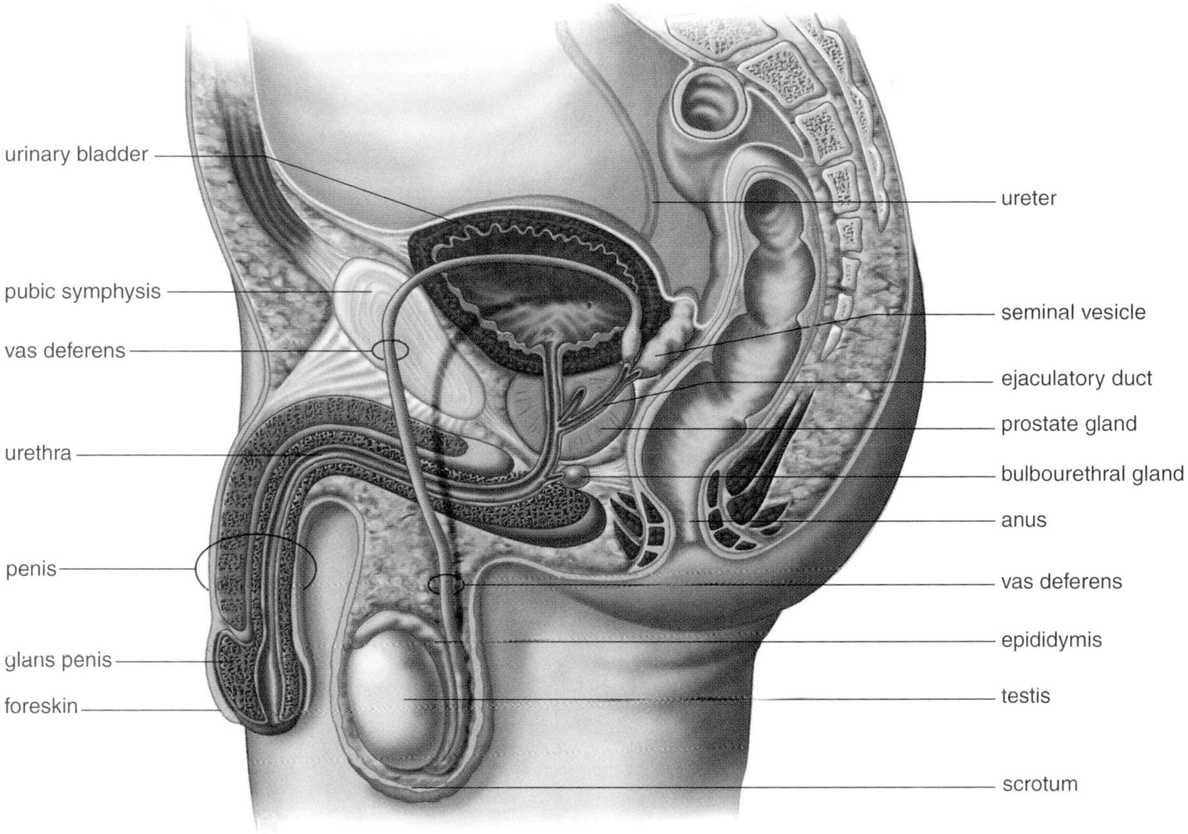

The male reproductive system
Figure 43.3

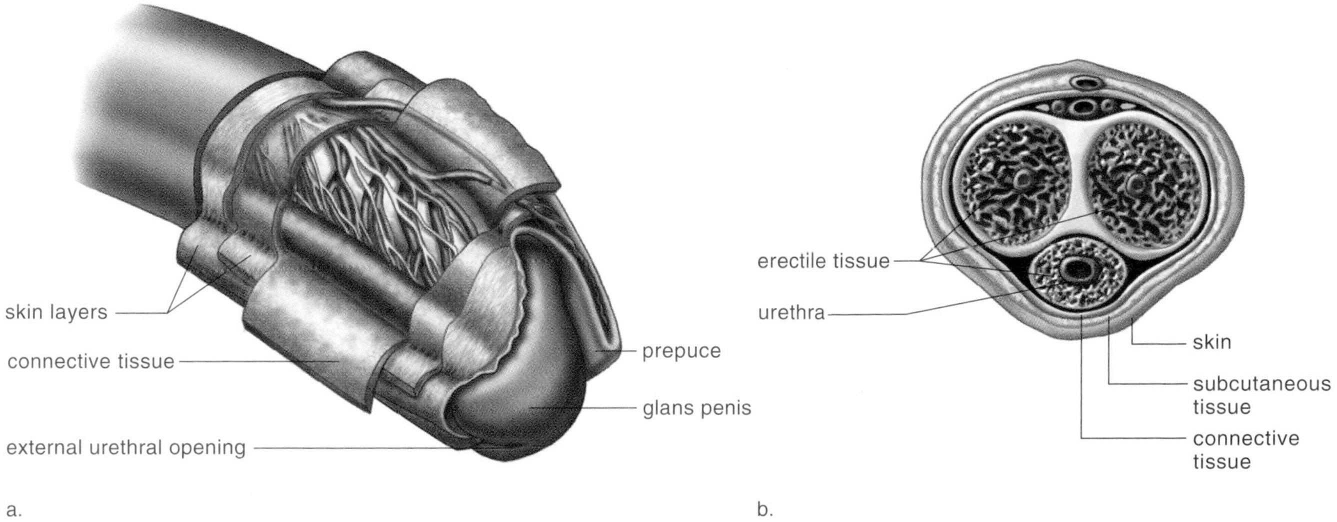

a.

b.

Penis anatomy
Figure 43.4

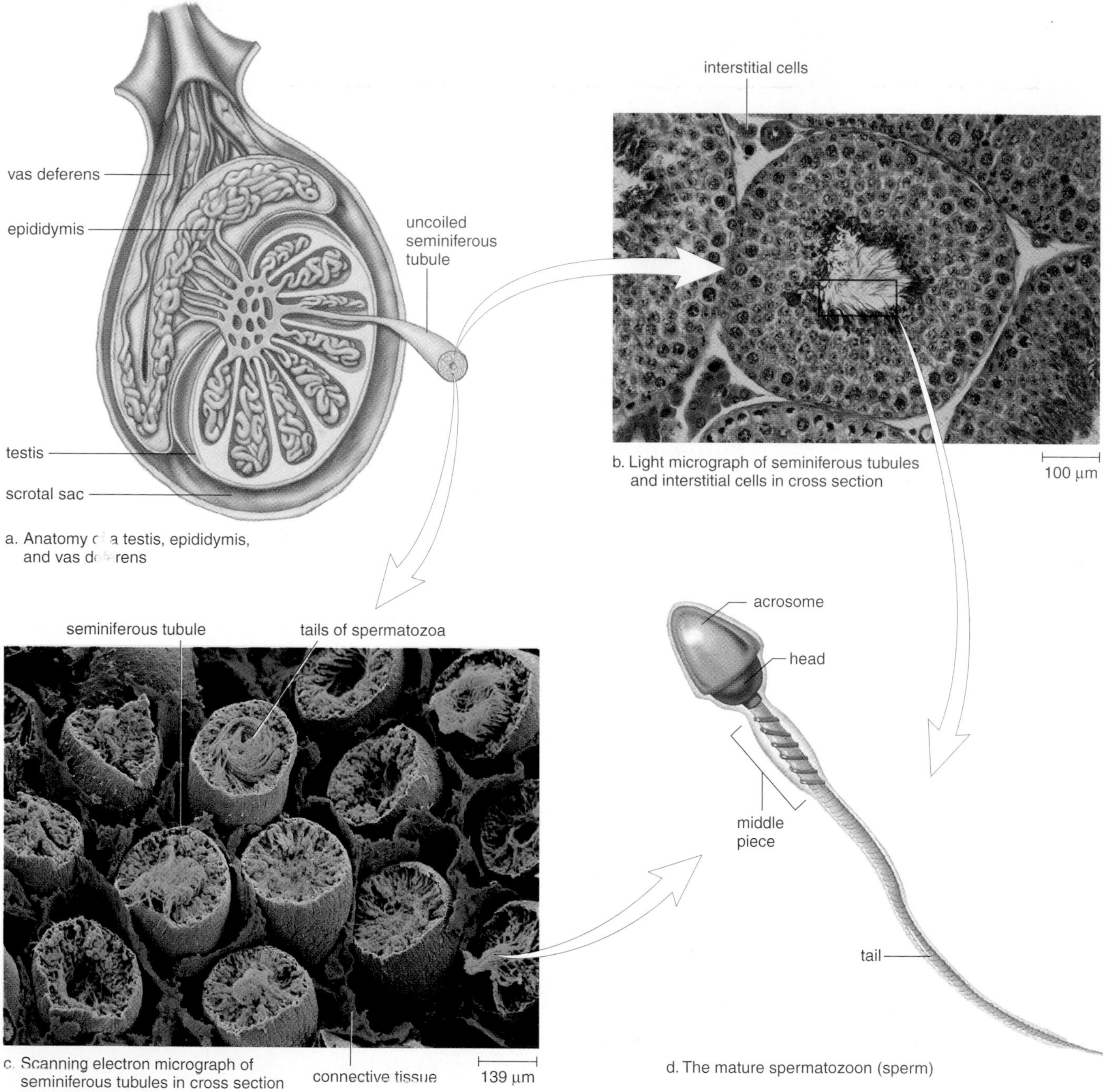

vas deferens

epididymis

testis

scrotal sac

a. Anatomy of a testis, epididymis,
   and vas deferens

uncoiled
seminiferous
tubule

interstitial cells

b. Light micrograph of seminiferous tubules
   and interstitial cells in cross section

100 μm

seminiferous tubule

tails of spermatozoa

c. Scanning electron micrograph of
   seminiferous tubules in cross section

connective tissue   139 μm

acrosome

head

middle
piece

tail

d. The mature spermatozoon (sperm)

Testis and sperm
Figure 43.5

Hormonal control of testes
Figure 43.6

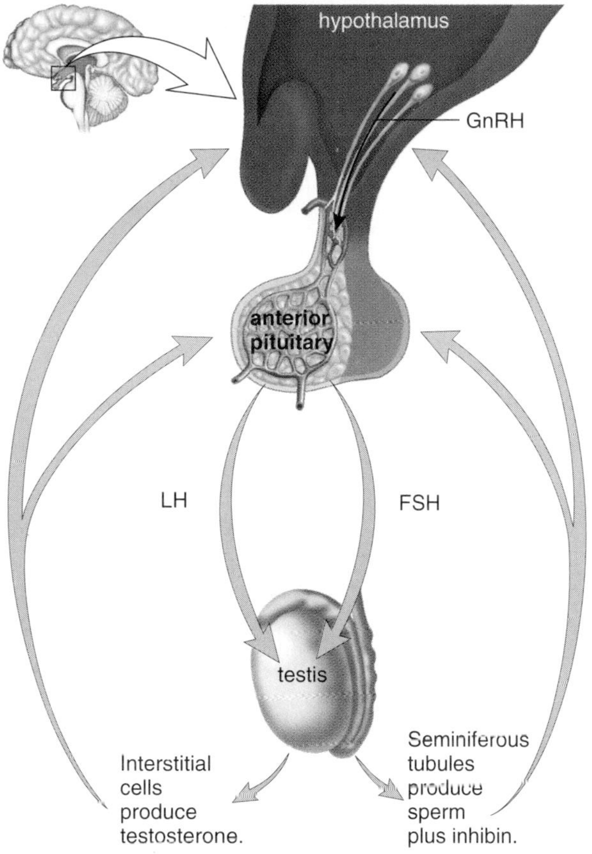

hypothalamus

GnRH

anterior pituitary

LH

FSH

testis

Interstitial cells produce testosterone.

Seminiferous tubules produce sperm plus inhibin.

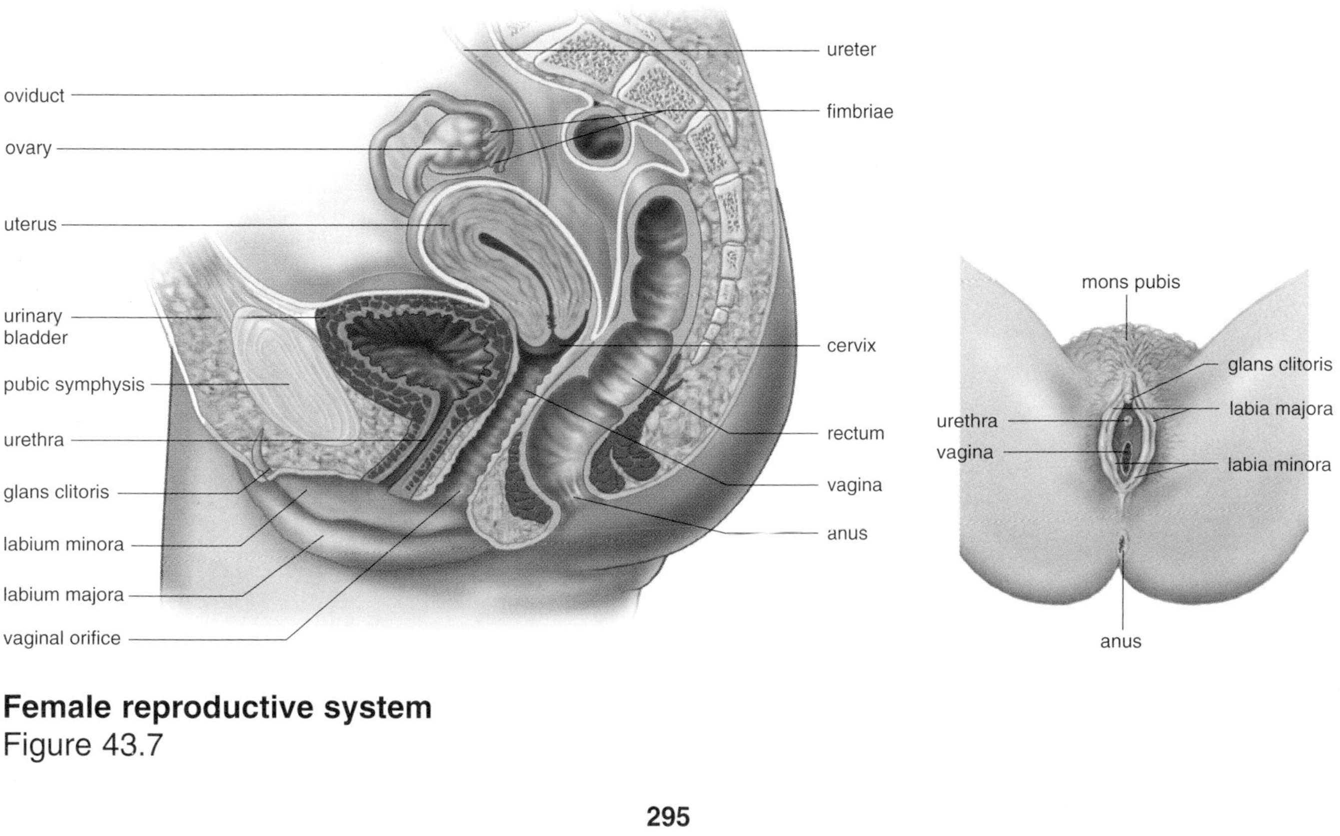

ureter

oviduct

fimbriae

ovary

uterus

urinary bladder

pubic symphysis

cervix

urethra

glans clitoris

rectum

labium minora

vagina

labium majora

anus

vaginal orifice

mons pubis

glans clitoris

urethra

labia majora

vagina

labia minora

anus

Female reproductive system
Figure 43.7

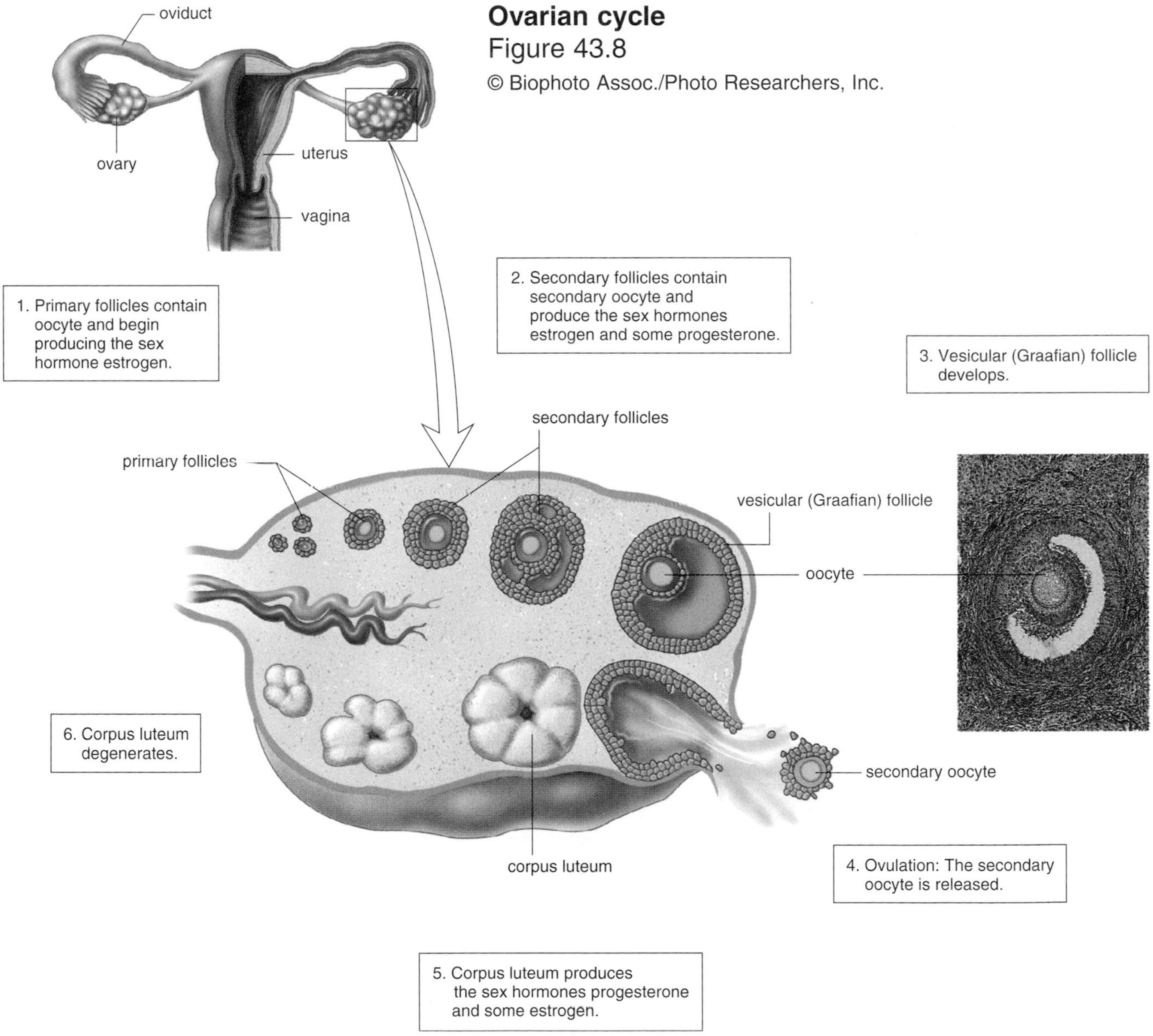

Ovarian cycle
Figure 43.8
© Biophoto Assoc./Photo Researchers, Inc.

oviduct

ovary

uterus

vagina

1. Primary follicles contain oocyte and begin producing the sex hormone estrogen.

2. Secondary follicles contain secondary oocyte and produce the sex hormones estrogen and some progesterone.

3. Vesicular (Graafian) follicle develops.

primary follicles

secondary follicles

vesicular (Graafian) follicle

oocyte

secondary oocyte

6. Corpus luteum degenerates.

corpus luteum

4. Ovulation: The secondary oocyte is released.

5. Corpus luteum produces the sex hormones progesterone and some estrogen.

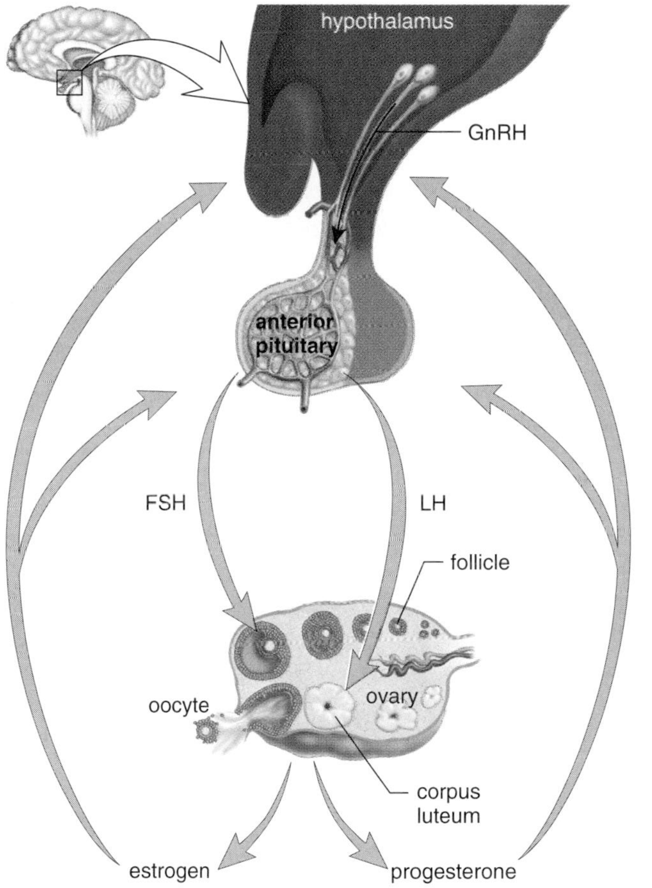

hypothalamus

GnRH

anterior
pituitary

FSH

LH

follicle

oocyte

ovary

corpus
luteum

estrogen

progesterone

Hormonal control of ovaries
Figure 43.9

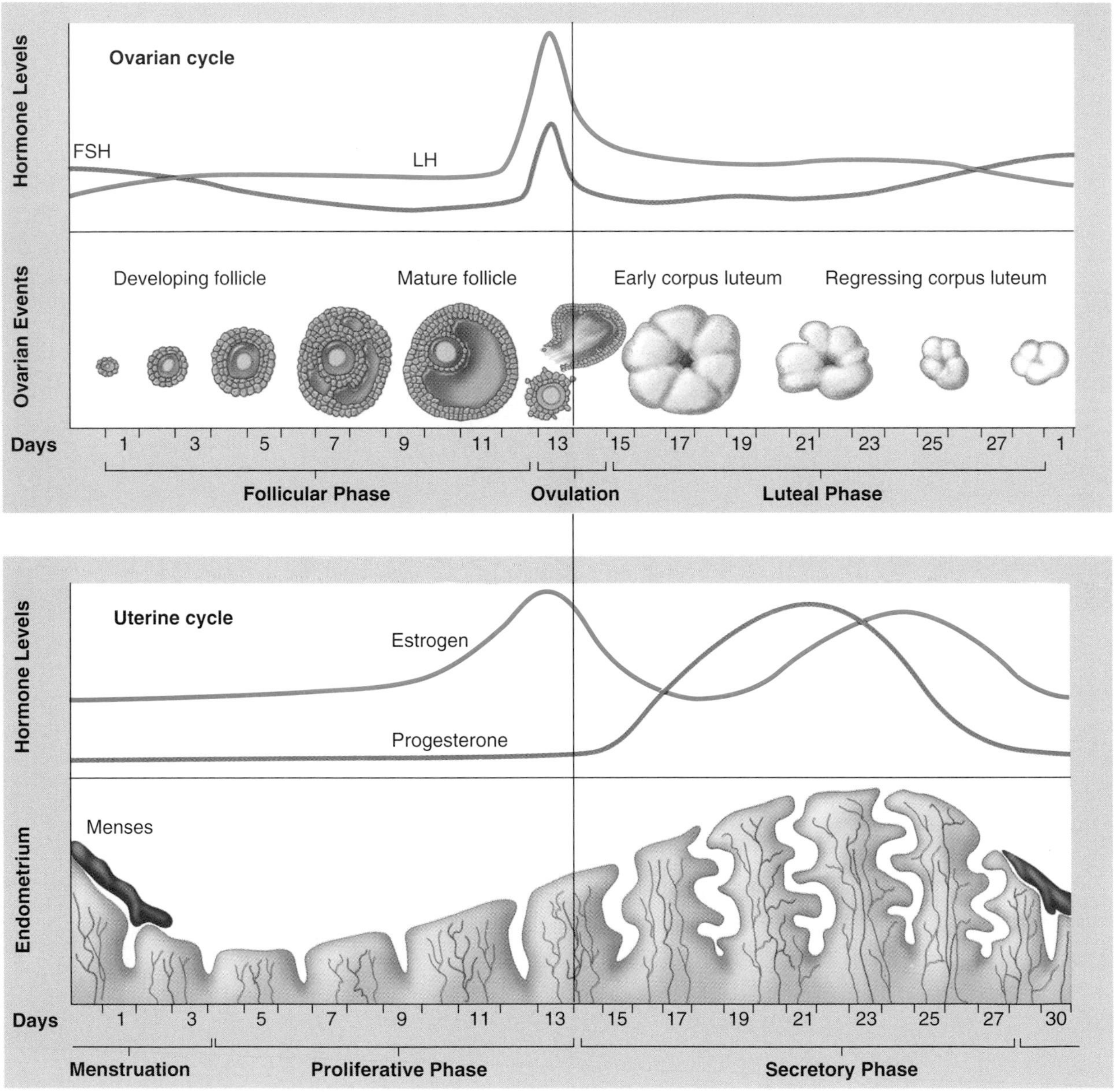

Female hormone levels during the ovarian and uterine cycles
Figure 43.10

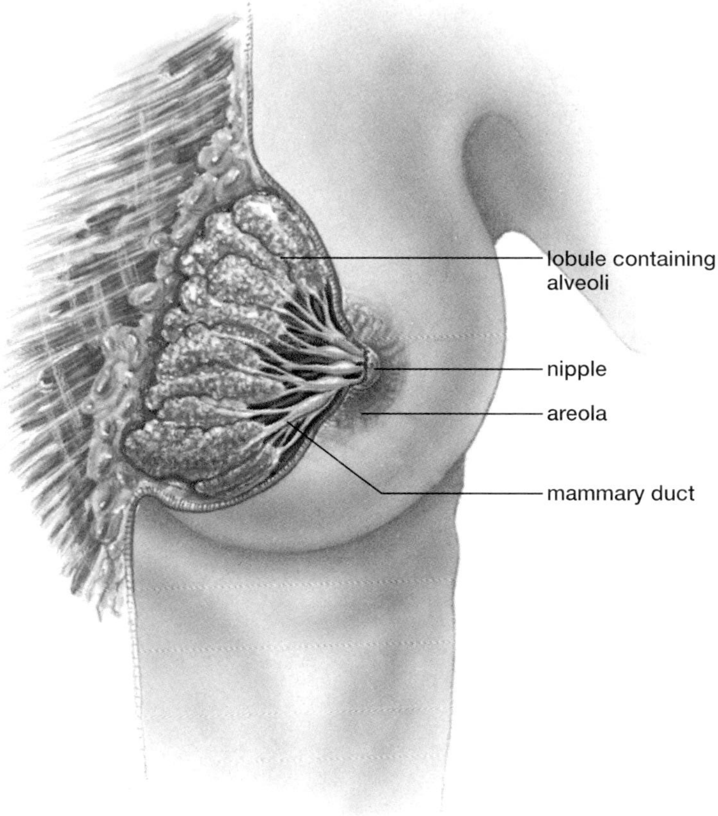

Anatomy of the breast
Figure 43.11

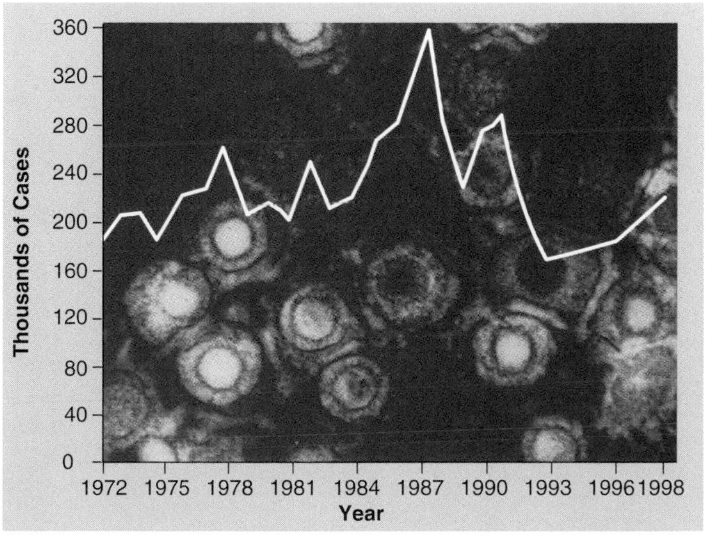

Genital warts
Figure 43.12
© CDC/Peter Arnold,Inc.

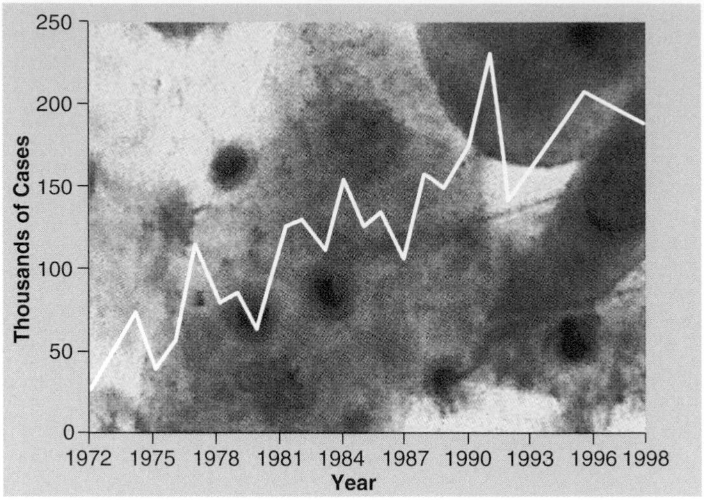

Genital herpes
Figure 43.13
© G.W. Willis/BPS

Chlamydial infection
Figure 43.14
© G.W. Willis/BPS

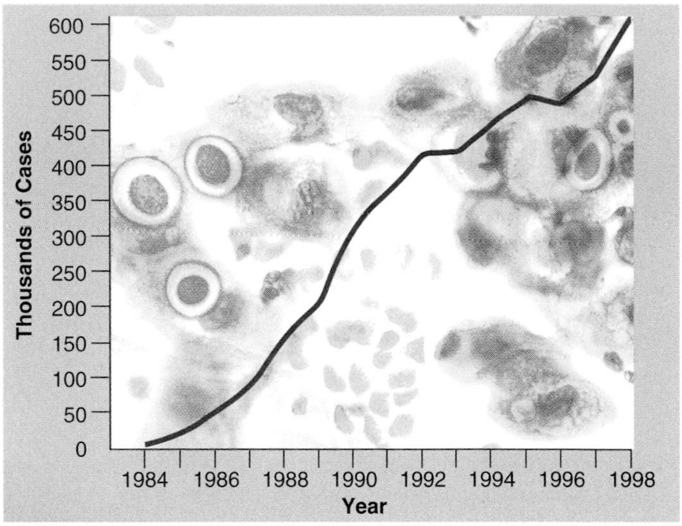

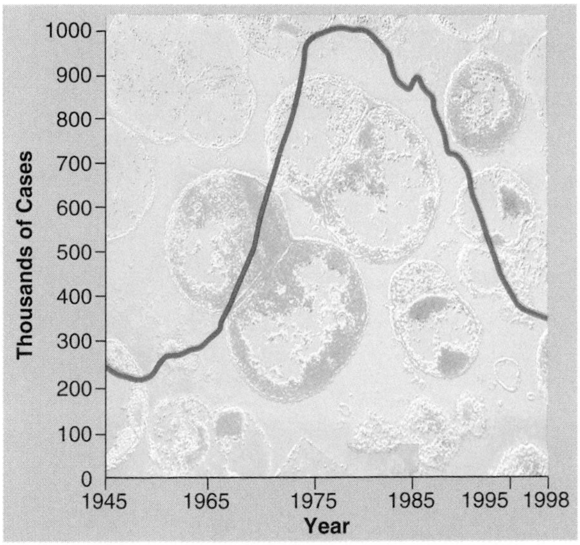

Gonorrhea
Figure 43.15
© CNR/SPL/Photo Researchers, Inc.

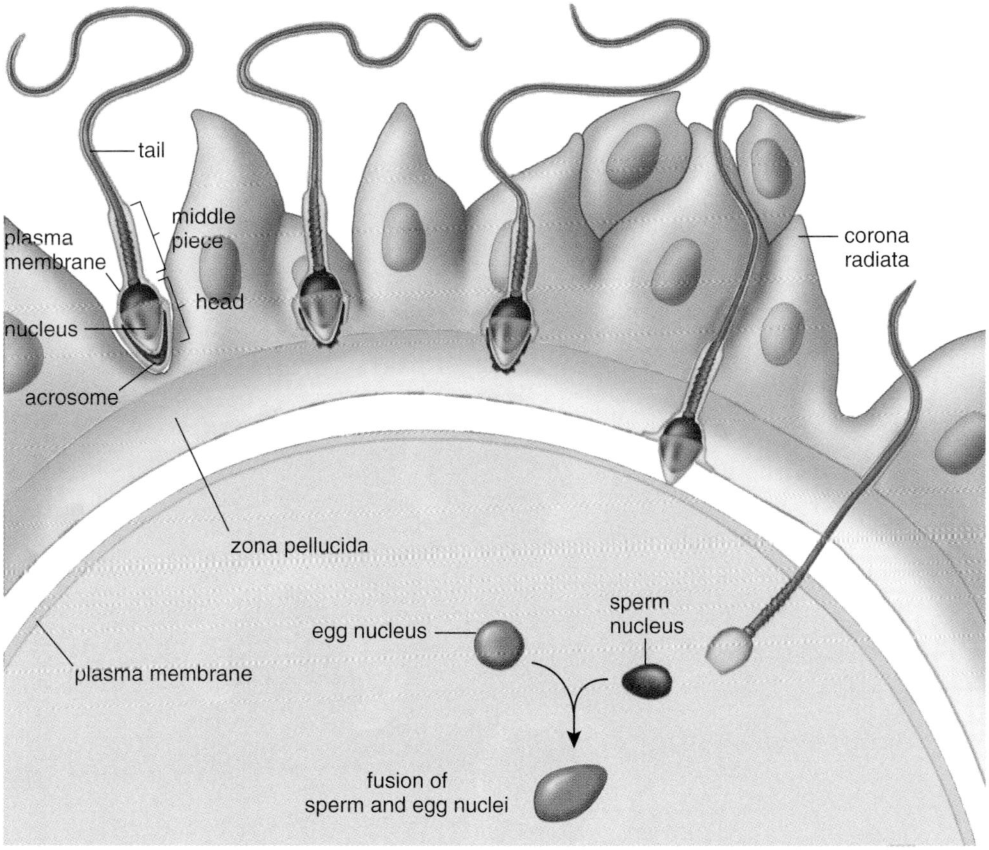

tail

middle
piece

plasma
membrane

head

nucleus

acrosome

corona
radiata

zona pellucida

plasma membrane

egg nucleus

sperm
nucleus

fusion of
sperm and egg nuclei

b.

Fertilization
Figure 44.1

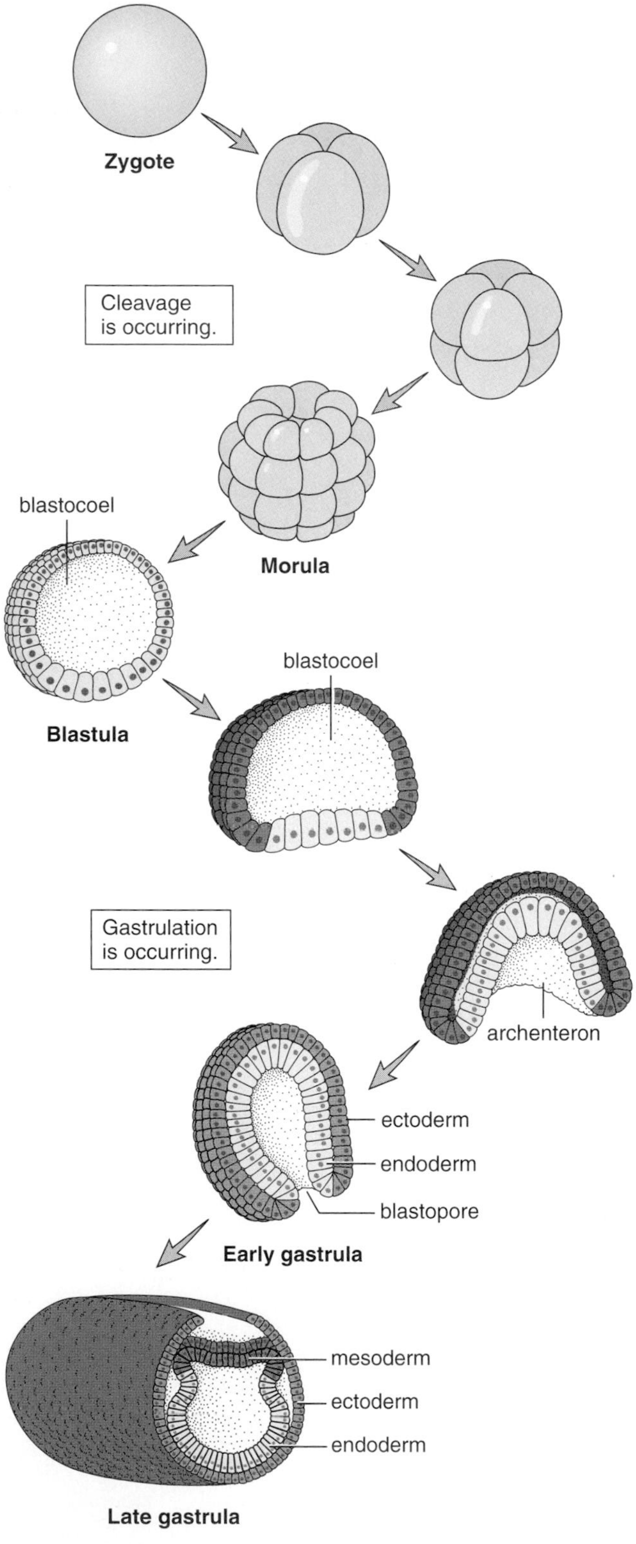
Zygote
Cleavage
is occurring.
blastocoel
Morula
Blastula
blastocoel
Gastrulation
is occurring.
archenteron
ectoderm
endoderm
blastopore
Early gastrula
mesoderm
ectoderm
endoderm
Late gastrula

Lancelet early development
Figure 44.2

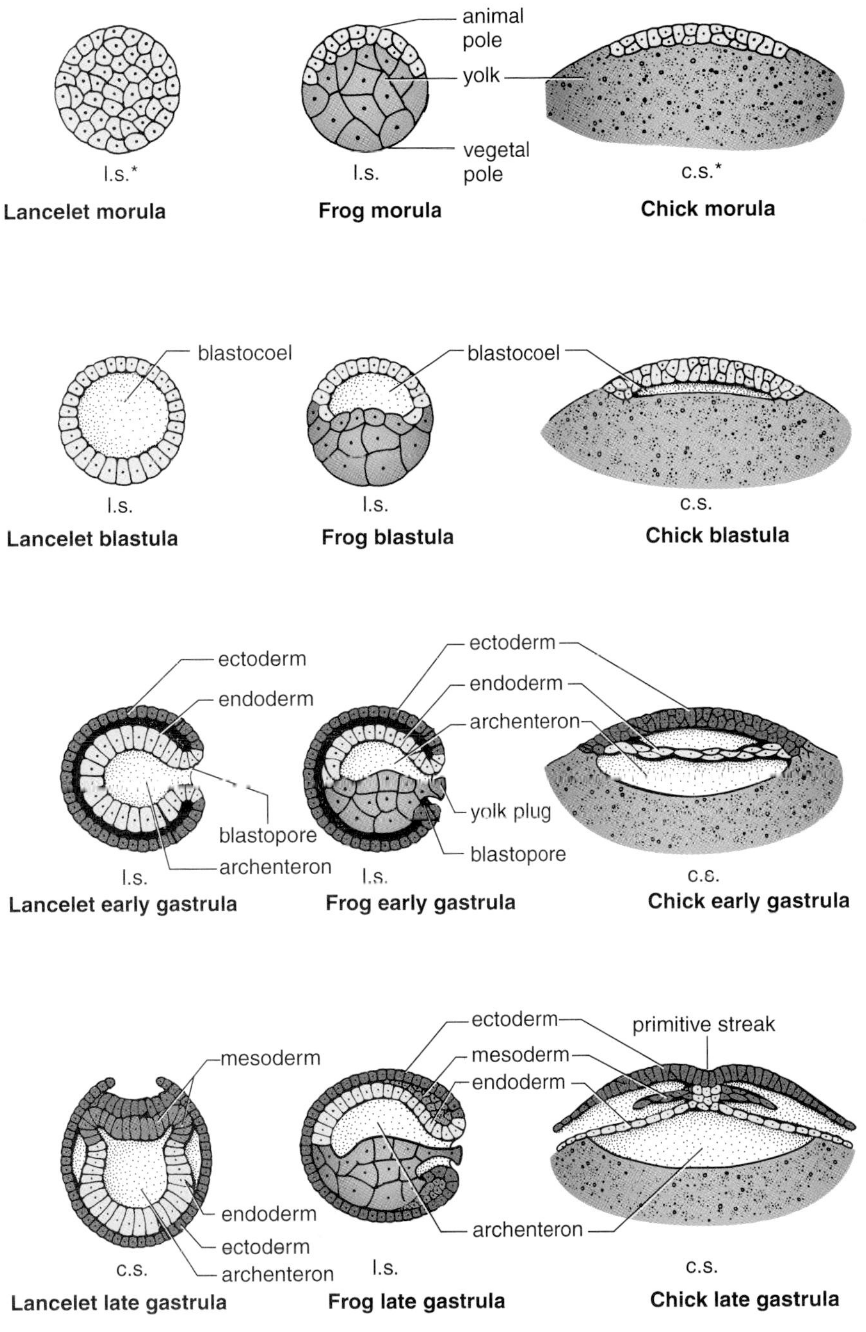

animal pole

yolk

vegetal pole

l.s.*

Lancelet morula

l.s.

Frog morula

c.s.*

Chick morula

blastocoel

l.s.

Lancelet blastula

blastocoel

l.s.

Frog blastula

c.s.

Chick blastula

ectoderm

endoderm

ectoderm

endoderm

archenteron

blastopore

archenteron

l.s.

Lancelet early gastrula

yolk plug

blastopore

l.s.

Frog early gastrula

c.s.

Chick early gastrula

mesoderm

endoderm

ectoderm

archenteron

c.s.

Lancelet late gastrula

ectoderm

mesoderm

endoderm

archenteron

l.s.

Frog late gastrula

primitive streak

c.s.

Chick late gastrula

*l.s. = longitudinal section; c.s. = cross section

Comparative animal development
Figure 44.3

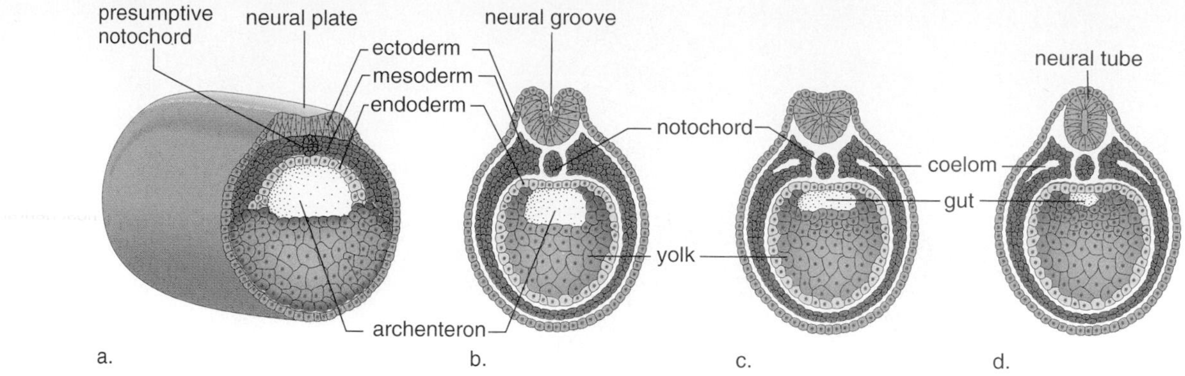

Development of neural tube and coelom in a frog embryo
Figure 44.4

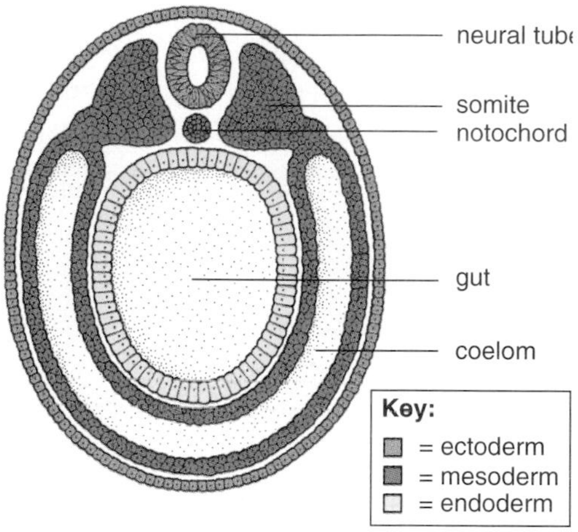

Chordate embryo, cross section
Figure 44.5

Key:
■ = ectoderm
■ = mesoderm
□ = endoderm

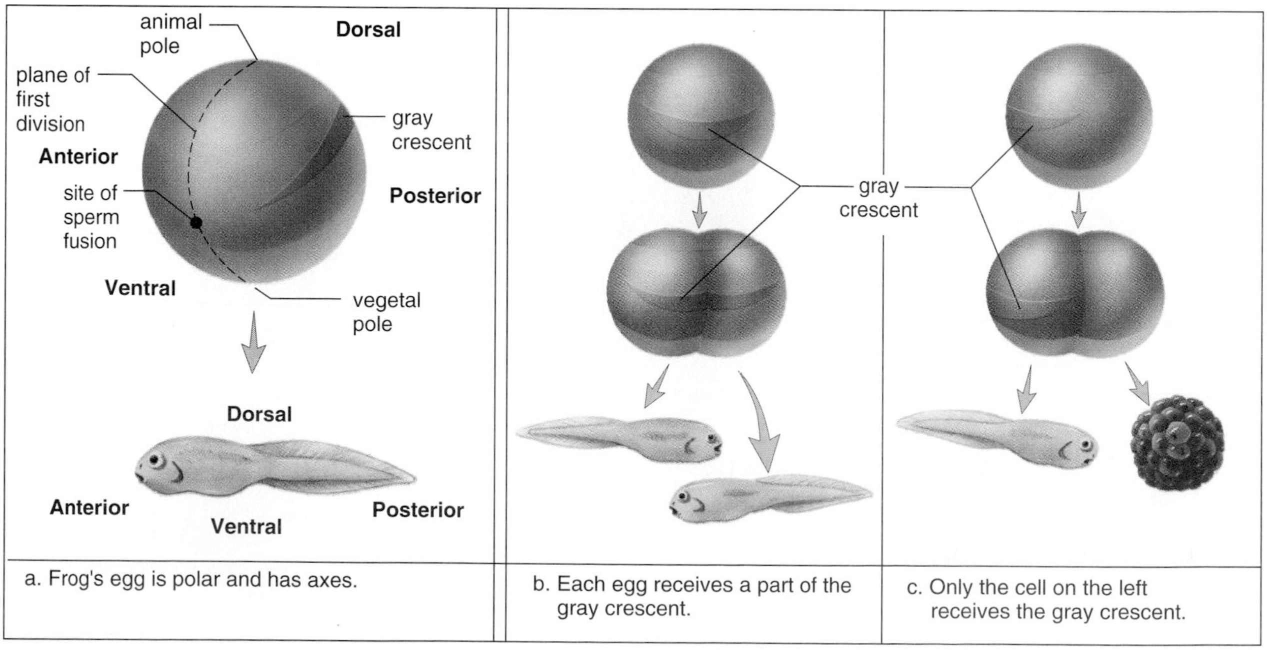

a. Frog's egg is polar and has axes.

b. Each egg receives a part of the gray crescent.

c. Only the cell on the left receives the gray crescent.

Cytoplasmic influence on development
Figure 44.6

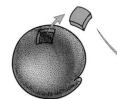
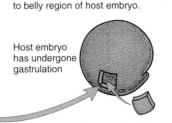
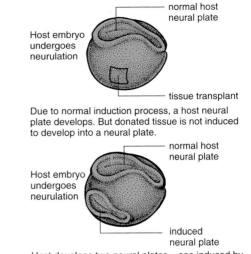

a. Presumptive nervous tissue is removed from a donor embryo.

Host embryo has undergone gastrulation

After removal of host tissue, donor presumptive nervous tissue is transplanted to belly region of host embryo.

Host embryo undergoes neurulation

normal host neural plate

tissue transplant

Due to normal induction process, a host neural plate develops. But donated tissue is not induced to develop into a neural plate.

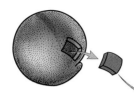

b. Presumptive notochord tissue is removed from a donor embryo.

Host embryo has undergone gastrulation

Donor presumptive notochord tissue is transplanted to a host embryo. Host belly tissue (which was removed) is returned to the host.

Host embryo undergoes neurulation

normal host neural plate

induced neural plate

Host develops two neural plates – one induced by host notochord tissue, the second induced by transplanted notochord tissue.

Control of nervous system development
Figure 44.7

gonad (8–16 divisions)

egg

cuticle (8–11 divisions)

gonad

vulva (10–13 divisions)

cuticle

intestine (3–6 divisions)

egg

vulva

nervous system (6–8 divisions)

intestine

sperm

pharynx (9–11 divisions)

nervous system

pharynx

a. Fate map

cell-death signal

Cell-death cascade

protease

nuclease

master protein (inactive)

b. Apoptosis

Development of *C. elegans*, a nematode
Figure 44.8

Pattern formation in *Drosophila*
Figure 44.10

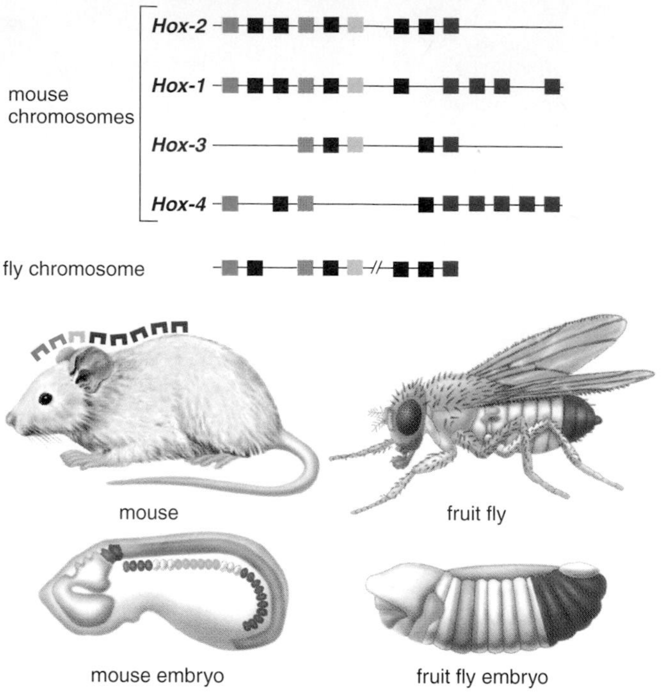

mouse chromosomes

Hox-2
Hox-1
Hox-3
Hox-4

fly chromosome

mouse

fruit fly

mouse embryo

fruit fly embryo

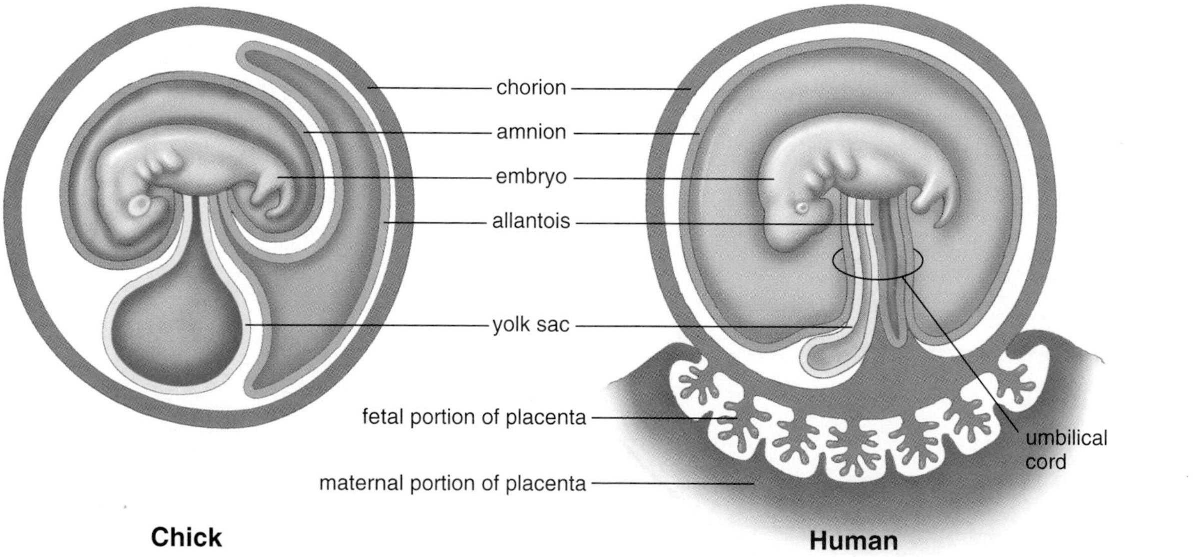

chorion
amnion
embryo
allantois

yolk sac

fetal portion of placenta
maternal portion of placenta

umbilical cord

Chick

Human

Extraembryonic membranes
Figure 44.11

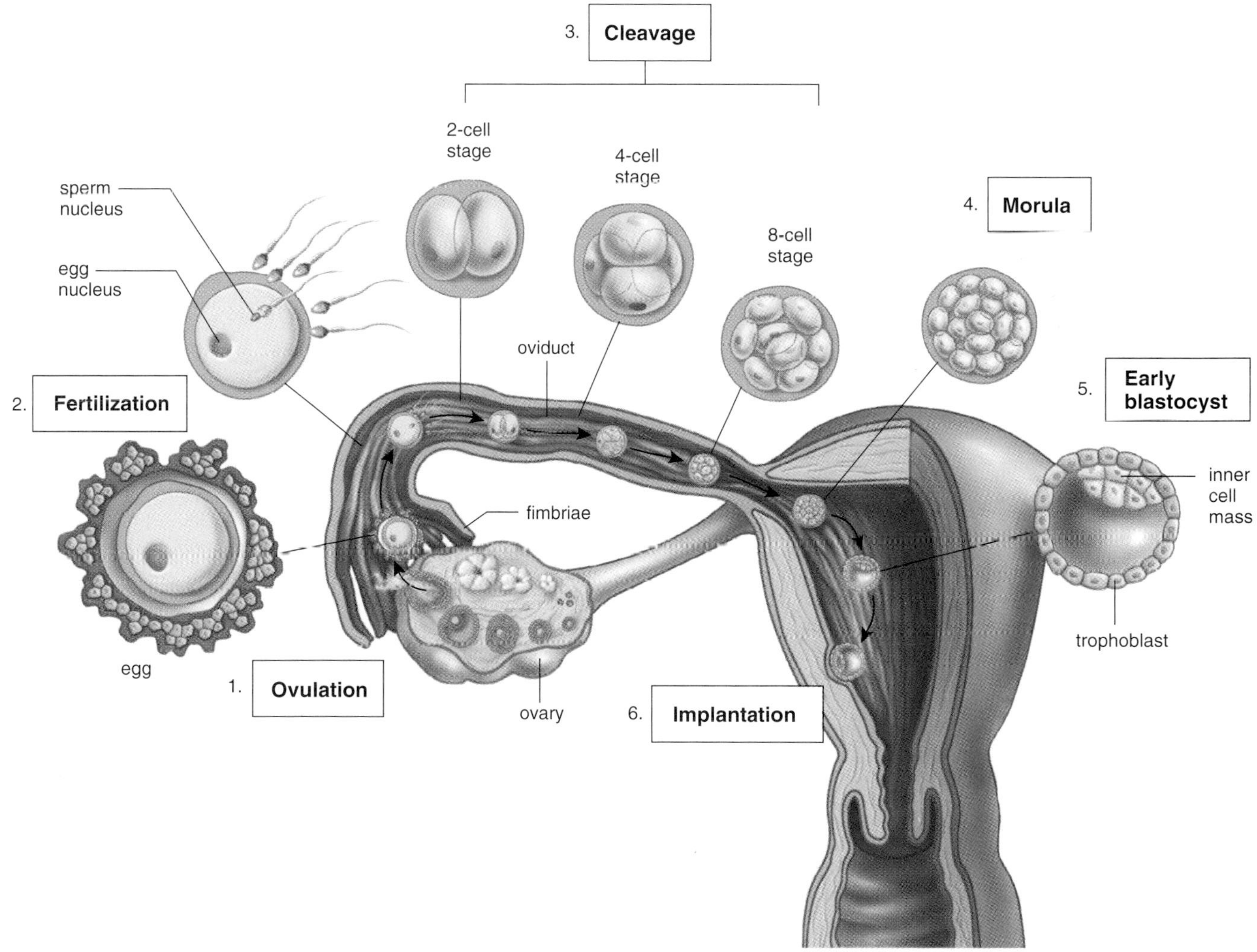

3. **Cleavage**

2-cell stage

4-cell stage

8-cell stage

sperm nucleus

egg nucleus

oviduct

4. **Morula**

2. **Fertilization**

fimbriae

5. **Early blastocyst**

inner cell mass

egg

1. **Ovulation**

ovary

6. **Implantation**

trophoblast

Human development before implantation
Figure 44.12

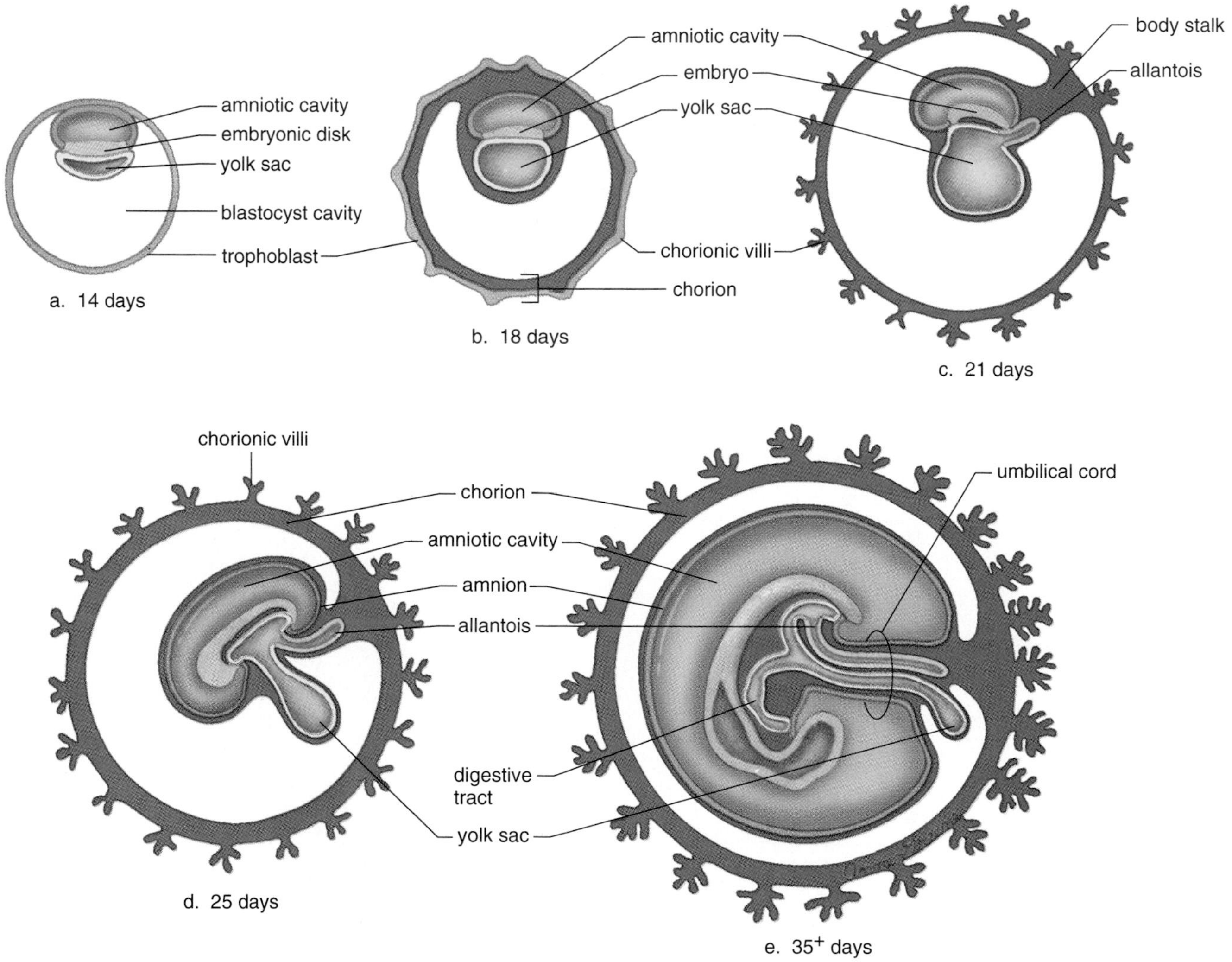
amniotic cavity
embryonic disk
yolk sac
blastocyst cavity
trophoblast
a. 14 days
amniotic cavity
embryo
yolk sac
chorionic villi
chorion
b. 18 days
body stalk
allantois
c. 21 days
chorionic villi
chorion
amniotic cavity
amnion
allantois
digestive tract
yolk sac
d. 25 days
umbilical cord
e. 35+ days

Human embryonic development
Figure 44.13

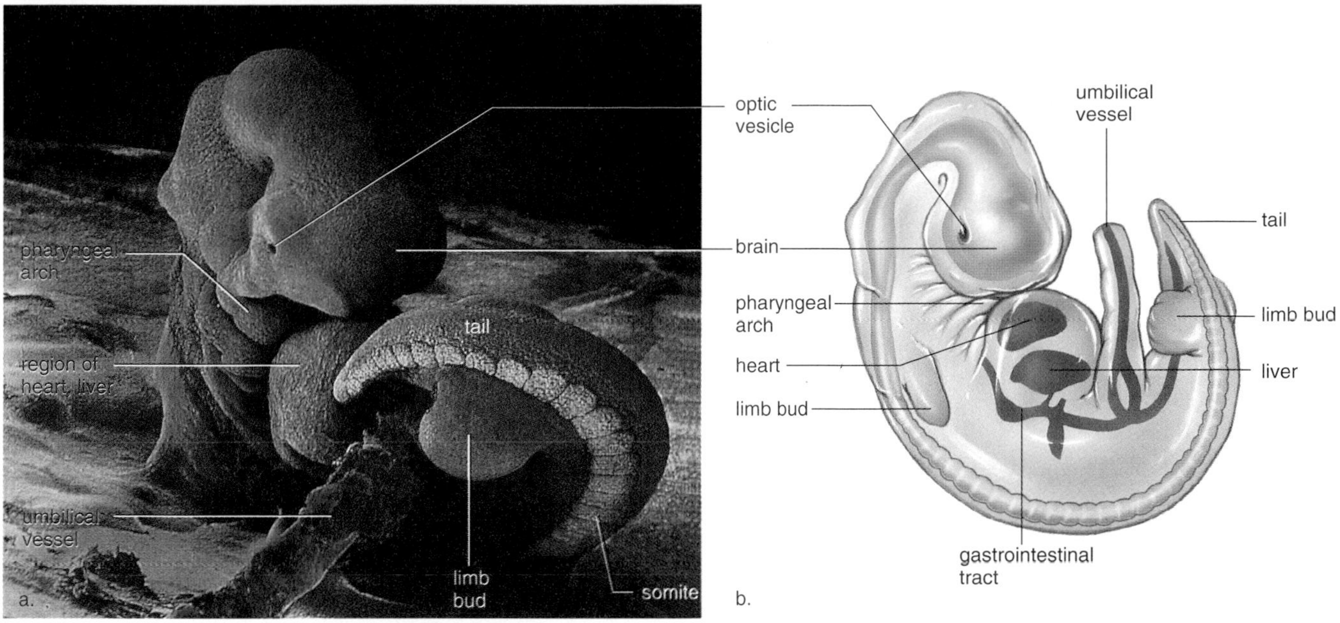

Human embryo at beginning of fifth week
Figure 44.14

a: © Lennart Nilsson *A Child is Born* Dell Publishing

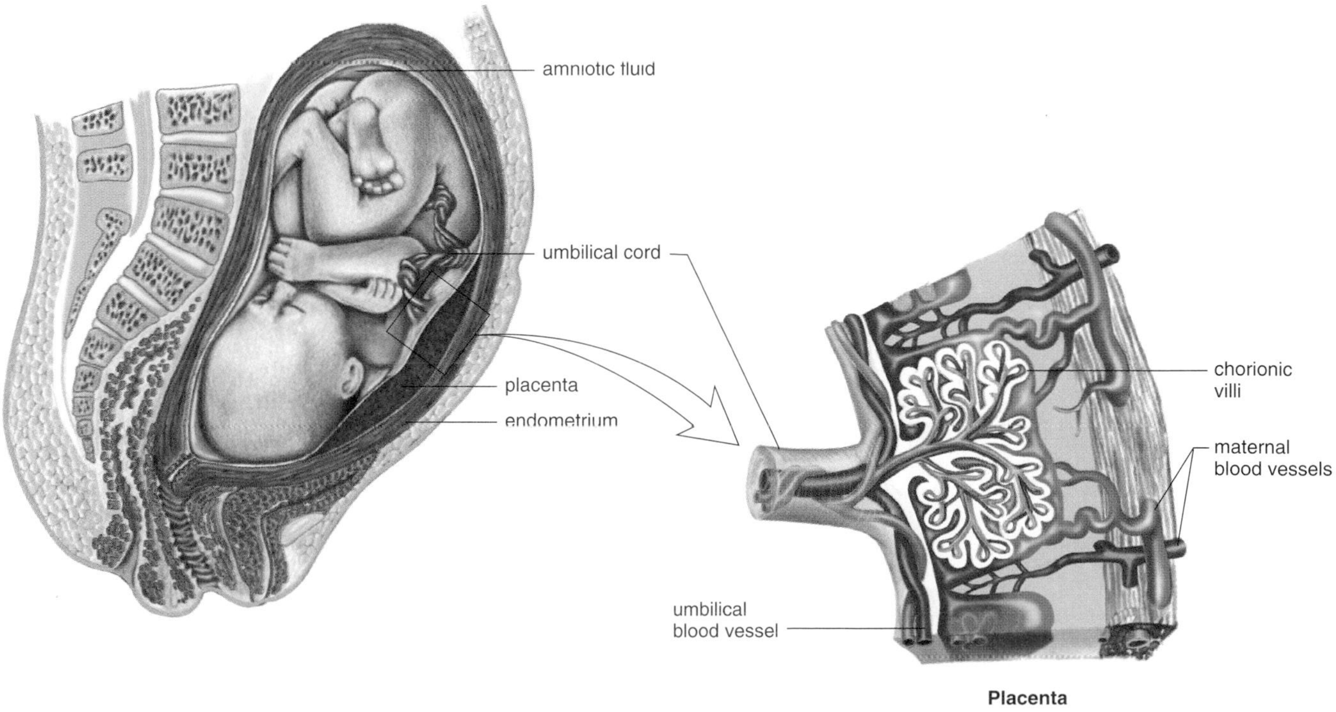

Anatomy of the placenta in a fetus at six to seven months
Figure 44.15

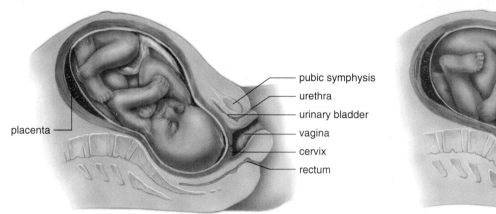

placenta

pubic symphysis
urethra
urinary bladder
vagina
cervix
rectum

a. 9-month-old fetus

ruptured
amniotic
sac

b. First stage of birth: cervix dilates

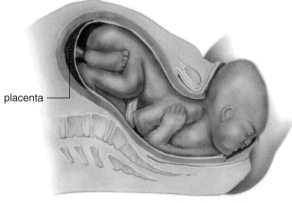

placenta

c. Second stage of birth: baby emerges

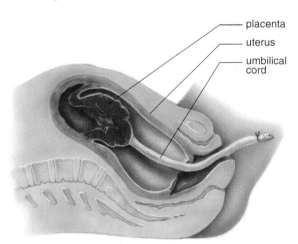

placenta
uterus
umbilical
cord

d. Third stage of birth: expelling afterbirth

Three stages of parturition
Figure 44.16

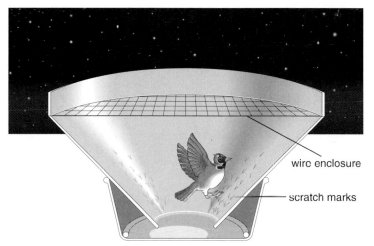

wire enclosure

scratch marks

longitudinal section of funnel cage

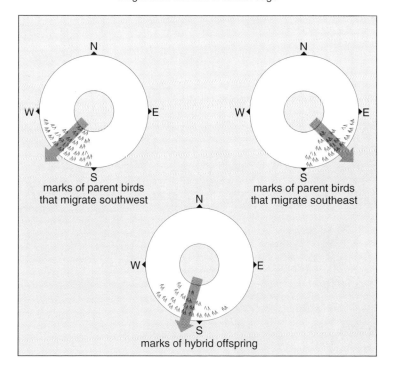

N

W ◄   ► E

S

marks of parent birds
that migrate southwest

N

W ◄   ► E

S

marks of parent birds
that migrate southeast

N

W ◄   ► E

S

marks of hybrid offspring

**Inheritance of migratory behavior
in Blackcap Warblers, *Sylvia***
Figure 45.1

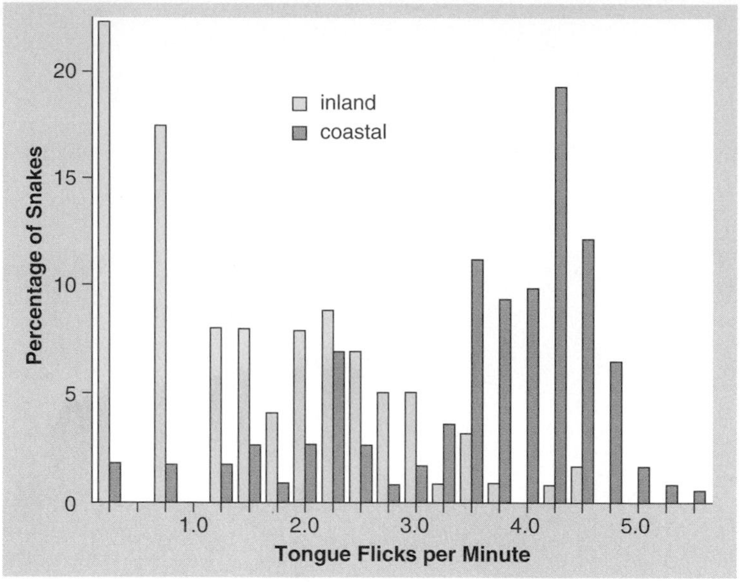

Feeding behavior in garter snakes,
Thamnophis elegans
Figure 45.2

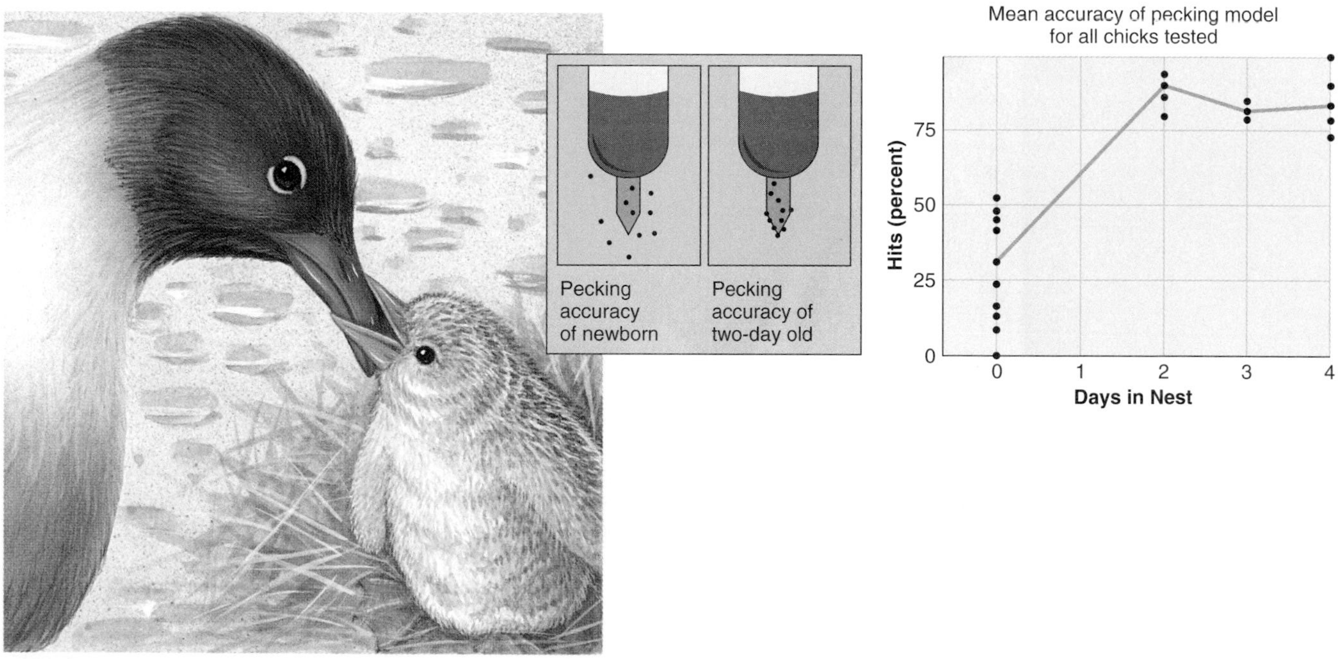

Feeding (pecking) behavior of Laughing Gull chicks, *Larus atricilla*
Figure 45.3

Isolated bird sings but song is not developed.

Bird sings developed song played during a sensitive period.

Bird sings song of social tutor without regard to sensitive period.

Song learning by White-crowned Sparrows, *Zonotrichia leucophrys*
Figure 45.4

male

female

Raggiana Bird of Paradise
Figure 45.5

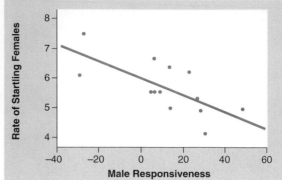
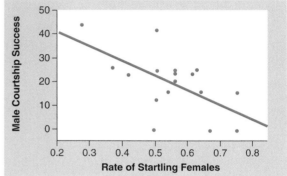

Robotic female bowerbird

Figure 45D

Courtesy Gail Patricelli/University of Maryland

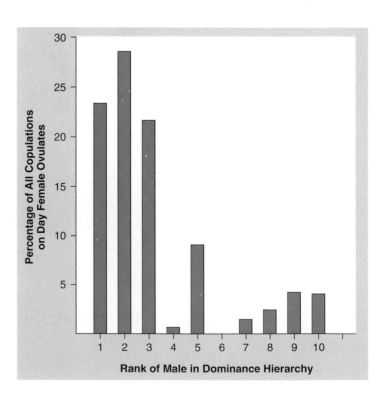

Female choice and male dominance among baboons

Figure 45.7

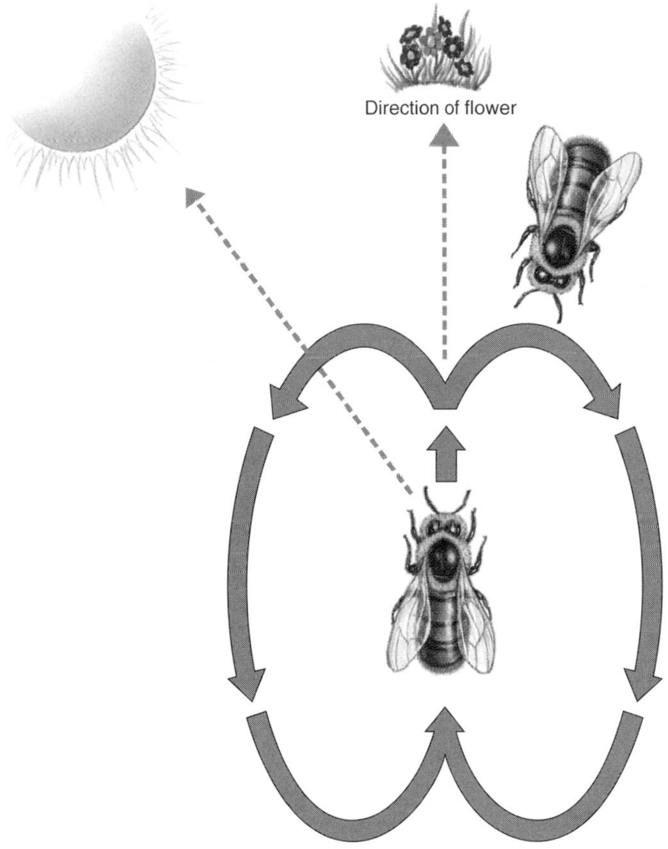

Direction of flower

b. Components of dance

Communication among bees
Figure 45.11

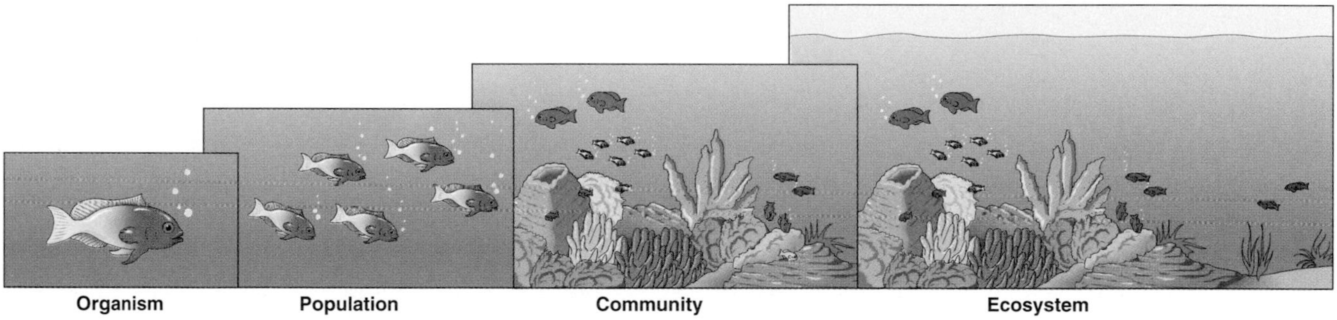

| Organism | Population | Community | Ecosystem |

Ecological levels
Figure 46.1

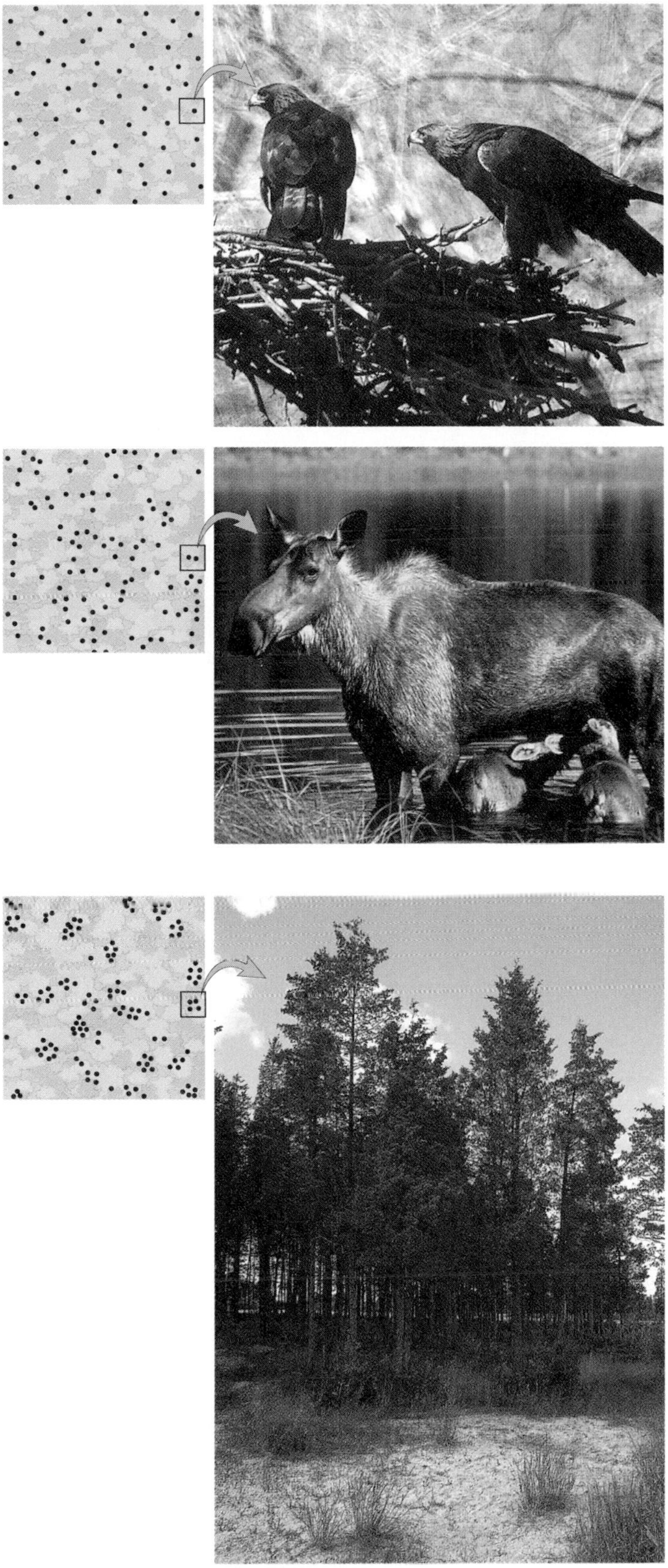

Patterns of dispersion within a population
Figure 46.2

a: © C. Palek/Animals Animals/Earth Scenes; b: © S.J. Krasemann/Peter
Arnold, Inc.; c: © Peter Arnold/Peter Arnold, Inc.

Model for exponential growth
Figure 46.4

Generation	Population Size
0	10.0
1	24.0
2	57.6
3	138.2
4	331.7
5	796.1
6	1,910.6
7	4,585.4
8	11,005.0
9	26,412.0
10	63,388.8

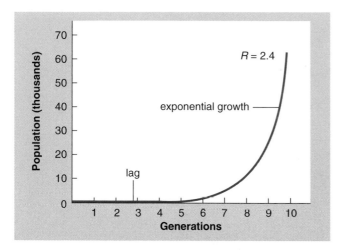

To calculate population size from year to year, use this formula:

$$N_{t+1} = RN_t$$

N_t = number of females already present

R = net reproductive rate

N_{t+1} = population size the following year

Growth of Yeast Cells in Laboratory Culture

Time (t) (hours)	Number of individuals (N)	Number of individuals added per 2-hour period $\left(\dfrac{\Delta N}{\Delta t}\right)$
0	9.6	0
2	29.0	19.4
4	71.1	42.1
6	174.6	103.5
8	350.7	176.1
10	513.3	162.6
12	594.4	81.1
14	640.8	46.4
16	655.9	15.1
18	661.8	5.9

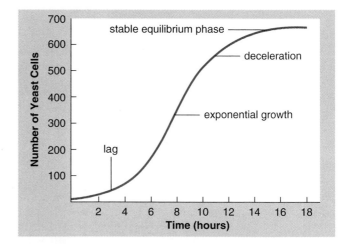

To calculate population growth as time passes, use this formula:

$$\frac{dN}{dt} = rN\left(\frac{K-N}{K}\right)$$

N = population size

dN/dt = change in population size

r = intrinsic rate of natural increase

K = carrying capacity

$\dfrac{K-N}{K}$ = effect of carrying capacity on population growth

Model for logistic growth
Figure 46.5

Survivorship curves
Figure 46.7

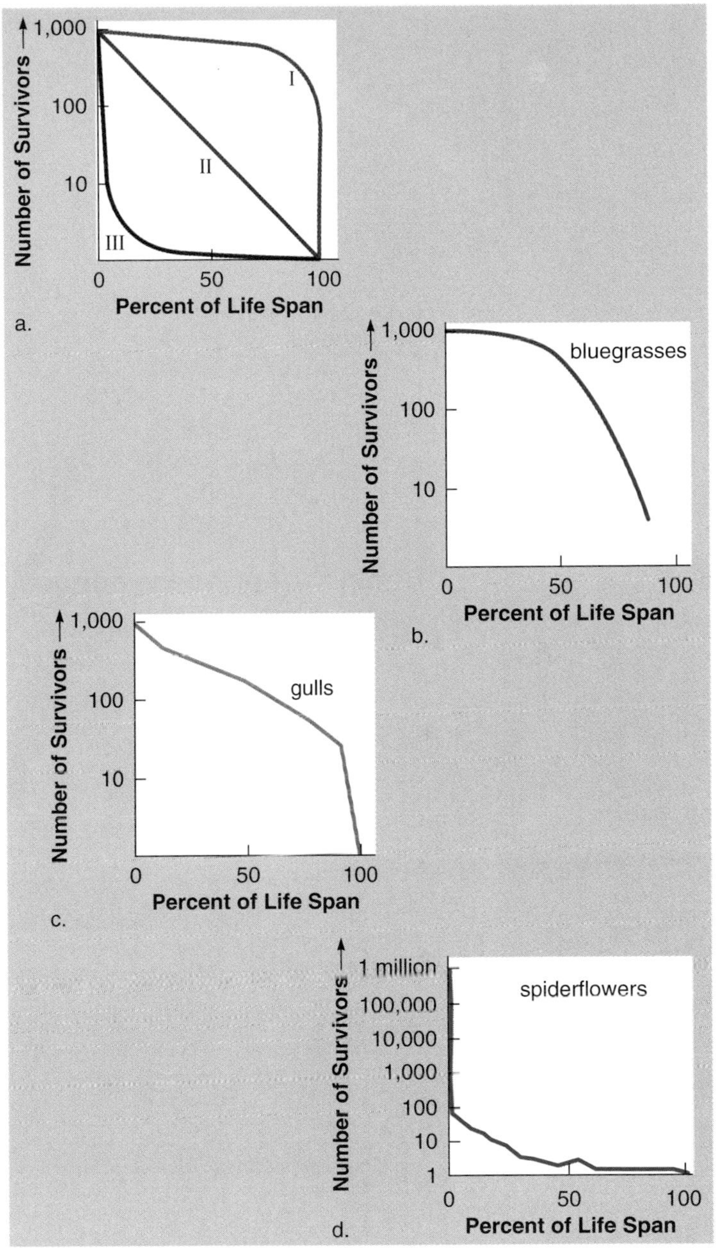

U.S. age distributions, 1910, 1960, and 2010 (projected)
Figure 46.8

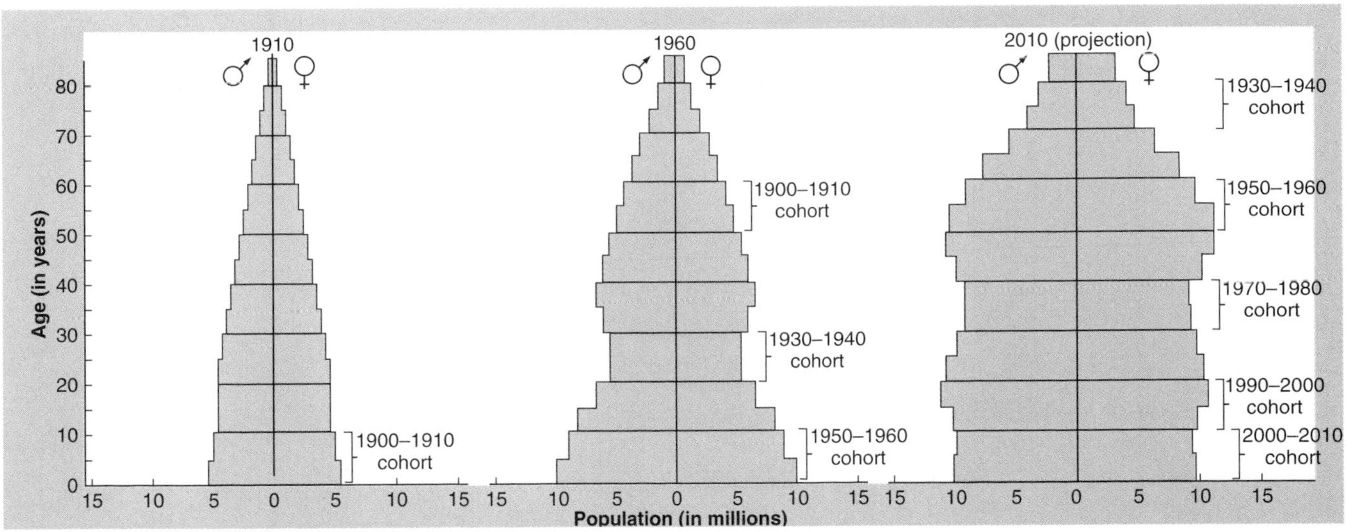

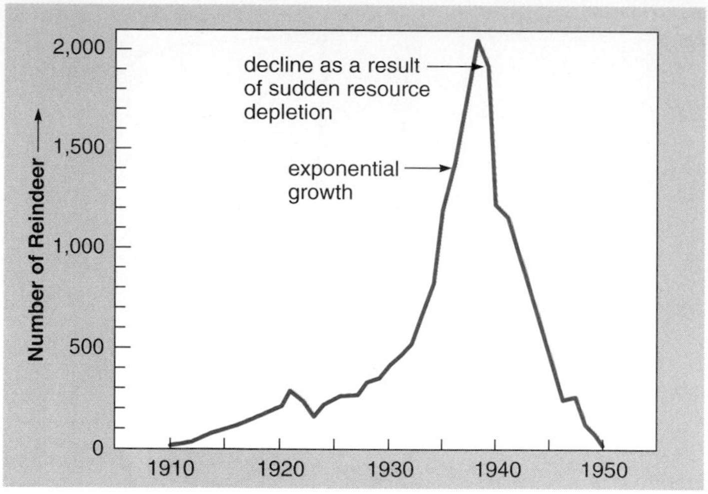

b.

Density-dependent effect
Figure 46.9

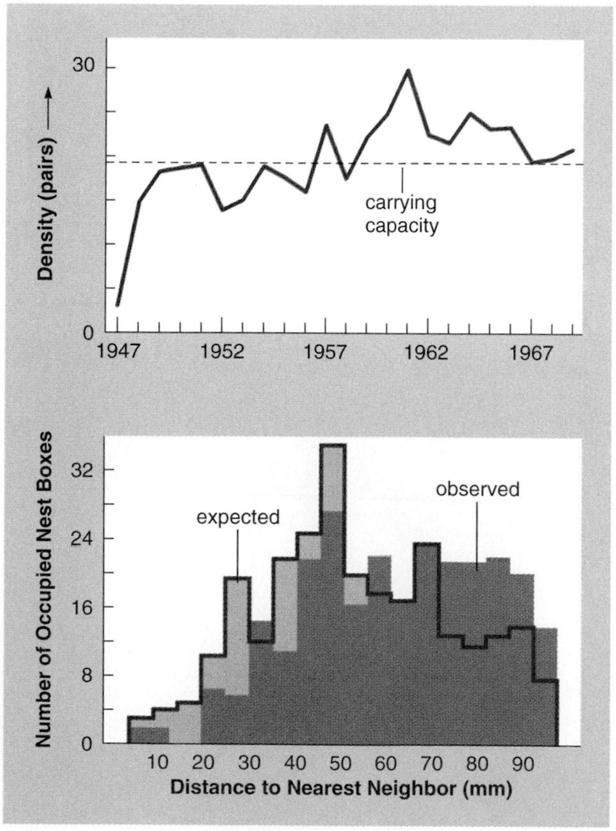

Great Tit, *Parus major*
Figure 46.11

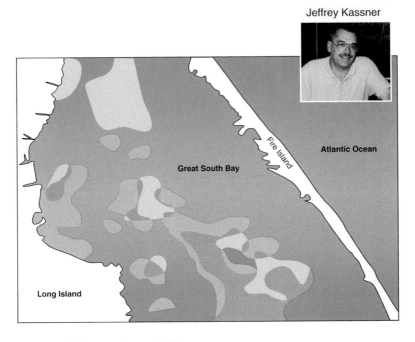

Jeffrey Kassner

Map of Great South Bay
Figure 46A
Courtesy Jeffrey Kassner

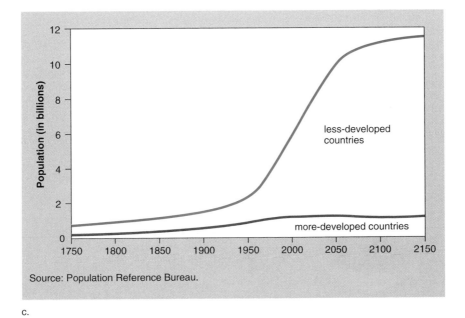

c.

World population growth
Figure 46.14

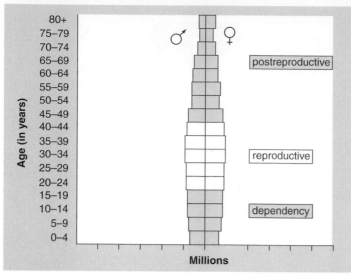

a. More-developed countries (MDCs)

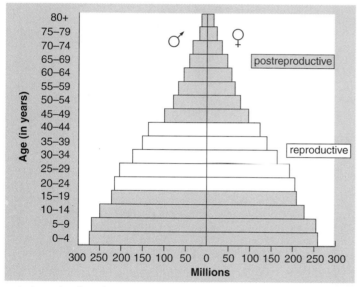

b. Less-developed countries (LDCs)

Age structure diagrams (1998)
Figure 46.15

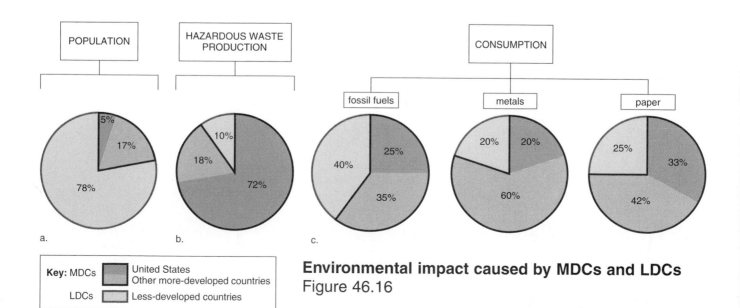

POPULATION

HAZARDOUS WASTE PRODUCTION

CONSUMPTION

fossil fuels

metals

paper

Key: MDCs
United States
Other more-developed countries
LDCs
Less-developed countries

Environmental impact caused by MDCs and LDCs
Figure 46.16

Community structure
Figure 47.1

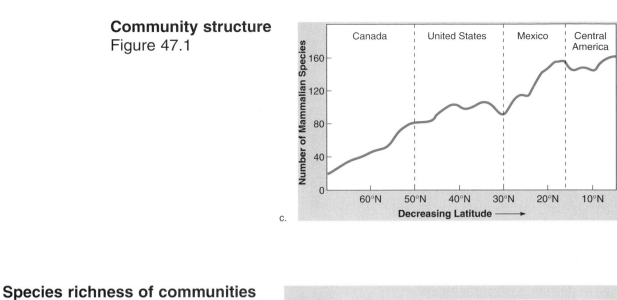

c.

Species richness of communities
Figure 47.2

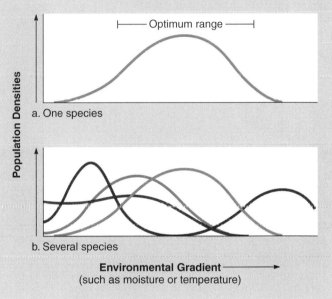

a. One species

b. Several species

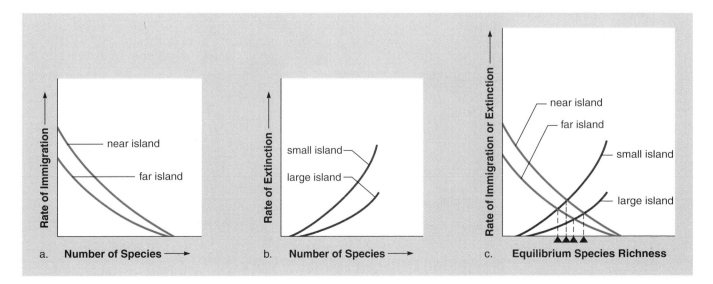

a.

b.

c.

Model of island biogeography
Figure 47.3

Flamingos feed on small molluscs, crustaceans, and vegetable matter strained from mud pumped through their bills by their powerful tongues.

Dabbling ducks feed by tipping, tail up, to reach aquatic plants, seeds, snails, and insects.

Avocets feed on insects, small marine invertebrates, and seeds by sweeping their bills from side to side in shallow water.

Oystercatchers pry open bivalve shells with their knifelike bills and probe sand for worms and crabs.

Plovers dart around on beaches and grasslands hunting for insects and small invertebrates.

Feeding niches for wading birds
Figure 47.4

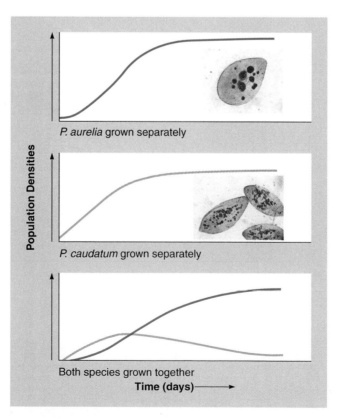

P. aurelia grown separately

P. caudatum grown separately

Both species grown together

Population Densities

Time (days)

Competition between two laboratory populations of *Paramecium*
Figure 47.5

© Ken Wagner/Phototake

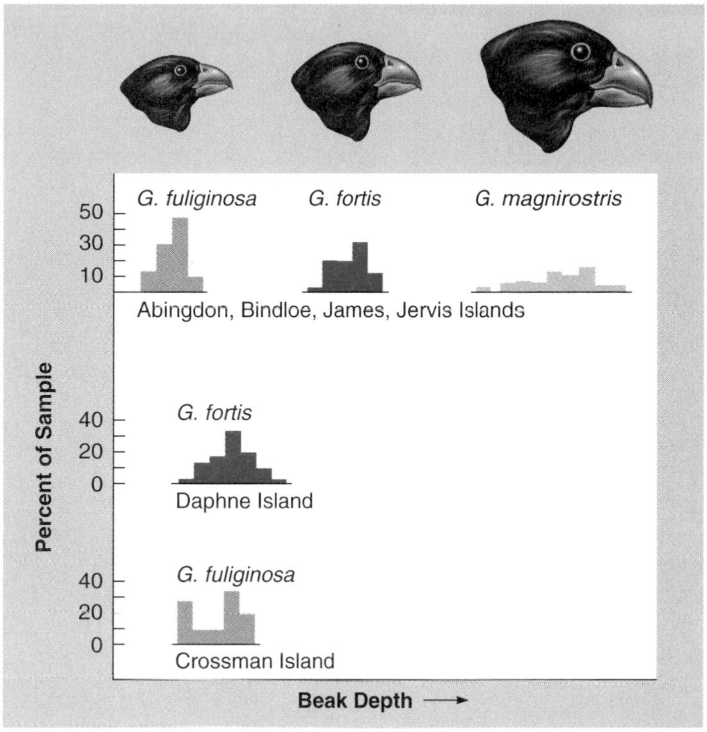

**Character displacement in finches
on the Galápagos Islands**
Figure 47.6

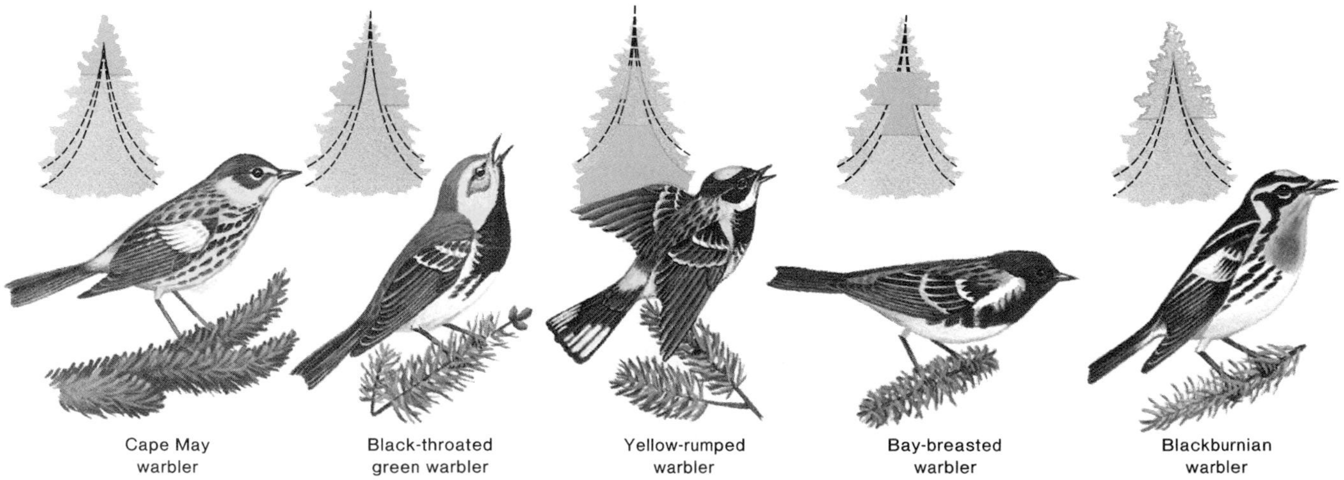

| Cape May warbler | Black-throated green warbler | Yellow-rumped warbler | Bay-breasted warbler | Blackburnian warbler |

Niche specialization among five species of coexisting warblers
Figure 47.7

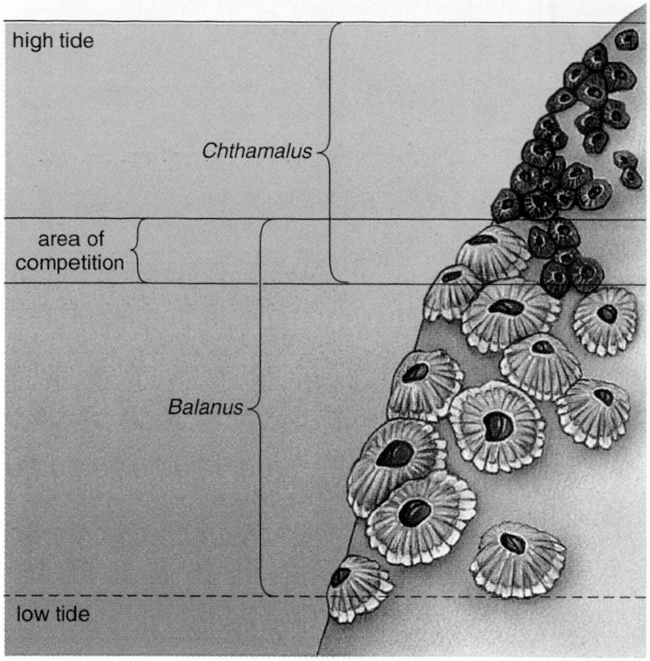

Competition between two species of barnacles
Figure 47.8

Predator-prey interaction between *Paramecium caudatum* and *Didinium nasutum*
Figure 47.9

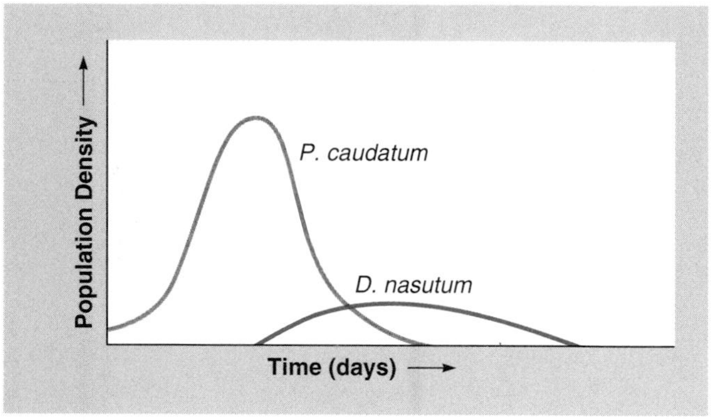

b. Population density of *Paramecium* and *Didinium* over time.

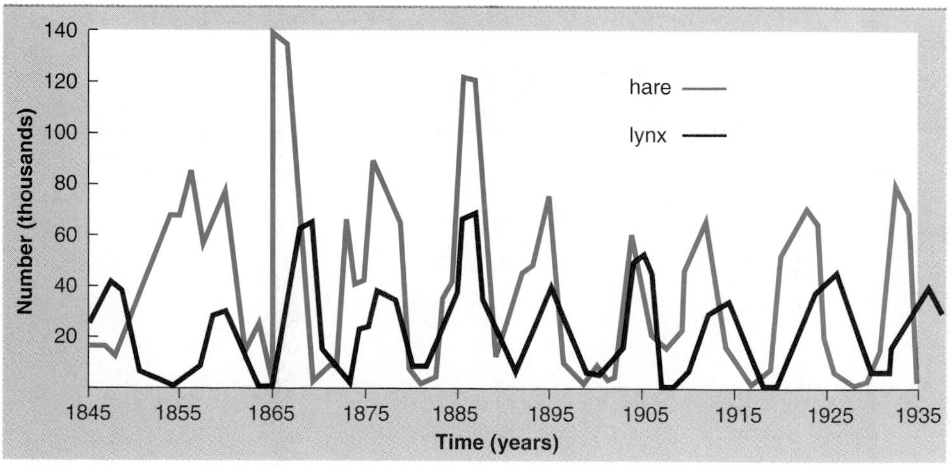

Predator-prey interaction between a lynx and a snowshoe hare
Figure 47.10

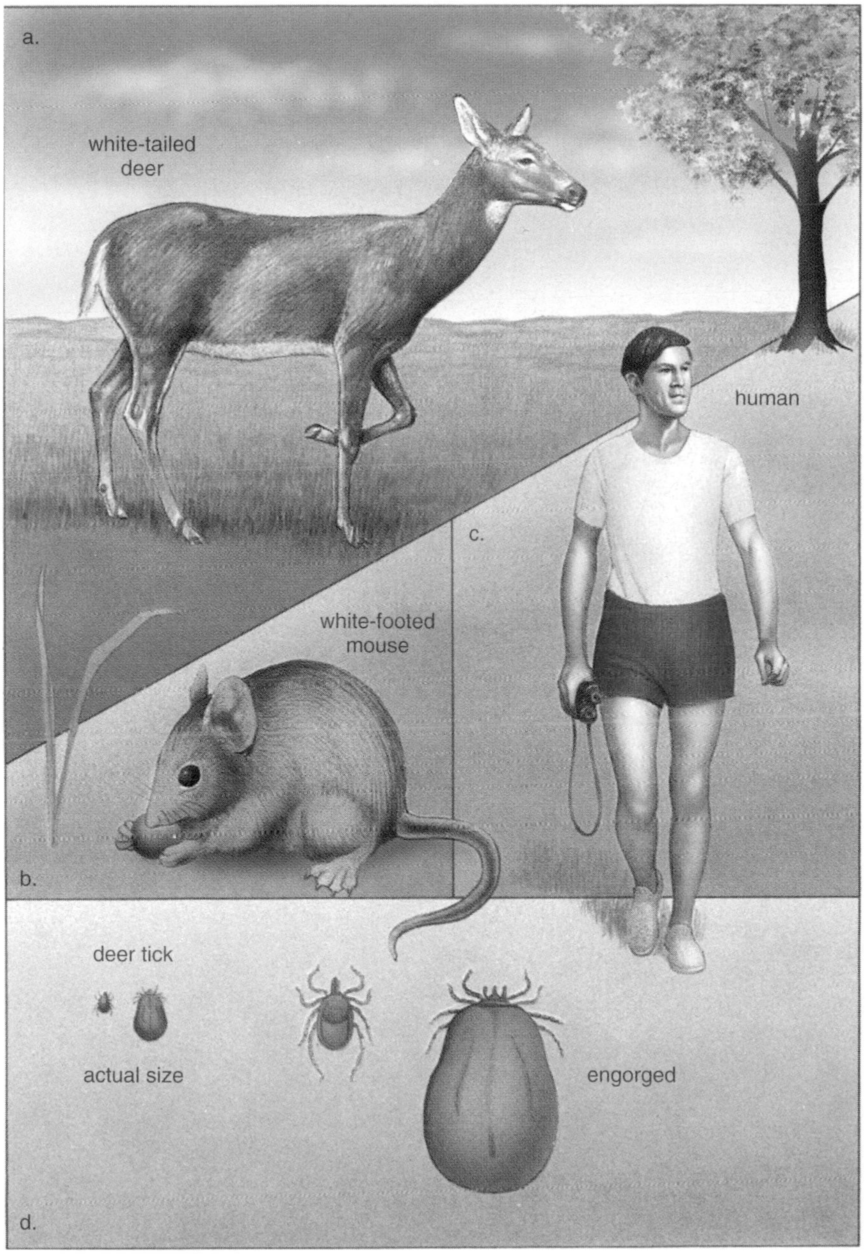
a.
white-tailed
deer
human
c.
white-footed
mouse
b.
deer tick
actual size
engorged
d.

The life cycle of a deer tick
Figure 47.14

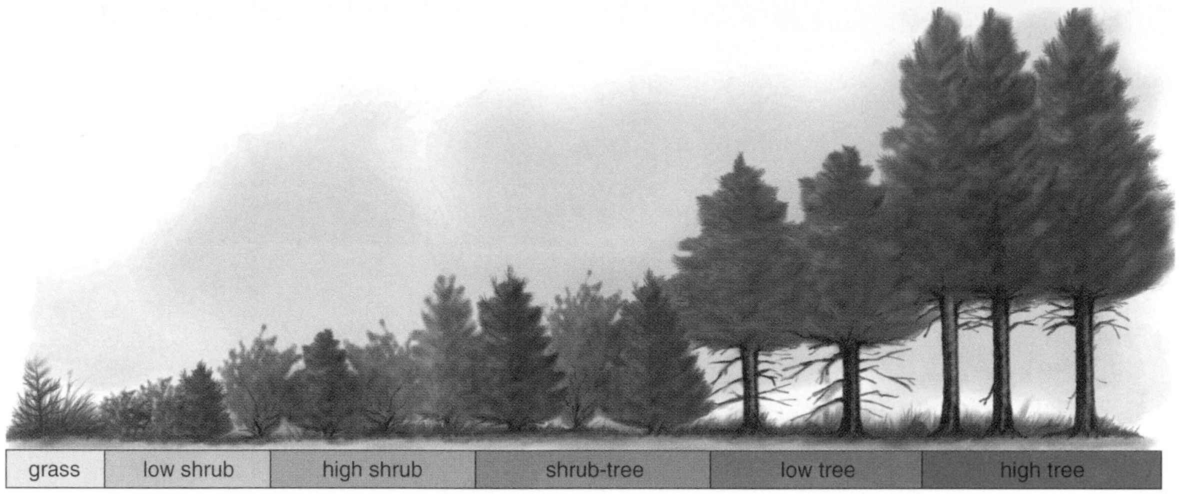

Secondary succession in a forest
Figure 47.19

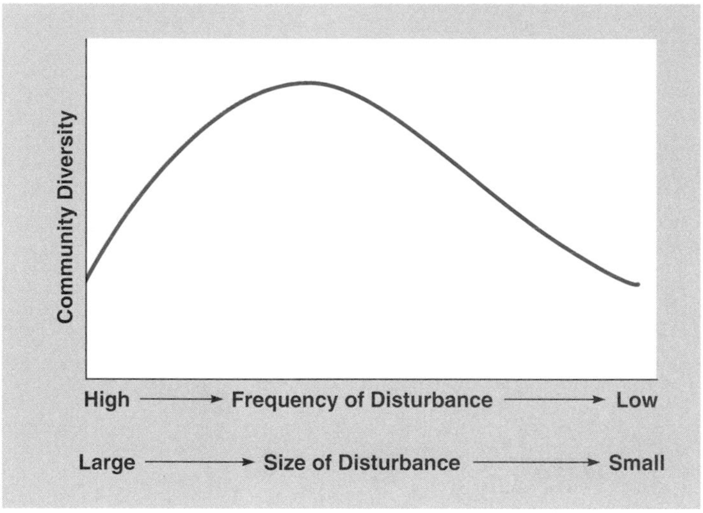

The intermediated disturbance hypothesis
Figure 47.20

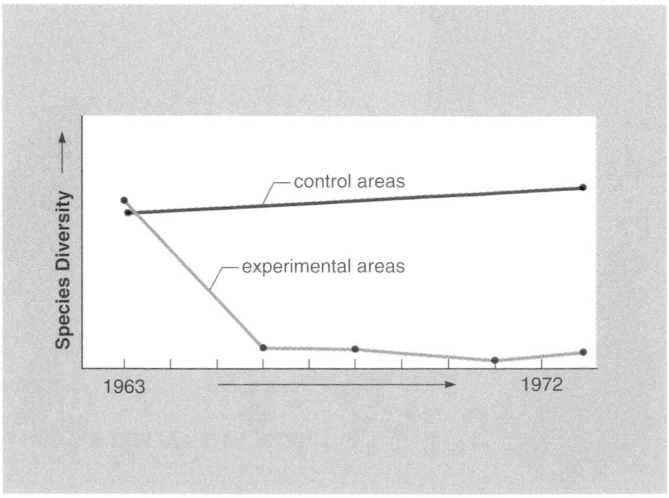

Effect of a keystone predator
Figure 47.21

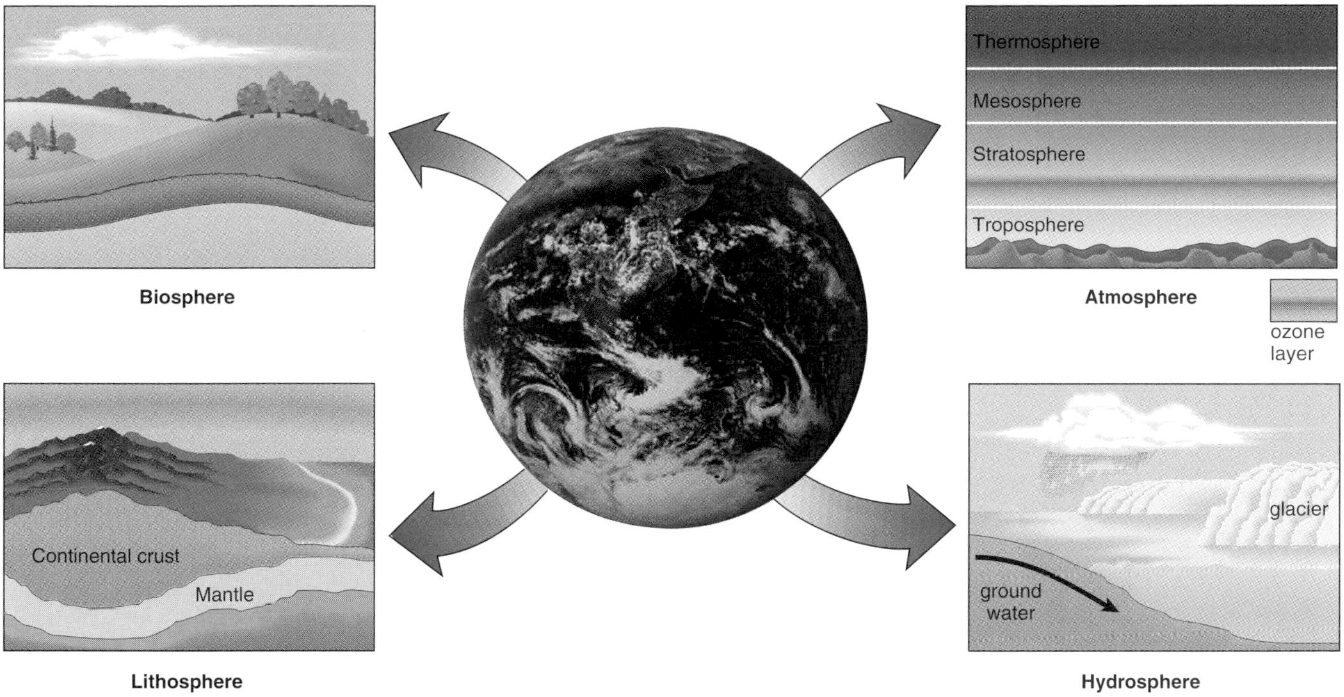

Biosphere
Figure 48.1
NASA

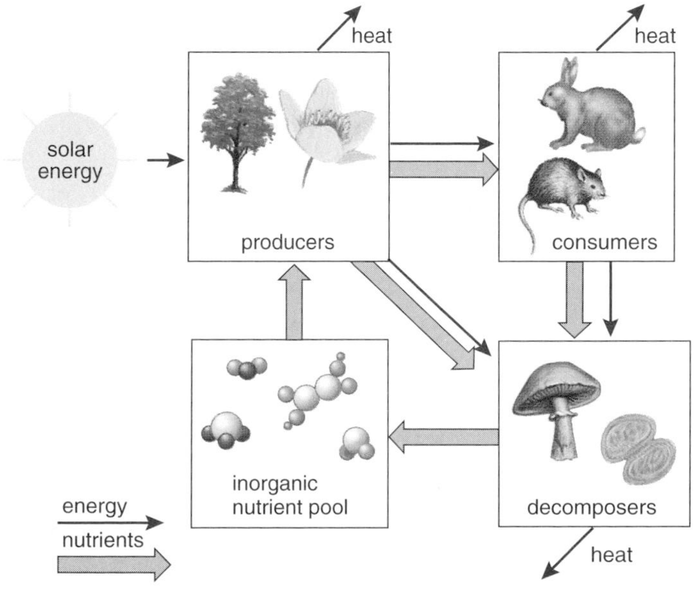

Nature of an ecosystem
Figure 48.3

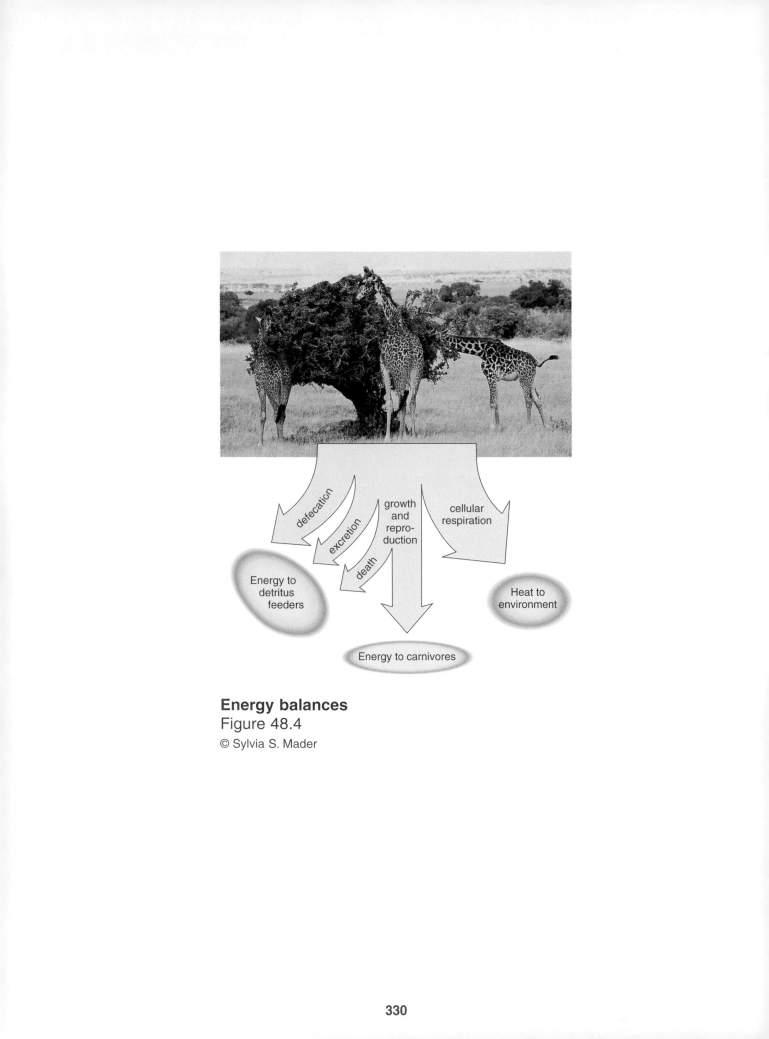

Energy balances
Figure 48.4
© Sylvia S. Mader

defecation
excretion
death
growth
and
repro-
duction
cellular
respiration
Energy to
detritus
feeders
Heat to
environment
Energy to carnivores

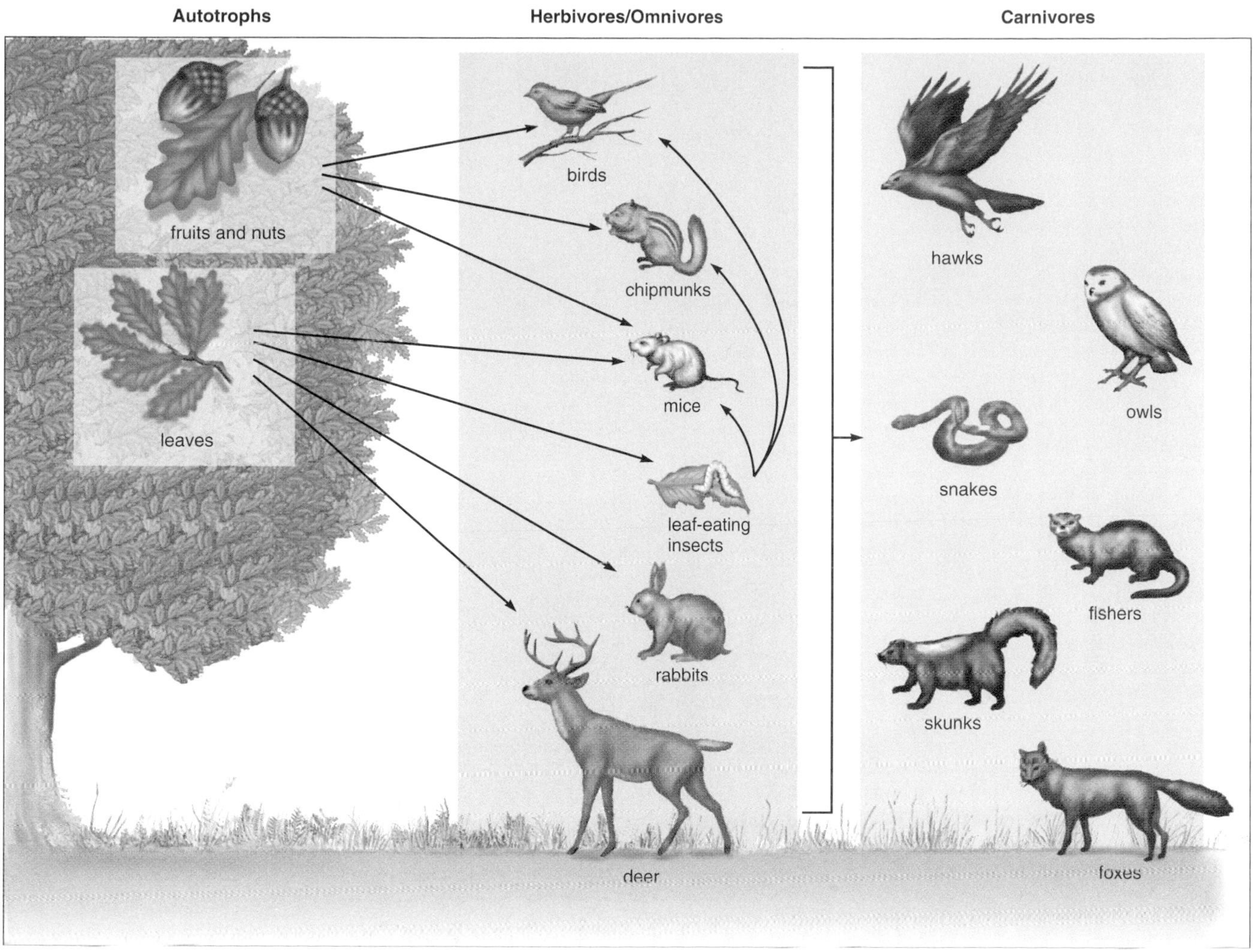

Grazing food web
Figure 48.5

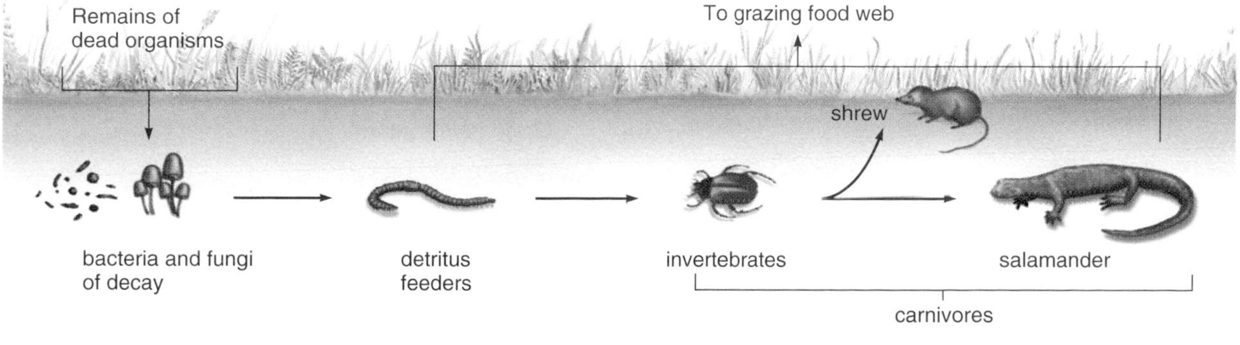

Detrital food web
Figure 48.6

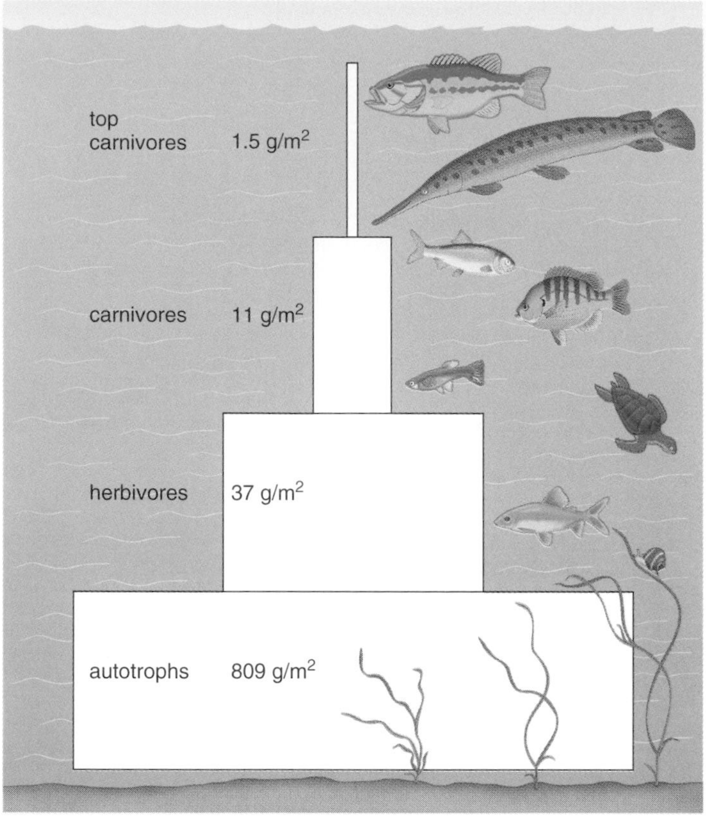

Ecological pyramid
Figure 48.7

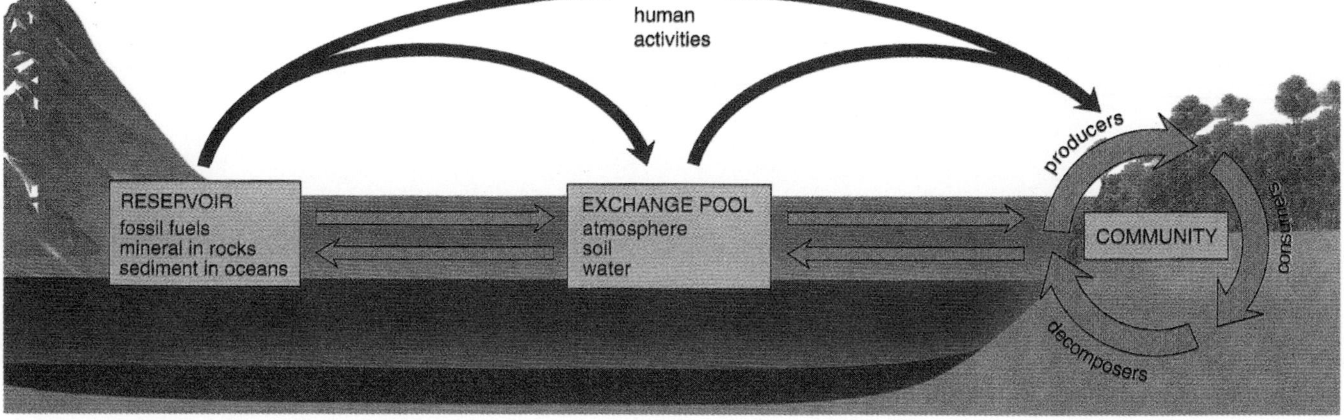

Model for chemical cycling
Figure 48.8

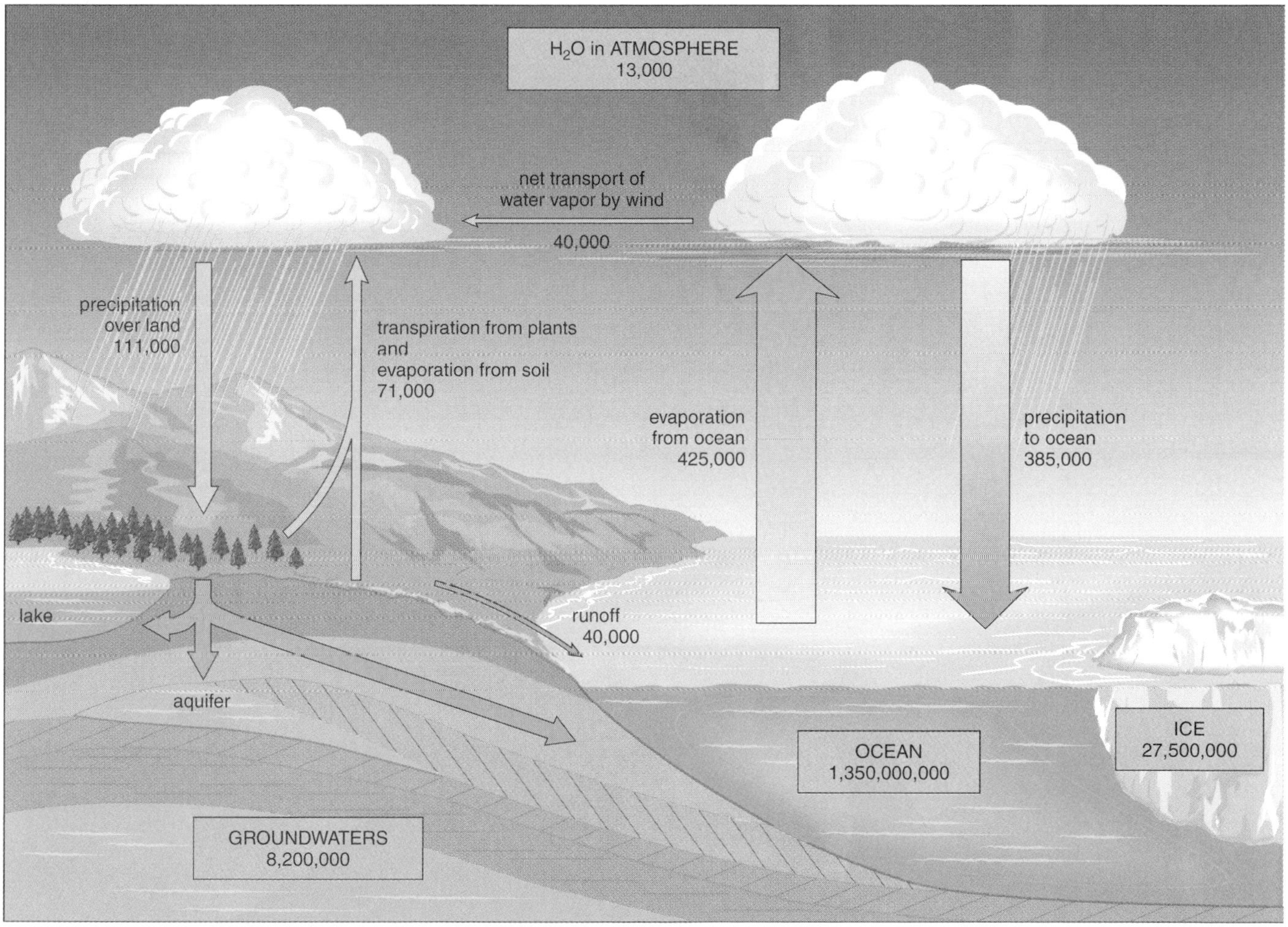

H_2O in ATMOSPHERE
13,000
net transport of
water vapor by wind
40,000
precipitation
over land
111,000
transpiration from plants
and
evaporation from soil
71,000
evaporation
from ocean
425,000
precipitation
to ocean
385,000
lake
runoff
40,000
aquifer
ICE
27,500,000
OCEAN
1,350,000,000
GROUNDWATERS
8,200,000

The hydrologic (water) cycle
Figure 48.9

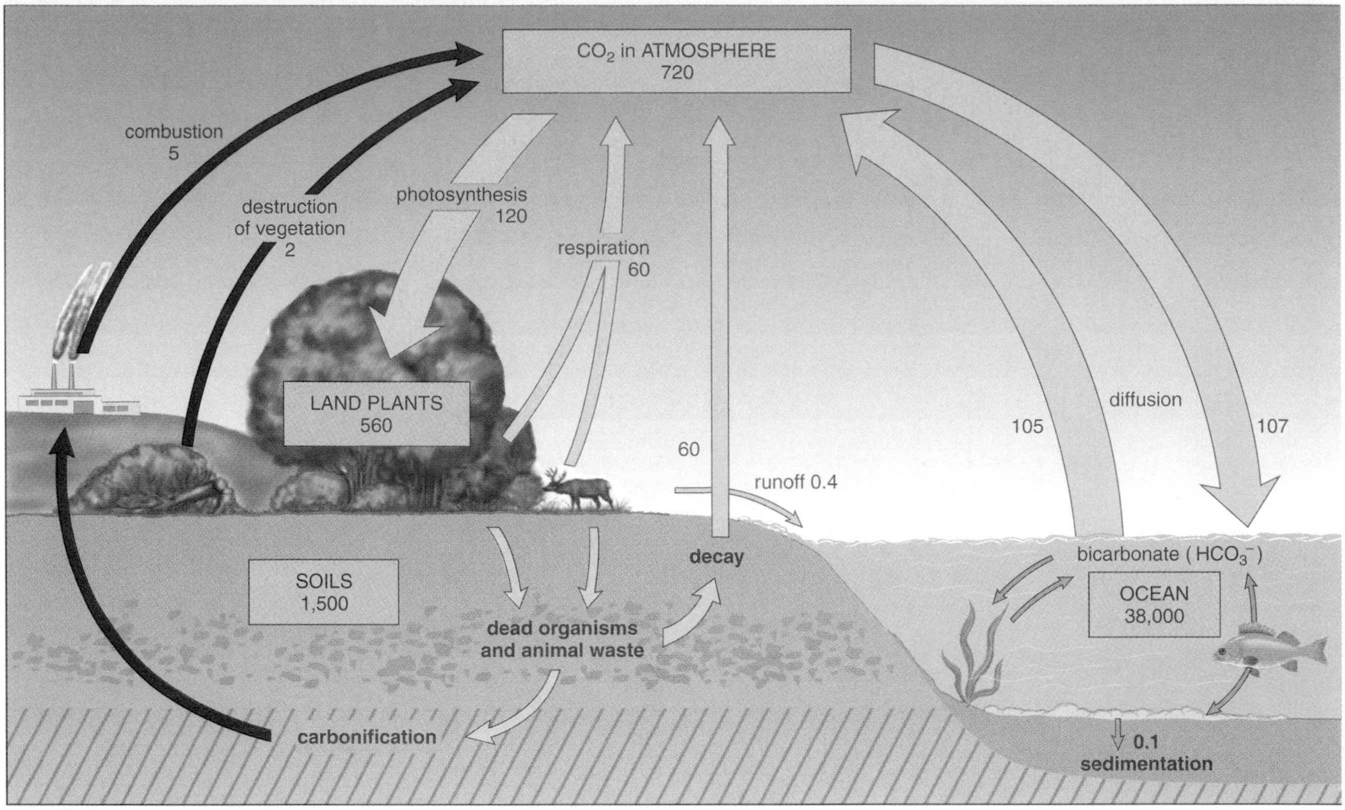

The carbon cycle
Figure 48.10

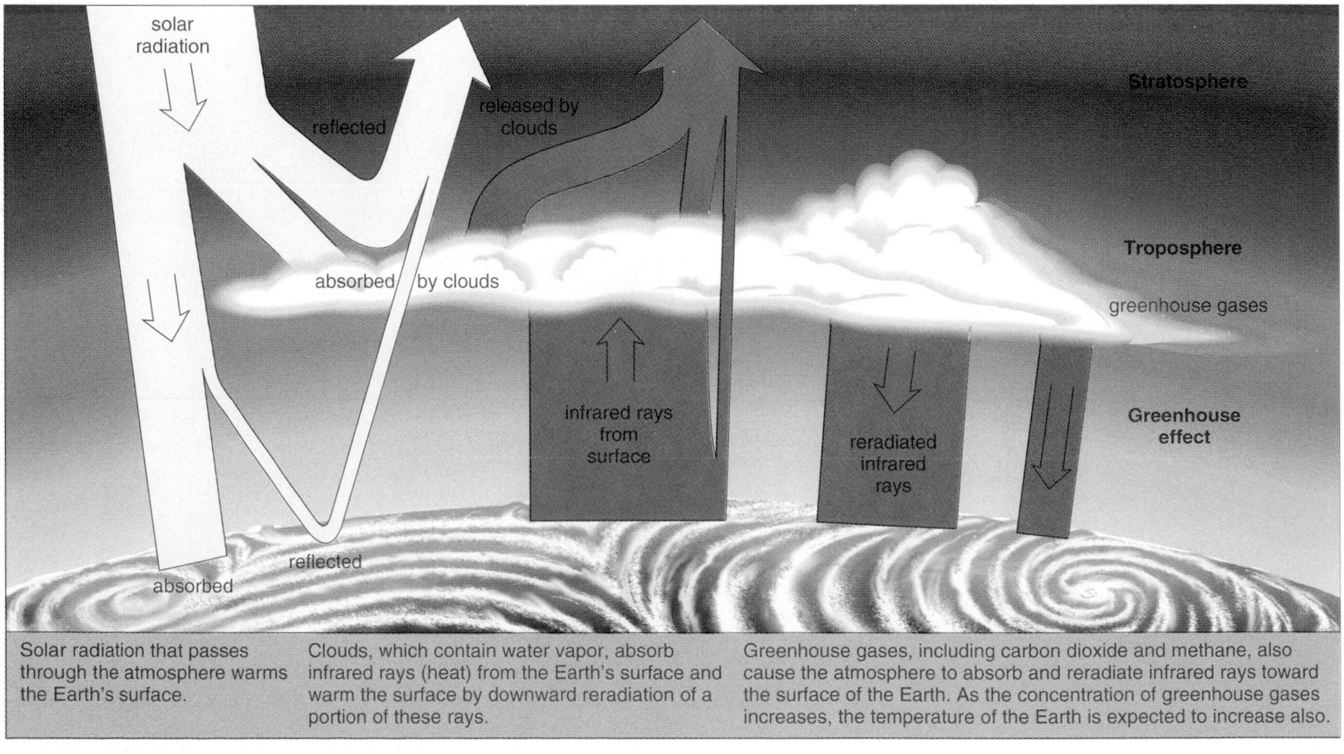

Solar radiation that passes through the atmosphere warms the Earth's surface.

Clouds, which contain water vapor, absorb infrared rays (heat) from the Earth's surface and warm the surface by downward reradiation of a portion of these rays.

Greenhouse gases, including carbon dioxide and methane, also cause the atmosphere to absorb and reradiate infrared rays toward the surface of the Earth. As the concentration of greenhouse gases increases, the temperature of the Earth is expected to increase also.

Earth's radiation balances
Figure 48.11

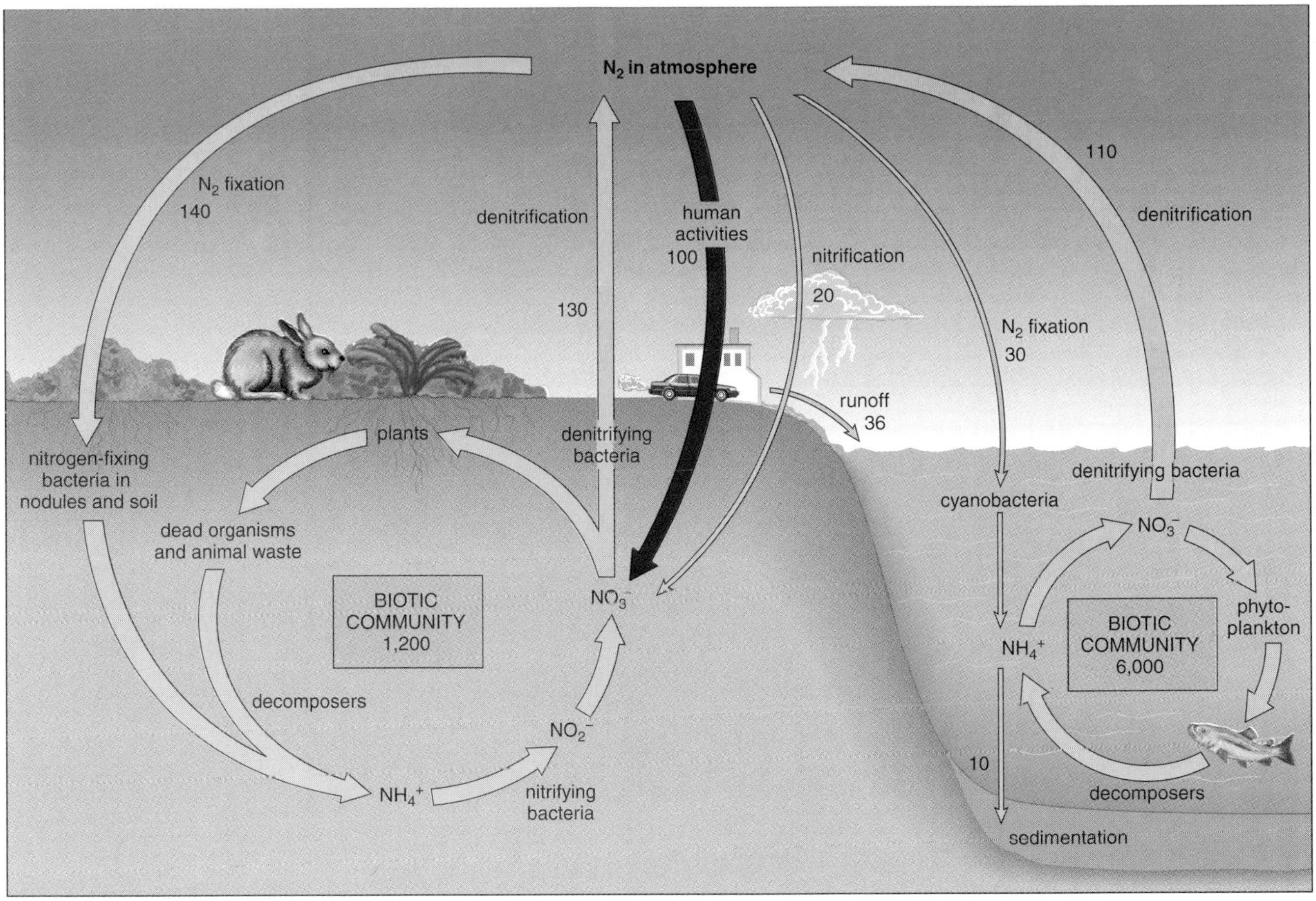

The nitrogen cycle
Figure 48.12

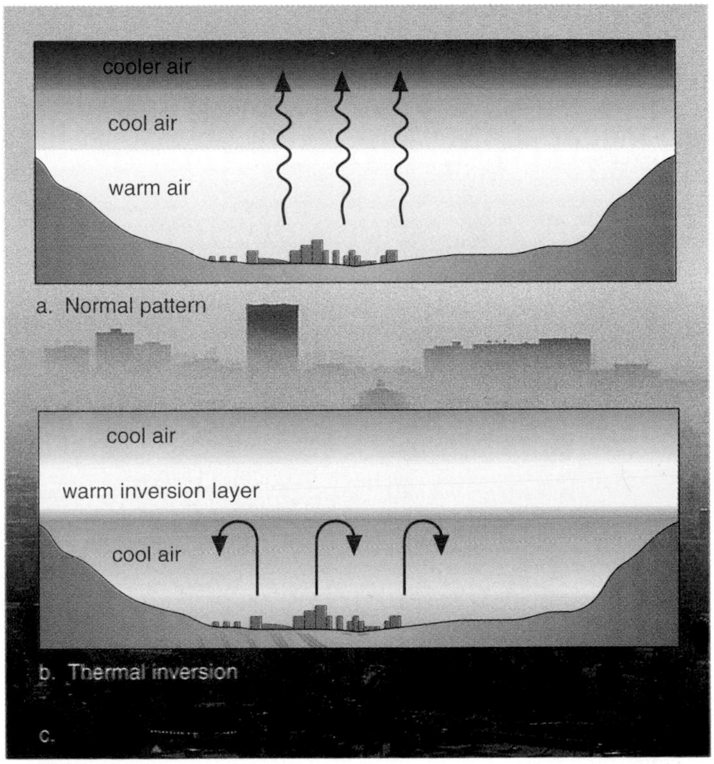

Thermal inversion
Figure 48.14

c: © Bill Aron/Photo Edit

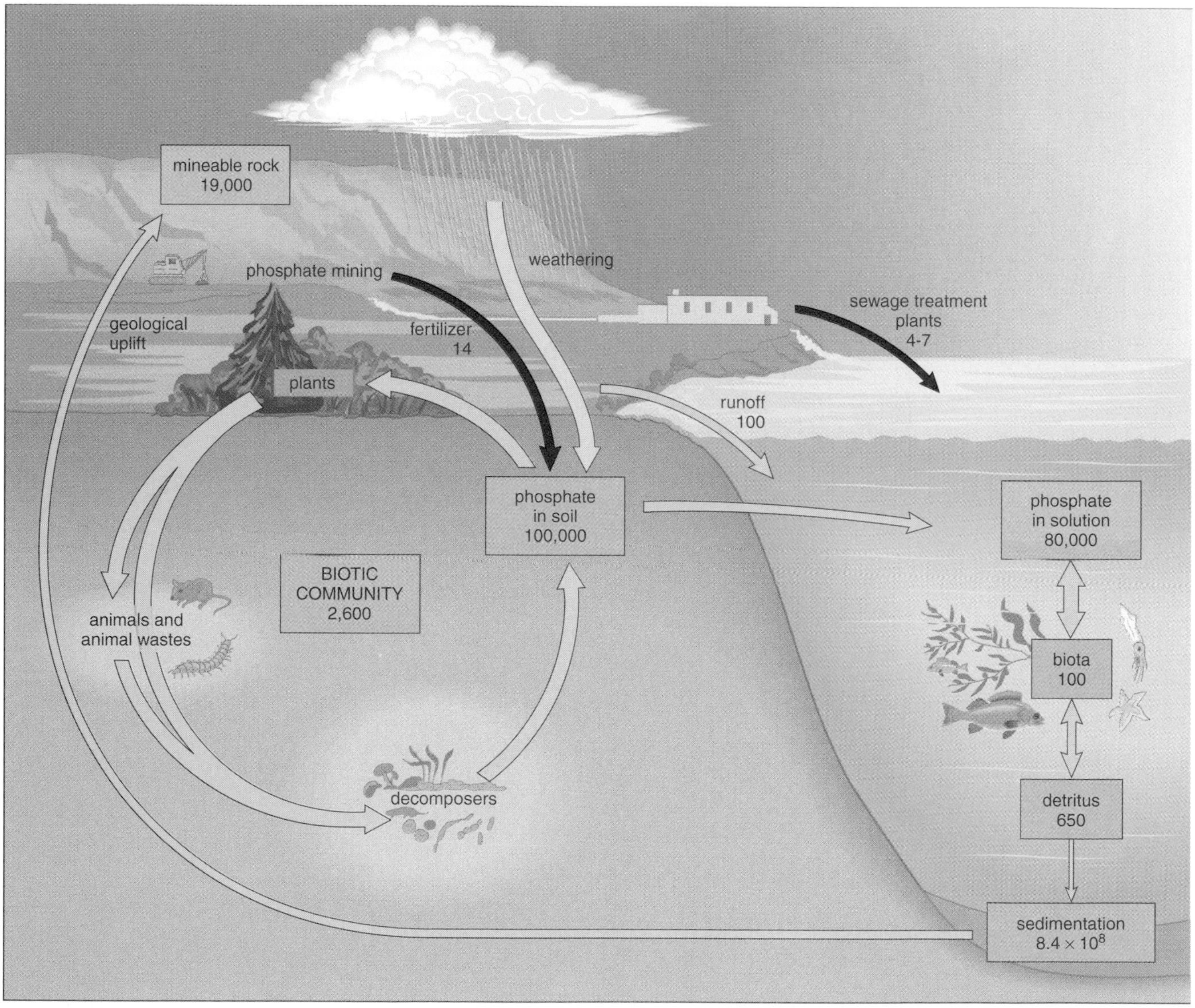
mineable rock
19,000
weathering
phosphate mining
geological
uplift
fertilizer
14
sewage treatment
plants
4-7
plants
runoff
100
phosphate
in soil
100,000
phosphate
in solution
80,000
BIOTIC
COMMUNITY
2,600
animals and
animal wastes
biota
100
decomposers
detritus
650
sedimentation
8.4 × 10^8

The phosphorus cycle
Figure 48.15

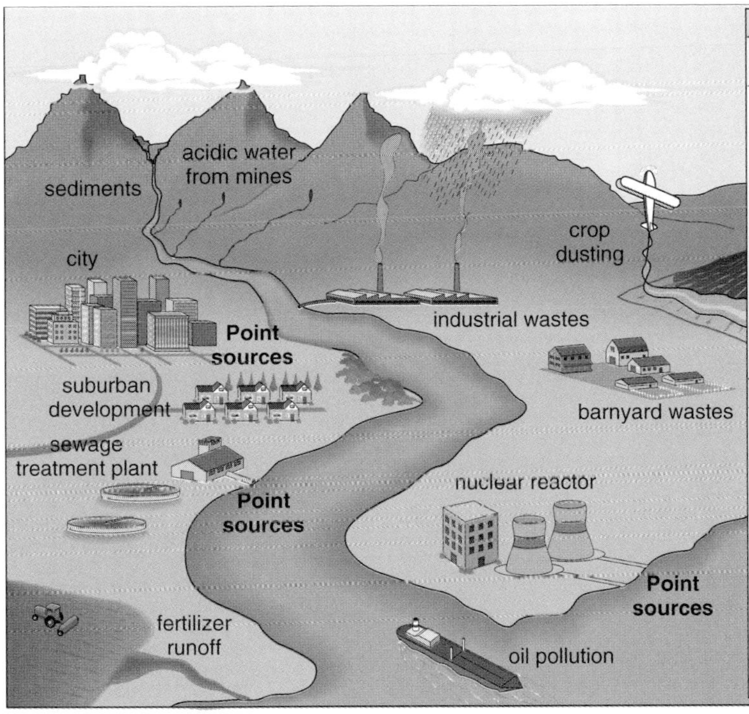

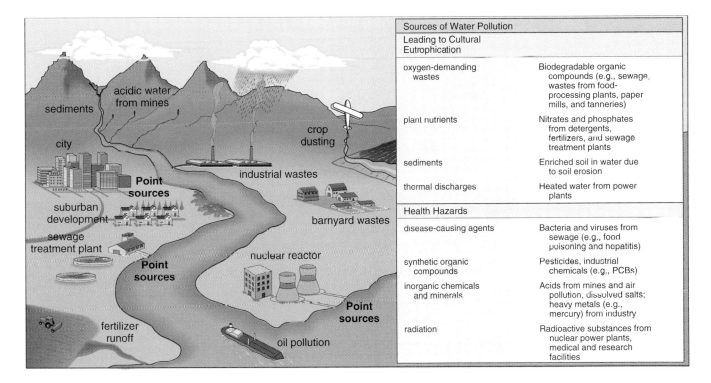

acidic water
from mines
sediments
city
crop
dusting
industrial wastes
Point
sources
suburban
development
barnyard wastes
sewage
treatment plant
nuclear reactor
Point
sources
Point
sources
fertilizer
runoff
oil pollution

Sources of Water Pollution	
Leading to Cultural Eutrophication	
oxygen-demanding wastes	Biodegradable organic compounds (e.g., sewage, wastes from food-processing plants, paper mills, and tanneries)
plant nutrients	Nitrates and phosphates from detergents, fertilizers, and sewage treatment plants
sediments	Enriched soil in water due to soil erosion
thermal discharges	Heated water from power plants
Health Hazards	
disease-causing agents	Bacteria and viruses from sewage (e.g., food poisoning and hepatitis)
synthetic organic compounds	Pesticides, industrial chemicals (e.g., PCBs)
inorganic chemicals and minerals	Acids from mines and air pollution; dissolved salts; heavy metals (e.g., mercury) from industry
radiation	Radioactive substances from nuclear power plants, medical and research facilities

Sources of surface water pollution
Figure 48.16

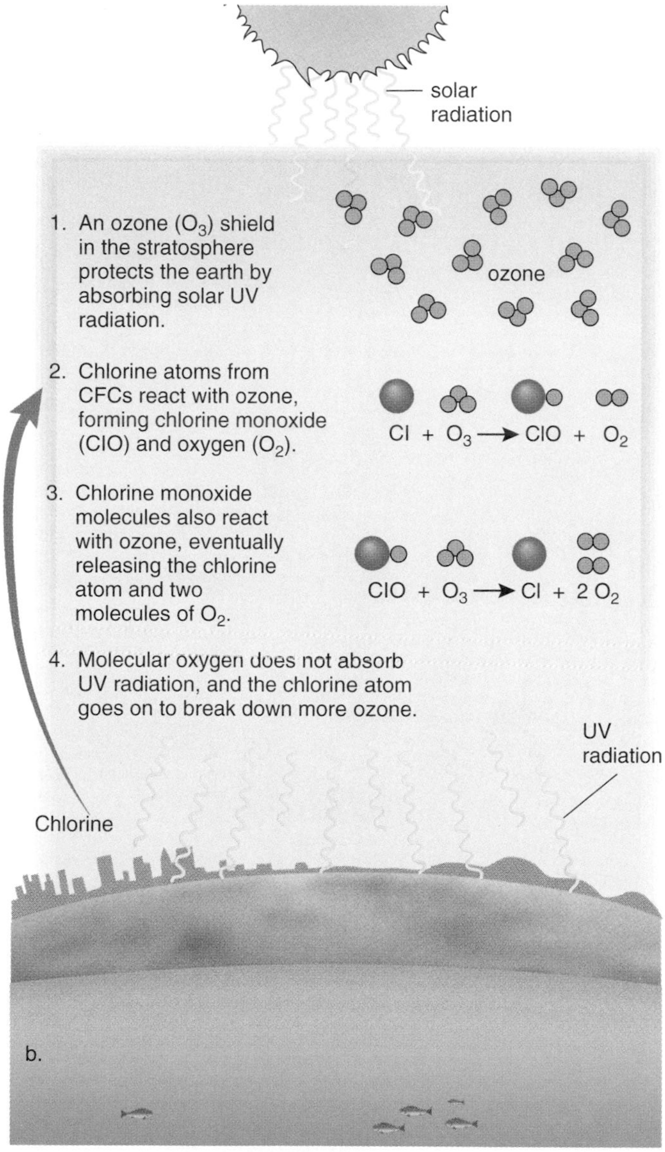

1. An ozone (O_3) shield in the stratosphere protects the earth by absorbing solar UV radiation.

2. Chlorine atoms from CFCs react with ozone, forming chlorine monoxide (ClO) and oxygen (O_2).

$$Cl + O_3 \longrightarrow ClO + O_2$$

3. Chlorine monoxide molecules also react with ozone, eventually releasing the chlorine atom and two molecules of O_2.

$$ClO + O_3 \longrightarrow Cl + 2 O_2$$

4. Molecular oxygen does not absorb UV radiation, and the chlorine atom goes on to break down more ozone.

Ozone shield depletion
Figure 48B

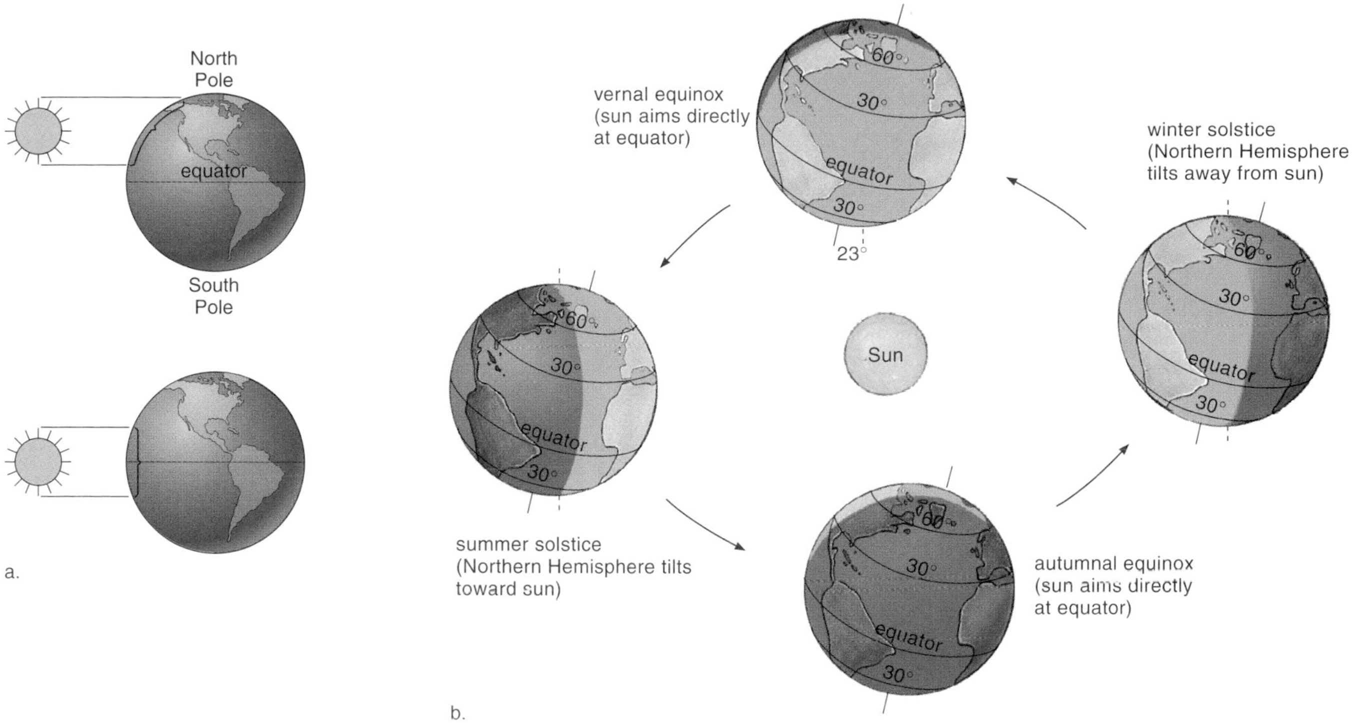

a.

b.

Distribution of solar energy
Figure 49.1

vernal equinox
(sun aims directly
at equator)

winter solstice
(Northern Hemisphere
tilts away from sun)

Sun

summer solstice
(Northern Hemisphere tilts
toward sun)

autumnal equinox
(sun aims directly
at equator)

Global wind circulation
Figure 49.2

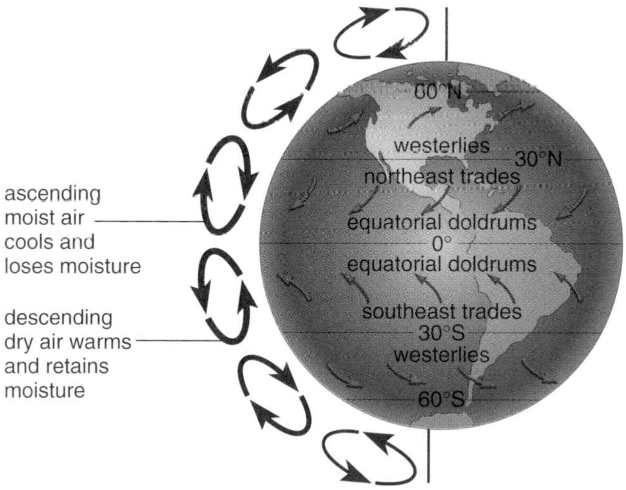

ascending
moist air
cools and
loses moisture

descending
dry air warms
and retains
moisture

60°N
westerlies
30°N
northeast trades
equatorial doldrums
0°
equatorial doldrums
southeast trades
30°S
westerlies
60°S

Formation of a rain shadow
Figure 49.3

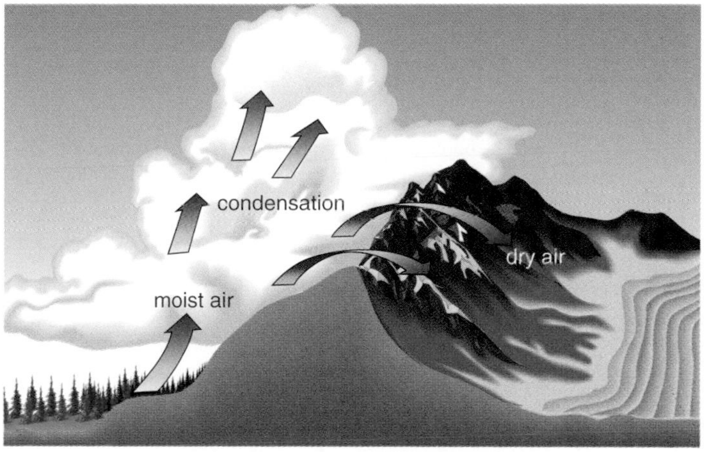

condensation

dry air

moist air

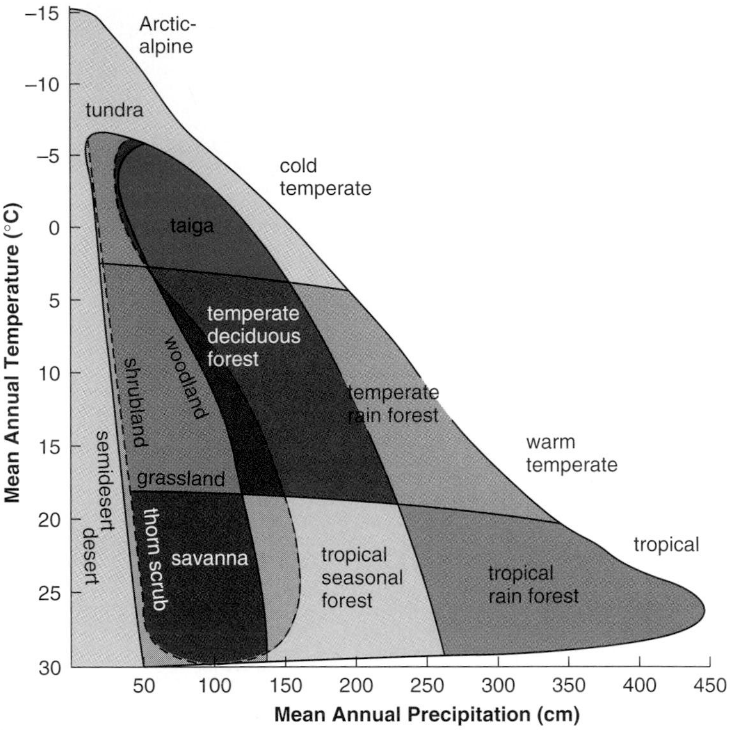

Arctic-alpine
tundra
cold temperate
taiga
temperate deciduous forest
woodland
shrubland
temperate rain forest
warm temperate
semidesert
grassland
thorn scrub
savanna
desert
tropical seasonal forest
tropical rain forest
tropical
Mean Annual Temperature (°C)
Mean Annual Precipitation (cm)

a.

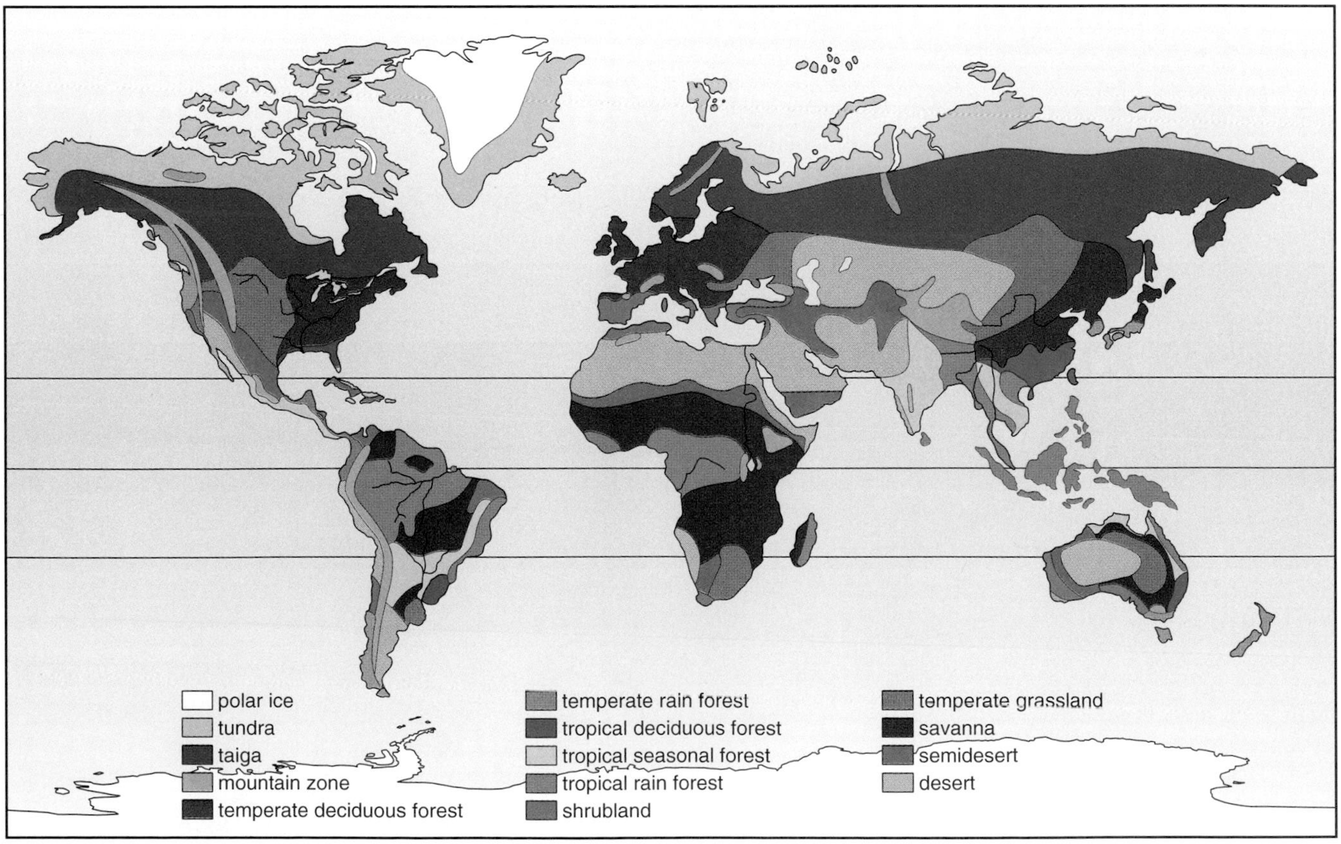

polar ice
tundra
taiga
mountain zone
temperate deciduous forest
temperate rain forest
tropical deciduous forest
tropical seasonal forest
tropical rain forest
shrubland
temperate grassland
savanna
semidesert
desert

b.

Pattern of biome distribution
Figure 49.4

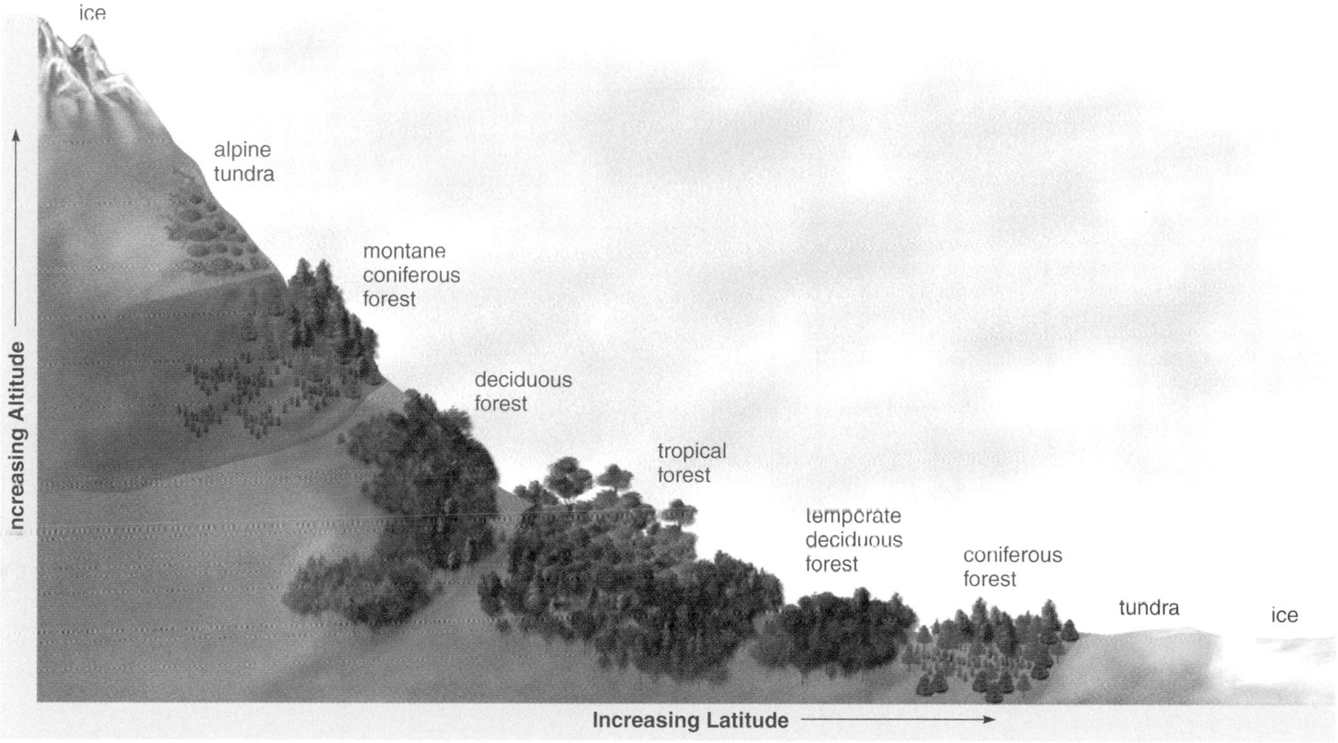
ice
alpine tundra
montane coniferous forest
deciduous forest
tropical forest
temperate deciduous forest
coniferous forest
tundra
ice
Increasing Altitude
Increasing Latitude

Climate and biomes
Figure 49.5

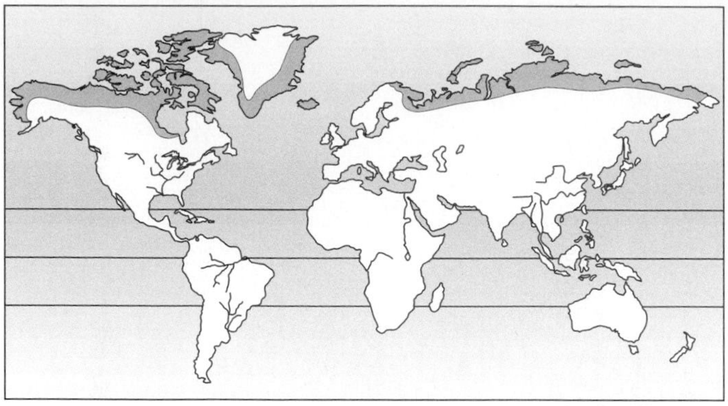

The tundra
Figure 49.6

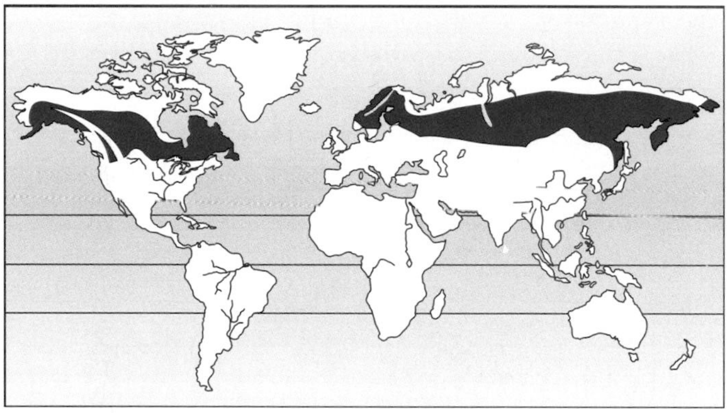

The taiga
Figure 49.7

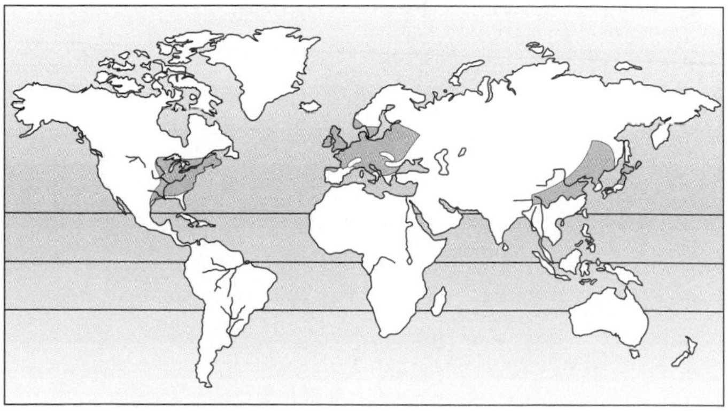

Temperate deciduous forest
Figure 49.8

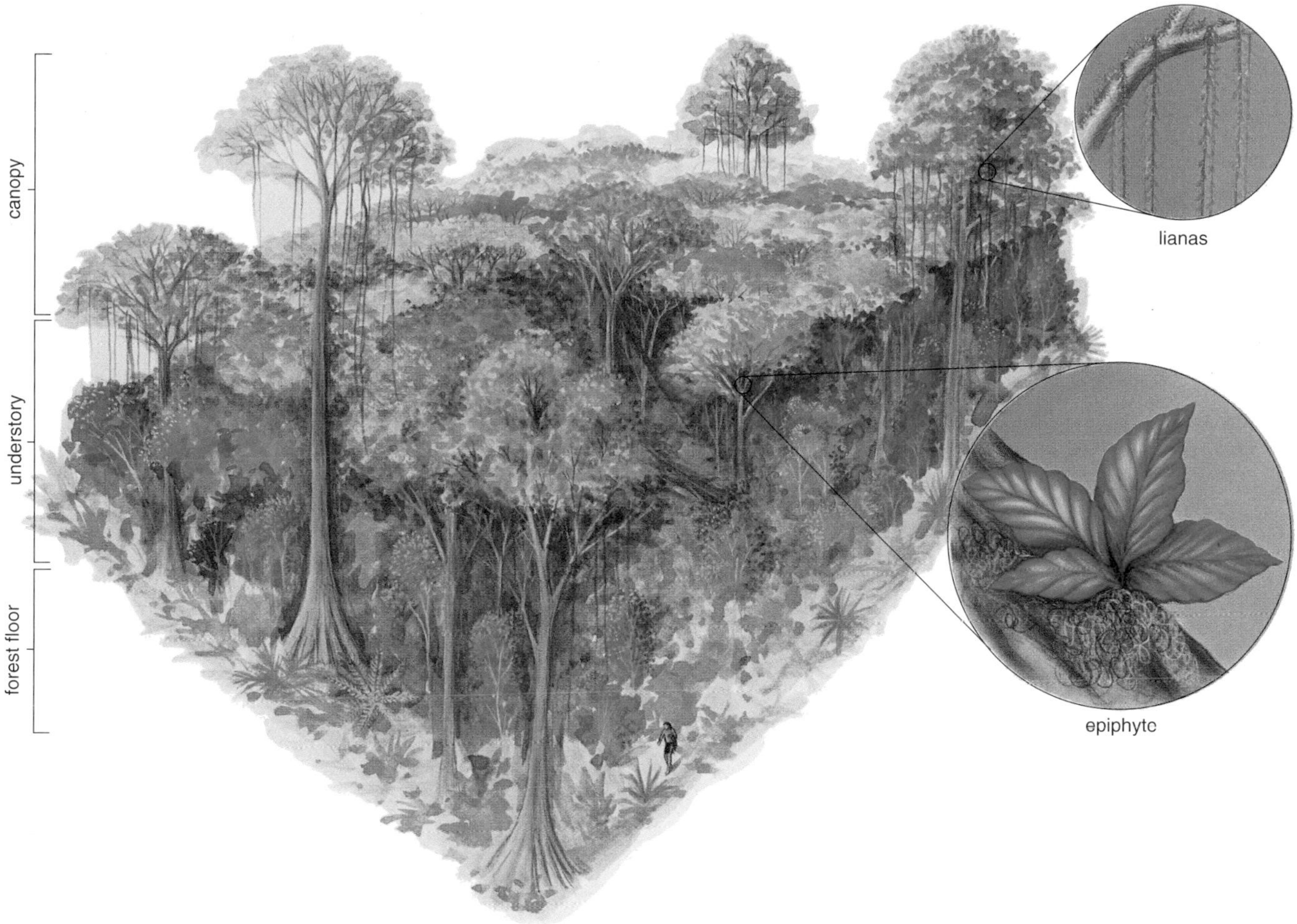

Levels of life in a tropical rain forest
Figure 49.9

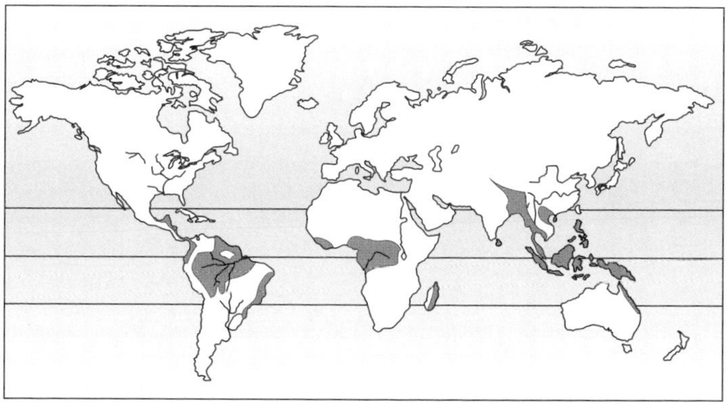

Animals of the tropical rain forest
Figure 49.10

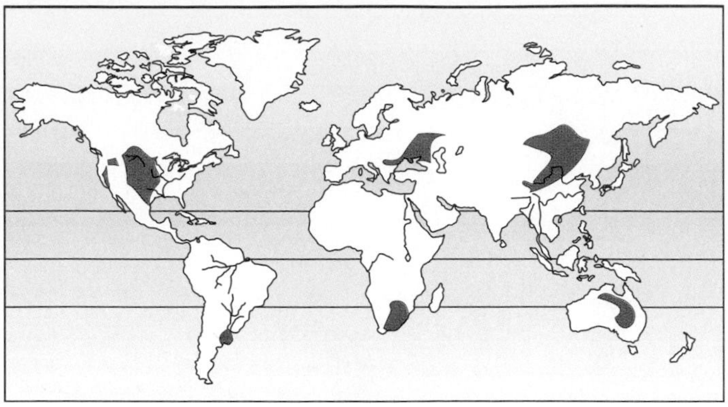

Temperate grassland
Figure 49.12

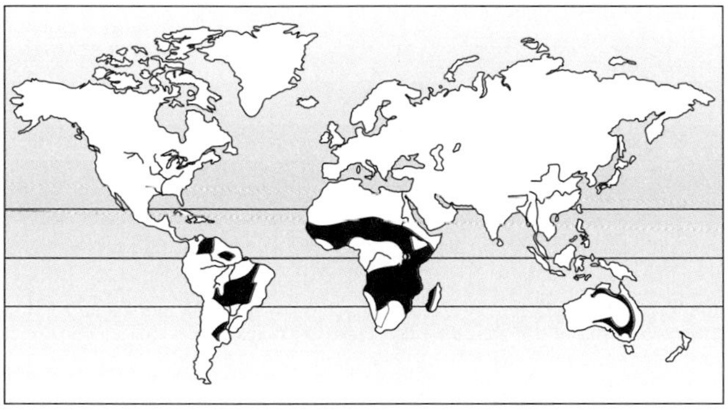

The savanna
Figure 49.13

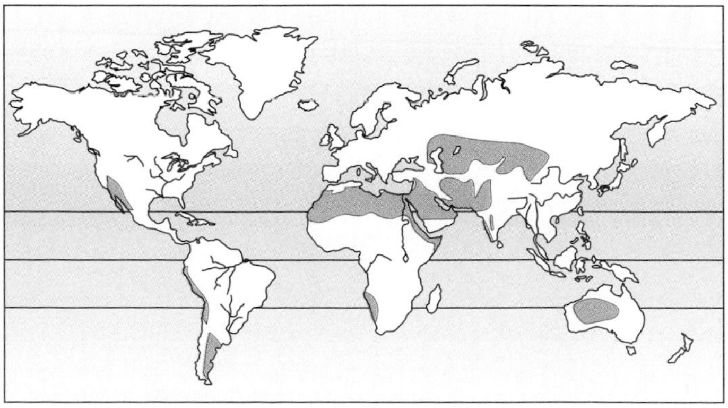

The desert
Figure 49.14

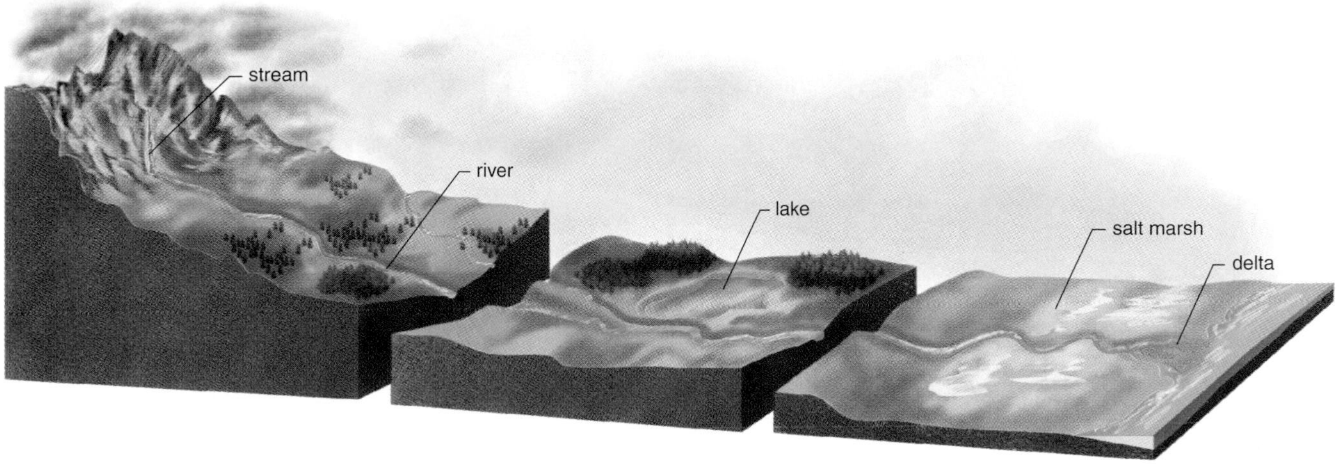

Freshwater and saltwater communities
Figure 49.15

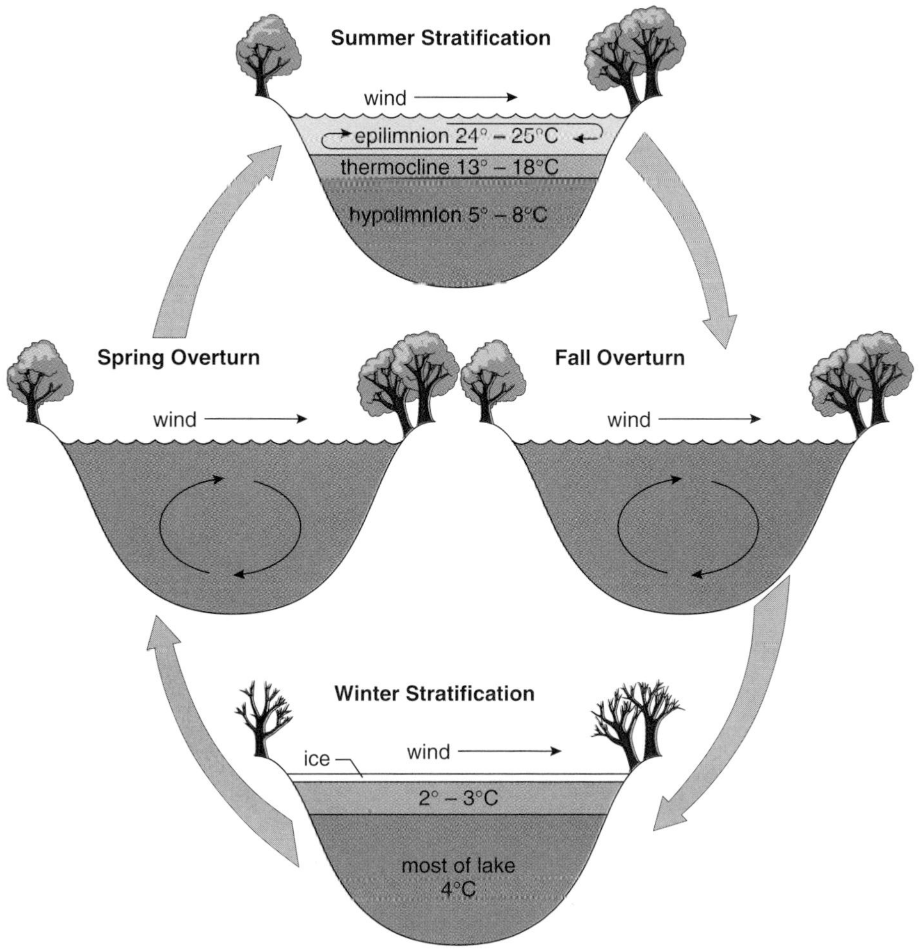

Lake stratification
Figure 49.17

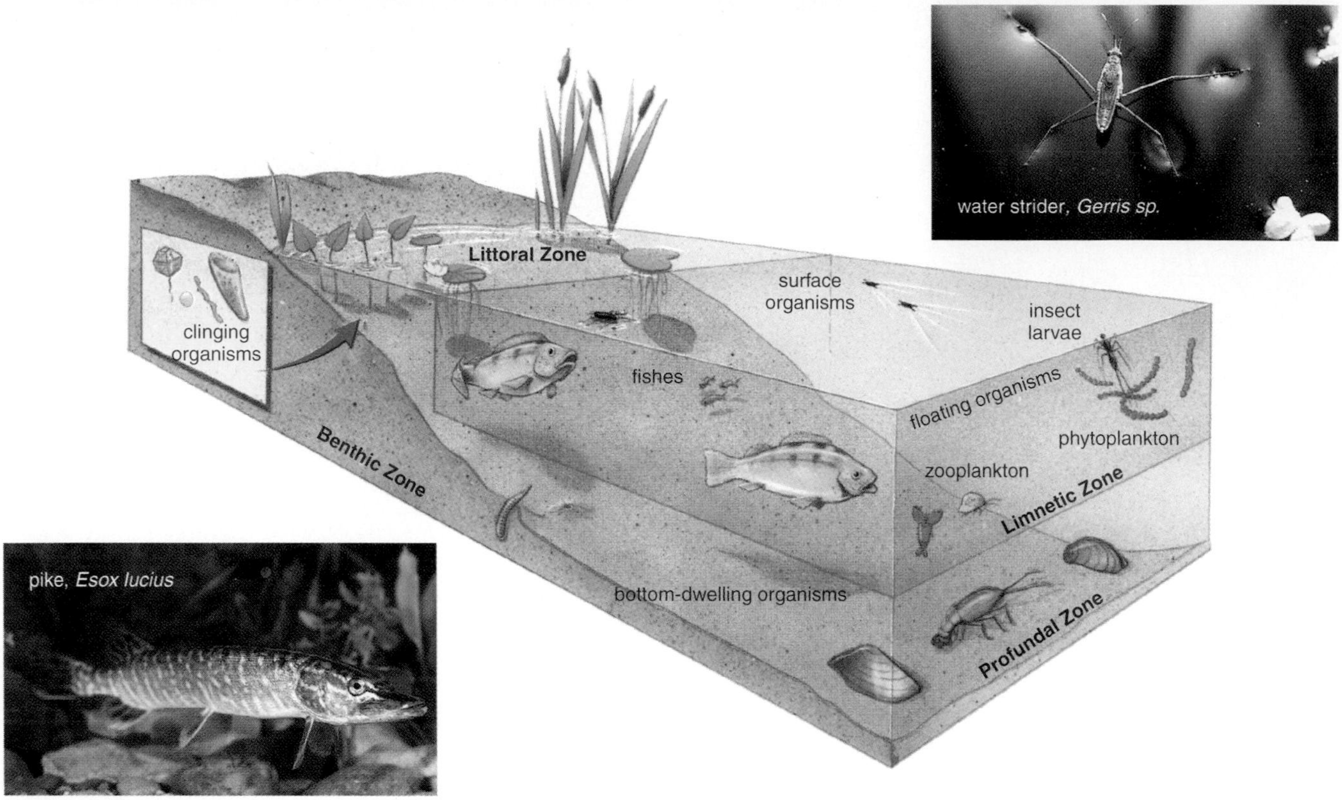

Zones of a lake

Figure 49.18

pike: © Robert Maier/Animals Animals/Earth Scenes; pond skater: © G.I. Bernard/Animals Animals/Earth Scenes

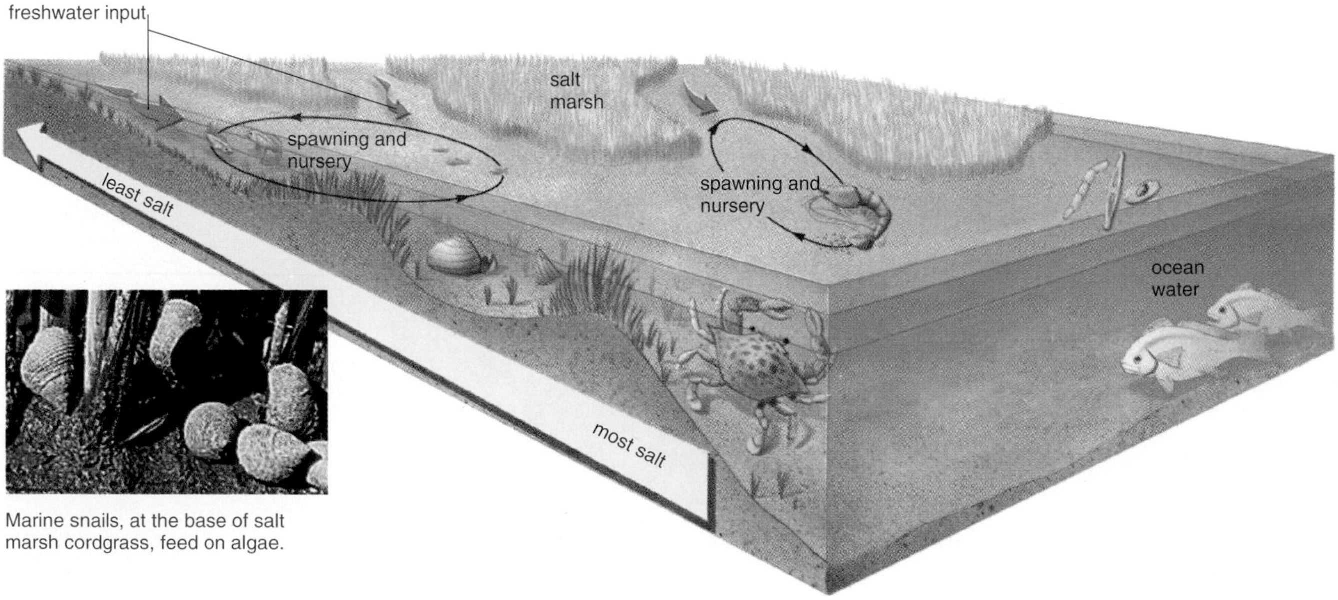

Estuary structure and function

Figure 49.19

b: © Heather Angel

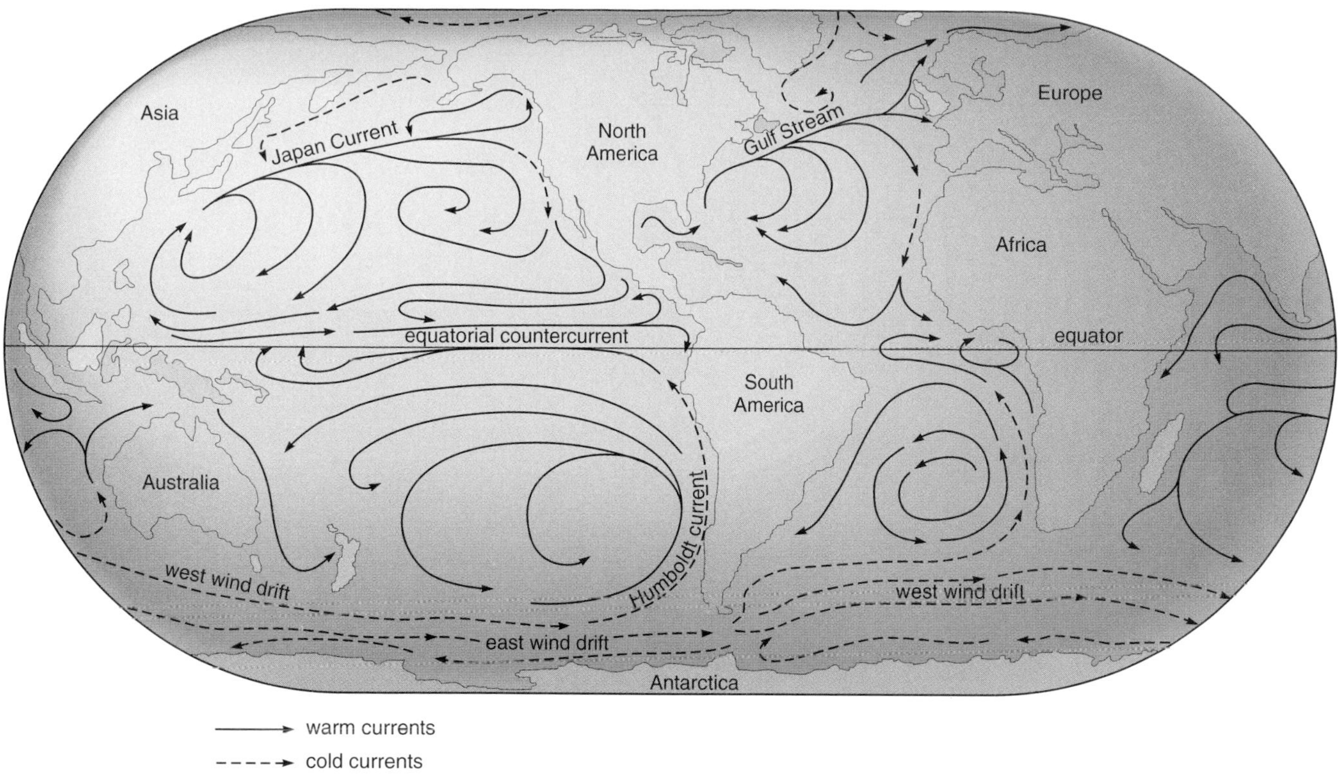

Ocean currents
Figure 49.22

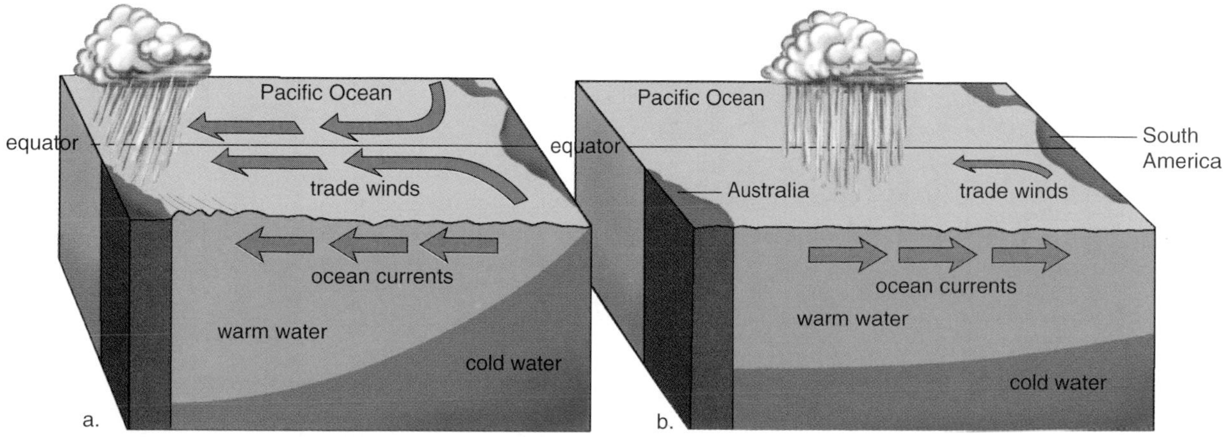

a. La Niña
Figure 49B

b. El Niño

347

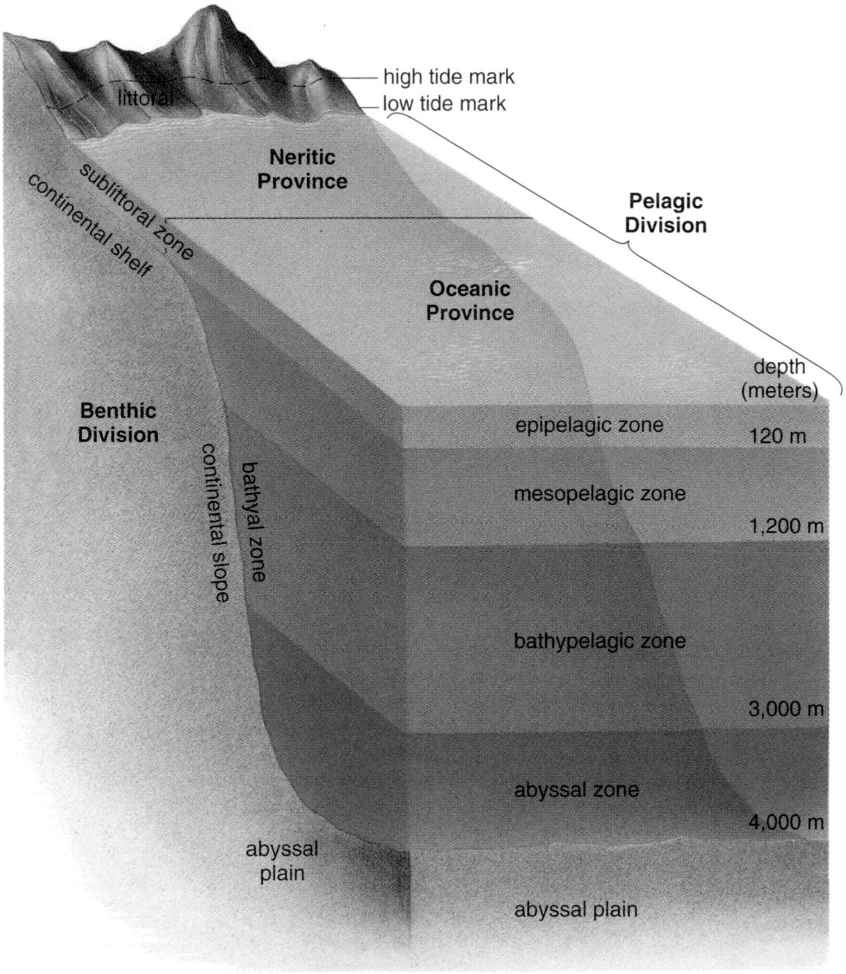
high tide mark
low tide mark
littoral
Neritic Province
Pelagic Division
sublittoral zone
continental shelf
Oceanic Province
depth (meters)
Benthic Division
epipelagic zone
120 m
mesopelagic zone
1,200 m
continental slope
bathyal zone
bathypelagic zone
3,000 m
abyssal zone
4,000 m
abyssal plain
abyssal plain

Marine environment
Figure 49.23

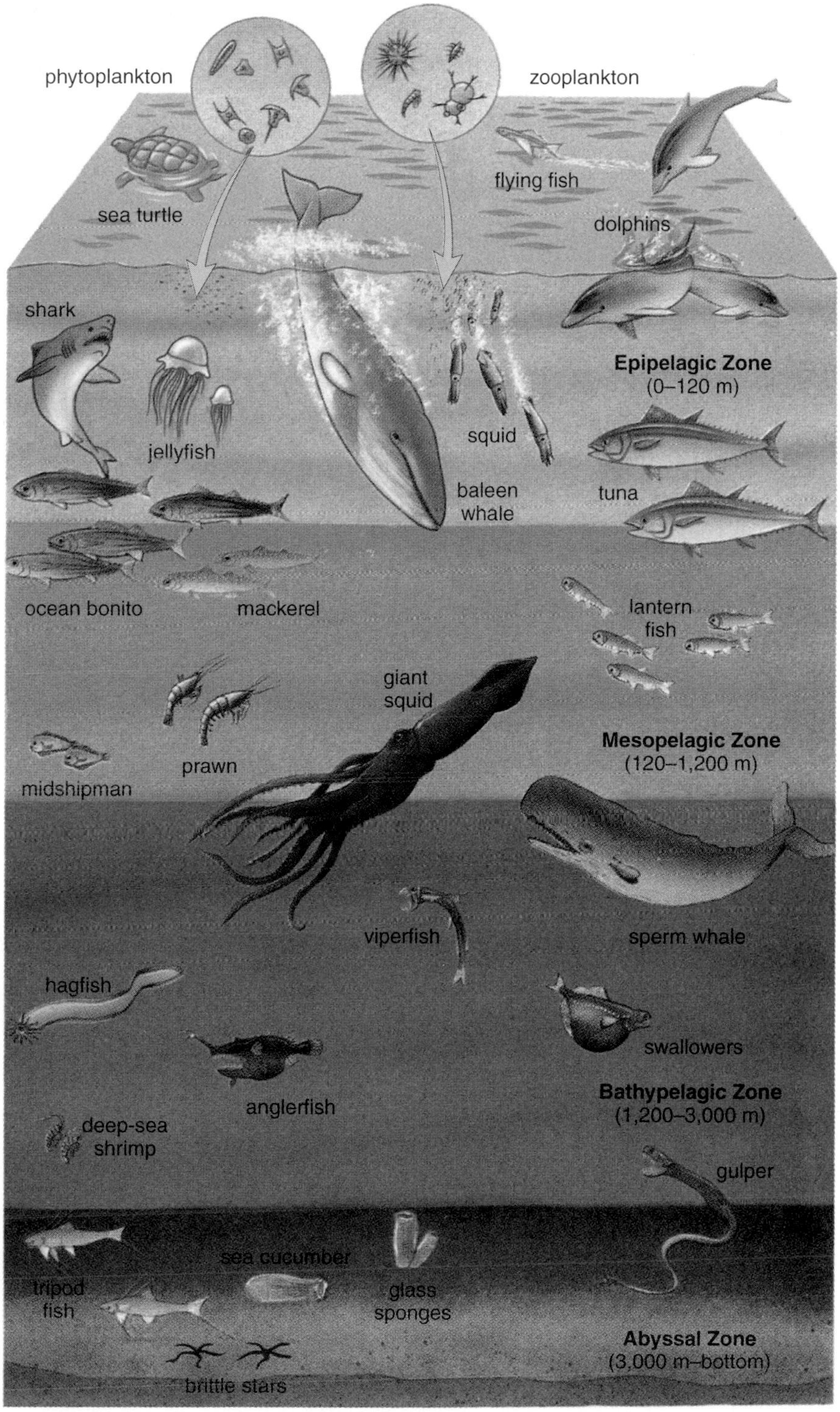
phytoplankton
zooplankton
sea turtle
flying fish
dolphins
shark
jellyfish
squid
baleen
whale
Epipelagic Zone
(0–120 m)
tuna
ocean bonito
mackerel
lantern
fish
giant
squid
prawn
midshipman
Mesopelagic Zone
(120–1,200 m)
viperfish
sperm whale
hagfish
swallowers
anglerfish
Bathypelagic Zone
(1,200–3,000 m)
deep-sea
shrimp
gulper
sea cucumber
tripod
fish
glass
sponges
brittle stars
Abyssal Zone
(3,000 m–bottom)

Ocean inhabitants
Figure 49.24

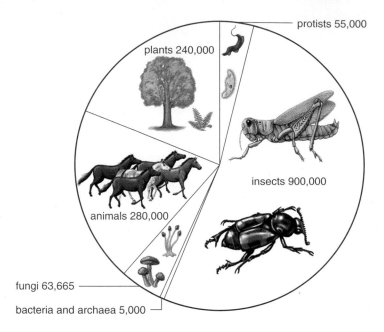

Number of described species
Figure 50.1

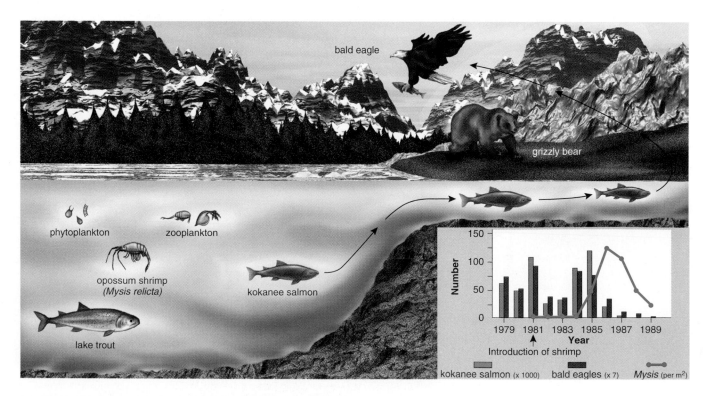

Eagles and bears feed on spawning salmon
Figure 50.2

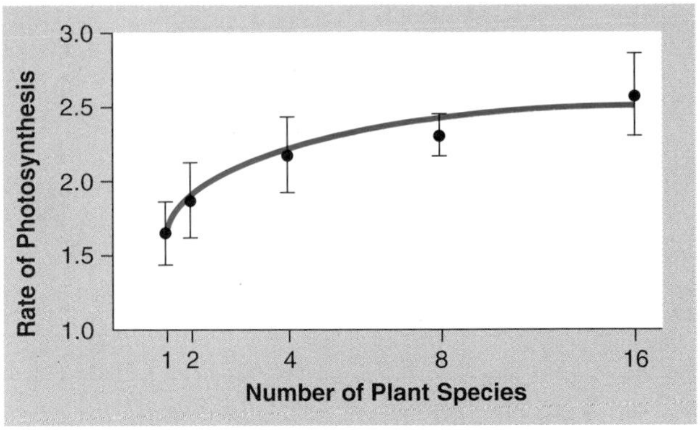

Indirect value of ecosystems
Figure 50.4

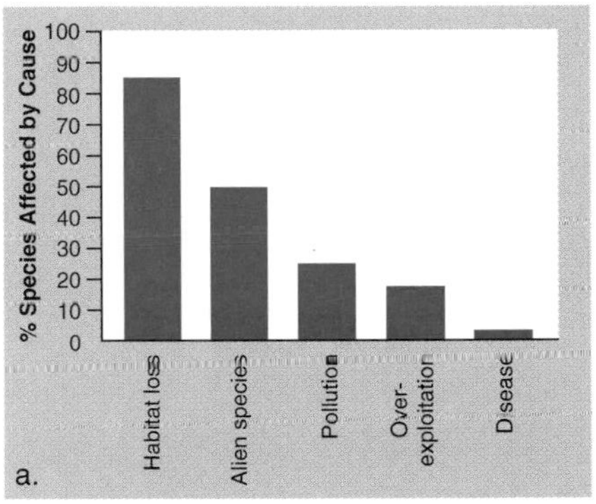

Habitat loss
Figure 50.5

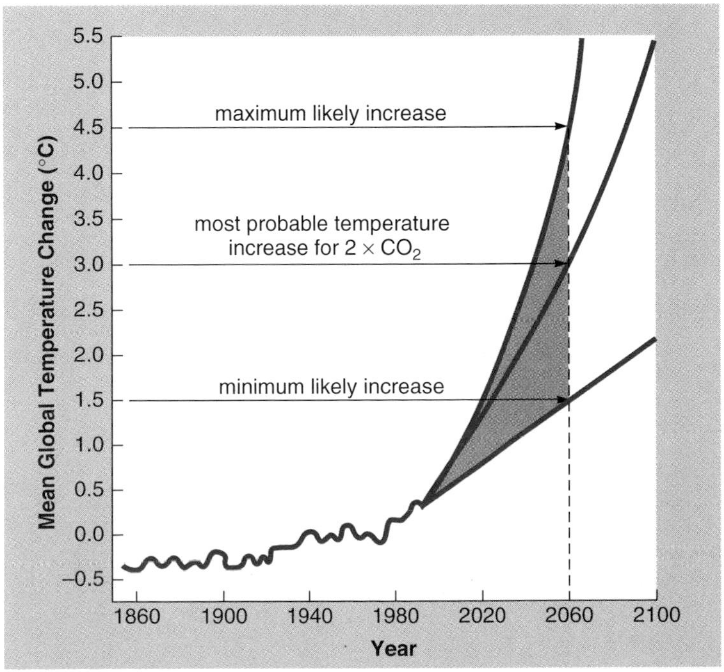

Global warming
Figure 50.7

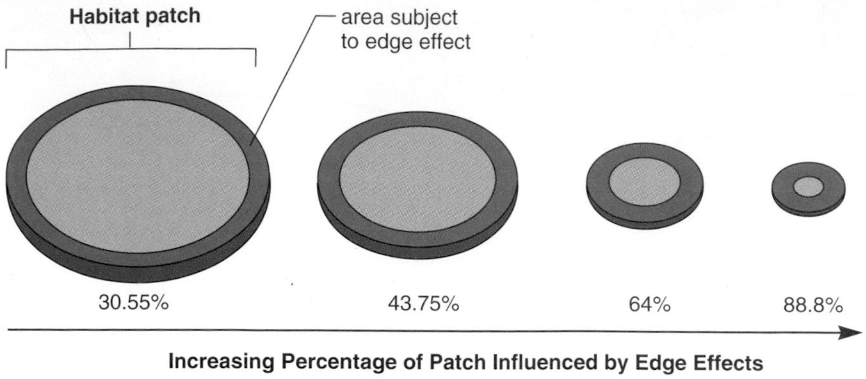

Increasing Percentage of Patch Influenced by Edge Effects

Edge effect
Figure 50.10

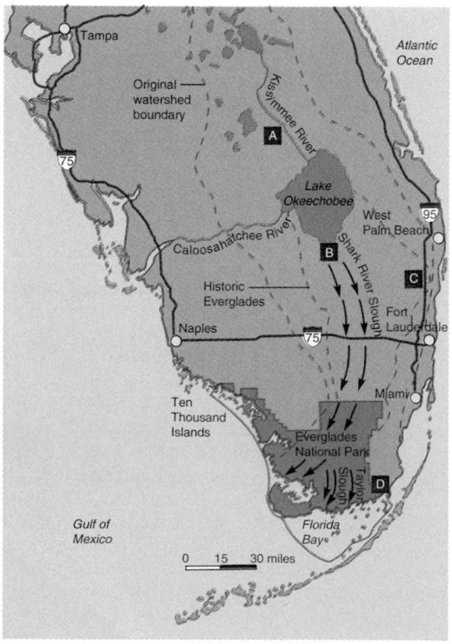

Restoration of the Everglades
Figure 50.11